Y0-CAM-290

Nutrition of pond fishes

NUTRITION OF POND FISHES

BALFOUR HEPHER

Formerly of Fish and Aquaculture Research Station, Dor, Israel

CAMBRIDGE UNIVERSITY PRESS

Cambridge

New York New Rochelle Melbourne Sydney

Published by the Press Syndicate of the University of Cambridge
The Pitt Building, Trumpington Street, Cambridge CB2 1RP
32 East 57th Street, New York, NY 10022, USA
10 Stamford Road, Oakleigh, Melbourne 3166, Australia

First published 1988

Printed in Great Britain at the University Press, Cambridge

British Library cataloguing in publication data

Hepher, Balfour
Nutrition of pond fishes.
1. Fish-culture 2. Fish ponds
I. Title
639.3′11 SH159

Library of Congress cataloguing in publication data

Hepher, Balfour, 1925–
Nutrition of pond fishes.
Bibliography: p.
Includes index.
1. Fishes–Feeding and feeds. 2. Fish ponds.
I. Title.
SH156.H47 1988 639.3′11 87-24968

ISBN 0 521 34150 7

US

Contents

List of Symbols

ADC	Apparent digestibility coefficient
ADP	Adenosine diphosphate
AMP	Adenosine monophosphate
ATP	Adenosine triphosphate
B	Biomass
BV	Biological value
C_g	Calorific value of weight gain
CoA	Coenzyme A
CSC	Critical standing crop
DE	Digestible energy
dW/dt	$\dot{W}$, Instantaneous weight gain; growth
E	Ivlev's index of electivity
E_f	Energy excreted in faeces
E_m	Energy excreted in urine and through the gills; metabolic energy excretion
E_{pg}	Apparent utilization efficiency of *ME* for growth
E_{pm}	Apparent utilization efficiency of *ME* for maintenance
EAAI	Essential amino acid index
EFA	Essential fatty acids
F	Fat content of fish (%)
F_n	Ingested natural food
F_s	Ingested supplementary feed
FCR	Feed conversion ratio
G	Dietary energy requirement for growth
g	Weight gain (grams); net energy requirement for growth
G_w	Average percentage daily gain ('specific growth')
G6PDH	Glucose 6-phosphate dehydrogenase
GDP	Guanosine diphosphate
GTP	Guanosine triphosphate
HUFA	Highly unsaturated fatty acids

K_1 Growth coefficient of the first order = ratio between energy gain and ingested energy
K_2 Growth coefficient of the second order = ratio between energy gain and digestible energy
M Dietary energy requirement for maintenance
ME Metabolizable energy
N Number of organisms
N_B Nitrogen content of fish body
N_E Endogenous nitrogen excretion
N_f Faecal nitrogen
N_I Nitrogen ingested
N_r Retained nitrogen
N_u Urinary nitrogen
NAD Nicotinamide adenine dinucleotide
NADH Nicotinamide adenine dinucleotide, reduced form
NADP Nicotinamide adenine dinucleotide phosphate
NADPH Nicotinamide adenine dinucleotide phosphate, reduced form
NE Net energy (maintenance + growth)
NPN Non-protein nitrogen
NPU Net protein utilization
P Production
p Ratio between metabolizable energy and ingested energy
p_t Ratio between net energy and ingested energy
PER Protein efficiency ratio
PPV Productive protein value
PR Protein requirement
PUFA Polyunsaturated fatty acids
Q Energy consumption; metabolism
Q_A Active metabolism
Q_B Basal metabolism
Q_M Total energy requirement for maintenance (including *SDA*)
Q_m Net energy requirement for maintenance
Q_R Routine metabolism
Q_S Standard metabolism
R Food requirement; ration
R_d Daily food intake
RQ Respiratory quotient
SDA Specific dynamic action
SG Specific growth coefficient
T Temperature
t Time (usually in days)
TCA Tricarboxylic acid cycle
W Weight of fish
$\bar{W}$ Average weight of fish

$\dot{W}$	dW/dt, Instantaneous weight gain; growth
W_t	Weight of fish at the time t
W_∞	Asymptotic maximum weight of the fish

Preface

Aquaculture has recently made a great stride forward. New areas of fish ponds have been developed and old culture systems have been rejuvenated and intensified. However, development of aquaculture is dependent on knowledge related to various disciplines, some associated with the fish itself, such as physiology and genetics, and others with the environment, such as limnology of the culture system and the life cycles of food organisms. The multi-disciplinary nature of aquaculture is strongly expressed in the subject of feeding fish in ponds. Here a number of disciplines converge: anatomy and physiology of fish feeding, animal nutrition, and ecology of fish feeding. Efficient diets and feeding charts for fish in ponds can be produced only by integrating all these disciplines. However most researchers concentrate on one of these aspects, and only a few see the whole picture. The intention of the book is to acquaint researchers, students and fish farmers with the holistic approach. No doubt more extensive and deeper reviews can be found in the literature than those presented here if each discipline is considered separately, but the aim is to combine them, so that the physiologist, for example, will understand some of the ecological problems and the ecologist will appreciate some of the nutritional difficulties.

My thanks are due to many who helped me write this book, by fruitful discussions, or by reading the manuscript and offering their critical remarks. Special thanks are due to Prof. H. J. Schoonbee of Rand Afrikaans University, South Africa, Prof. R. J. Roberts, Dr R. H. Stirling, Dr K. Jauncey and Dr M. Beveridge of Stirling University, Scotland, UK and Prof. A. Gertler of the Hebrew University of Jerusalem. I hope this work will benefit the development of aquaculture throughout the world.

Biographical note

Balfour Hepher was born in Haifa, Israel, in 1925. He received his MSc degree in Agriculture from the Hebrew University in Jerusalem in 1950 and then, at the same University, embarked upon a career of nearly forty years in aquaculture, with a PhD study on the dynamics of phosphorus in fishponds. After some years of work on pond fertilization he turned his attention to nutrition of pond fish, using both natural and supplemental foods. Later research interests included the ecological principles governing intensive systems and problems of fish metabolism, particularly carbohydrates.

Dr Hepher had a long association with the Fish and Aquaculture Research Station at Dor, Israel; he was its director from 1971 to 1975 and continued there as a Senior Scientist, actively involved in research and advising visiting aquaculturists. He lectured on aquaculture at the Universities of Jerusalem and Tel Aviv and fulfilled several other teaching commitments locally and internationally. He travelled extensively abroad, using his ability to assess innovations, such as intensive culture systems and techniques of demand feeding, in one country and to transfer them to another, thus improving aquaculture worldwide.

Dr Hepher's many professional activities included contributions to several national and international organisations. He was a member of FAO Task Forces, and acted as a consultant to the World Bank/UNDP on the integration of aquaculture and waste recycling, as well as advising several Third World countries, particularly in the Far East and Latin America, in different aspects of aquaculture. He published over 60 papers on fertilization, manuring, feeding and management of fishponds, and his editorial responsibilities included membership of the Boards of Bamidgeh (the Bulletin of Aquaculture in Israel) and the international journal Aquaculture. This, his second book, was in its final stages of production when he died on 4 February 1988.

Introduction

The yield of fish per pond area or per culture unit, and thus the profitability of fish farming, depends to a large extent on the amount of supplementary feed used. The more intensive the aquaculture system, the greater the importance of supplementary feeding, and the greater the proportion of the feed costs to total production costs. According to Collins & Delmendo (1979) the cost of supplementary feed in a flowing water system and in cage culture can reach 56.3–56.8% of the total production costs. Cost of feed in pond culture can also be high. According to calculations made by the Fish Breeders Association of Israel (September 1985, personal communication) the feed costs in a polyculture pond farm of an average yield of 3.5 t/ha, where common carp constitute about 50% of the yield, have reached 30.6% of the total production costs. This included depreciation and interest; when these items were excluded cost of feed was 37.0% of the direct total production costs. This emphasizes the importance of determining the optimum level of feeding. Feeding too little will result in the utilization of most of the feed for maintenance. Conversely, feeding too much will result in a low utilization rate and a waste of feed. In both cases feed conversion ratio, i.e. the amount of feed used for a unit gain in fish weight, will increase and the profitability of the fish farm will correspondingly decrease.

Determination of correct feeding level for pond fishes is more difficult than for most farm animals, since these fish obtain part of their food supply from natural sources. The fish farmer supplements this natural food to meet the nutritional requirements of the fish. In this respect the situation is analogous to cattle or sheep on a pasture. However, since the variety of food organisms in the pond is far larger than that in the pasture, and their life cycles much shorter than those of grasses, it is much more difficult to estimate the amount of food consumed by fish in ponds. The fact that the fish cannot be seen in the turbid water of the pond, which obscures their feeding behaviour, adds to the difficulty of estimating the amount of food they consume.

Three components are thus involved in the nutrition of pond fish: (a) the food requirement of the fish (R); (b) available natural food[1] (F_n); (c) supplementary feed (F_s), so that

$$R = F_n + F_s \quad (1)$$

Determination of supplementary feeding level is less complex in flowing-water fish culture systems such as trout raceways. Due to the rapid rate of water exchange, the production of natural food in such systems is negligible. All the nutritional requirements of the fish must, therefore, be supplied through feed from outside sources provided by the fish farmer, a situation similar to that of many farm animals. However, the situation is different for warmwater fishes cultured in ponds. Here the contribution of natural food cannot be ignored and must be taken into account when estimating the supplementary feeding level. For such an estimation, both the nutritional requirement and the proportion supplied by the natural food must be known. However, in the determination of these two parameters a number of difficulties are encountered, since our knowledge of the nutritional requirements of most warmwater fishes, though considerably advanced during the last decade, is still limited. Thus, for instance, in estimating the amount of food required to satisfy the nutritional requirements, one must take into account the utilization rate of various foods for the different metabolic processes, about which only meagre information exists. Little progress has also been made so far in estimating the amount of natural food available to fish under varying pond conditions. Since the total amount of natural food in a pond is limited, fish density will determine the average amount of natural food available per fish. Thus the interrelations between fish density, mean fish weight, and the available and ingested natural food are equally relevant.

The purpose of this book is to discuss some of the aspects mentioned above, with the hope that this will lead to improved understanding of the methods of determination of the supplementary feeding required by fish in ponds. This discussion is divided into two parts. The first deals with food requirements, i.e. the quantitative and qualitative requirements for maintenance and growth, and the utilization rates of foods for these purposes. The second part deals with food sources and feed conversion. Here the problems of estimating natural food consumption by fish, and interrelationships with fish density and weight, are discussed.

---

[1] Throughout this book a distinction is made between natural *food* and supplementary *feed*.

Part I

Food requirement

1

The balance of energy

The quantitative requirement of any food depends largely on its composition. The most efficient level of feeding is attained only when the correct supply of energy and essential nutrients is available in the proportions required by the fish for maintenance and growth. Any deviation from this 'ideal' composition will also change the quantitative food requirement. For the sake of clarity, it is necessary to separate the discussion of the quantitative aspects of feeding from the qualitative aspects. When dealing with the quantitative aspects it is usually assumed that the diet consumed corresponds to this 'ideal' composition. However, the effect of the composition of the diet actually consumed on the quantitative requirements must also be taken into account.

In the transformation of food energy to net energy available for metabolism and growth, a considerable portion is lost. Although some energy loss is unavoidable, it should obviously be minimized. This can only be done if the processes by which energy is lost have been studied and the amount lost can be estimated. Figure 1 presents graphically the transformation of food energy to net energy for maintenance and growth, and indicates the losses incurred during these transformations.

The digestible energy (*DE*) is part of the food energy which is absorbed by the fish, while the energy egested in the faeces (E_f) is lost. Some of the digestible energy is lost through the excretion of catabolites in the urine and through the gills (E_m). This loss is largely the result of catabolism of proteins and their deamination when amino acids are utilized for energy rather than for synthesis of new tissue proteins. The energy assimilated in the body is the metabolizable energy (*ME*), but part of it is again lost through a number of processes. The loss of energy through these processes, grouped together under the name 'specific dynamic action' (*SDA*), is worth some further elaboration.

According to the laws of thermodynamics, the transformation of energy from one form to another always results in entropy – the 'loss' of energy as heat. The transformation of food energy to net energy within the body is carried out stepwise through catalytic processes with the minimum possible heat loss, but it cannot be entirely eliminated. This loss, known in nutrition as 'heat increment', is related to the biochemical oxidation processes and, therefore, varies with different feedstuffs. McDonald *et al.* (1966) give the following example of the efficiency of metabolic oxidation of glucose to energy. The total energy content of glucose at 37 °C is 703 kcal/mol. Through its glycolysis to 2 mols of pyruvic acid, a net 8 mols of ATP are produced. An additional 30 mols of ATP are produced in the final oxidation of the pyruvic acid to carbon dioxide and water. The total production of ATP from ADP is thus 38 mols, each of which yields approximately 12.5 kcal of energy when broken by hydrolysis under physiological conditions. Thus, each mol of glucose oxidised yields a total of 475 kcal as free energy, and the efficiency of energy utilization from glucose by the body is: $475/703 \times 100 = 67.5\%$.

In endotherms (= homeotherms), which derive their body temperature from their own metabolic activity, the heat produced by the oxidation of food (whether through entropy or due to utilization of free energy in maintenance processes of which the final result is heat), is utilized, at least

Figure 1. Flow chart of energy transformations and losses in fish.

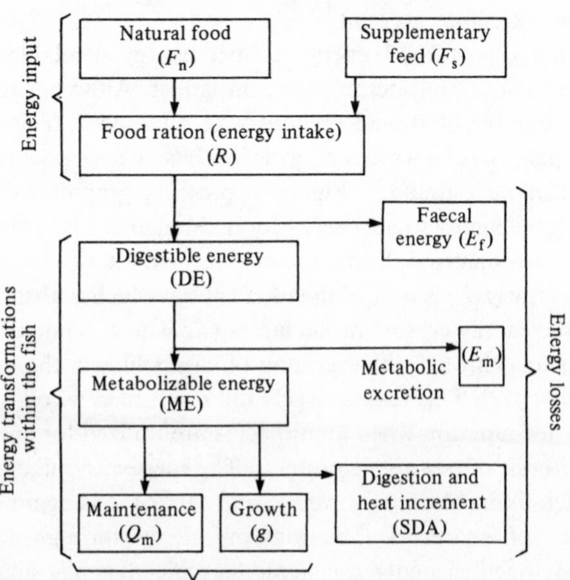

in part, to warm up their bodies. This is of special importance when the feeding level is low, and the amount of energy for heat production is limited. In ectotherms (= poikilotherms), including the fish, which obtain their heat from the environment and which cannot maintain a body temperature different from the ambient temperature, the situation is different. Here the heat produced through metabolic activity is simply lost. The unutilized portion of the food energy which turns into 'heat increment' should, therefore, be considered as an unavoidable 'tax' on energy consumption.

For the vital processes of life, such as blood circulation, respiration, osmoregulation, suspension and balancing in water, and other essential functions, the fish requires a certain amount of energy. This is maintenance metabolism. Since the fish must sustain maintenance metabolism even when food energy is not available during fasting, the required energy has to be obtained by utilizing its own tissues, especially fat reserves, and the fish will consequently lose weight. It is possible to determine the amount of energy required for the essential maintenance metabolic processes under such conditions by several methods: (a) by determining the calorific value of the tissues lost during starvation; (b) by determining the heat produced by metabolic processes (direct calorimetry); (c) by determining the amount of oxygen utilized in these processes (indirect calorimetry). These methods are discussed in greater detail in Chapter 4. It should be emphasized, however, that all these methods measure the total heat production which includes the heat increment, i.e. the unutilizable portion lost as heat. Since it is difficult to distinguish between the utilizable and unutilizable energy at this stage, the total consumed energy is used as a base line for further calculations as 'maintenance metabolism' (Q_m).

In practice, it should be taken into account that when depriving fish of food, it takes the fish some time to adapt to the lower supply of energy until they reach maintenance metabolism level. Brown (1946c, 1957), working with brown trout (*Salmo trutta*), observed that the reduction of the amount of food toward the maintenance level resulted initially in weight loss, followed by weight gain. Similar results (Figure 2) have been obtained by Hepher *et al.* (1983) working on red tilapia (*Oreochromis* sp.). From their results it seems that after a certain adaptation period less energy is required for the metabolic processes, since the rate of loss of body tissues was higher during the first week of fasting and lower during the subsequent weeks, when adaptation had been reached. These studies support results of various other experiments by Hickman (1959) on *Platichthys stellatus*, Muir *et al.* (1965) on *Kuhlia sandvicensis*, and Brett (1965) on *Salvelinus fontinalis*. In all these experiments the rate of metabolism during the initial period of starvation was higher than during the subsequent period, showing an adaptation to the deprivation of external energy source. According to Beamish (1964a) this adaptation is associated with reduction of spontaneous activity by the fish (p. 127).

The minimum amount of food a fish must eat to provide for the essential metabolic processes, while maintaining its body weight, is 'maintenance requirement' (M). The 'tax' or heat increment on this source of energy is higher than that on tissue energy. While the main nutrient utilized from the tissues is fat, the nutrients most abundant in food are protein and carbohydrate, which are utilized to a lesser extent than fat. In addition to the levy for maintenance as heat loss, there is an ancillary energy requirement associated with the consumption of food and disposal of the metabolites which pass through the fish. These require extra energy. Both the loss of energy due to lower utilization of food nutrients and the additional energy requirement due to feeding are included in the 'specific dynamic action' (SDA). Thus:

$$M = Q_{m} + SDA \tag{2}$$

Figure 2. Changes in average body weight of red tilapia (*Oreochromis* sp.) of about 15 g starved or fed at 0.5, 1.0 and 3.0% of their body weight per day (from Hepher *et al.*, 1983).

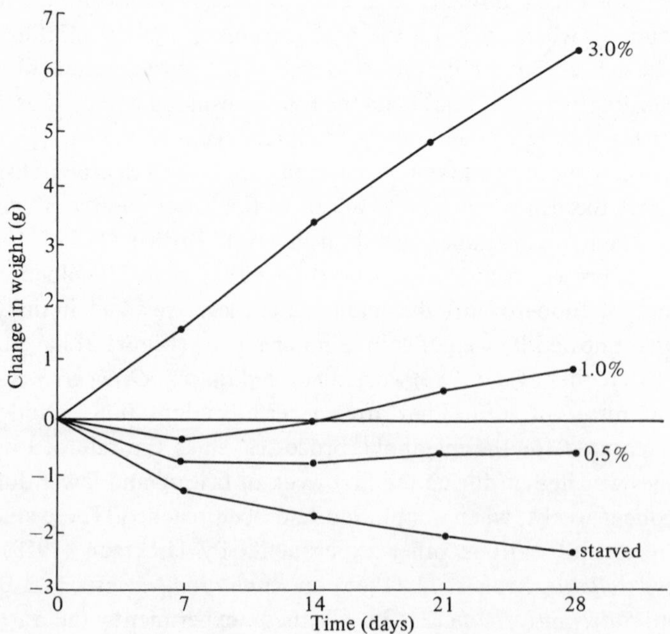

The food energy locked in the newly synthesized fish tissues (= growth, g) can be easily measured. However, it should be remembered that here, too, food energy is not transformed completely into tissue energy, and that some is lost. This loss is both for handling of the extra food consumed to promote the growth, and for the synthesis of the tissues. The amount of food energy required for growth (G) is thus:

$$G = g + SDA \tag{3}$$

Determining the food ration

For the determination of food ration the balance of energy must be calculated, taking into account both the final net energy (NE) required by the fish for maintenance and growth and the losses during transformations:

$$R = NE + SDA + E_m + E_f \tag{4}$$

The ratio between the net energy and the food ration gives the utilization rate of the food. The practical aspects of food utilization are discussed in Chapter 10. However, for a better understanding of the relationships between the components of equation (4) a discussion on those between the net energy and the SDA, or in other words, on the utilization rate of the metabolizable energy will be helpful.

Utilization efficiency of metabolizable energy (ME) depends on the process for which the energy is required – whether for maintenance or for growth. Anabolic processes and the synthesis of tissues usually require some extra energy, which may result in a smaller portion of the metabolizable energy being deposited as body tissues than that utilized as free energy for maintenance processes. Utilization of ME for maintenance is therefore discussed separately from utilization for growth.

It must be remembered that the calculation of the utilization efficiency of ME for maintenance metabolism as measured on starved fish includes some utilized heat increment. It is more appropriate, therefore, to call the utilization efficiency for maintenance 'apparent utilization efficiency'. According to Kleiber (1961) and Windell (1978), the partial apparent utilization efficiency of food ME for maintenance (E_{pm}) can be expressed by the equation:

$$E_{pm} = \frac{\text{Energy requirement for maintenance as supplied by body tissues}}{\text{Energy requirement for maintenance as supplied by food } ME} \tag{5}$$

Partial efficiency for maintenance can also be determined from feeding experiments in which two levels of feeding are employed. The one is at, or somewhat below, maintenance requirement and the other is lower. In this case E_{pm} can be calculated from the ratio between the tissue energy which was saved from loss through feeding at the higher level and the extra food energy added to save it:

$$E_{pm} = \frac{\text{Difference in energy utilized from body tissues}}{\text{Difference in food } ME \text{ ingested}} \tag{6}$$

By multiplying this value by 100 the ratio is expressed as a percentage of metabolizable energy ingested.

Harper (1971, quoted from Beamish, 1974) stated that if each food ingredient is provided separately to an endothermic animal, the apparent partial efficiencies for maintenance are 95% for carbohydrates, 87% for lipids and only 60% for proteins. Table 1 presents some data on partial apparent efficiencies of food *ME* for maintenance for two simple-stomached animals as given by McDonald *et al.* (1966). E_{pm} for fish is discussed in greater detail in Chapter 4. It should be mentioned here, however, that E_{pm} in fish is much lower than in the higher vertebrates mentioned in Table 1, and is usually about 50%, with a smaller variation between food ingredients (Hepher *et al.*, 1983).

Partial efficiency for production of body tissues is lower than that for maintenance. It will, naturally, depend on the efficiencies of the anabolic processes related to the synthesis of body proteins and fats. In general, these processes are more complicated than the catabolic processes and require extra energy. Brody (1945), Kleiber (1961) and Warren & Davis (1967) call the ratio between growth and metabolizable energy used to attain it, after maintenance metabolism requirement has been satisfied, 'utilization efficiency of energy for growth' (E_{pg}) and also 'net efficiency'.

$$E_{pg} = \frac{G}{ME - M} \tag{7}$$

Table 1. *Utilization efficiency of metabolizable energy from various feedstuffs used for maintenance metabolism by dog and rat*

Feedstuffs	Animal	Utilization efficiency
Glucose	Dog	95%
Starch	Rat	88%
Olive oil	Dog	95%
	Rat	99%
Casein	Rat	76%

Source: From McDonald *et al.* (1966).

From theoretical calculations on the efficiency of synthesis processes through the appropriate metabolic pathways it was found (McDonald *et al.*, 1966) that from 100 kcal of glucose only 70 kcal of fat can be produced. In protein synthesis the energy cost of linking amino acids together is relatively small. If the amino acids are present in the right proportion, the efficiency of protein synthesis is about 80%. However, if the amino acids are not present in the correct proportions, some are deaminated and the efficiency falls. In relating the figures given above to the efficiency with which metabolizable energy is utilized, one should add the loss of energy due to *SDA*. Maximum theoretical partial efficiency for tissue production then declines to about 66% of the metabolizable energy for the production of fat from glucose, and to about 61% for production of proteins from protein precursors.

The estimation of the actual partial efficiency for growth *in vivo* is somewhat complicated since the resulting tissues may contain protein and fat in varying proportions. Such an estimation therefore requires analysis of the added tissues. When adult farm animals are fattened, the main tissue produced is fat and it is relatively easy to determine the efficiencies of different feeds for fat production. Such values, taken from McDonald *et al.* (1966) are presented in Table 2. In general, there is an agreement between these values and the calculated theoretical values given above, with variations due to the animals and the feedstuffs. Comparing the values in Table 2 with those in Table 1 shows that the net efficiencies for fattening of higher vertebrates are indeed lower than the partial efficiencies for maintenance.

The partial efficiencies for both maintenance and for growth may change with the feeding level. In most cases, however, such changes are slight. If we assume that the relative efficiencies for maintenance and

Table 2. *Utilization efficiencies (%) of metabolizable energy of various feedstuffs for the production of fat by pig and fowl*

Feedstuffs	Pig	Fowl
Starch	76	57
Ground nut oil	86	78
Protein	62	55
Barley	87	—
Corn	74	—

Source: From McDonald *et al.* (1966).

growth do not change with feeding level, Figure 3 represents graphically the relationships between metabolizable energy and heat increment in animals when *SDA* for growth is higher than for maintenance. In such conditions, for calculating the food ration the *SDA* must be treated separately from other energy losses, Also, *SDA* for maintenance must be distinguished from *SDA* for growth (see also Ursin, 1967). The losses through faecal and metabolic excretion can be grouped together (p) and taken as a function of the food ration:

$$p = \frac{R - (E_f + E_m)}{R} \tag{8}$$

and then:

$$pR = ME = M + G = \frac{Q_m}{E_{pm}} + \frac{g}{E_{pg}} \tag{9}$$

or

$$R = \frac{M + G}{p} = \left[\frac{Q_m}{E_{pm}} + \frac{g}{E_{pg}}\right] \times \frac{1}{p} \tag{10}$$

Most of the conclusions which led to the development of equation (10) were based on values found in higher vertebrates, mainly farm animals. Are these also true for fish? Losses of energy through *SDA* in fish are considered by many authors to be related to the food consumed, without distinguishing between the different functions of maintenance and growth. The *SDA* is then taken as a certain ratio of food energy (e.g. Beamish, 1974). Kerr (1971a) and Kitchell *et al.* (1974) state that there is a linear

Figure 3. Relationship between metabolizable energy intake and heat production (after McDonald *et al.*, 1966).

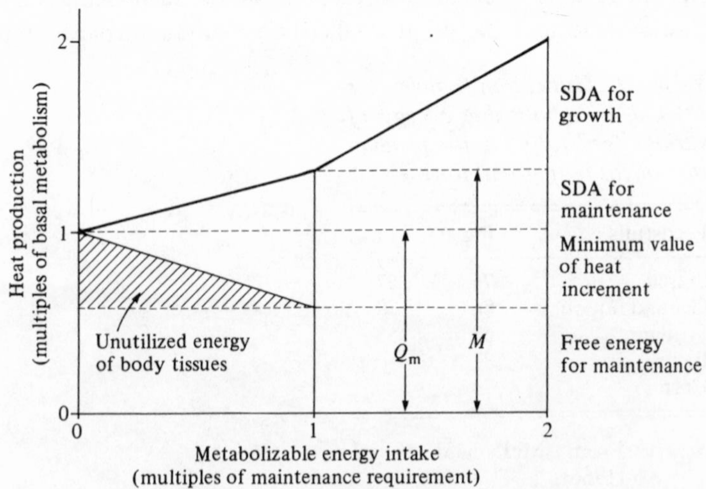

relationship between *SDA* and food consumption when the latter exceeds the maintenance metabolism. This approach is expressed graphically in Figure 4, after Smith (1980), which represents the food energy partition in the body as a function of feeding level. If *SDA*, like other energy losses, is indeed proportional to the ingested food energy, all losses of energy can be accumulated and the total energy loss (p_t) can be expressed as a fraction (or percentage) of the ingested food:

$$p_t = \frac{R - (E_f + SDA)}{R} \tag{11}$$

and then:

$$p_t R = NE = Q_m + g \tag{12}$$

Food ration is then expressed as:

$$R = \frac{Q_m + g}{p_t} \tag{13}$$

The last two equations were given (though with different notations) by Winberg (1956), who also gave a rough estimate for p_t as 0.8.

By replacing the symbols in these equations with actual values obtained for pond fishes, it should be possible to determine their requirement for food energy. Some attempts have been made at estimating these values for various fish species. Table 3 gives, for example, some estimates of p_t. These seem to support Winberg's rough estimate, and indeed, the equations and

Figure 4. Partition of ingested energy in a growing fish at various feeding levels (after Smith, 1980).

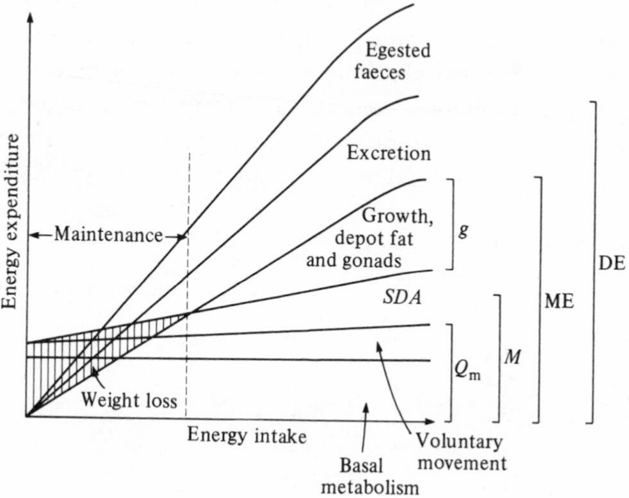

the estimate for p_t were quoted and subsequently used by a number of authors (e.g. Paloheimo & Dickie, 1966a). However, later studies carried out by more reliable methods suggest that Winberg's estimate is too conservative on the one hand and too general on the other. Losses of food energy, especially in growing fish, seem to be higher than the 20% estimated by Winberg. The undigestible component of the food egested by fish usually amounts to at least 14–15% and even more for certain feeds and fish species (see also Chapter 2 and Appendix I). Thus, McComish (1970, quoted from Kitchell *et al.*, 1974) estimated this loss for bluegill sunfish (*Lepomis macrochirus*) at 20%, Wissing (1974) with white bass (*Morone chrysops*) at 27–39%, and Tandler & Beamish (1981) with largemouth bass (*Micropterus salmoides*) at 28% of the ingested food. McComish (1970, quoted from Kitchell *et al.*, 1974) and Beamish (1974) estimated the losses through metabolic excretion at 3–8% of the ingested food and losses by *SDA* to an additional 13–14%. However, other estimates of *SDA*, discussed in greater detail in Chapters 4 and 5, give much higher values. Brett & Groves (1979) estimated the metabolizable energy fraction of the food in herbivorous fish to amount to about 40% of ingested energy. For omnivorous fish they consider a range of 25–30%, and only with carnivorous fish does the loss approach Winberg's (1956) generalized estimation of 20%. Even higher losses of food energy were found by Fischer (1970) in grass carp (*Ctenopharyngodon idella*). He estimated the assimilated energy at about only 13% of the ingested energy. Thus, it can be seen that p_t fluctuates at a very wide range. The variations in the estimates of p_t, as in the specific sources of energy losses in fish, are not surprising. It must be remembered that each of the processes contributing to the energy loss is affected by a considerable number of factors. Some of these factors are related to the fish itself, such as the species, weight, activity, physiological condition and genetic characteristics, some to environmental conditions, such as water temperature, water quality, and the presence of toxicants and catabolites, whilst others are related to the nature of the available food, its composition, quantity, etc. Since food

Table 3. *Estimates of p_t in fish by various researchers*

Author	Fish species	p_t	Remarks
Winberg (1956)	General	0.8	
Davis & Warren (1965)	*Cottus perplexus*	0.82	May not acccount for metabolic excretion
Brocksen *et al.* (1968)	*Salmo clarki*	0.819	May not acccount for metabolic excretion
Brocksen & Bugge (1974)	*Salmo gairdneri*	0.712–0.848	May not acccount for metabolic excretion
Staples & Nomura (1976)	*Salmo gairdneri*	0.717	
Davis (1967)	*Carassius auratus*	0.9	

efficiency may be affected by a multitude of factors, it is obvious that the use of general mean values for estimating food requirements, without taking into account the actual nutritional state of the fish, may lead to an erroneous estimation of the food ration. For the values inserted in the above equations to be meaningful, the effect of the factors mentioned above should be known and considered.

In this book a discussion on the factors affecting food requirement and utilization is attempted. Since one must take into account the arguments discussed above that there is a difference in partial efficiencies for maintenance and growth, these parameters are discussed separately in Chapters 4 and 5.

2

Ingestion, digestion and absorption of food

The alimentary canal

The food utilized by fish in natural habitats and ponds is very diverse, as also are their feeding habits. Some are carnivores, omnivores or detritivores. Some feed from the bottom, while others feed from the water column. As a result of their evolution, teleosts have developed special morphological and physiological characteristics to match their food and feeding habits. These adaptations are principally characterized by the morphology of the alimentary canal, but degree of relationship between the morphology of the alimentary canal and food taken varies. Different species with the same type of diet may differ in the structure of the alimentary system, but functional adaptations related to the nature of food and feeding habits usually remain similar. Morphological adaptations of fish to their feeding habits have been reviewed extensively by Kapoor *et al.* (1976a); some of these will be discussed here briefly.

Location of food

For most pond fishes chemoreception is most important for locating food, since sight is frequently quite poor in these fishes, especially in the turbid water of the pond. The gustatory system of the fish was well reviewed by Kapoor *et al.* (1976b). In teleosts taste buds are commonly found on the body and even on the fins. The number of taste buds may vary greatly from one part of the body to another, but the greatest number are found in the region most closely associated with food contact, namely, mouth, pharynx, lips, gill rakers, gill arches and barbels. Atema (1971) summarized the distribution of taste buds in the yellow bullhead (*Ictalurus natalis*). He found five taste buds per square millimetre over the body with slightly higher counts on the dorsal part, 10/mm^2 on the nasal area and lips, and 10–25 mm^2 on the barbels. According to Atema (1971) and Finger (1983) the taste buds differentiate into two specific taste systems. One is innervated by the facial nerve, subserving all the taste buds on the body skin, lips and anterior part of the mouth; the other is innervated by the vagal and glosso-pharyngeal nerves and contains all the taste buds on the posterior part of the mouth and gill arches. The functions of these two taste systems were determined by selective ablations removing either the sensory area of the facial lobe or that of the vagal lobe. These operations showed that the facial taste system functions in accurate localization of the food, and triggers the pick-up reflex. The vagal taste system controls the swallowing reflex. Without its control the fish cannot swallow its food.

The barbels, which can be found in a number of fish species, such as various cyprinids and catfishes, serve as important sense organs. The number of barbels is different for each species. Catfishes have eight particularly large barbels. Taste buds are pear-shaped, embedded in the epidermis. They are generally more numerous on the distal parts of the barbels. The barbels are also rich in sensory buds sensitive to touch. The fishes with barbels may well find their food by a combination of taste and touch (Rajbanshi, 1966, 1979).

In most fishes the olfactory organ is an important chemical receptor, reacting to chemical changes in water. Nostrils are located on each side of the upper part of the snout in front of the eyes. Unlike the situation in higher vertebrates, the nostrils do not normally communicate with the mouth cavity, except in a few species of fishes, but form two separate nasal blind sacs. There are usually two nostrils to a nasal sac on each side of the snout. Water enters the forward nostril and after irrigating the sac is expelled through the rear nostril. This circulation may depend in some species (e.g. Anguillidae) on currents created by the beating of cilia in the nostrils, whereas in others (e.g. Cyprinidae) water is deflected through the front nostril by a special skin fold between the two nostrils, as the fish swims. In some fishes, such as cichlids, each sac has only a single nostril. In these, water circulation depends on compression and relaxation of the nasal sacs, which correspond to the respiration movements. The nasal sacs are lined with an undulating and pleated sensory epithelium that is linked to the brain through the olfactory nerves. Bakhtin & Filiushina (1974) conducted an electron microscopy study of the olfactory epithelium goblet cells and found ultrastructural diversity in their secretion. Some secrete mucopolysaccharides and some lipoproteins. According to these authors the different mucous secretions form on the surface of the epithelium complexes which transport molecules of odorous substances to the receptor membrane.

The mouth and buccal cavity

The position and size of the mouth show a close relation to the location and size of food items. The mouth is usually situated at the tip of the snout and parallel to the longitudinal axis of the fish in species that are primarily adapted to seizing food located in front of them, such as the common carp. In species that obtain food primarily from below, such as some loaches (Cobitidae), the mouth is often located inferiorly, on the lower side of the snout. When the fish generally feeds upwards in the water column, as with plankton filter feeders (e.g. silver carp), the mouth is superior, located on the upper side of the snout and directed upwards.

The relative size of the mouth appears to depend on the size of the food particles ingested. Aleev (1963) gives the relative length of the mandibles of various fish to characterize the dimensions of their mouths. He showed that species that feed on plankton, plant food, or small benthic organisms generally have smaller mouths than the predatory carnivores.

In many fish species the mouth can protrude forward. The fore-lip or premaxilla is connected by a flexible membrane-like tissue to the skull and can be moved by a pedicel which slides against the frontal part of the skull. When protruded the length of the mouth can extend from 5 to 25% of the length of the head (Harder, 1975). The forward extension of the mouth facilitates the grasping of food and prey and increases the efficiency of grabbing or sucking of food from the bottom, while enabling a hydrodynamic movement when the mouth is retracted.

Fish have three kinds of teeth: mandibular, buccal and pharyngeal. The more active predators, such as the pikes (*Esox* spp.) and the salmonids, have strong jaws and sharp mandibular teeth on them. These teeth serve to bite and catch their prey. Others, such as some tilapias, scrape their food from rocks or feed on invertebrates, and have only small mandibular teeth on their jaws, or more often only rough, scraper-like lips. Cyprinidae have no teeth at all on their jaws.

Some fish have buccal teeth in the mouth to hold the prey and prevent its escape if, as is usually the case, it is swallowed alive. But the most common are the pharyngeal teeth. These are situated on the modified fifth gill arch, which does not carry gills. Pharyngeal teeth can be very sharp to cut the food, as in the grass carp (*Ctenopharyngodon idella*), but more often they crush and grind the food, as in common carp and goldfish (King, 1975). It is obvious that in fish it is better for the grinding to be done in the pharynx than in the mouth, as it is in mammals, since the food particles may pass over the gills and gill rakers thus obstructing respiration. Common carp have molar-like pharyngeal teeth which grind the food against a horny pad located on the floor of the skull.

Fish which feed on small particles must filter their food and separate it from the water. This is done inside the mouth by the modified gill rakers. These are arranged in two rows on each gill arch like the teeth of a comb, forming with the rakers of the neighbouring arches an efficient screen which guards the delicate gills from harm by sharp particles and filters food particles. The gill rakers can be more or less densely spaced according to the size of the food particles usually eaten by a particular fish species. In some plankton filtering fishes the gill rakers have developed into very fine sieves which, in conjunction with a curtain of mucus, trap the microscopic organisms. The silver carp (*Hypophthalmichthys molitrix*) has a specially fine sieve which can filter plankton as small as 20–25 μm (Antalfi & Tölg, 1971). The concentrated slurry then flows along special grooves to the pharynx.

Many mucous cells are found in the buccal cavity walls. These seem to have been developed from ordinary epithelial cells (Sinha, 1975). The mucus the cells secrete help to coagulate and bind small food particles and to smooth the swallowing of large prey. The number of mucous cells has been found to be related to the food habits of the fish. Sinha (1975) found a low count of mucous cells in the carnivorous *Clarias batrachus* and high count in the herbivores *Labeo rohita*.

From the mouth the food passes into the pharynx and then to the oesophagus. This is a short, straight muscular tube between the mouth and the digestive tract. It has a ciliated epithelial lining which is rich in mucus secreting cells. In predaceous species such as pike and perch, the oesophagus is capable of a great degree of stretching, which makes it possible to swallow large prey. The powerful oesophageal sphincter serves to prevent the entry of water from the respiratory system when food is swallowed.

Stomach and intestine

From the oesophagus the food passes into the digestive tract proper. A number of fishes, among them many cyprinids, lung fishes and cyprinodonts, have no stomach or acid secretion associated with it. The food passes into the intestine where digestion occurs. Though the lack of stomach has been considered by many authors a primitive stage of evolutionary development, Barrington's (1957) opinion is that it is in a later, more specific, development that these fish species have lost the stomach, which existed in earlier developmental stages. As a support to this theory Barrington points out that the lack of a stomach appears in various unrelated groups of fish, while the same genus may include species with and species without a stomach.

The intestine is a simple tube, often long and coiled, especially in herbivores. Where the stomach is absent the fore-gut extends to the posterior end of the abdominal cavity, where it coils first. The fore-gut is wider than the rest of the intestine and serves as a pseudo-stomach (Figure 5). There are no clear boundaries between the oesophagus and the fore-gut or between the fore-gut and the hind-gut, except that the latter is narrower. The intestine is lined throughout with a simple mucoid columnar epithelium which shows a zigzag pattern along the intestine. The mucoid epithelium of the alimentary tract in various vertebrates has been shown to contain mucous substances of marked chemical diversity (Western, 1971), presumably of physiological significance in the digestive processes, and also playing a protective role against mechanical and chemical injuries, including auto-digestion. Yamada & Yokote (1975) found that the intestinal mucus of the Japanese eel is composed of elaborate mucosaccharides, which seems to be true for most fishes.

Electron microscopy studies in some animals, including fish, have shown that there are sub-microscopic cytoplasmic projections (microvilli) and folds in the epithelium. Most of the digestion takes place between the microvilli or on the cell membranes. Ugolev (1960, 1965), who reviewed this form of digestion, differentiates between cavital digestion, which is produced by enzymes secreted into the gastrointestinal cavity, and

Figure 5. A ventral view of the intestine of a stomachless fish (a cyprinid) (according to Sarbahi, 1951).

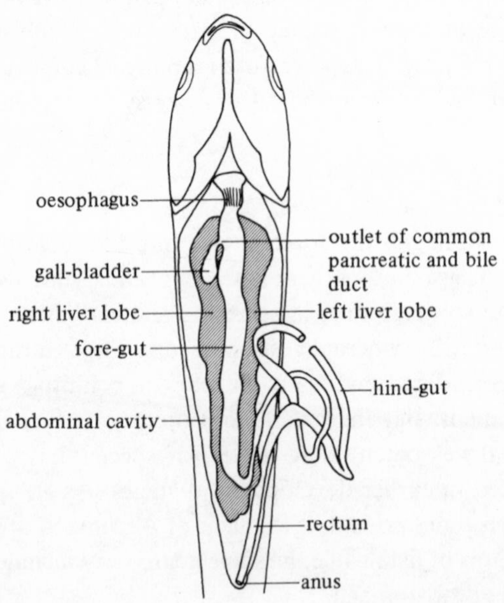

membrane digestion, which is due to enzymes adsorbed from the chyme and enzymes structurally associated with the membranes of the intestinal cells. These latter enzymes are responsible, according to Ugolev, for the final hydrolysis of the nutrients, the products of which are absorbed immediately into the cells. Berman (1967) called this 'parietal digestion'. He found that the level of parietal digestion on the epithelium of a one-year-old common carp was 20 times higher than that in the intestinal lumen, but in a two-year-old fish it was only nine times higher.

Variations in the ratio between the activity of the parietal digestion and that in the lumen also occur with season. Kuzmina (1977) determined this ratio with regard to α-amylase in 12 species of fish. In most species parietal digestion, as compared to cavital digestion, was highest in summer. Differences were found also between those species.

Spannhof & Plantikow (1983) found that parietal digestion is also important in trout, although not to the same degree as mentioned above for the common carp. They diluted intestinal juices from rainbow trout (*S. gairdneri*) with 43% glycerol at a ratio of 1:50 and added pieces of intestine to the mixture. Amylase activity in the mixture was measured and compared with mixtures containing no intestinal tissues. They found that the amylase activity was higher in the former by 12–39%. Since the effect of the parietal digestion is obviously proportional to the surface area of the intestine, the difference in the effects of parietal digestion in common carp and rainbow trout may possibly be associated with the relatively smaller intestinal surface area of the trout. In addition to the activities mentioned above, the intestinal mucosa is engaged in the active transport of nutrients into the blood.

The relatively narrow and often long and coiled intestine which extends from the fore-gut of stomachless fishes occupies most of the abdominal cavity. It may be differentiated into three portions: a proximal portion (mid-gut 1), a distal portion (mid-gut 2) and a short posterior intestine, or rectum, which ends at the anus. Though these are not differentiated much in their morphological characteristics, they seem to vary somewhat in their physiological function. According to Noaillac-Depeyre & Gas (1974) the absorption of lipid occurs essentially in the fore- and first mid-gut portion, while the absorption of protein occurs in the distal portion of the mid-gut.

The liver varies in shape and size among different fish species. In most warmwater pond fishes the liver is a relatively large, elongated, brown coloured organ composed of two lobes which extend along the fore-gut. The lobes unite at their anterior and posterior ends. The gall-bladder, which contains a greenish-yellow bile, is found between the right lobe and the fore gut. The pancreas usually does not appear as a localized organ but rather as diffused islands of secretory tissue. These are variable in location even within a single species. They can be located mainly on the liver and around the hepatic portal vein, in which case they form a compounded organ – the hepatopancreas In other cases the islands of pancreatic tissue may be scattered not only on the liver but also on the intestine, the mesentery of the intestine and the pyloric caeca, and around the spleen and gall-bladder. The pancreatic duct joins the bile duct to form a common duct leading to the fore-gut or, in fish having a stomach, to the duodenum just behind the stomach. In most warmwater pond fishes the endocrine components of the pancreas, the islets of Langerhans, consist of small cells also diffused alongside the enzyme secreting cells.

The stomach, when present, is situated just behind the oesophagus. The stomach wall consists of a number of layers, characteristic for most vertebrates (Kapoor *et al.*, 1976a). Lining the stomach is a typical columnar epithelium rich in gastric glands which secrete both acid and enzymes. Only one type of secretory cell has been histologically identified in the gastric glands of teleosts and no physiological division of secretory function was found (Western & Jennings, 1970; Kapoor *et al.*, 1976a; Shafi, 1980). However, a distinction on the basis of staining reaction has been made in some fish between 'neck cells', generally mucus producing, and 'granular' enzyme secreting cells (Kapoor *et al.*, 1976a). The food in the stomach is usually completely surrounded by a sheath of mucus which prevents a direct physical contact between the food and the gastric mucosa (Western, 1971).

In its simplest form (e.g. in tilapia) the stomach is clavate, but often it is sigmoid consisting of a descending 'cardiac' portion (or 'corpus'), and an

Figure 6. Basic anatomy of the digestive tract of a salmonid fish (after Ellis *et al.*, 1978).

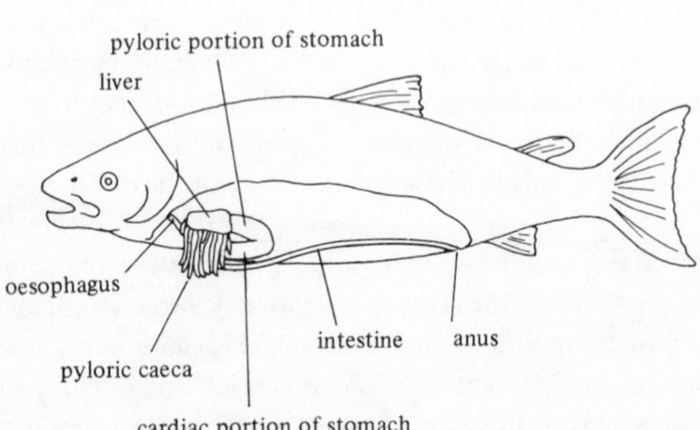

ascending 'pyloric' portion nearest to the intestine (Figure 6). The cardiac portion is egg-shaped and is easily distended when full with food. Some fish have a muscular, gizzard-like stomach which also grinds the food, e.g. the grey mullet (*Mugil* spp.). Verma *et al.* (1974) studied the morphological variations in the stomachs of 15 species of fish. They suggested that the size of the stomach is correlated with the feeding habits of the fish. In most of the herbivores the stomach forms a prominent bulbous sac-like structure. Also in active predators, e.g. *Clarias batrachus* and green snakehead (*Ophicephalus punctatus*), which take a large amount of food at any time, the stomach is well developed and bulbous, while in the case of some less active or burrowing fishes the stomach is relatively smaller and forms a tubular structure.

Unlike higher vertebrates, many fish have a number of blind tubular outgrowths around the pylorus close to where the intestine leaves the stomach. These are the pyloric caeca. They resemble the appendix in man, although they are further forward in the digestive tract. Though the number of pyloric caeca is not always constant in a species, the variation of their number among species has been used in identification of some species. Fish that lack a stomach also lack pyloric caeca. *Mugil* spp. have 7–12 caeca according to species, in *Salmo* spp. their number is 40–50, in *Oncorhynchus* spp. 50–60, and in the marine tunas there are several hundreds. Pyloric caeca are histologically almost identical with the intestine and, as in the intestine, are lined by a single columnar epithelium rich in secretory cells (Lawrence, 1950). Various functions have been suggested for the pyloric caeca, e.g. an accessory food reservoir; a digestive function supplementing that of the stomach and the intestine; absorption of carbohydrates and fats; resorption of water and inorganic ions; and to increase intestinal surface area (Kapoor *et al.*, 1976a). Some of the enzymes secreted in the pyloric caeca are specific. For example, Yoshinaka *et al.* (1973) found a collagenase in the pyloric caeca of yellowtail (*Seriola quinqueradiata*). This enzyme was capable of breaking down connective tissues of the prey which were not affected by regular proteinases. After a long period of starvation the pyloric caeca may degenerate. This may lower digestibility, as was found by Windell (1967) in bluegill sunfish (*Lepomis macrochirus*). Fish starved for 25 days showed noticeably atrophied pyloric caeca and the digestibility dropped by about 50%.

The overall length of the intestine, relative to the length of the fish body, varies among different fish species and seems to be related to their feeding habits. That of herbivores is usually longer than that of carnivores, probably because most of the digestion in the latter takes place in the stomach, and because digestion of food of animal origin is easier than that of plant food which includes hard cell walls. The longest intestine relative to body length occurs in microalgae filter feeders. The relative length of the intestine of omnivores may vary. According to Al-Hussaini (1947) and Kapoor *et al.* (1976a) this variation depends upon the proportion of indigestible to digestible materials in their diet. Hsu & Wu (1979) studied this aspect in eight species of fish having different food habits. The relative length of gut ($RLG = \frac{\text{length of gut}}{\text{length of body}}$) found by them are given in Table 4. The values given in this Table fit well with this concept. Al-Hussaini (1947) and Kapoor *et al.* (1976a) conclude that, in general, the lengths of the intestine in carnivores, herbivores and microphages are one half, two and five times their respective body lengths, while that of omnivores varies between two and six times their body length. The relative length of the gut may change within the species with the change in their feeding habits. Sinha & Moitra (1975) found in *Labeo rohita*, and Lassuy (1984) in damselfish, *Stegastes lividus*, that the *RLG* of the fry, which feed on zooplankton, is lower than that of herbivorous adult fish. Lassuy (1984) concludes that juvenile fish compensate for the lower efficiency in assimilation of plant food due to the shorter intestine and its small diameter, by including a higher percentage of animal material in their natural diets.

The relative length of the intestine also seems to be related to the duration of the intestinal digestion. Svob & Kilalic (1967) measured the digestion periods in the intestine of various cyprinids by X-rays, and

Table 4. *The relative length of the gut (RLG) of fish species having different feeding habits*

Fish species	*RLG*
Carnivores	
Japanese eel (*Anguilla japonica*)	0.46
Formosan snakehead (*Channa maculatus*)	0.57
Formosan catfish (*Clarias fuscus*)	0.68
Herbivores	
Grass carp (*Ctenopharyngodon idella*)	2.16
Microphages	
Silver carp (*Hypophthalmichthys molitrix*)	5.28
Tilapia (*Sarotherodon mossambicus*)	6.29
Omnivores	
Common carp (*Cyprinus carpio*)	2.04
Goldfish (*Carassius auratus*)	5.15

Source: From Hsu & Wu (1979).

found differences among the species which were proportional to the number of intestinal coils and length. It should be remembered, however, that one of the main factors affecting the digestion is the total surface area of mucosal folds, and this is not necessarily proportional to the length of the intestine (Barrington, 1957).

The digestive enzymes

Digestion is the process by which food in the digestive tract is split into simpler compounds capable of passing through the intestinal walls to be absorbed in the blood stream. Proteins are hydrolysed into free amino acids or short peptide chains of several amino acids, carbohydrates are broken down into simple sugars, and fats into fatty acids and glycerol. These processes are carried out by the digestive enzymes while the food is transferred along the digestive tract. The mucus secreted in the mouth and pharynx of fish differs from that of higher vertebrates in that it seems to lack digestive enzymes. Enzyme secretion in the oesophagus has been found in a number of fish species (Sarbahi, 1951; Nagase, 1964; Kawai & Ikeda, 1971) but the digestion proper starts in the stomach, or fore-gut in fish lacking a stomach.

Digestive enzyme activity may change with fish age, physiological state and season. Kitamikado & Tachino (1960a) found that amylolytic activity in young rainbow trout is quite high and increases as the fish grows, reaching a peak at a weight of 100 g. Over this weight the activity decreases and in large fish (about 1000 g) the amylolytic activity is considerably lower than in young fish. An even more pronounced difference was found for the proteolytic activity of the rainbow trout. In its first developmental stages this activity is considerably lower than in the adult fish and is again reduced with age (Kitamikado & Tachino, 1960b). Decreasing proteolytic and amlolytic activity with age was also found by Morishita *et al.* (1964) and Stroganov & Buzinova (1969a, b) in a number of other fishes. Kawai & Ikeda (1971) could not find α-glucosidase and lactase in small common carp (6.5 g), though they were found in larger fish. Ananichev (1959) found that amylase activity of burbot (*Lota lota*), pike-perch (*Stizostedion vitreum*) and bream (*Abramis brama*) was at a maximum during the period of preparation for spawning. Onishi *et al.* (1974) studied the changes in activity of digestive enzymes in pyloric caeca of rainbow trout during sexual maturation. They found that the activity ratio between amylase and protease increased, which seems to indicate a higher utilization of carbohydrates during this period. Seasonal differences were also found in the activities of the digestive enzymes. Ananichev

(1959) found the maximum activities of digestive enzymes to coincide with periods of intensive food intake. Thus, for burbot (*Lota lota*) it was in winter, while for pike-perch (*Stizostedion vitreum*) and bream (*Abramis brama*) it was in summer. Chepik (1964) found that the activity of the digestive enzymes of common carp was higher in spring than in any of the other seasons. The adjustment of enzyme activity to the seasonal food intake may be correlated with the effect of temperature on the enzyme activity, as found by Hofer (1979a, b) for roach, *Rutilus rutilus*, or can be achieved independently of temperature as was found by him for the rudd, *Scardinius erythrophthalmus*.

Each digestive enzyme has its highest activity at a certain optimum temperature. Jančařík (1949), Bondi & Spandorf (1953) and Kitamikado & Tachino (1960b) determined the optimum temperature for protease activity in common carp and rainbow trout at 38–40 °C (somewhat higher – 45 °C – for trout's trypsin activity). A higher optimal temperature of about 50 °C was determined by Chiu & Benitez (1981) for the activity of intestinal amylase in milkfish (*Chanos chanos*). Above the optimal temperature there is a sharp drop in enzyme activity. However, since this temperature is above the lethal temperature for most fishes, this is not applicable to live fish. Thus, within the range of temperatures found in fish ponds, in many fish the higher the temperature the higher the enzymatic secretion and activity (Schlottke, 1938–9; Nordlie, 1966; Smit, 1967; Trofimova, 1973). However, in many other fish there is a clear temperature compensation. When fish are acclimated to lower temperature they secrete more enzymes than those acclimated to higher temperatures, when both are held at the same temperature. Due to this compensation Smit (1967) found only small variation in enzyme secretion of acclimated catfish, *Ictalurus nebulosus* within the temperature range of 20–30 °C. This may be the cause of the different effect of temperature on the digestive enzyme activities in various species. Morishita *et al.* (1964) compared the activities of the digestive enzymes of a number of fish species. They found that those of salmonids were more active at lower temperatures than those from warmwater fish studied (*Seriola quinqueradiata, Anguilla japonica* and *Plecoglossus altivelis*).

The secretion of acid

In the stomach there is usually a secretion of hydrochloric acid which lowers the pH. As in the higher vertebrates, the acid is formed by a reaction between carbonic acid (H_2CO_3) and sodium chloride (NaCl), both supplied to the acid secretory cells) through the circulatory system. The reaction results in the production of mono-sodium carbonate ($NaHCO_3$)

and hydrochloric acid (HCl). In fish, contrary to the case in mammals, the same cells secrete both the acid and the digestive enzymes (Barrington, 1957).

Acid secretion increases considerably after feeding, but it may continue, although at a very limited rate, even in fish with empty stomachs or fasting. Holstein (1975) studied the gastric acid secretion in the cod (*Gadus morhua*), using a catheter which drained the stomach. Unstimulated acid secretion was very low, not exceeding 8 µmol H^+/kg/h. Western (1971) studied the change in gastric pH after feeding of bullheads, *Cottus gobio*, and *Enophrys bubalis* and found a similar pattern in both. Gastric pH, which was neutral before feeding, became progressively more acidic after ingestion of food, reaching pH 2.0 some 30 h later. Gastric acidity was rigidly maintained at this level while food remained in the stomach, but after emptying, the gastric pH slowly rose to neutrality again. Moriarty (1973) found that acid secretion in *Oreochromis niloticus* has a diurnal cycle which corresponds to the feeding cycle. It increases during the early morning, reaching a peak in the late morning when feeding is at its maximum, and declines to its lowest level during the night. The concentration of acid in the stomach can be quite high. Bowen (1976) found that the pH in the stomachs of actively feeding *Oreochromis mossambicus*, examined immediately after capture, was consistently below 2.5 and usually in the range 1.25 – 1.5, which is lower than that found in mammals. Moriarty (1973) found a pH of 1.4 – 1.6 in the stomach of *O. niloticus*. In such acid conditions prevailing in the stomach of tilapia, the green chlorophyll is converted into brown phaeophytin. Moriarty attributes the ability of *O. niloticus* to digest blue-green algae to the high acidity in its stomach. Blue-green algal cells have tough walls and usually are not easily digested by most fish. Since the cell wall is rich in pectin, it was assumed that the digestion of these algae is dependent on the presence of enzymes which break down pectin, such as pectinesterase or polygalacturonases. However, Fish (1960) could not find these enzymes in the stomach of *O. mossambicus*, even though this fish can also digest blue-greens (Fish, 1960; Tropical Fish Culture Research Institute, Malacca, 1961). This supports Moriarty's assumption that it is mainly the high acidity which contributes to the blue-greens' digestibility. Blue-greens can also be digested by some other fish species such as milkfish (*Chanos chanos*), pearlspot (*Etrophs suratensis*), grey mullet (*Mugil cephalus*), and *Haplochromis nigripinnis* (Pillay, 1953; Hickling, 1970; Moriarty, 1976). This may also be associated with a low gastric pH. Digestion of blue-green algae is not common to all tilapias. Fish (1951) found that *Tilapia esculenta* from Lake Victoria can digest diatoms but not blue-green or green algae. A somewhat higher pH than that mentioned above for the tilapias has been reported for other species of fish, e.g. between 2 and 4 in the stomach of channel catfish (Page *et al.*, 1976), and between 2 and 6 in the stomach of *Dicentrarchus labrax* (Alliot *et al.*, 1974).

In fish lacking a stomach there is also no acid secretion. The pH of the entire digestive tract is neutral or slightly basic. Al-Hussaini (1949) found a pH of 6.12–7.72 in the intestine of common carp and *Rutilus* sp. Cockson & Bourn (1973) found the pH of the intestinal fluid of the cyprinid *Barbus paludinosus* to be 5.8 and 7.8 for the fore-gut and hind-gut respectively. The pH of the intestine is probably associated with the secretion of the bile, which is neutral or slightly acid.

In fish possessing a stomach the intestine behind the duodenum also has a basic pH. Western (1971) studied the digestion by *Cottus gobios* and *Enophrys bubalis* and found that in non-feeding fish of both species the mid-gut is maintained at neutral or slightly alkaline pH. The pH of the intestine increases in feeding fish, though it fluctuates due to the influx of acidified chyme from the stomach and with the secretion of bile. According to Page *et al.* (1976) the pH of the bile of channel catfish is 6.1–7.5. In the majority of the cases the pH of the intestine is maintained within the range 7.0–9.0 (Alliot *et al.*, 1974, for *Dicentrarchus labrax*; Page *et al.*, 1976, for channel catfish). The fluctuations in pH were most marked in the anterior part of the intestine and the pyloric caeca where the mixing of the fluids and the chyme takes place, and were reduced in the posterior intestine. The pH of the intestinal contents after mixing is higher than that of either the chyme or the bile, indicating that a further source of alkali, discharged simultaneously into the intestinal lumen, is involved in neutralizing the chyme. This source is most probably the exocrine pancreas and the intestinal mucosa.

Proteinases

Pepsin is the major proteolyptic enzyme in the fish stomach. It has been extracted in a crystalline form from a number of fishes and found to be more active than mammalian pepsin. Thus, Norris & Mathies (1953) have found that the crystalline pepsin of some tunas (*Thunnus albacares, T. thynnus*, and *T. germo*) is more active than that of the pig when tested against the same unit weight of protein. Also Ananichev (1959) points out that the affinity of fish pepsin for its substrate is about 150 times greater than that of mammalian pepsin. Fish pepsin acts in acid conditions as in higher vertebrates. The pH at which maximum proteolytic activity occurs may vary somewhat among fish species. In *Ictalurus* sp. it is 3–4 (Smith, 1980), in rainbow trout (*S. gairdneri*) 2.5–3.5 (Kitamikado & Tachino,

1960b; Kapoor *et al.*, 1976a) and about 2.0 in some marine flat fishes (Yasunaga, 1972).

Moriarty (1973) found pepsinogen in extracts of the stomach wall of *Oreochromis niloticus*, but its proteolytic activity was low and, according to this author, the digestive process in the stomach of this fish is due mainly to the high acid concentration. Hsu & Wu (1979) compared the pepsin activity of another tilapia, *O. mossambicus*, with that of several carnivorous fishes: Formosan snakehead (*Channa maculatus*), Japanese eel, (*Anguilla japonica*) and Formosan catfish (*Clarias fuscus*), and found the former to be much lower. The ratios of the activities between these species were 1.0:27.1:10.5:8.5 respectively.

In fishes, both with and without a stomach, a mixture of proteolytic enzymes is secreted into the intestine. Since in most cases these enzymes were not isolated, the combined proteolytic activity in the intestine is often referred to as 'trypsin activity'. Some attempts were made, however, to isolate and identify these enzymes. Overnell (1973) found trypsin, chymotrypsin, and carboxypeptidase A and B in the mesentery of the pyloric caeca of cod (*Gadus morhua*). Their origin seemed to be pancreas cells which were diffused in the mesentery. Kalač (1975) isolated and partly purified the proteolytic enzymes from the extracts of pyloric caeca homogenates of mackerel (*Scomber scombrus*), using electrophoresis and polyacrylamide gels. He detected five enzymatic protein bands. Two had a specific trypsin activity. Also Jany (1976) found 'trypsin and chymotrypsin-like' proteases, as zymogens, in the hepatopancreas of *Carassius auratus*, but could not find elastase. Cohen (1981) isolated and purified the enzymes from the pancreas of the common carp. She found that the pancreas secretes chymotrypsin, trypsin, elastase and carboxypeptidase B. The compositions of the chymotrypsin and trypsin from common carp were similar, but not identical, to those of cattle and pigs. The trypsin in carp is richer in the acidic residues of amino acids, especially glutamine/glutamate and tryptophan, than bovine and porcine trypsins, making it anionic. The composition of the carboxypeptidase B was different from that of mammals. Yoshinaka *et al.* (1984), on the other hand, found in the pancreas of the catfish *Parasilurus asotus* two carboxypeptidase B enzymes which were very similar to those of the higher vertebrates. Each of the enzymes found by Cohen (1981) in the common carp pancreas consisted of enzymatically active fractions (isoenzymes) which were of identical molecular weight. Chymotrypsin had two such fractions and trypsin four. The activity of carp chymotrypsin was found to be higher than that of the bovine homologue. Ananichev (1959) states that the affinity of fish trypsin for its substrate is 14 times higher than mammalian trypsin. Also Ooshiro (1971) compared purified proteinase from pyloric caeca of a mackerel with that of mammals and found the former to be much more active than the latter. The purified mackerel proteinase hydrolysed no less than 70% of the peptide bonds of casein, while crystalline bovine trypsin hydrolysed only 15%. Indeed, Dabrowski (1983) found that free amino acid concentration in the gut content of common carp appeared to be several times higher than that in mammals.

The relative activity of chymotrypsin and trypsin may differ in different fish species. Cohen (1981) and Jonas *et al.* (1983) found a higher activity of chymotrypsin than trypsin in common and silver carp, but in sheatfish (*Silurus glanis*) the activity of trypsin was about four times that of chymotrypsin.

All of the intestinal proteolytic enzymes are active at a pH range of 6–11. Bondi & Spandorf (1953, 1954) state that in common carp proteolytic enzymes secreted by the pancreas are more active at pH 7 than at 8.2, but Cohen (1981) showed that the maximum activity of these enzymes is reached at pH 8.0–8.5. Jany (1976) has shown that unlike their bovine counterparts, the trypsin and chymotrypsin of *Carassius auratus* are unstable below pH 6.

The pancreas is the principal protease secreting organ. However, other organs, such as the liver, intestinal mesentery, spleen, gall bladder and even the bile itself show a clear proteolytic activity. It is logical to assume that since, pancreatic tissue is diffused among these organs, part of the activity, if not all, is associated with this diffused tissue. Other important sources of proteolytic activity are the intestine and the pyloric caeca (Sarbahi, 1951; Yasunaga, 1972; Hsu & Wu, 1979). 'Trypsin activity' was detected in extracts from the intestinal wall and mucous epithelium along the entire length of the intestine. Polimanti (1912) identified the intestinal proteolytic enzymes as erepsin, which today is considered in mammals primarily as an intracellular enzyme in the absorptive mucosal epithelium. Jančařík (1964) and Sarbahi (1951) found that the proteolytic activity of the intestinal lumen juice is higher than that of the intestinal wall extract. This is probably the result of the addition of pancreatic enzymes to the intestinal enzymes. Most authors agree that the proteolytic activity of the combined mixture of the pancreatic enzymes with those of the intestine is higher than that of each of them alone (Babkin & Bowie, 1928; Beauvalet, 1933; Jančařík, 1949; Bondi & Spandorf, 1953, 1954; Uchida *et al.*, 1973; Smith, 1980). Jančařík (1949) found that mixing a low concentration (0.05%) of intestinal juice with extracts of the intestinal wall, liver or pancreas activated their proteolytic enzymes. Such activation was also found, though to a lesser degree, when intestinal wall extract was mixed with extracts of the liver and pancreas. Smith (1980) found that the proteolytic activity of the mixture was 10 times higher than the activity of each separately. All these authors conclude that, as in all other vertebrates, the mucous epithelium of the intestine secretes enterokinase which activates the proenzymes (zymogens) secreted by the pancreas into the intestinal cavity. This activation is an irreversible process in which the proenzymes are undergoing a limited proteolysis into trypsin which, in turn, converts more trypsinogen into trypsin in a self-induced autoactivation process. The trypsin thus produced also converts chymotrypsinogen, proelastase and procarboxypeptidase into their respective active forms. The zymogens have a certain activity of their own, but this activity is enhanced considerably after their activation by the enterokinase (Neurath & Walsh, 1976; Cohen, 1981). Yamashina (1956) showed that the mammalian enterokinase is a glycoprotein which activates trypsinogen through the cleavage of the same lysin-isoelectric bond as does trypsin during autoactivation. Neurath *et al.* (1970) have found that the activation of the dogfish (*Squalus suckleyi*) trypsinogen is accelerated by calcium (Ca^{2+}) ions. Yoshinaka *et al.* (1981a, b) showed that in rainbow trout starved for eight weeks activities of trypsin and chymotrypsin in the intestine were very low, but the activities of their zymogens stored in the pancreas were high, to provide a ready supply of trypsin and chymotrypsin when required.

Jančařík (1949) found a proteolytic activation of liver, pancreatic, and intestinal wall extracts when these were mixed with low concentrations of extracts of crustacean food organisms (e.g. *Daphnia* sp.). In this work, as well as in later studies, Jančařík (1964, 1968) supports the theory suggested by Schaeperclaus (1961) and Wunder (1936) on the role of exogenic enzymes, contributed by food organisms, in the digestion of food by fish. According to this theory, the addition of exogenic enzymes may increase the digestibility of food ingredients, and this was considered to be a major advantage of natural food over supplementary feed. Jančařík suggested that the exogenic proteolytic enzymes do not add to the activity of the endogenic ones, but rather *activate* them. It is doubtful whether such activation is really required in healthy adult fish, but the effect of exogenic enzymes contributed by zooplankton may have some importance in the feeding of fish larvae, in which the mechanism of enzyme secretion may not be fully developed. Dabrowska *et al.* (1979) added extracts of fish hepatopancreas and intestine to an artificial diet given to common carp larvae. Growth and survival of these larvae were better than those fed on an artificial diet without the extracts, but less than that of larvae fed on natural food. It stands to reason that exogenic enzymes will be effective, if they are, more in stomachless fishes, since they may be destroyed in the acidic medium of the stomach.

Bondi & Spandorf (1953, 1954) found a high proteolytic activity of the pancreatic extract of common carp, but low activity of intestinal extract, in breaking down protein (casein). However, when the activities of the same extracts were examined on peptone, which is a product of partial breakdown of protein, the activity of the intestinal extract was higher than that of the pancreas. Mixing these two extracts did not result in a higher breakdown of peptone compared with that by the intestinal extract alone. The authors concluded that the two extracts complement each other and act on different parts of the protein at different stages in its breakdown. These results support those of Schlottke (1938–9). He found that pancreatic proteinase enters the intestinal cavity only when the latter is full of food. Pancreatic proteinase could not be found in the empty intestine. However, aminopeptidases and dipeptidases, formed in the mucoid membrane of the intestine itself, could be found at that time. These results were in line with later findings on the existence of two kinds of proteinases: exopeptidases which break down peptide bonds joining terminal amino acids, and endopeptidases which act, in addition, upon more centrally situated bonds of the protein chain (Smith, 1951; Barrington, 1962). According to Barrington (1957) both of these groups exist in fish pancreas. Indeed, Cohen (1981) found in the pancreas of common carp, on the one hand, chymotrypsin, trypsin and elastase which are endopeptidases, each acting on specific peptide bonds, and on the other hand, carboxypeptidase which complements the breakdown of the peptides produced by the former enzymes. However, in addition to these pancreatic enzymes the intestine also secretes enzymes which act in a similar way on oligopeptides, such as tri- and dipeptidases (Sastry, 1977; Ash, 1980; Boge *et al.*, 1981), aminopeptidases (Sastry, 1972; Jany, 1976; Bouck, 1979; Richard *et al.*, 1982; Plantikow & Plantikow, 1985). These enzymes, which are associated with the brush border of the microvilli of the midgut, complement the pancreatic enzymes in the hydrolysis of the proteins and play a dominant role in the final phase of protein digestion.

Lipases

Lipolytic activity has been found in extracts of digestive organs of many fish. Within the wide class of enzymes catalysing the hydrolysis of various esters, distinction is made between lipases and esterases on the basis of their relative preferential specificity. The natural substrates for lipases are triglycerides of long chain fatty acids, whereas esterases act on simple esters of low molecular weight acids. Both lipases and esterases have been identified in fish. However, both groups are usually non-specific in their action and the nature of either the fatty acid or the alcohol residue has mostly but a secondary effect influencing only the rate of hydrolysis of the ester. This hydrolysis rate can be affected by other factors, such as the presence of surface active substances or temperature to a much greater degree, making the distinction between lipases and esterases even more doubtful. Ball *et al.* (1937, quoted from Bier, 1955) have found that the hydrolysis of the higher, but not the lower, saturated triglycerides is very dependent on temperature. At low temperature the pancreatic lipase of mammals would appear to be esterase and with the increase in temperature the maximum rate of hydrolysis shifts toward longer chain fatty acids. This shift may be of special importance in ectotherms where body temperature may change widely. These considerations should be taken into account when discussing the lipolytic enzymes found in fish by various authors.

Esterase and lipase were found along the entire length of the digestive tract of most fish examined. This includes the stomach, pyloric caeca, and the whole length of the intestine (Al-Hussaini & Kholy, 1953; Kitamikado & Tachino, 1960c; Nagase, 1964; Sastry, 1974; Goel, 1975; Swarup & Goel, 1975). Patankar (1973) studied the esterase activity in the stomach of four fish species having different food habits. He concluded that the carnivore *Osteoletus ruber* showed more lipase activity in its stomach and epithelial cell layer than the plankton feeder *Opisthopterus tardor*, the herbivore *Labeo rohita* and the omnivore *Sarotherodon mossambicus*, and that there was a correlation between the food habits of the fish and its stomach esterase. Specific variation may also exist in the amount and nature of lipase in the pancreas. Keddis (1957) could not find lipase or esterase in the pancreas of *Oreochromis niloticus*. Moriarty (1973) could not find lipase in the pancreas of the same fish species, but found esterase. On the other hand, Goel (1975) found a strong lipase activity in the pancreas of the Indian carp *Cirrhinus mrigala*.

Lipase activity in fish seems to be greater than in mammals. Ananichev (1959) found that the affinity of lipase from fish such as burbot (*Lota* sp.), pike-perch (*Stizostedion vitreum*) and bream (*Abramis brama*) for its substrate is 40 times higher than that of mammalian lipase. The action of lipases depends to a very high degree on the presence of surface active substances such as the bile salts. The role of the bile salts has been studied very little in fish. The two major bile acids in the gall-bladder of fish are cholic acid, conjugated with taurine to form taurocholic acid, and taurochenodesoxycholic acid. They account for about 85% and 14% respectively, of the total bile acids of the rainbow trout (Denton *et al.*,

1974); 83% and 15% respectively, in channel catfish (*Ictalurus punctatus*), and 84% and 16%, respectively, in blue catfish (*Ictalurus furcatus*) (Kellogg, 1975). These acids and their salts serve as detergents which emulsify the lipids and ease the action of the lipolytic enzymes. The emulsification also enables the direct absorption of some lipids, while their esterification is brought about in the epithelium of the intestinal wall (Lovern, 1951; Bergot & Fléchon, 1970).

Carbohydrases

The digestion of carbohydrates seems to be affected and stimulated by the stomach acid juice (Moriarty, 1973), but the most important factor in the hydrolysis of carbohydrates is the carbohydrases. These enzymes are of a special interest in fish since not all fish species digest carbohydrates with the same efficiency. Carnivorous fish such as the salmonids digest some carbohydrates, notably starch, less efficiently than omnivorous or herbivorous fish.

Amylases are the enzymes responsible for the hydrolysis of starch into glucose. It is generally held that the digestive amylase of animals is α-amylase which acts on the 1,4′-α-glucosidic bond and converts starch into a mixture of glucose and maltose (while β-amylase acts on the non-reduced part of the starch molecule and forms maltose). Amylases have been found in most omnivorous fish such as cyprinids (Sarbahi, 1951; Bondi & Spandorf, 1954) and in herbivorous fish such as tilapias (Al-Hussaini & Kholy, 1953; Fish, 1960; Nagase, 1964; Moriarty, 1973) and milkfish (Chiu & Benitez, 1981). However, there was some controversy about their presence in carnivores. Nagayama & Saito (1968) stated that the amount of amylase in the digestive tract of carnivorous fish such as salmonids, eels or yellowtail is negligible, whereas other studies have found higher amounts of amylases in these fishes (Kitamikado & Tachino, 1960a – in rainbow trout; McGeachin & Debnham, 1960 – in largemouth black bass, *Micropterus salmoides*; Ushiyama *et al.*, 1965 – in chum salmon, *Oncorhynchus keta*). An important fact to be noted here is that the digestion of starch in carnivores is inversely related to its content in the feed. The higher the starch content, the lower its digestion (Phillips *et al.*, 1948; Kitamikado *et al.*, 1965; Nose; 1971; Spannhof & Kuehne, 1977). According to Singh & Nose (1967), rainbow trout absorbed 69% of the starch when its content in the feed was 20%, but when the content of starch increased to 60%, only 26% of it was absorbed. A somewhat higher digestion has been found by the same authors for dextrin, which is a partially hydrolysed starch. At 20% dextrin 77% was absorbed, but at 60% dextrin only 45% was absorbed. Similar results have been obtained by Cowey & Sargent (1972) for the digestion of dextrin by plaice (*Pleuronectes platessa*) and by Kitamikado *et al.* (1965) for the yellowtail (*Seriola quinqueradiata*). According to Falge *et al.* (1978) the reduced digestibility with increased starch content is related to actual reduction in gland stimulation and enzyme secretion. If this is so, it is contradictory to the situation found in tilapia and common carp where the amylolytic activity increases after ingesting a feed rich in starch (Nagase, 1964; Kawai & Ikeda, 1972). Indeed, Spannhof & Plantikow (1983) have found that in rainbow trout the maximum volume of intestinal juice produced after feeding with a carbohydrate diet was more than double that produced after feeding a protein-rich diet, and amylase activity was 15 times higher than pre-feeding activity. Most authors suggested that in carnivorous fish only a limited amount of amylase is secreted and therefore its activity is restricted to digesting only low concentrations of starch. Kitamikado & Tachino (1960a) found that the amylolytic activity in rainbow trout was lower than that in common carp, but higher than that in Japanese eel (*Anguilla japonica*). Similar results were obtained by Hofer & Sturmbauer (1985). While the activity of amylase in common carp was 616 mU/ml, that in rainbow trout was only 21 mU/ml. Ushiyama *et al.* (1965) found the amylolytic activity of the intestine of chum salmon (*O. keta*) to be 1/411 that of common carp. Cockson & Bourn (1972) compared the amylolytic activity of a herbivore – *Tilapia shirana* to that of a omnivore – *Clarias mossambicus*, in which prey constitutes a large proportion of its food. They found this activity to be much higher in the former than in the latter. Kapoor *et al.* (1976a) attempted to express these differences as the ratio of the activity of the amylase and that of the proteolytic trypsin-like enzymes in different fish species. They found that in carnivores this ratio is about 0.125, compared with 1.1–3.4 in herbivores and omnivores. The carnivores showed the highest proteolytic and lipolytic activities, while grey mullet (*Mugil* sp.), which feed on plankton and detritus, showed the lowest activities (though only a small difference was found between the proteolytic activity in a planktonivore tilapia and the carnivore perch (*Perca* sp.). The above relationships are also well presented in a comparative study by Shimeno *et al.* (1979) made on common carp and yellowtail. Their results are given in Table 5.

The limited amylolytic activity in carnivores could have been explained by a more localized amylase secretion. Fish (1960) states that in the tilapia *Oreochromis mossambicus*, which is principally herbivorous, the amylase activity is dispersed along the entire intestine, but in the carnivorous perch (*Perca* sp.) it is concentrated in the pancreas only. Others also have reported to have found amylolytic activity in extracts of the pancreas, liver, stomach, intestine and pyloric caeca of herbivores and omnivores (Vonk, 1937; Al-Hussaini, 1949; Nagayama & Saito, 1968; Sarbahi, 1951; Kawai & Ikeda, 1971). Jančařík (1964) claimed that various parts of the digestive tract of common carp are involved in secretion of amylase, and Cockson & Bourn (1973) state that the amylolytic activity of the posterior part of the intestine of *Barbus paludinosus* is equal to that of the anterior part. Here again, however, one should remember, as has been justly pointed out by Barrington (1957), that the pancreatic tissue is dispersed in these organs and the amylolytic activity may be a result of 'contamination' by this tissue. Vonk (1937) found that in common carp the amount of amylase in the pancreatic tissue was about 50–100 times as large as in the gut extracts. He considered amylase detected in the latter to have originated from the pancreas. Bondi & Spandorf (1954) have pointed out that most of the amylolytic activity in common carp was found in the pancreatic extracts. Intestinal extracts did not show amylolytic activity at all and seemed to be lacking in amylase, though maltose was hydrolysed slowly by these extracts (about 30% during three days). This controversy has been resolved by Yamane (1973a, b) who studied the localization of amylase activity in common carp, tilapia (*Oreochromis mossambicus*), and bluegill sunfish (*Lepomis macrochirus*), histochemically, by a substrate film method instead of the tissue homogenization procedure. He found that, although weak amylolytic activity was detected in various parts of the digestive organs, this activity appeared only on the luminal surface of the tract and not in the mucosa or submucosa. Also, no activity was found in the cells of the liver. However, strong amylase activity was found in the pancreatic exocrine cells. Yamane's conclusions were that amylases are synthesized and secreted by the cytoplasm of the pancreatic exocrine cells alone. This is probably true for both carnivorous and non-carnivorous

Table 5. *Digestive enzyme activities (in activity units[a] per 100 g body weight) in common carp and yellowtail*

Enzyme	Common carp	Yellowtail
Amylase	1040 ± 700	12.5 ± 2.1
Pepsin	not found	112.0 ± 15.0
Trypsin	64.4 ± 39.0	52.2 ± 10.9

Note: [a] Activity unit = the amount of extract required to hydrolyse 1 μmol of substrate or coenzyme in one minute.

Source: From Shimeno *et al.* (1979).

fish and therefore localization of amylase secretion cannot be a satisfactory explanation for its limited activity in carnivores, but must arise from either its limited production in the pancreas, or its subsequent inactivation. Onishi *et al.* (1973a, b) measured the sequence of the amylase levels in the intestinal contents and hepatopancreas of common carp after feeding. They found the activities of amylase and protease in the intestinal contents to increase gradually after feeding, reaching a maximum after five hours. Concurrently with this increase, amylase levels in the hepatopancreas decreased sharply after feeding, probably due to the secretion of the accumulated enzyme into the intestinal lumen. The enzyme was replenished rather slowly in the hepatopancreas. It reached a minimum concentration two hours after feeding and returned to a pre-feeding level only after 24 h. Protease concentration was also reduced in the hepatopancreas after feeding, but recovery was much more rapid than for amylase. This may suggest a limited amylase secretion in the pancreas. However, a different explanation was given for reduced amylase activity in carnivorous fish by Spannhof & Plantikow (1983). They found that the amylase activity in the intestinal juice was different when 'soluble starch' or crude potato starch were ingested. In the latter case amylase activity of the intestinal juice was only 28% that when 'soluble starch' was ingested. By eluting the insoluble chyme residues from fish which received potato starch, about 70% of the amylase activity in the intestinal juices of the fish receiving 'soluble starch' was recovered. The authors concluded, therefore, that the amount of amylase secreted is constant, regardless of the form of starch, but in the case of crude starch the amylase was bonded to the starch and thus rendered inactive. Spannhof & Plantikow (1983) have also shown *in vitro* that amylase is strongly absorbed by the starch from which it can be eluted by repeated washing. This absorption of the amylase reduces the activity of the enzyme, which acts as a free mobile enzyme, and thus inhibits the hydrolysis of the starch.

Another explanation was given by Hofer & Sturmbauer (1985). They found wheat and some other grains to contain albumins which inhibit the α-amylase of carp and trout. However, in contrast to Spannhof & Plantikow (1983) quoted above, they found that in carp the amount of amylase actually secreted when native wheat was consumed was three to four times higher than when fed extruded wheat in which the amylase inhibitor was destroyed (Sturmbauer & Hofer, 1986). Thus, according to these authors, the carp is able to regulate its amylolytic activity and to make up for the inhibition by the albumins.

The optimum pH for α-amylase and maltase activity in most fish examined was in the range 7–8 (Kuzmina & Nevalennyy, 1983).

Other carbohydrases beside amylase, such as glucosidases, maltase, sucrase, lactase, cellobiase and others, have been found in most fishes, both carnivorous and non-carnivorous (Tunison *et al.*, 1941; Phillips *et al.*, 1948; Sarbahi, 1951; Kitamikado & Tachino, 1960a; Kawai & Ikeda, 1971). Bondi & Spandorf (1953, 1954) found maltose to be rapidly hydrolysed to glucose by pancreatic extract of common carp. In contrast to the digestion of starch, carnivorous fish digest glucose and disaccharides well. Table 6 presents data on digestibility of various carbohydrates by brook trout (*Salvelinus fontinalis*), as given by Phillips *et al.* (1948). It can be seen that the digestibility decreases with the increase in the molecular weight of the carbohydrate, starch being the lowest. Higher digestiblity of oligosaccharides such as dextrin or hydrolysed starch than that of raw starch was also found by others (e.g. Singh & Nose, 1967). This may explain the differences in the efficiencies of the various carbohydrates in the food of chinook salmon (*Oncorhynchus tshawytscha*) as found by Buhler & Halver (1961). They found that sucrose could replace glucose, but fructose was only 80% as efficient as glucose, galactose only 60%, and glucosamine less than 50%.

Cellulase was not found in trout (Kitamikado & Tachino, 1960a; Lindsay & Harris, 1980) nor in most fish examined. Chiu & Benitez (1981) could not find cellulase even in milkfish (*Chanos chanos*), although this fish is a herbivore which feeds mainly on naturally occurring algae. Cellulose included in trout diets was not digested (Bergot, 1981) and had a growth depressing effect (Hilton *et al.*, 1983). Also, Bondi & Spandorf (1953) could not find any cellulolytic activity in the extracts of pancreas and intestine of common carp. However, *in vivo* experiments showed a considerable digestion of cellulose by this fish, which they explained by the

Table 6. *Digestibility (in %) of various carbohydrates by brook trout (Salvelinus fontinalis)*

Carbohydrates	% Digestibility
Glucose	99
Maltose	93
Sucrose	73
Lactose	60
Starch (cooked)	47
Starch (raw)	38

Source: Data from Phillips *et al.* (1948).

presence of rich intestinal bacterial flora breaking down cellulose. This theory is supported by Stickney & Shumway (1974) who studied the occurrence of cellulase in the stomachs of 62 different fish species. Only 16 of these species, mainly estuarine and freshwater fishes, showed some cellulase activity, while the others showed none. No correlation was found between the food habits as reported in the literature and the presence of high or low cellulase activity in those fish which demonstrated activity. The authors attempted to determine the source of the cellulase activity in the channel catfish (*Ictalurus punctatus*) and found that after fingerlings of this species were exposed to streptomycin solution for 24 h they showed no cellulase activity. From this they concluded that the cellulase activity, at least in channel catfish but possibly also in other fishes, originates from alimentary tract microflora rather than from cellulase secreting cells within the fish. Another extensive survey on the presence of cellulase in fish was conducted by Lindsay & Harris (1980), who studied 138 marine and freshwater fishes representing 42 species. The majority of omnivores and piscivorous fishes demonstrated no cellulase activity or only a slight amount. In those fishes exhibiting cellulase activity, its level varied within and between the species; it was higher in fishes feeding extensively on invertebrates. From these results the authors also concluded that the fish examined do not seem to possess an endogenous cellulase or even maintain a stable cellulolytic microflora. The cellulolytic activity, according to Lindsay & Harris (1980) seems to be the result of cellulase enzymes produced within the invertebrates and bacteria ingested by the fish and which became closely associated with the gut wall of the fish. In that they lent support to the 'exogenic enzyme' theory (p. 31). The lack of cellulase in the digestive tract of salmonids may explain why Buhler & Halver (1961) found a decrease in growth of chinook salmon (*O. tschawytscha*) with the increase in the content of α-cellulose replacing dextrin in the diet.

Other digestive enzymes

Other enzymes can also be found in the digestive tract of fish. Chitinase was found in the stomachs of some fishes (Okutani & Kimata 1964a, b; Micha *et al.*, 1973; Pérès *et al.*, 1974; Danulat & Kausch, 1984). Lindsay (1984a) determined the activity of chitinolytic enzymes in the stomach or fore-gut of 29 Northern European marine and freshwater fishes. He found no correlation with a chitin eating (i.e. crustacean eating) habit. However, he observed that species that were able to disrupt their prey mechanically had low activity, while species that ingested their prey whole had high activity. He concluded that the primary function of gastric chitinase in fish may be to disrupt chemically the chitinase envelope of the prey.

Trout secrete a gastric chitinase, the activity of which is amongst the highest recorded in fish (Micha *et al.*, 1973). Lindsay (1984b) also found in trout other endogenous gastric chitinolytic enzymes such as chitobiase and lysozyme. However, the digestibility of purified chitin included in rainbow trout diets is considerably lower than could be expected given the amount of chitinase secreted by the stomach of this fish (Lindsay *et al.*, 1984a). In a later study, Lindsay (1984b) found that fish gastric chitinase, although attacking soluble chitin derivatives (such as glycol chitin) at random, was able to hydrolyse only the apexes of the chitin microfibrils. Chitinase absorbed into the chitin substrate away from the apical area of the microfibrils, does not find a susceptible substrate for hydrolysis and this results in a depressed hydrolytic rate. Gastric lysozyme was rapidly absorbed onto both chitin and cellulose but chitobiase was not absorbed by any of these. Thus diets containing cellulose or chitin may seriously interfere with the function of the gastric lysozyme. This resembles the effect of starch on the activity of amylase (p. 37).

Alkaline and acid phosphatases have been detected in the digestive tract of fishes by a number of authors (Al-Hussaini, 1949; Western, 1971; Whitmore & Goldberg, 1972a, b; Sastry, 1975; Goel, 1975). Both alkaline and acid phosphatases have been found in the border microvilli of fish intestine and in different tissue layers of the stomach wall (Western, 1971; Sastry, 1975). However, they were associated primarily with microvilli of the absorptive cells and did not occur in the other, non-absorptive, tissues (Western, 1971). The activities of these enzymes are lowest in the oesophagus and fairly low in the rectum (Sastry, 1975). In the stomach, a high acid phosphatase activity is accompanied by a low activity of alkaline phosphatase, whereas in the intestine the alkaline phosphatase activity is higher than the acid phosphatase. In higher vertebrates it has been shown that epithelia involved in active transport of nutrients contain high concentrations of phosphatases. These enzymes catalyse the splitting of inorganic phosphate from organic phosphate. Alkaline phosphatase is active in alkaline media and acid phosphatase in acid media. Presumably the phosphorylation of glucose would be carried out by hexokinase on the outside of the epithelial cell nearest the lumen of the intestine while dephosphorylation would have to take place in the inside of the cells of the same region, since it has been shown that epithelial cells accumulate the sugar (Davison, 1970). However, acid alkaline phosphatase is associated not only with active transport of glucose but also of protein and lipid (Tuba & Dickie, 1955). Utida & Isono (1967) found that in the Japanese eel (*Anguilla japonica*) alkaline phosphatase was also associated with the active transport of water and sodium ions.

Gastrointestinal flora

The issue of cellulose digestion and that of other hardly digestible food ingredients emphasizes the importance of intestinal bacterial flora which assist in digesting these ingredients as well as taking part in other nutritional processes in the gut.

The gastrointestinal bacterial flora of fish appears to be simpler than that of endotherms. While the digestive tract of endotherms is populated mainly by obligate anaerobes, the predominant bacteria isolated from most fish guts have been aerobes or facultative anaerobes (Shewan, 1961; Aiso *et al.*, 1968; Sera & Ishida, 1972; Matches *et al.*, 1974; Trust *et al.*, 1979; Kakimoto & Mowlah, 1980). Among these, *Vibrio* and *Aeromonas* are usually dominant. The number and characteristics of the bacteria present may be different in different species or conditions they are kept in. Mattheis (1964) examined the intestines of 24 rainbow trout, 13 common carp and seven brown trout (*S. trutta*) and isolated from these fish 219 bacterial strains. The intestine of brown trout from natural waters contained a relatively small number of bacterial species as compared with the intestine of rainbow trout from a hatchery, where a larger number of bacterial species (28) has been found. The largest number of bacterial species was found in the intestine of common carp, where *Pseudomonas* and *Areomonas* were dominant. Sakata *et al.* (1980a) and Sugita *et al.* (1981, 1982a, b) found high numbers of aerobic and facultative anaerobes in which *Vibrio* and *Aeromonas* predominated in the intestine of *Tilapia zillii, Oreochromis niloticus* and *O. mossambicus*. Although the number of bacterial species in the intestines of salmonids may be low, the actual concentration can be quite high. Trust & Sparrow (1974) found as many as 10^8 viable enteric organisms per gram weight of the alimentary tract of freshwater salmonids. Large numbers of bacteria were also found in the intestines of common carp, grass carp and tench by Syvokene & Grigorovie (1974). Data on the bacterial concentrations found in the faeces of common carp by Dabrowski & Wojno (1978) are presented in Table 7.

The fact that the enteric bacteria found in the intestines of fish are very often of the same kinds as those found free in the surrounding water or in the food of the fish led some researchers to suggest that these organisms are only transient residents of the fish gastrointestinal tract, and that fish do not have a true enteric bacterial population. However, recent studies have shown the presence of specific intestinal bacteria, mainly obligate anaerobes, in the guts of many fishes. The first to report on the presence of obligate anaerobes in the intestine of fish were Trust *et al.* (1979) who found significant numbers of them among the resident bacterial flora of grass carp and goldfish, but not in rainbow trout. Sakata *et al.* (1980b,

1981) also found high numbers of enteric bacteria in various freshwater fishes (six species), reaching 10^6–10^9/g intestinal content. They found obligate anaerobes predominating over the facultative anaerobes in most cases, especially in the intestines of *Oreochromis niloticus* and ayu (*Plecoglossus altivelis*).

In addition to digestion, intestinal bacteria also take part in other nutritional functions such as non-protein nitrogen (*NPN*) utlization and vitamin production. Teshima & Kashiwada (1967, 1969) found 209 different groups of bacteria in the intestine of common carp, some of which seem to be able to synthesize vitamin B_{12} and nicotinic acid. Lesauskiene *et al.* (1974) cultured intestinal bacteria of common carp and white amur (*Ctenopharyngodon idella*) and found that they liberated a considerable amount of methionine into the culture media. This may indicate that intestinal bacteria can also produce essential amino acids to compensate partly for deficiency of these in the diet.

Absorption

After the food is digested, the resulting simpler compounds are transported through the intestinal wall and absorbed into the blood. This transport of nutrients may be partly by diffusion, but in part it is an active process mediated by carriers in the cell membranes. Evidence for three separate transport systems has been obtained by Buclon (1974): one for D-glucose, one for neutral acyclic L-amino acids, and a third for diamine-L-amino acids. As in mammals, sodium has been found to be essential for the evolution of the transport current and the transport, therefore, depends on external Na^+ concentration (Smith, 1969; Smith & Ellory, 1971; Buclon, 1974: Boge & Pérès, 1983). In marine fishes Cl^- has also been found to affect transportation (Boge & Pérès, 1983).

Smith (1970) used perfused intestines of goldfish and found that 18 amino acids were transported actively through the mucosa against their

Table 7. *The concentration of intestinal bacteria in the faeces of common carp*

Bacterial groups	Concentration (10^5/g faeces)
Heterotrophic	70–2700
Ammonification	25.6–436
Urea decomposing	1.3–140
Citrate decomposing	0.6–98.4

Source: Data from Dabrowski & Wojno (1978).

respective concentration gradients. Stokes & Fromm (1964), studying the intestinal transport of D-glucose in a perfused segment of trout intestine, and Boge *et al.* (1979, 1981), who studied the intestinal absorption of glycine by rainbow trout, found transport sites in both mid-gut and hind-gut, but maximum transport rates were higher in the mid-gut than in the hind-gut. From their study they have concluded that glycine is transported in an active, saturable, process.

Amino acid transport is competitive, and the transport of one amino acid may be inhibited in the presence of other amino acids (Kitchin & Morris, 1971). Thus, Ingham & Arme (1977) have found that L-valine and L-methionine are competitive inhibitors of L-leucine uptake. The presence of L-valine, however, even at a relatively high concentration (100 mM) did not reduce the uptake of L-methionine to the level which was expected from its K_1 constant. The authors suggest that the portion of L-methionine inaccessible to L-valine inhibition may represent uptake through a second site. They concluded that transport of neutral amino acids can be carried out through more than one site and by more than one system. There is, however, competitive inhibition of uptake between amino acids in rainbow trout, indicating that at least part of the transport is shared.

Transport mechanism may change with temperature. Stokes & Fromm (1964) found the transport rate of D-glucose through perfused segments of trout intestine to be greatly increased at higher water temperatures. Escoubet *et al.* (1974) found a 37% increase in the rate of intestinal absorption of glycine (but not of glucose) by rainbow trout when water temperature was increased from 11 to 20 °C. Higher intestinal absorption of glycine was also recorded by Pérès *et al.* (1974) *in vitro* with rainbow trout.

However, intestinal absorption of most amino acids, although not all, is subject to temperature adaptation (Smith, 1970). Cold adapted fish could absorb these amino acids at a higher rate than warm adapted ones. Thus, Kitchin & Morris (1971) and Smith & Kemp (1971) found that intestinal absorption of valine, measured *in vitro*, using goldfish acclimated to different temperatures, was lower at high acclimated temperature, while the absorption of phenylalanine and methionine was not affected. Kitchin & Morris (1971) suggested that at high temperatures, both the number and turnover rate of the carrier for valine decreased, but not those for methionine, which were temperature independent. Smith & Kemp (1971) found that with the adaptation of goldfish to higher temperature (change from 16 to 30 °C), a change in the membrane fatty acid composition occurred. The proportion of stearic acid increased, while that of docosa-hexaenoic acid decreased. These changes were complete after 48 h and seem to have affected the transport of valine, but not that of phenylalanine and methionine. Smith (1970) suggested that this is related to the lipophilic properties of the amino acids. The different physical properties of the individual amino acids may give rise to selective amino acid transport through the membrane, which changes when lipid composition of the membrane changes.

Nutrients other than amino acids can affect the membrane transport of amino acids. Read (1967) found that galactose inhibited the uptake of cycloleucine by dogfish intestinal tissue in a 10 min incubation (but not in a two minute incubation) and Hokazono *et al.* (1979) found that 10 mM glucose inhibited the transport of L-lysine through the intestinal wall of rainbow trout. Read's (1967) studies on the source of this inhibition led him to conclude that the uptake of the sugar, being sodium dependent, causes a highly localized alteration of the infracellular concentration of sodium, and thus affects the cycloleucine uptake, which is also sodium dependent.

From the above discussion it is clear that different free amino acids may be transported and absorbed at different rates. Dabrowski (1983) states that the rate limiting factor in the utilization of amino acids by fish is not digestion but rather absorption. Most of the amino acids, excluding histidine and lysine, were less efficiently absorbed by common carp than the non-essential amino acids. A high increase in free amino acid content in the intestine of the common carp led to only a small increase in the absorption rate, and as a result free amino acids were removed in the faeces in concentrations higher than usually found in higher vertebrates. Differences in the absorption rate observed in some fishes may affect the availability of the amino acids at the proportions required for biosynthesis and thus reduce their utilization for growth. The problem seems to be more acute when fish such as common carp are fed free crystalline amino acids (p. 202). In natural conditions, however, this seems to be corrected by the ability of some fish species to absorb oligopeptides (Boge *et al.*, 1981) and even macromolecules of protein (Boge & Pérès, 1983). Noaillac-Depeyre & Gas (1973) have found that common carp have, in the distal segment of their mid-gut, a special ultrastructure of the enterocytes which enables the absorption of macromolecules. They found that molecules of horseradish peroxidase (MW 40 000) are transported through the epithelium of an adult common carp mid-gut, apparently without modification, either directly into the blood circulation or to be accumulated in supranuclear vacuoles of the epithelium, where they are decomposed by lytic enzymes. Noaillac-Depeyre & Gas (1973) pointed out that the structural arrangement enabling the absorption of macromolecules has been found in newborn mammals, but there it disappears at about the 21st day of postnatal development. Grabner (1985) and Grabner & Hofer (1985) simulated *in vitro* the digestion processes by rainbow trout and common carp. Digesting in this way a single cell protein of methanophilic bacterium, broad bean and soya bean has shown that 50–60% of the end products were oligopeptides with a molecular weight of less than 40 000 (less than 25 000 in the broad and soya beans), and only 30–40% were free amino acids. This seems to support the suggestion that most of the protein absorption is that of oligopeptides. It should be noted, however, that the free amino acid fraction contained about 80% essential amino acids, while the peptide fraction only about 40%. This form of absorption may increase the total amino acid absorption, especially of some amino acids which individually are poorly absorbed, as found in humans by Silk *et al.* (1973).

Different carbohydrates may also be absorbed at varying rates. Furuichi & Yone (1982b) studied the intestinal absorption of glucose, dextrin and α-starch by red sea bream (*Chrysophrys major*) and found that during the first two hours after ingestion, glucose is absorbed faster than dextrin, and the latter faster than α-starch. The different rates of absorption may affect the synchronization with the glycolytic enzymes secretion, and thus reduce the utilization rate of carbohydrates (p. 83).

Digestion rate

Digestion rate is the time it takes to digest the food, i.e. the rate of passage of the food through the digestive tract. This has been studied mainly in fish with a stomach, where the amount of food in the stomach was determined at various times after feeding. The stomach evacuation period has also been called 'digestion rate' or 'gastric digestion rate'. However, since the stomach evacuation may result in the passage of partly digested chyme into the intestine, and not only through true digestion, this is referred to here as 'gastric evacuation rate' and the term 'digestion rate' is reserved for the passage of the food through the entire digestive tract.

Gastric evacuation is a complex process, and is very incompletely understood for fish. Gastric digestion and evacuation are affected by a number of factors such as the quantity and quality of the food, the rate of secretion of gastric juice, gastric mobility, and the capacity of the intestine to accept chyme from the stomach. Still, at a given temperature there is a constant decrease in gastric content with time. Job (1977) claimed that when the gastric evacuation rate was expressed as the percentage of the food eaten leaving the stomach per hour, the highest evacuation rate was about eight hours after feeding. However, this does not agree with results obtained by Hunt (1960) and Pandian (1967) who found that, at a constant temperature, there is an almost linear relationship between percentage of stomach evacuation and time. More precise studies, which were based on the amount of organic matter (i.e. fresh weight of food less water, ash and chitin) or dry weight, rather than wet weight of food, have shown that the relationship is curvilinear (Fabian *et al.*, 1963; Kitchell & Windell, 1968; Beamish, 1972). Molnar & Tolg (1962a) and Fabian *et al.* (1963) measured the gastric evacuation rate by X-rays. They, as well as Brett & Higgs (1970) and Elliott (1972), showed that the evacuation rate is exponential with time. When expressed as the amount remaining in the stomach as per cent of body weight, the logarithms of these values form a linear regression with time (Figure 7). This means that the gastric evacuation rate is proportional to the amount of food remaining in the stomach, and when the latter gets smaller, so does the evacuation rate.

The digestion rate of fish with a stomach is usually slower than in fish without a stomach. Rozin & Mayer (1961) fed goldfish (*Carassius auratus*) pellets at a constant rate during one hour per day. By substituting pellets containing carmine on a given day they determined the digestion rate. At the test temperature (24.5 °C) the median time of the first appearance of red faeces was seven hours after ingestion; the great majority of the red faeces was excreted from eight to 24 h after ingestion of food. In contrast, in some of the carnivorous fishes periods of five, and even up to 18 days have been recorded as digestion rates (Barrington, 1957).

Figure 7. Gastric evacuation rates of sockeye salmon (*Oncorhynchus nerka*) at three temperatures after acclimation (from Brett & Higgs, 1970).

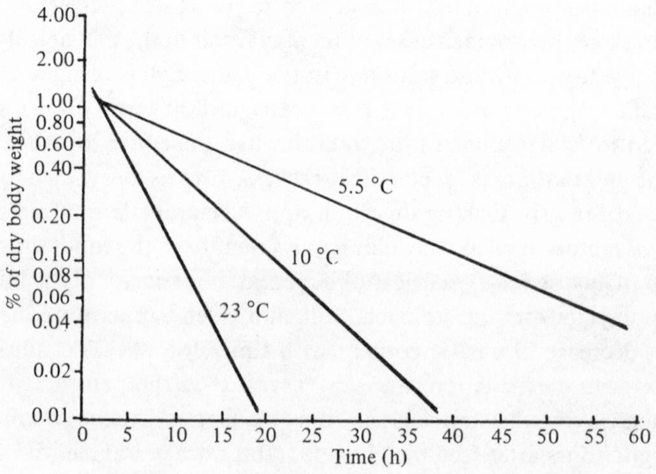

Digestion rate seems to be independent of the size of the fish (Garber, 1983; Talbot *et al.*, 1984). However, other factors may affect it. Food composition is one of these factors. Different digestion rates have been recorded in fish fed different foods, both natural and supplementary. Ranade & Kewalramani (1967) measured the rate of food passage through the intestines of Indian major carps (*Labeo rohita, Cirrhinus mrigala* and *Catla catla*). The algae *Microcystis, Anabaena, Scenedesmus* and *Chlorella* were retained in the intestine for a longer time than the algae *Spirogyra* and *Ulothrix*. Also Kitchell & Windell (1968) and Windell & Norris (1969a, b) have found differences in gastric evacuation rates in pumpkinseed sunfish (*Lepomis gibbosus*) and rainbow trout fed various feeds. Spannhof & Plantikow (1983) measured the digestion rate by rainbow trout of protein-rich and starch-rich diets and found that the rate of chyme passage of the latter through the intestine was greater. Following the feeding with the starch-rich diet, 64–72% of all faecal matter was voided within six hours of feeding, compared with only 50% for the protein-rich diet.

Gastric evacuation rate in rainbow trout increased as meal size increased beyond 1.1–1.5% of body weight (Windell & Norris, 1969a; Tyler, 1970). Although Elliot (1972), working with brown trout (*Salmo trutta*), and Kitchell & Windell (1968), working with pumpkinseed sunfish (*Lepomis gibbosus*) found that the digestion rate is directly correlated with the amount of ingested food, many other studies show that this relationship is not linear (e.g. Garber, 1983). Hunt (1960) found that when meal size tripled the digestion period slightly more than doubled. Beamish (1972) found that when largemouth bass (*Micropterus salmoides*) were fed at 2% of their body weight the stomach was evacuated completely after 14 h, but when they were fed 8% of their body weight, evacuation time increased to almost 27 h. The effect of ration size on food retention in the intestine was similar. When the fish were fed 4% of their body weight, the main defecation period was 12–36 h after feeding and lasted until 48 h, but when the fish were fed larger amounts most defecation took place after 12–48 h and lasted until 72 h after feeding.

Of all the factors affecting digestion rate, temperature seems to be the most important. The higher the temperature, the higher the gastric evacuation rate (Figure 7), and the shorter the time food stays in the stomach (Shrable *et al.*, 1969; Brett & Higgs, 1970; Edwards, 1971, Elliot, 1972; Job, 1977; Fänge & Grove, 1978; Ross & Jauncey, 1981). Molnar & Tölg (1962a, b) and Fabian *et al.* (1963) found a semi-logarithmic relationship between gastric evacuation rate and temperature in four carnivorous fish species within the temperature range of 5–25 °C.

The relationship between passage of food through the entire gut (digestion rate) and temperature is similar to those mentioned above for gastric evacuation rate. Phillips *et al.* (1961) reported that a food capsule containing red dye passed more quickly through the digestive tract of brook trout when the water temperature was higher. Similarly, Ross & Jauncey (1981), using a radiographic method in tilapia hybrids, and Fauconneau *et al.* (1983), using chromic oxide in rainbow trout, found that digestion rate was affected by temperature in a similar way to the gastric evacuation rate. This is also true for fish lacking a stomach. Scheuring (1928) found a semi-logarithmic relationship between temperature and the time between feeding and defecation in *Misgurnus fossilis* (Cypriniformes, Cobitidae). A strong effect of temperature on digestion rate of common carp, grass carp (*Ctenopharyngodon idella*) and silver carp (*Hypophthalmichthys molitrix*) was also found by Maltzan (1935) and Okoniewska & Kruger (1979). Peters *et al.* (1972) studied the digestion rate of three fish species – *Lagodon rhomboides, Leiostomus xanthurus*, and *Menidia menidia*, at different temperatures, after a single unrestricted feeding of commercial pelleted feed. Digestion rate was measured by a serial slaughter method. At lower temperatures (6, 12 and 18 °C) there was a 'latent' period immediately following ingestion during which the contents of the alimentary tract did not decline. After adjustment to this latent period, digestion rates were described by a double log regression which showed the time for complete evacuation as a function of temperature and amount of food.

Digestibility coefficient

During the passage of food through the digestive tract not all of it is digested and absorbed. The undigested portion is egested as faeces. The absorbed portion is determined by difference between the ingested and egested nutrients, and usually expressed as a percentage of the amount ingested as 'apparent digestibility coefficient':

$$ADC = \frac{\text{nutrient ingested} - \text{nutrient egested}}{\text{nutrient ingested}} \times 100 \qquad (14)$$

Digestiblity coefficient can be determined for the food dry matter, but since the digestibility can be different for protein, carbohydrate, and lipid, it is usually determined separately for each of these nutrients.

Determining food digestibility

Determination of the digestiblity coefficient by the balance method is commonly used for many farm animals. Since the end of the last century it has been used for fish (Homburger, 1877, quoted from Knauthe, 1898). However, because the use of this method for fish has met a number of constraints, only limited information has been gathered since then on the digestibility coefficients of various feedstuffs for fish.

The balance method is based on as accurate as possible determinations of the amounts of ingested and egested nutrients. However, with fish the food is given in water and faeces are egested into water, and nutrients of both food and faeces may be dispersed and dissolved in water. Since the lost nutrients are treated as if they are absorbed by the fish, an error is caused in the digestibility coefficient which is biased toward higher values, according to the losses of nutrients in water. Tunison *et al.* (1942) pointed out these losses when reporting their study on the digestibility coefficients of feeds by brook trout (*Salvelinus fontinalis*). To illustrate them, they gave the following example: From 64.1 g of starch offered to the trout, 1.9 g remained uneaten and recovered as particles from water. However 9.4 g were found in solution in the water. 23.4 g of the starch were found in the particulate faeces which were egested into water and recovered, but 12.8 g were found in solution. This was also true for proteins: 6–41% of the proteins offered to the trout remained uneaten, 60–93% of this portion was found in solution or in a very fine suspension. Mann (1948) also showed that the food loses nutrients during its immersion in water (Table 8). This bias in digestibility coefficient which can result in the balance method can be seen from results obtained by Smith & Lovell (1971, 1973). They have collected the faeces from different parts of the intestine of channel catfish (*Ictalurus punctatus*) and from the water of the aquarium after faeces have been voided. These results are given in Table 9. Windell *et al.* (1978) have studied this more thoroughly. They determined the effect of the time the faeces remain in water before they are collected

Table 8. *Weight of food and content of nutrients in sweet lupin and rye after immersion in water for varying periods (in % of initial weight and contents) for coarse and fine particle sizes*

	Sweet Lupin				Rye			
	Weight		Protein content		Weight		Sugar content	
Immersion period	Coarse	Fine	Coarse	Fine	Coarse	Fine	Coarse	Fine
10 min	100	95	100	100	100	100	100	66
20 min	95	83	93	90	92	86	88	58
2 h	80	80	90	82	90	80	50	33

Source: Data calculated from Mann (1948).

on the loss of nutrients, and found that the major loss is incurred during the first hour of immersion. During this time about 21% of the dry matter, 12% of the protein and 4% of the lipids were lost, increasing the digestibility coefficients by 11.5, 10 and 3.7% respectively. Within 16 h the losses of nutrients from the faeces reached 31, 12 and 9.8%, respectively, and the increase in digestibility coefficients were 17, 10 and 8.2%.

Many early studies on food digestibility by fish have used the balance method in spite of its drawbacks. In some cases special efforts have been made to reduce the error as much as possible by holding the fish first in one aquarium for feeding for a scheduled time (5–60 min in different studies), and then transferring the fish to another aquarium in which the faeces are collected. The water from both aquaria is filtered to collect food residues or faeces. This procedure was adopted originally by Morgulis (1918) and later by many others (e.g. Tunison *et al.*, 1942; Hanaoka *et al.*, 1948; Furukawa *et al.*, 1953; Bondi *et al.*, 1957).

In order to overcome the drawbacks of the balance method discussed here, many researchers have recently been using the 'inert indicator' method. In this method a certain amount of inert substance, which is not digested by the fish, is introduced into the examined food. A number of inert substances can serve this purpose such as lignin, chromic oxide (Cr_2O_3), and various radio-isotopes such as ^{144}Ce (Peters & Hoss, 1974); the most common is chromic oxide, which was used in most studies on fish. The concentrations of the nutrient and the inert indicator are determined in both the food and the faeces and the apparent digestibility coefficient (*ADC*) of the nutrient is calculated according to the following formula:

$$ADC\,(\%) = 100 - \left[\frac{\%\text{ indicator in food}}{\%\text{ indicator in faeces}} \times \frac{\%\text{ nutrient in faeces}}{\%\text{ nutrient in food}} \times 100 \right] \quad (15)$$

The advantage of this method over the previous one is that there is no need to determine the amount of food ingested nor the amount of faeces egested. It is sufficient to analyse a sample of both and determine the percentage contents of nutrient and indicator. Nevertheless, even this method is not free of constraints and drawbacks. If the faeces are to be sampled from the water, after being egested, the analysis is subject to the same errors as the balance method due to dispersion and solution of nutrients in water. In some studies the faeces are extracted from the rectum by stripping (e.g. Singh & Nose, 1967). However, since digestion and absorption continue all along the digestive tract, it makes a significant difference which part of the intestine the faeces were stripped from. Austreng (1978) obtained higher digestion coefficients for all nutrients when samples were stripped only from the hindmost part of the rectum, behind the anal fin, than from samples stripped from the entire abdomen. Errors may also be introduced due to the inclusion in the sample of sperm, ova or mucus (Windell, 1978). The digestibility coefficients obtained by stripping the hindmost part of the abdomen in Austreng's (1978) study were higher than those sampled by extracting the faeces by dissection of the same part. The difference may be due to the inclusion of ova, sperm, or mucus. The frequent handling of the fish for stripping may also cause, according to the author's experience, stress which is sometimes expressed in diarrhoea and incomplete food digestion. Some researchers have sampled faeces from the posterior 2–3 cm of the intestine by incision, after sacrificing the fish. Beside the need to sacrifice the fish in each experiment, it has been claimed that this practice results in low values of digestibility coefficients since digestion continues until the faeces are actually voided (see Table 9).

The most disturbing sources of error of the inert indicator method are, however, related to the indicator used. For this method to be reliable it must meet the following conditions: (a) the indicator must be evenly distributed in the feed and in the faeces, so that samples will be representative; (b) the indicator must not be absorbed through or adsorbed onto the intestinal mucosa; (c) the rate of passage of the indicator through the digestive tract must be equal to that of the food nutrients. These conditions are not always met. Knapka *et al.* (1967) showed that in a mammal (donkey) the excretion of chromic oxide is not always complete and can vary during the day. Digestibility coefficients

Table 9. *Apparent digestibility coefficients of protein by Ictalurus punctatus at two protein levels and for faeces collected at different parts of the digestive tract and from the aquarium*

	Faeces collected from:				
Protein level[a]	Stomach	Fore-gut	Hind-gut	Anus	Aquarium
20%	61.6	65.4	75.0	80.0	96.7
40%	61.4	72.2	86.5	91.6	98.3

Note: [a] Per cent of the feed.

Source: From Smith & Lovell (1971, 1973).

based on chromic oxide (Cr_2O_3) were significantly lower than those based on other indicators. Also, in fish it was found that chromic oxide does not always move through the digestive tract at the same rate as the food (Bowen, 1978). In experiments conducted by the author on common carp it was found that different samples of faeces collected during the day had variable proportions of chromic oxide and the nutrients examined. At certain times, usually toward the end of the experiment, only the mucous sheath of the faecal pellets, empty of food, was excreted. This sheath was very often coloured green with chromic oxide. Since some of the chromic oxide seems to be retained in the fish or slowed down in passage through the intestine, faecal samples may contain less indicator than they should, causing erroneous digestibility coefficients, usually biased toward lower than true values. Henken *et al.* (1985) determined digestibility of dry matter, crude protein and gross energy by African catfish (*Clarias gariepinus*) using the balance method and the chromic oxide indicator method under identical conditions. The latter method gave lower values although the chance of leaching of nutrients was equal for both. Tacon & Rodrigues (1984), who compared apparent digestibility coefficients of rainbow trout using chromic oxide at three dietary concentrations (0.5, 1.0 and 2%), found that with 2% inclusion level nutrient digestibility coefficients were significantly higher than those for fish fed the lower chromic oxide levels. They explained this by proposing that chromic oxide at the higher inclusion level passed through the gastrointestinal tract at a faster rate relative to the digesta.

Very few determinations of digestibility in fish have been carried out using indicators other than chromic oxide. Tacon & Rodrigues (1984) tried polyethylene and acid washed sand as indicators, but obtained poor results since all digestibility coefficients determined using these indicators were significantly lower, and displayed a wider variation. Hirao *et al.* (1960) and Yamada *et al.* (1962) used ^{32}P in ammonium phosphomolybdate, which is insoluble in water, but their results were so unacceptably low that it seemed that some of the phosphate had been absorbed by the fish. Moriarty & Moriarty (1973) used $NaH^{14}CO_3$, incorporated in various algae, to determine the digestibility of these algae by *Oreochromis niloticus*. It is obvious, however, that this method is limited to live food organisms which can incorporate the isotope from the medium prior to their ingestion.

Several research workers have attempted the use of indigestible indigenous components of the diet as inert reference indicators. The advantage of this approach is that it can be applied to analyses of food digestibility by fish in their habitats. Such dietary components that have been used as indicators include silica (Hickling, 1966), cellulose (Buddington, 1979), magnesium (Klekowski & Duncan, 1975), hydrolysis-resistant organic matter (HROM) – essentially cellulose and chitin (Buddington, 1980; De Silva & Perera, 1983), crude fibre (Tacon & Rodrigues, 1984) and hydrolysis-resistant ash (HRA) (De Silva & Perera, 1983). Bowen (1981) who has used the last indicator, states that his preliminary comparisons with some of the other indigenous indicators showed that a small but significant fraction of both total ash and HROM were assimilated and therefore are not suitable as indicators.

In spite of the constraints and drawbacks of the inert indicator method, most researchers used this method for measuring digestibility in fish as it seems to reduce the difficulties associated with the balance method. One may, however, conclude that a reliable method for *in vivo* determination of digestibility in fish has still to be developed.

A very common method for a rapid determination of digestibility of various feedstuffs for farm animals is the *in vitro* test. This is done under standard conditions using pure enzymes extracted from the animals, such as pepsin for protein digestion. This, of course, requires a prior correlation between the *in vitro* and *in vivo* digestibility determinations. Such tests can increase considerably the amount of information on the digestibilities of various feedstuffs for formulating fish diets. However, only a little work has been done along these lines. Bondi & Spandorf (1953) found that soybean oil meal is digested by pancreatic extract of common carp better than meat meal, and that wheat bran is digested less than pure starch. More recent work by Grabner (1985) and Grabner & Hofer (1985) developed a method to simulate, *in vitro*, the proteolytic digestion processes by rainbow trout and common carp at temperatures, pH, water content, enzyme concentrations and digestion rates close to those found in these fishes. However, only a limited attempt has been made to correlate these results with *in vivo* digestibility coefficients in order that this information could be applied to diet formulation.

Apparent and true digestibility

In addition to the dietary protein, a certain amount of protein of endogenous origin appears in the digestive tract, where it is digested along with that derived from the diet. Part of this endogenous protein consists of the enzymes and mucoproteins secreted into the digestive tract; the rest consists of tissue protein from desquamated mucosa cells. The mucosa is one of the few tissues in which a continuous renewal of cells occurs. However, digestion of this endogenous protein is not complete and part of it is voided in the faeces. Also, in many fishes the faeces are voided as pellets in a mucous sheath which may contain protein. Since the source of this protein is the metabolic processes, it is called 'metabolic faecal protein'. The inclusion of the metabolic protein with the dietary protein in the faeces causes a bias of the digestibility coefficients toward lower values. Such coefficients, calculated without taking into account the metabolic protein are, therefore, 'apparent digestibility coefficients' to be distinguished from the 'true digestibility coefficients'.

If one considers the metabolic faecal protein to be proportional to the amount of food ingested, and not affected by the composition of the food, it may be expected that the apparent digestibility coefficient of protein will increase with increase in protein content of the food. Nose (1967), discussing this point, provided the following example. He found that the metabolic faecal protein in trout amounts to 3.6 g/100 g food. If it is considered, for the sake of this example, that the true digestibility coefficient of the protein is 100% and the amount of food ingested is 100 g, the apparent protein digestibility coefficient obtained with different protein concentrations would be:

Protein in food (%)	Metabolic protein (g)	'Digestible' protein (g)	Apparent digestibility coefficient (%)
100	3.6	96.4	96.4 × 100/100 = 96.4
50	3.6	46.4	46.4 × 100/50 = 92.8
10	3.6	6.4	6.4 × 100/10 = 64.0

Nose (1967) points out that in many cases, where the effect of protein concentration on its digestibility is discussed, the increase in digestibility with increase in concentration may have resulted from the inclusion of metabolic faecal protein, and the true digestibility value would not have shown much variation, as was indeed found by Ogino & Chen (1973a). It should be mentioned, however, that in some cases, such as the yellowtail (*Seriola quinqueradiata*), an increase in true digestibility with increased protein content in feed has been found. In any case, it is important to remember that the higher the protein content in the diet, the smaller the difference between apparent and true protein digestibilities. Ogino & Chen (1973a) have shown that the difference is small when the amount of protein ingested is over 150 mg/100 g fish body weight per day.

In order to determine the true digestibility, the amount of metabolic faecal protein should be first determined and subtracted from the protein in the faeces. In spite of its importance, only a few studies have been made to determine the metabolic faecal protein in fish and the factors affecting it. Metabolic faecal protein is usually expressed in relation to either food ingested or fish body weight. Since these two parameters are correlated, the variations in the two methods of expression are usually small. Metabolic faecal protein can be determined by measuring the amount of nitrogen (from which crude protein is calculated by multiplying by 6.25) excreted in the faeces when the fish is fed a non-protein diet. This, however, may be a problem in carnivorous fish which do not accept such feed. Some researchers have determined, therefore, the metabolic faecal nitrogen in faeces under fasting conditions (e.g. Tunison *et al.*, 1942). It is clear, however, that this introduces a certain bias and that the metabolic nitrogen determined under fasting conditions is lower than that determined when the fish is fed. Kim (1974) determined the metabolic faecal nitrogen of common carp by finding the regression between the protein level in the feed and the nitrogen excreted in the faeces per 100 g feed ingested. By extrapolation to zero protein in the feed (Y-intercept of the regression) he could determine the metabolic faecal nitrogen (Figure 8). Ogino *et al.* (1973) compared this method with that of feeding a non-protein diet. The results are given in Table 10. It can be seen that the difference between them is small, being somewhat lower in the Y-intercept method. However, it can also be seen that temperature may affect the

Figure 8. Relationship between dietary nitrogen (%) and faecal nitrogen (g N/100 g dry diet) in common carp (from Kim, 1974).

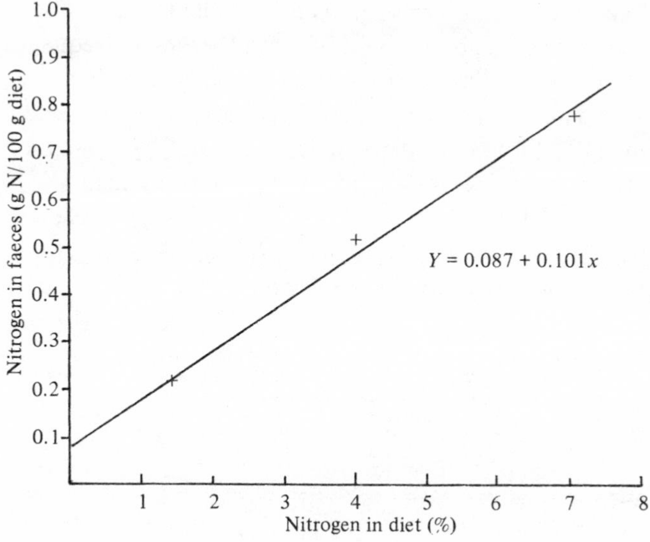

metabolic faecal nitrogen excretion, which increases with an increase in temperature. On the whole, it is clear that for omnivores and herbivores, feeding a non-protein diet and measuring the faecal nitrogen is the simplest way to determine metabolic faecal protein.

Factors affecting digestibility

Digestion of food depends on three main factors: (a) the ingested food and the extent to which it is susceptible to the effects of the digestive enzymes; (b) the activity of the digestive enzymes; (c) the length of time the food is exposed to the action of the digestive enzymes. Each of these main factors is affected by a multitude of secondary factors, some of which are associated with the fish itself, such as its species, age, size, and physiological condition; some associated with the environmental conditions, such as water temperature; and some are related to the food, viz., its composition, particle size and amount eaten. The more important of these factors and their effect on digestibility deserve a further discussion.

Fish species Digestibility coefficients may vary among fish species, due both to differences in the digestive system and its digestive enzymes, and to the different foods consumed. In spite of these differences and the lack of pepsin in fish without a stomach, variations in the digestibilities of proteins and lipids among species are small. Much more pronounced are variations in digestibility of carbohydrates, especially starch. As has already been mentioned, carnivorous fish digest starch to a much lesser degree than omnivorous and herbivorous fish. While the digestibility of raw starch in salmonids can be as low as 38–55% (Bergot & Breque, 1983; see also Table 6), depending on its concentration in the diet, that of common carp remains high even at high concentrations (see Appendix I). Chiou & Ogino (1975) found the digestibility of α-starch by common carp to be 84% when its concentration in the diet was as high as 48%.

Table 10. *Metabolic faecal nitrogen egested by common carp as determined by two methods and in two environmental temperatures*

Method	Temperature (°C)	Fish weight (g)	mg N/100 g diet/day	mg N/100 g weight/day
Non-protein diet method	20	156–213	144±27.7	3.3±0.82
	27	133–215	180±29.4	4.4±0.70
Y-intercept method	20	73–125	123	2.7
	27	50–150	151	3.3

Source: From Ogino *et al.* (1973).

Differences in digestibility of carbohydrates can also occur among different strains of carnivorous fish within the same species, as was demonstrated by Refstie & Austreng (1981) for rainbow trout.

It should be noted that in addition to the low digestibility, carnivores have other physiological limitations to utilization of carbohydrates (see Chapter 3). Tunison *et al.* (1939) and Phillips *et al.* (1948) stated that digestible carbohydrates in trout feed should not exceed 12% since a higher content causes an accumulation of glycogen in the liver, associated with severe physiological disturbances and sometimes death of fish. Also Edwards *et al.* (1977) reported that increasing dietary carbohydrate level from 17 to 35% depressed the growth of rainbow trout. Refstie & Austreng (1981) fed rainbow trout with a diet containing 41.6% nitrogen-free extracts for 282 days with no apparent pathological signs or increased mortality, but this diet reduced growth and produced larger livers than diets lower in carbohydrates. Anyhow, it is clear that there are limitations on the amount of carbohydrate in carnivorous fish diets, even when it is digestible.

Fish age It has already been mentioned that enzymatic activity may vary with fish age, and that proteolytic and amylolytic activities of trout in its first developmental stages are lower than in later stages. This may, of course, affect digestibility coefficients. Kitamikado & Tachino (1960b) and Kitamikado *et al.* (1964b) found that the digestibility by rainbow trout of casein and protein from fish meal and beef liver were much lower at the early developmental stage (under 10 g) than in later stages (10–100 g), during which it did not vary much. This was especially noticeable with fish meal. Also Windell *et al.* (1978) found a difference in digestibility of proteins between small rainbow trout (18.6 g) and larger ones, but only at a relatively low water temperature (7 °C). At higher temperatures this difference was not noticed.

Physiological conditions Stressed fish, due either to excessive handling or to disease, may have a disturbed digestibility. Job (1977) stated that tilapia (species not given) which were fished in natural water bodies and transferred into tanks showed increased defaecation until acclimated. A long period of starvation may also affect enzyme secretion and digestibility. Starvation generally reduces the hydrolytic capacity of the intestine (Risse, 1971) by reducing the activity of the digestive enzymes (Ananichev, 1959; Overnell, 1973). Schlottke (1938–9) found that secretion of enzymes into the intestine of common carp that have fasted for a long period was higher after the second feeding than after the first one.

Parallel to the seasonal variations in digestive enzymes activity (p. 25), seasonal variations in digestibility may also occur. Chepik (1964) found that digestibility by common carp was highest in spring and rather sluggish in winter.

Water temperature Contrary to endotherms, where enzymes are active in a more or less constant temperature, the digestive enzymes of ectotherms are affected by external temperature. Increasing the temperature may increase both enzyme secretions and enzyme activity (p. 26). Temperature may also affect the rate of absorption of digested nutrients through the intestinal wall (p. 43). However, temperature also affects the rate of passage of food through the digestive tract. The higher the temperature, the more rapid is the transport of food and the shorter its exposure time to the digestive enzymes (p. 47). It is clear that the effect of temperature on digestion is compounded. On the one hand it affects enzyme activity and absorption and on the other it is balanced by the effect on the time the food stays in the digestive tract. To these, one should also add the effect of temperature compensation. Due to this compensation lower relative amounts of enzymes are secreted by high temperature acclimated fish than by low temperature acclimated ones (Smit, 1967). Only a few studies have been made on the effect of temperature on digestibility coefficient, which is the resultant of the above mentioned factors. Cho & Slinger (1979) showed that the digestibility coefficient of a compounded test diet by rainbow trout did not vary at temperature range of 9–15 °C, and increased slightly when temperature increased to 18 °C (Table 11). Shcherbina & Kazlauskene (1971) quoted Krazinkin (1935, 1952) who found that the digestibility of total nutrients by common carp increased somewhat with increase in temperature, but digestibility of protein remained unchanged

Table 11. *Effect of temperature on the digestibility of nutrients of a compounded test diet by rainbow trout*

Temperature	Digestibility coefficients (%)			
(°C)	Dry matter	Energy	Crude fat	Protein
9	67.9	75.2	89.8	91.3
12	68.2	75.0	88.2	90.0
15	67.4	74.4	89.4	90.0
18	69.8	77.4	92.6	93.5

Source: From Cho & Slinger (1979).

or even decreased slightly. Also Brizinova (1949, 1953, quoted from Shcherbina & Kazlauskene, 1971) found that the digestibility of protein from natural food (Tendipedidae larvae) by common carp and grass carp decreased with increase in temperature. The results of the experiment carried out by Shcherbina & Kazlauskene (1971) are given in Table 12. These results also show that digestibilities of carbohydrates and lipids by common carp increased with temperature, but those of cellulose and the absorption of minerals decreased.

It should be emphasized that although the digestibility coefficient may not change much with increased temperature, digestion rate usually does (p. 47). This allows the fish to eat more so as to supply the increased energy requirement due to the increase in temperature. This has been demonstrated quite clearly for salmon by Brett *et al.* (1969).

Water salinity Not much is known of the effect of salinity or other factors related to water composition on digestibility. MacLeod (1977) found that the digestibility of dry matter, energy and protein by rainbow trout fell linearly with increasing water salinity from freshwater to sea water (32.5% salts). It is not known whether this is a direct effect of the salt content or an indirect stress effect.

Food composition Nutrients of different foodstuffs may be digested to a different degree. This is often related to the source and composition of the foodstuff. Foods of plant origin are usually digested to a lesser degree than foods of animal origin. Plants have thicker and more resistant cell walls which are more difficult for the digestive enzymes to penetrate. Also the high cellulose content may affect their digestibility. Cellulose, which itself is hardly digested, may envelop and protect other, more digestible, nutrients such as protein and carbohydrate from the digestive enzymes.

Digestibility may be affected by the form of the food and its method of processing. Sandbank & Hepher (1978) have shown that algae dried in the

Table 12. *Digestibility of sunflower oil meal by two-year-old common carp at different temperatures*

Temperature	Digestibility coefficients (%)				Absorption (%)		
(°C)	Hydrolysable carbohydrate	Crude fat	Crude protein	Crude cellulose	Ca	Mg	P
16–17	19.0	78.6	81.7	53.9	27.0	39.0	40.5
22	47.9	78.2	80.0	49.6	33.7	46.3	35.6
26–27	30.6	86.4	75.1	46.5	11.6	30.3	21.8

Source: From Shcherbina & Kazlauskene (1971).

sun were less digestible than those dried on a steam drum (Table 13). Rapid drying on the steam drum seems to cause the bursting of the algal cell wall and expose the inner contents to the action of the enzymes, while slow drying in the sun may cause the shrinkage and hardening of the walls, protecting the cell contents from the enzymes.

Milling can increase digestibility. Mann (1948) determined the digestibility, by common carp, of sweet lupin and rye of different particle sizes. The results given in Table 14 show that the smaller the particle size, the higher is the digestibility. Experiments at the Tropical Fish Culture Institute, Malacca (1961) have shown that milled algae were digested better by *Oreochromis mossambicus* than non-milled algae. Kitamikado *et al.* (1964b) sifted fish meal and fed rainbow trout with the fine particle alone. They found that these particles were digested better than unsifted meal. It should be remembered, however, that the smaller particles may lose more nutrients into the water by leaching (p. 293, Table 8).

Cooking of the food by a method involving an extrusion process (see p. 293) can make it more digestible. This is especially true for carbohydrates, which become gelatinized, and thus more digestible by salmonids (R. R. Smith, 1971; Bergot & Breque, 1983; Vens-Cappell, 1984). Lovell (1984) has shown that grinding and cooking, as in the extrusion process, improves the digestibility of protein and starch (but not of fat) from practical diets by channel catfish, as can be seen in Table 14.

Tunison *et al.* (1942, 1943, 1944) have shown that the digestibility of protein by trout depends on the content of carbohydrates in the feed: the higher the carbohydrate content, the lower the protein digestibility. Their explanation of this is that the undigested portion of the carbohydrates passes more rapidly through the alimentary canal, carrying with it some of the proteins. According to Falge *et al.* (1978) the increase in carbohydrate content of the diet actually reduces the activity of the proteolytic enzymes, although, as they themselves point out, this activity remains still quite high. It is doubtful, therefore, whether the digestibility of the protein is

Table 13. *Digestibility coefficients (%) by common carp of unicellular algae dried in the sun as compared to those dried on a steam drying drum*

Algal species	Sun-dried	Drum-dried
Ankistrodesmus sp.	79.0	85.0
Scenedesmus sp.	60.8	68.4

Source: From Sandbank & Hepher (1978).

really affected by carbohydrate in this way. Watanabe *et al.* (1979) showed that an increase in lipid content of feed from 5 to 23% increased the digestion of total energy by rainbow trout. This was due to an increase in the digestion of protein from 98.4 to 98.9%; of carbohydrates from 50.7 to 58.5%; and also that of the lipids themselves from 74.7 to 87.5%.

Nose (1967) showed that the digestibility of lipids depends on their composition and saturation level (Table 15). It decreased with increase in number of carbon atoms in the fatty acid chain, and increases with the number of double bonds.

Various feeds may contain digestive enzyme inhibitors which reduce digestibility. Thus, for instance, raw soybean meal contains trypsin inhibitor, which has to be destroyed thermally before the soybean can be used for feeding fish (p. 211). Hofer & Sturmbauer (1985) found in raw wheat a strong α-amylase inhibitor, which also can be destroyed by heating.

Feeding level and frequency One of the most important questions, from the point of view of applied fish nutrition, is whether the amount of feed ingested affects its digestibility. From the literature, however, it seems that

Table 14. *The effect of grinding and cooking on the digestibility of feedstuffs by common carp and channel catfish*

(a) Digestibility by common carp of sweet lupin and rye at two levels of grinding

	Sweet Lupin		Rye	
Grinding level	Protein	Carbohydrate	Protein	Carbohydrate
Coarse	84.0	50.0	84.4	33.0
Fine	93.1	71.0	88.0	60–75

(b) Digestibility by channel catfish of nutrients from compounded feed[a] processed in various ways

	% Digestibility			
Process	Protein	Starch	Fat	Gross energy
Coarsely ground	82	64	89	62
Finely ground	84	70	90	65
Extruded	86	80	90	72

Notes: [a] Contains 12% fish meal, 50% soybean meal and 30% corn.

Source: (a) From Mann (1948).
(b) From Lovell (1984).

there is no such direct effect. Bondi *et al.* (1957) found that digestibility by common carp of 30–60 g does not change much with the increase in the amount of feed eaten (Table 16). Also Windell *et al.* (1978) could not find variations in the digestibility of protein and lipid by rainbow trout with increasing feeding levels from 0.4 to 1.6% of body weight per day, although, as could be expected for carnivorous trout, digestibility of carbohydrate, and as a result that of dry matter, decreased significantly at the higher feeding level (1.6% of body weight per day). On the other hand, Henken *et al.* (1985) found that apparent digestibility of protein (and dry matter) by African catfish (*Clarias gariepinus*) was negatively correlated with feeding level. These conflicting results indicate that further study on this problem would be beneficial.

The stability of digestibility coefficients with increased amounts of food may be explained as resulting from two factors: (a) increased enzyme secretion with increased food uptake into the stomach and intestine (Liebich & Mann, 1950; Windell, 1967); (b) longer retention of food in the stomach and intestine. It is known that in mammals gastric secretion is evoked by distension of the stomach wall, which activates the

Table 15. *Digestibility coefficients by rainbow trout of lipids having different composition*

Lipid composition	Digestibility
10:0	98
12:0	96
14:0	87
15:0	81
16:0	78
16:1	89
17:0	82
18:0	66
18:1	93
18:2	95
18:3	91
18:4	92
20:4	82
20:5	93
22:1	89
22:6	88

Notes: [a] For the method of designation of fatty acid composition see footnote to p. 217.

Source: From Nose (1967).

neuro-hormonal mechanism (Gomazkov & Krayukhin, 1961). Distension of the stomach wall also evokes the secretion of gastric juice in fish. This has been proven by inserting into the stomach inert materials such as foam plastic, inert starch or glass beads (Smit, 1967; Western & Jennings, 1970; Norris *et al.*, 1973). It would thus appear that the larger the bulk of the ingested food, the greater is the enzyme secretion. Also gastric and intestinal retention time increases with increasing feeding level (p. 47). These factors may be the reason why digestibility coefficients did not vary much among the different feeding levels in Beamish's (1972) experiments. However, since both increased enzyme secretion and food retention seem not to be linearly related to the amount of food eaten, digestibility may become lower when very large amounts of food are ingested. No doubt this subject is worth a further study.

Feeding frequency also does not seem to affect digestibility greatly. Hudon & De La Noue (1984) did not find any difference in apparent digestibility of dry matter, protein and energy when feeding frequency was increased from two to six times per day.

From the foregoing discussion it can be seen that digestibility in fish is regulated quite efficiently, and except for the effect of food composition and form, digestibility is quite stable at varying conditions. Appendix I presents digestibility coefficients of various feedstuffs by different fish species as obtained from the literature. These were determined by different methods (balance or inert indicator), at various temperatures and under various conditions. These factors may have caused errors in the individual determinations. However, it is hoped that due to the large number of citations, the errors will balance each other, giving an average value which should represent the correct digestibility coefficient. These values do not, however, provide a solution to the problems presented in the foregoing discussion.

Table 16. *Digestibility coefficients (%) by common carp of rice meal fed at different levels*

Feeding level (g/fish/day)	Nutrients			
	Proteins	Lipids	Fibres	*NFE*[a]
2	86.0	87.5	86.8	88.0
4	81.0	85.4	87.1	79.8
6	88.3	94.1	93.7	93.3

Note: [a] Nitrogen free extracts.

Source: From Bondi *et al.* (1957).

3

Energy pathways

In order to obtain a fuller understanding of the effects of absorbed nutrients on maintenance and growth for the purpose of better evaluating their nutritional value, it is important to know the metabolic pathways of these nutrients in the fish body. These may proceed in two different directions: catabolism and anabolism. In catabolism the relatively large molecules of nutrients absorbed into the blood are broken down and transformed into smaller molecules of simpler compounds. This is done in two stages: (a) transformation of absorbed nutrients into a limited number of intermediate compounds of relatively smaller molecular weight, releasing about one third of the energy in the nutrient as free energy; (b) final oxidation of the intermediate compounds to carbon dioxide and water, releasing the remaining two thirds of the energy. In anabolism, complex constituents of the tissues and body fluids are formed from simple precursors, which are usually the intermediate products of the first stage of catabolism. The rather intricate web involved in the transformation of matter and energy is called 'intermediate metabolism'. Processes of intermediate metabolism are similar in most animals, though variations may exist. The metabolic pathways of each of the major nutrients – carbohydrates, lipids and proteins – with specific regard to fish, will be discussed here separately, as was done in the case of digestibility (see Chapter 2). It should be borne in mind, however, that these pathways converge at crucial points and should, therefore, be considered as one comprehensive system.

Metabolism of carbohydrate

When digested, carbohydrates are converted mainly into glucose which is absorbed into the blood. Some studies have shown that, shortly after feeding on carbohydrates, the glucose level in the blood of fish increases considerably (Phillips *et al.*, 1948; Taggart, 1974, quoted from

Tiemeier & Deyoe, 1980). Nagai & Ikeda (1971a, b) did not find such an increase in blood glucose after feeding common carp on a high carbohydrate (90%), low protein (10%) diet. On the contrary, this diet caused the blood glucose level to decrease. Nagai & Ikeda explained this phenomenon as an accelerated turnover of blood glucose through oxidation, due to lack of proteins which usually supply part of the energy required by the fish.

The utilization of glucose for energy is carried out in two stages. The glucose is first converted, through a chain of reactions of the

Figure 9. The Embden–Meyerhof–Parnas and the pentose phosphate shunt pathways for the metabolism of glucose. (after Hoar, 1975). Underlined compounds – dietary precursors; heavy lines – main process direction; broken lines – process includes intermediate metabolites not given in the diagram.

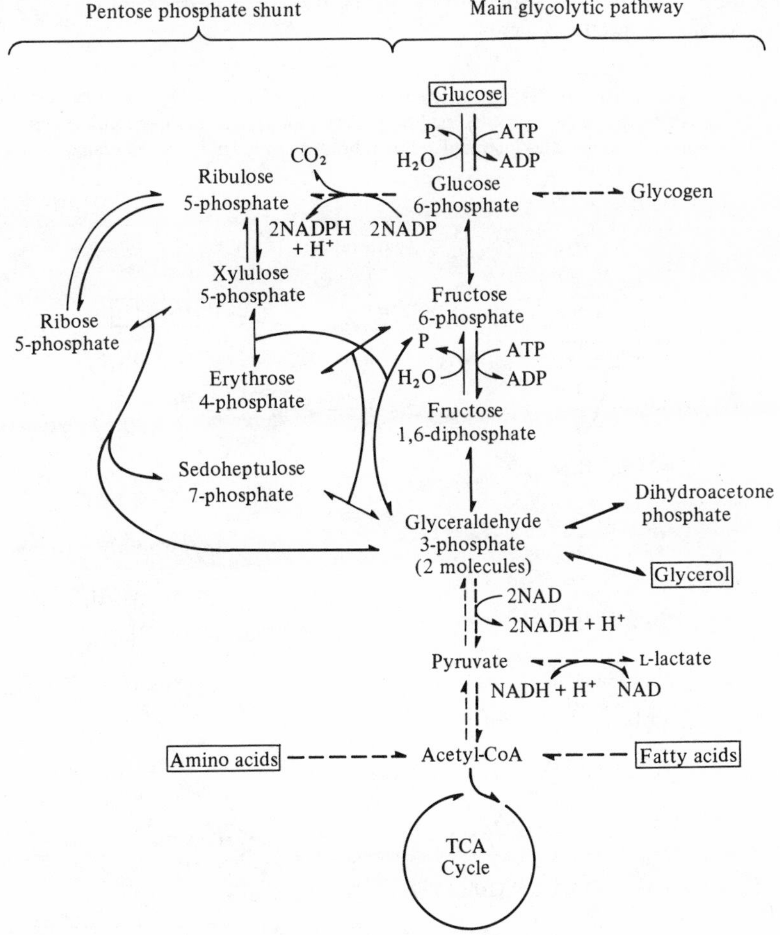

Embden–Meyerhof–Parnas glycolytic pathway, into pyruvic acid. These processes are carried out in the cell cytoplasm and do not require oxygen. The second stage is carried out through another series of reactions of the Krebs or tricarboxylic acid cycle (TCA cycle), which is sometimes also called the 'citric acid cycle', in the cell mitochondria. In these processes pyruvate is converted to acetyl coenzyme A which is further oxidized to carbon dioxide and water, producing reduced nicotinamide adenine dinucleotide (NADH-H^+) and flavin adenine dinucleotide ($FADH_2$), which through oxidative phosphorylation in the cytochrome system convert adenosine diphosphate (ADP) into the energy-rich adenosine triphosphate (ATP). The processes of oxidative phosphorylation are aerobic and require oxygen. The two stages of the glucose conversion are described in Figures 9 and 10. Part of the energy of glucose is released by oxidation in the first stage of glycolysis, but most is released through the processes of the TCA cycle.

Figure 10. The tricarboxylic acid cycle (after Hoar, 1975). Underlined compounds – dietary precursors; heavy lines – main process direction; broken lines – process includes intermediate metabolites not given in the diagram.

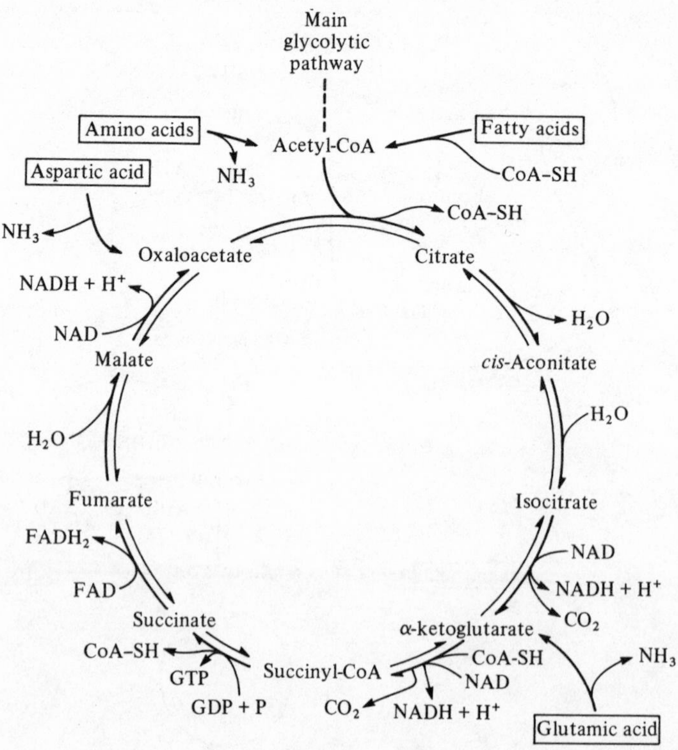

Most of the glucose catabolized in animal tissues proceeds through the glycolytic sequence producing pyruvate, most of which is, in turn, oxidized via the TCA cycle. There are, however, some alternative metabolic pathways by which glucose can be catabolized. One of these is the 'pentose phosphate shunt' also called 'hexose mono-phosphate shunt' or 'Warburgh–Dickens pathway'. Glucose 6-phosphate is the precursor for this pathway, which leads to the formation of two special products: reduced nicotinamide adenine dinucleotide phosphate (NADPH) and ribose 5-phosphate. The first product is used for a number of biosynthetic processes, and especially the synthesis of fatty acids and steroids, and the second is necessary for the synthesis of nucleotides and nucleic acids.

In order to obtain a better understanding of the metabolism of glucose, it is important to determine the relative role of the pentose phosphate shunt and the factors affecting it. A number of methods have been used for this determination. One is to measure the relative activity of the major enzymes of each of the pathways, such as that of phosphoglucomutase (PGM) and phosphoglucoisomerase (PGI), both active in the central glycolytic pathway, and glucose 6-phosphate dehydrogenase (G6PDH) or 6-phosphogluconate dehydrogenase (6PGDH), active in the pentose phosphate shunt. At times substances inhibiting the activity of some of these enzymes were used and the effect on glucose metabolism was studied. Thus, Ekberg (1958) used iodoacetate, which is a strong inhibitor of glyceraldehyde 3-phosphate dehydrogenase, an enzyme acting on one of the important links of the central glycolytic pathway (transforming glyceraldehyde 3-phosphate to 3-phosphoglyceroyl phosphate) and therefore inhibits metabolism through this pathway, but not through the pentose phosphate shunt. Some researchers used radioactively labelled carbon, ^{14}C, to label dietary glucose for the same purpose. These made use of the fact that in the central glycolytic pathway metabolism of the carbon atom C-1 of the glucose molecule is equal to that of C-6, and the ratio between these two carbons in the respired carbon dioxide is 1:1, but in the pentose phosphate shunt metabolism acts on C-1 alone. Thus every deviation from the ratio C-6:C-1 = 1 in the respired carbon dioxide of fish fed glucose labelled either at C-1 or C-6 indicates the participation of the pentose phosphate shunt in metabolism. Though this method has been criticised by Katz & Wood (1960), it has been used extensively in studies of fish. In order to follow better the fate and transformation of glucose, labelled intermediate compounds were sometimes administered, such as acetate-1-^{14}C, and the labelled carbon was traced in the respired carbon dioxide or glycogen (Hochachka, 1962; Nagai & Ikeda, 1972, 1973).

Studies carried out by the methods mentioned above have shown that in fish under normal conditions glycolysis proceeds mainly through the central glycolytic pathway, and the ratio of C-6:C-1 in the respired carbon dioxide approaches 1 (Brown, 1960). However, metabolism through the pentose shunt also occurs to a certain extent (Shimeno & Takeda, 1972, 1973). Liu *et al.* (1970) studied glycolysis in the fish *Cichlasoma bimaculatum* with the aid of glucose, acetate and gluconate labelled at various carbon atoms. Their results show that only 66% of the glycolysis is carried out through the central pathway and the rest through other pathways such as the pentose shunt (19%) and the glucuronic acid pathway (22%).

The level of the shunt metabolism may vary among different organs of the fish. Bachand & Leray (1975) found that the pentose shunt activity in the erythrocytes of yellow perch, *Perca flavescens*, is as efficient as that of the central pathway. This may be true also for other fish species, but from determinations of the G6PDH enzyme activity, Shimeno & Takeda (1973) showed that the highest pentose shunt activity in fish is in the liver or hepatopancreas. In common carp this activity was about 90% of the G6PDH activity of the whole body. G6PDH activity and thus HMP shunt metabolism was of relatively little importance in fish muscle, whereas the activities of PGI, PGM, and phosphofructokinase enzymes were high in the muscle, which may indicate metabolism through the central glycolytic pathway.

Activity of the pentose phosphate shunt also shows variation between fish species. Shimeno & Takeda (1973) found that the activity of G6PDH was distributed in livers of all of the fish species tested by them, but varied according to species. The highest activity was observed in the liver of freshwater eel, which was more than twice that found in the liver of common carp and three times that in the liver of rainbow trout. Yamauchi *et al.* (1975) found the activity of G6PDH in the liver of brook trout (*Salvelinus fontinalis*) to be 10 times as high as that in the liver of rainbow trout (*Salmo gairdneri*).

It seems that the activity of the pentose phosphate shunt pathway in the liver is associated primarily with the production of NADPH and ribose 5-phosphate which are eventually transported from the liver to the target tissue, while the predominance of the glycolysis through the central pathway in the muscle supplies the necessary energy for movement. Variations in the activity of the shunt among different species may be the result of variations in requirements for these metabolites, which may in turn be affected by external factors such as food and water temperature.

There is some controversy regarding the effect of fasting or food composition on the rate of metabolism through the pentose shunt. Buhler

& Benville (1969) state that the specific activity of G6PDH in yearling rainbow trout remained unchanged when the fish starved for a period as long as eight weeks, and also when starved fish were fed various diet compositions. Hochachka (1961), on the other hand, obtained quite different results. He found that the activity of the shunt decreased during starvation and that glycolysis is carried out mainly through the central pathway. Also, the activity of G6PDH found in brook trout (*Salvelinus fontinalis*) from natural ponds or streams was lower than that from the hatchery trout which received a high carbohydrate diet. This seems to indicate an effect of food composition on the glycolytic pathway. Still other studies by Nagayama *et al.* (1972b) show that during starvation of common carp the concentration of PGM and PGI enzymes, active in the central glycolytic pathway, decreased, while that of G6PDH and glucose 6-phosphatase, active in the pentose shunt, increased. This would mean, in contrast to Hochachka's results, an increase in the activity of the pentose shunt. It is obvious that more study is required to clear this controversy.

More important and confirmed variations in glucose metabolic pathways are those related to anoxia and especially to acclimation to low temperatures. Hochachka (1967), reviewing some of the aspects of metabolic adaptation to temperature, points out the changes occurring in glycolysis in muscle and liver tissues during cold acclimation, and the difference between the two. In both the muscle and liver of cold acclimated fish glycolysis through the central glycolytic pathway (measured by oxidation of glucose-6-^{14}C) was much higher than that in warm acclimated fish. TCA cycle activity (measured by acetate-1-^{14}C oxidation), however, was essentially unchanged, and in the liver it even decreased somewhat. This indicated that in cold acclimated fish extramitochondrial is favoured over mitochondrial metabolism. Hochachka postulated a shift in the specific isozyme pattern in which one type is favoured under one metabolic regime (e.g. cold compensation) while another is dominant under a different regime. This shift is presumably induced by hormones.

Changes in glycolysis in the liver, as found by Hochachka (1967) were more complex than in muscle. While in muscle of cold acclimated fish there was only a slight activation of lipogenesis (acetyl-1-^{14}C incorporation into lipid) and a slight corresponding increase in pentose phosphate shunt participation, in the liver of the cold acclimated fish lipogenesis was strongly activated and was about three times higher than that of warm acclimated fish. In association with this higher role of fatty acid synthesis, a clear rise in the shunt activity appeared (Kanungo & Prosser, 1959b; Hochachka, 1961, 1969; Hochachka & Hayes, 1962; Yamauchi *et al.*, 1975). The role of the pentose phosphate shunt in fish acclimated to cold temperatures is indicated in an experiment by Ekberg (1958) who used iodoacetate to inhibit glucose metabolism through the central glycolytic pathway. He used a homogenate of gills from goldfish acclimated at two temperatures, 10 and 30 °C, and then measured the oxygen requirements of the homogenates with and without iodoacetate. The oxygen requirement of the homogenate of the fish acclimated at 10 °C was inhibited by the iodoacetate by 53%, while that of the fish acclimated at 30 °C was 77%. Similar results were obtained by Hochachka & Hayes (1962) for brook trout. Inhibition of glucose metabolism by iodoacetate in fish acclimated at 15 °C was 88.1%, but only 66.7% in those acclimated at 5 °C. Other methods used by Hochachka & Hayes have also indicated an increase in the pentose phosphate shunt in fish acclimated to cold temperatures. From these studies it seems that the utilization of the shunt pathway when the temperature decreases is temporary and associated with the reorganization of the cells, especially the synthesis of cell membranes with phospholipids containing a higher percentage of highly unsaturated fatty acids. The pentose shunt seems to have a role in this lipogenesis by providing the necessary NADPH for the transformation of acetyl-CoA to fatty acids (Yamauchi *et al.*, 1975). Hochachka & Hayes (1962) traced the labelled carbon of fat added to the diet of trout and found 76.9% of its total activity in the muscle of cold acclimated fish (4 °C) after 15 min, compared with only 10.5% in fish acclimated at 15 °C.

It is thus obvious that cold acclimation is a rather complicated process. An interesting example is described by Hochachka (1967) to illustrate the complexity of the control of the reaction sequence involved in cold acclimation of energy saturated cells (Figure 11). At low temperatures the energy requirements for growth are very small, causing the ratio between ATP and AMP to increase. Since ATP exerts a restraining influence on the TCA cycle, the activity of this cycle is inhibited. However, since glycolysis in the muscle continues, the lactose concentration in the blood and liver rises, which also causes the increase in pyruvate concentration. Though the TCA cycle activity is reduced, the concentration of citrate, which is affected by the concentration of pyruvate, is increased. This has a number of consequences: (a) Citrate increases the substrate affinity of acetyl-CoA-carboxylase, the enzyme which activates the first stage of lipogenesis from acetyl-CoA. This results in an increased lipogenesis in cold acclimated fish. (b) Hydrogen formed by reducing NADP produced in the pentose phosphate shunt is utilized in some of the stages of lipogenesis. (c) Citrate is also a negative modulator of the enzyme phosphofructokinase, but does not affect fructose 1,6-diphosphatase, the enzyme which transforms fructose diphosphate to fructose 6-phosphate.

This causes the accumulation of the latter as well as glucose 6-phosphate, thus contributing to increased participation of the pentose phosphate shunt in glucose metabolism and increased supply of NADPH. (d) Increase in glucose 6-phosphate concentration also facilitates the synthesis of glycogen, and its concentration in the liver increases. This reaction sequence can explain the increase in lipid and glycogen contents in the liver of cold acclimated fish.

Muscle metabolism

In fish muscle, the primary energy-yielding pathway for carbohydrates is the glycolysis of the glycogen reserves. The blood-borne glucose is an alternative energy source. The enzyme hexokinase, which regulates the transfer of glucose to glucose 6-phosphate, is found with such

Figure 11. Organization of energy metabolism in liver during cold adaptation. Broken lines connect effectors (boxed in) with enzymes that they modulate. Effector activation is expressed with heavy arrows; effector inhibition is marked with a heavy cross. TCA cycle intermediates are drawn as if they were formed by the cycle, although at least three of these, citrate included, can be formed extramitochondrially (from Hochachka, 1967).

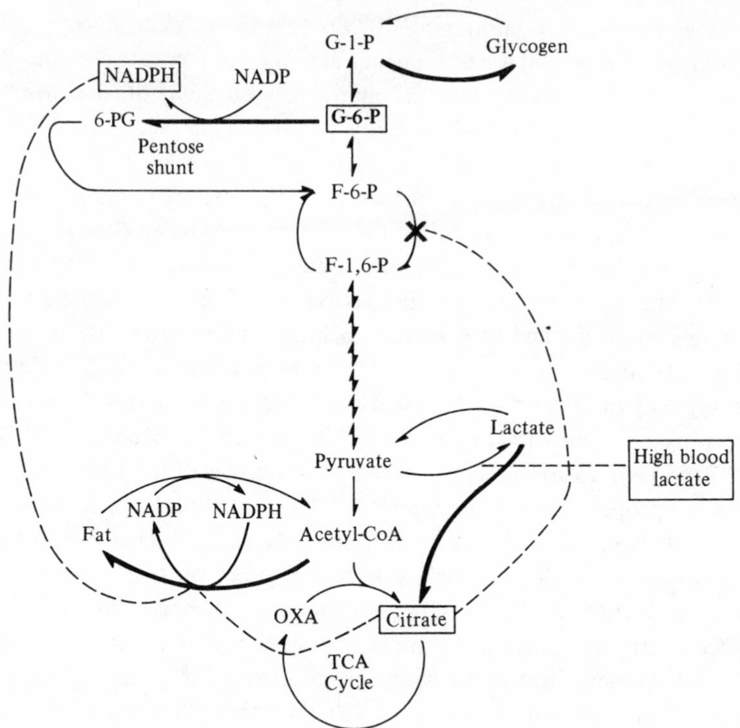

low activities in fish muscle, relative to other glycolytic enzymes, that it is unlikely that free glucose serves as an important source of energy for fish muscle. Glucose metabolism is carried out mainly in the liver and some other internal organs such as heart, kidney and spleen. The hexokinase activity in these internal organs is considerably higher than in the muscle. MacLeod *et al.* (1963) found that activity of hexokinase in the heart of trout was about 60 times as high as that in the muscle. Nagayama *et al.* (1972a) studied the activity of this enzyme in various organs of different fish species and found its concentration to be in the following order: heart > spleen > kidney > gills > liver > muscle > intestine. Also, the concentration of the enzyme glucose 6-phosphatase, which hydrolyses glucose 6-phosphate, is found mainly in liver and kidneys but to a very small extent in the muscle (Hochacka, 1969; Shimeno & Ikeda, 1967).

In order to understand better the utilization of the different energy sources by fish, the muscle, which requires a large portion of the energy for swimming, must be considered. As in other mammals, the muscles in fish are classified into three groups: (a) smooth, involuntarily controlled muscle, which appears mainly in the intestine and blood vessels; (b) striated, voluntarily controlled muscle, which forms the bulk of the fish flesh and serves mainly for swimming; (c) cardiac muscle. The striated skeletal muscles are composed of two types: (i) White muscle, having broad fibre cells which can contract rapidly. This constitutes the larger part of the skeletal muscle, in many fish 80–95% of the swimming musculature. (ii) Red muscle, usually as a thin superficial layer below the skin, forming thicker bands along both sides of the body at the level of the lateral lines and in certain other locations. It consists of narrow fibre cells which contract more slowly than those of white muscle. They are red due to the large number of blood vessels in them (Boddeke *et al.* 1959; Harder, 1975). The difference between these two types of muscle – white and red – is much more defined in fish than in higher vertebrates, where the two kinds of muscle fibres are mixed. In fish such a mixture of muscle fibres occurs only in the intermediate layer between the red and white muscles – the pink muscle (Johnston *et al.*, 1977; Johnston & Moon, 1980). Love (1970) presents a comprehensive review on the role of these muscles. Their relative proportion varies among species according to their swimming habits. Pelagic fish which swim continuously have a higher proportion of red to white muscle than fish that rest at times on the bottom (Greer-Walker & Pull, 1975). This proportion may change even in the same species with the change in swimming habits. Lewander *et al.* (1974) found that eels in their silver phase which are preparing for spawning migration, have a higher proportion of red muscle than the yellow phase. According to Johnston & Goldspink (1973a) the average weight of the superficial red muscle of crucian carp (*Carassius carassius*) accounts for about 7.4% of the total muscular mass.

Many studies have shown that there is a division of labour between the red and white muscles, where red muscle is responsible for the less energetic swimming efforts and white muscle for the more vigorous swimming activity, such as chasing prey or avoiding a predator. Extracellular recordings from the myotomal muscles of fish taken during swimming show that only red muscle fibres are active when fish swim slowly or at moderate speeds, whereas during vigorous movement the fast white muscle fibres are active (Bone, 1966; Tsukamoto, 1981). Since vigorous swimming is not the regular method of movement, the red muscle is generally the more active component of the muscular mass (Buttkus, 1963). Like cardiac muscle, red muscle has the ability to function continuously. This is essential for pelagic fish which swim continuously, hence the higher proportion of red muscle in these fish.

Glycogen levels in both muscle and liver of fish appear to be substantially lower than those recorded for mammals (Black *et al.*, 1961). Some researchers have found that the concentrations of glycogen, lipid and protein are higher in red muscle than in white (Bone, 1966; Love, 1970; Krishnamoorthy & Narasimhan, 1972; Lin *et al.*, 1974). However, the higher glycogen concentration in red muscle seems to result from its lower utilization as compared to white muscle. Black *et al.* (1961, 1962, 1966) found that during two minutes of vigorous swimming activity by rainbow trout, the glycogen concentration in white muscle was reduced by 50%. Wittenberger & Diaciuc (1965) showed that after intermediate swimming effort the glycogen concentration of white muscle of common carp had been reduced by 30%, but it did not change in red muscle. Fraser *et al.* (1966) found that during a rest period the glycogen concentrations of the red and white muscles of *Gadus morhua* were similar, but following vigorous activity the glycogen concentration in the red muscle was much higher, due to lesser degradation of the glycogen compared with the white. Similar results have been obtained by Wittenberger & Diaciuc (1965) and Bone (1966).

Johnston & Goldspink (1973a) studied the quantitative utilization of glycogen by white and red muscles of crucian carp during short periods of swimming. Glycogen utilization was found in the superficial red muscle. The $\log_{10}$ of total glycogen utilization rate was linearly related to swimming speed. Statistically significant glycogen utilization by white muscle, however, occurred only at speeds in excess of about three bodylengths per second. At these speeds most of the glycogen was utilized by white muscle and the superficial red muscle accounted for only 15–20% of the total glycogen utilization. Although the highest rate of glycogen utilization by white muscle occurs during 'burst' swimming, there is now much evidence which shows that utilization of glycogen in this tissue also occurs at less than 'burst' swimming speeds (Black *et al.*, 1962; Johnston & Goldspink, 1973b, c; Jones, 1982).

Thus, it seems that most of the energy in white muscle comes from glycogenolysis. The glycogen is transformed through the central glycolytic pathway into pyruvate. Under aerobic conditions pyruvate undergoes an oxidative decarboxylation to acetyl-CoA, which is further oxidized through the TCA cycle. However, during anaerobic conditions pyruvate is converted into lactic acid. Such conditions seem to be quite common in the tissue of a vigorously active fish. Brett (1970) points out that a 15 cm long salmon can swim at maximum speed of 150 cm/sec (10 bodylengths/sec) for 20 sec, after which it tires. During this vigorous movement oxygen requirement increases 100 times over the resting level. However, an experiment carried out by Wittenberger & Diaciuc (1965) showed no change in uptake of oxygen by white muscle due to swimming effort. The white muscle then goes into a temporary 'oxygen debt' while accumulating lactic acid (Black, 1955; Huckabee, 1958; Caillouet, 1964). This is accompanied by a decrease in muscle pH (Kutscher & Ackermann, 1933; Braekkan, 1956; Black *et al.*, 1962), followed by fatigue. Beamish (1966) noticed in several marine fishes that when they swim at a speed exceeding four bodylengths per second, they cannot endure this for more than about five minutes. According to Black (1958) and Wendt & Saunders (1973), the accumulation of lactic acid during hyperactivity and the change in muscle pH may be lethal. Exercised white muscle also contains a high concentration of creatine, which indicates utilization of creatine phosphate as an energy source for muscle contraction under anaerobic conditions. Creatine phosphate then reacts with ADP to form ATP, which is the main form of directly available energy in the muscle. According to Hamoir (1955) red muscle is able to resynthesize phosphate compounds in aerobic conditions at a higher rate than white muscle. Anaerobiosis may contribute to the fish energy budget at the start of swimming or when swimming speed changes, but normally most of the energy for fish movement is supplied by aerobic metabolism. Much indirect evidence suggests that at least 70–80% of the energy for critical speed of salmonids is supplied by aerobic metabolism. Aerobic metabolism can provide only about 50% of the thrust power for critical speed in cyprinids, but critical speeds, at which the fish is fatigued after a certain period of time, are higher in cyprinids than in salmonids (Jones, 1982).

Anaerobiosis and oxygen debt also can develop due to low water oxygen concentration (Heath & Pritchard, 1965; Jones, 1982). In both cases the oxygen debt is expressed by increased blood lactic acid concentration, decreased blood pH (Caillouet, 1968), and by high respiratory quotient (*RQ*), which is the ratio between the volume of carbon dioxide produced and oxygen consumed during respiration (Kutty, 1968; Peer & Kutty, 1981). Oxygen debt created by vigorous movement is made up subsequently by rapid ventilation during some hours of rest. According to Brett (1964) the recovery period following fatigue of a young sockeye salmon is two to four hours long.

Lactic acid formed by glycolysis in muscles of mammals is transported by the blood to the liver, where it has a dual fate: it is either oxidized to carbon dioxide and water, or undergoes a rapid regeneration back into glucose and glycogen for further use in the muscle. Through these processes there is a homeostasis of blood lactate level. It may continue to rise for a short time following a severe muscular effort, but it drops very rapidly thereafter. In fish, this process is much slower (Black *et al.*, 1962,

Figure 12. Changes in muscle, blood and liver levels of lactate in rainbow trout during 15 min of strenuous exercise and 24 h of recovery (from Black *et al.*, 1962).

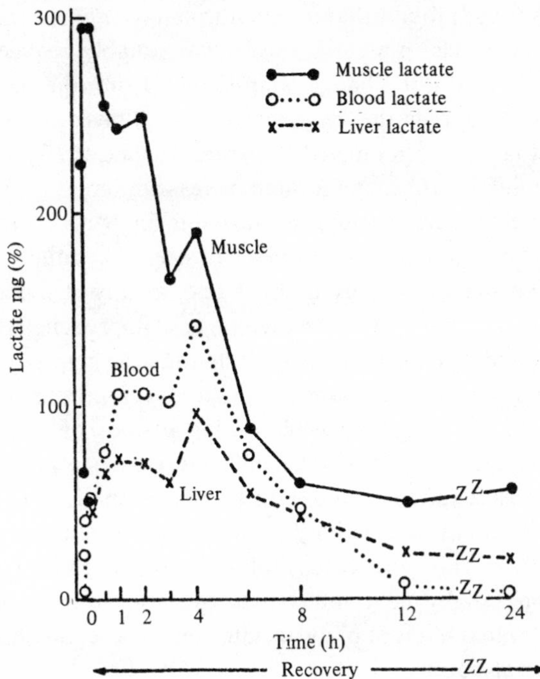

1966), and replenishment of glycogen reserves takes place at a much slower rate than in mammals (Bilinski, 1974). Black *et al.* (1961) found that muscle lactate in rainbow trout increased more than three-fold during the first two minutes of severe excercise and $4\frac{1}{2}$ times during the first nine minutes. During recovery after vigorous swimming, muscle lactate begins to decline upon cessation of the activity, but does not return to the pre-exercise levels until the sixth to eighth hour of recovery. Blood lactate continues to rise for two to four hours following cessation of the exercise and does not approach the pre-exercise level for eight to twelve hours (Figure 12). Accumulation of lactate in white muscle is often not proportional to the amount of energy generated by the muscle. Driedzic & Hochachka (1975) examined the possibility that lactate was not the sole end product of anaerobic metabolism and that amino acids were also utilized by white muscle as an anaerobic energy source. However, after an anaerobic work period induced by hypoxia, they could not detect any other product, nor did they detect a mobilization of the free amino acid pool. They concluded, therefore, that white muscle appears to depend solely on glycogenolysis and that the lower than expected lactate levels in the muscle may be due to transfer of some of this compound out of the tissue, despite its poor vascularization. This transfer of lactate from the muscle is not rapid enough during vigorous activity and lactate is accumulated. Beamish (1968) found that in Atlantic cod (*Gadus morhua*) swimming at moderate speeds, muscle glycogen was notably reduced, although blood lactate did not change appreciably from the non-exercising level. At higher speeds the depletion of muscle glycogen was more pronounced and lactate accumulated in the muscle and the blood. Also, Driedzic & Kiceniuk (1976) found a rapid increase in blood lactate concentration following fatigue, reaching a maximum in two to three hours, and then slowly returning to resting level. However, they found no increase in blood lactate level, or only a slight one, at any sustained swimming speeds, though in many cases the exercise level approached the critical velocity. These findings appear to suggest that muscle glycogen is utilized at any swimming velocity, but at moderate speeds it is presumably being subjected to complete oxidation, while during periods of 'burst' activity and final speed increment, the removal of lactate is too slow and it is accumulated in the muscle and the blood. Low participation of the liver in mobilizing glycogen for muscle activity is evident from studies by Wendt & Saunders (1973) who showed that following induced exercise, young Atlantic salmon usually have much reduced concentrations of muscle glycogen and elevated levels of blood lactate and glucose, but little or no reduction in liver glycogen.

According to Cowey & Sargent (1972) the accumulation of lactate in the blood of fish is associated with a low level of lactic acid dehydrogenase, the enzyme responsible for the metabolism of lactic acid. Black *et al.* (1961) suggested that this is also associated with low body temperature in fish. This is supported by Barrett & Robertson-Connor (1964) who found that in yellowfin tuna (*Thunnus albacares*) and skipjack (*Ketsuwonus pelamis*), which have a higher body temperature than other fish species, blood lactate returns to normal levels already after two hours of recovery.

In contrast to the case of white muscle, the role of red muscle was not at first clear, and some authors suggested that it acts as an inner organ supporting the metabolism of white muscle, similar to the role of the liver. This theory was stimulated by the fact that red muscle contains high concentrations of B vitamins and lipids, similar to the composition of the liver (Braekkan, 1956; Mori *et al.*, 1956; Tsuchiya & Kunii, 1960; Stirling, 1972). It is now apparent, however, that the high lipid content of red muscle is associated with the fact that lipid is the main source of energy for this muscle. George (1962) and George & Bokdawala (1964) found that lipase activity in red muscle is higher than in white. Their conclusion was that red muscle is adapted to aerobic metabolism, and that lipids are their main source of energy. Bilinski (1963) and Driedzic & Hochachka (1978) found that the capacity of red muscle to oxidize fatty acids is much higher (according to the latter authors up to 10-fold) than that of white muscle. This was supported by a number of other physiological studies. The number of mitochondria, which are the sites of fatty acid oxidation in the cell, is larger in red muscle than in white (Buttkus, 1963; Bone, 1966; Nishihara, 1967; Lin *et al.*, 1974). Jones & Randall (1978) state that the density of capillaries in red muscle is 10 times higher than in white muscle. Somero & Childress (1980) found that the activity in red muscle of citrate synthase, an enzyme associated with aerobic metabolism through the TCA cycle, is almost 10 times that in white muscle, while the activity of lactate dehydrogenase (LDH), which is associated with the anaerobic activity of glycolysis, is only one-fifth as high in red muscle. Activities of transaminases, which convert amino acids into oxaloacetate and pyruvate, are also found to be higher in red muscle than in white (Wittenberger & Giurgea, 1973), which indicates a more active catabolism of protein to provide energy in red muscle.

While red muscle is adapted mainly to aerobic metabolism it can also function anaerobically. In contrast to white muscle, where anaerobic metabolism is associated with glycogenolysis only, resulting in lactate as sole end product (Driedzic & Hochacka, 1975), evidence has been obtained (Johnston, 1975a, b) that in red muscle anaerobic pathways other than glycolysis are operated in which L-alanine and succinate have been observed to accumulate as anaerobic end products.

It can thus be summarized that white muscle, which is responsible for more vigorous movements, utilizes muscle glycogen as its main energy source. This muscle tires rapidly due to exhaustion of the glycogen reserves and accumulation of lactic acid. The liver does not support the larger glycogen requirement by regenerating the lactic acid to glycogen or mobilizing liver glycogen. For vigorous movement large amounts of oxygen are required. Since these cannot be supplied at the same rate as required, glycogen metabolism proceeds anaerobically causing oxygen debt, which is made up later during a rest period. The red muscle, on the other hand, is used for normal movement and utilizes lipids and protein as its main source of energy, which is produced by aerobic pathways. Oxygen is supplied by the numerous blood vessels in this muscle (Bokdawala & George, 1967a, b; Wittenberger, 1967; Rayner & Keenan, 1967; Gordon, 1968; Cowey & Sargent, 1972). It may well be that the relative requirement of carbohydrates, compared with protein and lipids as sources of energy, depends to some extent on the proportion of energy expenditure by the two kinds of muscle.

Regulation of carbohydrate metabolism

Metabolism of glucose is largely controlled by hormones. In endothermic animals there is a pronounced homeostasis of blood glucose; it is controlled mainly by the hormones insulin and glucagon secreted by the pancreas. Such hormones are also found in fish, though here the homeostasis of glucose is less pronounced. Insulin and somatostatin cause the glucose level to decrease (hypoglycaemia) and its rapid utilization in the tissue or transformation into glycogen which is accumulated especially in the liver. When insulin is deficient the blood glucose concentration increases (hyperglycaemia), glucose metabolism is impaired and a diabetic condition develops. Glycogenesis is also depressed, which results in glycogen concentration; lipogenesis, which usually proceeds from glucose or acetate in normal animals adequately nourished by carbohydrates, is impaired; and energy is supplied mainly by increased gluconeogenesis from lipids and proteins. It seems that insulin affects glucose metabolism in several ways. It affects the first step in conversion of glucose to glucose 6-phospate. Probably the chief determinant of this reaction is hexokinase, and insulin may modify its activity. It also causes a more rapid transfer of glucose, as well as amino acids and other metabolites, through cell membranes (Butt, 1975). The exact pathway in which insulin exerts its effect in the cell is not clear. It has been suggested (Saltiel & Cuatrecasas,

1986; Saltiel *et al.*, 1986) that insulin activates an endogenous selective phospholipase which hydrolyses a membrane glycolipid, resulting in the release of a complex carbohydrate–phosphate substance containing inositol and glucosamine that regulates cAMP phosphodiesterase and perhaps other enzymes. Insulin has been shown to cause a decrease in hepatic cAMP and to prevent the increase of cAMP produced by glucagon (Jefferson *et al.*, 1968) and therefore it seems that the action of insulin may also be mediated in this way.

Insulin is produced in the β-cells of the pancreatic islets of Langerhans (while the α-cells produce glucagon). Histological, immunological and biological evidence demonstrates the presence of insulin in all groups of fish, and the hormone has been purified from the pancreatic islets of several species (Epple, 1969; Lewander, 1976). Its primary structure has been determined for several species of fish. These molecules are similar in being composed of two chains, but differ among fish species by a few amino acids, particularly in the A chain and at the terminal of the B chain (Fontaine, 1975).

In higher vertebrates glucagon, also secreted by the pancreas, causes glucose level to increase (hyperglycaemia) by affecting phosphorylation in the liver, transforming glycogen to glucose. Sutherland & Cori (1948) found glucagon to accompany insulin even in highly purified pancreatic preparations. Phosphorylase is present in the liver in active and inactive forms. Glucagon causes an increase in the active form. Adrenalin (epinephrine) has a similar effect, but unlike glucagon, in addition to the liver, it also affects various other organs, such as skeletal muscle. Glucagon, adrenalin and glucocorticoids also stimulate gluconeogenesis, converting pyruvate to glucose. In contrast to insulin, the action of these hormones is mediated by cAMP. Both cAMP and its dibutyryl derivative stimulate phosphorylase activity (and lipolysis) and inhibit glycogen synthetase (Butt, 1975).

Only relatively little information is available on glucagon in fish. α-Cells similar to mammalian glucagon-producing cells were found in the pancreas of all fishes except cyclostomes (Epple, 1969; Fontaine, 1975). Falkmer (1961, 1966) observed a hyperglycaemic response in *Cottus scorpius* after an injection of an alcoholic insulin-free extract of the islets. An application of histochemical procedure for the detection of tryptophan has also suggested the existence of this hormone in the α-cells of teleosts (Thomas, 1970). Klein & Lange (1972), using an immunofluorescence method, have shown the presence of glucagon in the α-cells of *Xiphophorus helleri*. Ince & So (1984) used a mammalian glucagon radioimmunoassay to measure immunoreactivity in a perfused pancreas of European eel (*A. anguilla*), and found glucagon-like immunoreactivity. Noe & Bauer (1970) reported on the glucagon precursor in the pancreatic islets of the anglerfish (*Lophius americanus*), but there does not appear to be any direct report on the identification of glucagon in fishes. The difficulty in identifying glucagon may be due to its high biological and immunological species specificity. Nevertheless, the injection of mammalian glucagon into fish has a strong hyperglycaemic effect. It decreases the glycogen synthetase just as it does in mammals (Murat & Plisetskaya, 1977).

Adrenocorticoid hormones, especially cortisol, which in teleosts is the principal plasma corticosteroid (Cavin & Singley, 1972; Fryer, 1975; Peter *et al.*, 1978; Truscott, 1979) also have a hyperglycaemic activity. Dave *et al.* (1975) found that during prolonged starvation of the European eel (*A. anguilla*) there is an increase in the level of blood cortisol. This seems to increase the rate of gluconeogenesis, using amino acids as an energy source for the starving fish. Adrenocorticoid hormone secretion is regulated through the pituitary. Butler (1968) hypophysectomized freshwater North American eel (*Anguilla rostrata*) and found a depletion in liver glycogen and increased gluconeogenesis. Injecting hydrocortisone inhibitors also decreased liver glycogen, while injecting fish with cortisol increased the utilization of body protein for energy through gluconeogenesis (Storer, 1967).

Through the effect of glycaemic hormones the blood glucose level in endotherms remains quite constant. It increases during the first hour after ingesting a carbohydrate-rich diet, but returns to its normal level within two to three hours. This, however, is not the case with fish, especially carnivores such as trout and salmon. Many studies (e.g. Phillips *et al.*, 1945, 1948; Palmer & Ryman, 1972; Cowey *et al.*, 1977a, b; Bergot, 1979a) have shown that the blood glucose level increases rapidly and considerably after consumption of a carbohydrate diet and its subsequent decrease is slow. Phillips *et al.* (1945) found that when brook trout of an average weight of 12 g were fed 25 mg glucose, blood glucose increased 110% over the normal level. The higher the glucose in the food, the higher was its concentration in the blood (Phillips *et al.*, 1948; Bergot, 1979b). Blood glucose returned to its initial level only after 24 h. Similar phenomena were observed by Furuichi *et al.* (1971) and Furuichi & Yone (1971b) for red sea bream, Thorpe & Ince (1974) for northern pike (*Esox lucius*), and by Ince & Thorpe (1974) for European eel (*Anguilla anguilla*). In the last two cases a dose of 0.5 g/kg glucose was injected, which resulted in hyperglycaemia which returned to normal levels after 24–48 h. Differences among fish species in this respect were demonstrated by Yone (1979) who fed common carp, red sea bream, and yellowtail a gelatin capsule containing glucose at a rate of 167 mg/100 g body weight. Blood glucose increased in common carp from 40 mg% to 180 mg% within an hour. In red sea bream the same increase was noticed in two hours and in yellowtail blood glucose increased from 120 mg% to 210 mg% in three hours. Blood glucose did not return to its normal values even after five hours, but it was then much higher in yellowtail than in red sea bream, and the latter was higher than in common carp. The sluggish reduction in the serum glucose is a result of its low utilization. Lin *et al.* (1978) used labelled glucose to determine the utilization rate in coho salmon (*Oncorhynchus kisutch*). They found it to be 0.35–0.43 mg/min/kg body weight, which is much lower than that usually observed in higher vertebrates.

In view of the above results, most authors concluded that with respect to blood glucose, fish behave like diabetic mammals. Indeed, when insulin was injected into fish prior to feeding glucose, blood glucose increased to a lesser degree and returned to normal after 3–15 h (Phillips *et al.*, 1945; Yone, 1979), and in red sea bream it did not increase at all in comparison with control fish which did not receive glucose (Furuichi & Yone, 1971b). Moreover, injecting insulin to fish without feeding glucose caused hypoglycaemia (Cowey *et al.*, 1977a; Yone, 1979; Ottolenghi *et al.*, 1982) which persisted for a long period. Kuzmina (1971a, b) injected insulin into crucian and common carp. The degree of hypoglycaemia and the rate of its appearance were dependent on the dose injected. When 12 unit/kg were injected it appeared after three days, and at 60 unit/kg it appeared after three hours, but in both cases it continued for five days. Thorpe & Ince (1974) injected northern pike with bovine insulin at 10, 25, and 50 unit/kg. This resulted in a hypoglycaemia and death 27–72 h after injection. When, however, they injected a dose as low as 2 unit/kg of cod fish insulin, a hypoglycaemia was produced which persisted for up to seven days, while bovine insulin at the same dose produced only a transient hypoglycaemia. Ince & Thorpe (1974) obtained similar results, though with a somewhat weaker hypoglycaemia, by injecting the same doses of cod and bovine insulin into European eel. It seems that fish are more sensitive to fish insulin than to bovine insulin.

There is a certain controversy as to the fate of the glucose after insulin injection. According to Kuzmina (1971b), as in higher vertebrates, liver and muscle glycogen increase, indicating glycogenesis. Also Cowey *et al.* (1977a) found an increase in liver glycogen following insulin injection into rainbow trout. However, Furuichi & Yone (1971b) and Ablett *et al.* (1983) found no glycogenesis in liver or muscle after injection of glucose and insulin, and that the latter seems to have stimulated glucose metabolism in other pathways. Tashima & Cahill (1968), who studied the effects of glucose and insulin administered to *Opsanus tau*, and Ince & Thorpe (1976) studying the same problem in the pike (*Esox lucius*) concluded that one of the prime effects of insulin is on the incorporation of glucose into skeletal protein, i.e. providing α-keto acids for the synthesis of non-essential amino acids. Tashima & Cahill (1968) also observed the transformation of glucose into skeletal muscle glycogen, and Ince & Thorpe (1976) found that it was transformed into liver and muscle lipids.

From the above, as well as from other experiments, it seems that fish are deficient in the insulin regulation system and therefore their glucose metabolism is impaired, resembling that of diabetic mammals (Nagai & Ikeda, 1973). Little is known on regulation of insulin secretion in fish. Phillips *et al.* (1948) supposed that there is a deficiency in the amount of insulin secreted, and related this to the relatively small number, in fish, of pancreatic Islets of Langerhans which secrete insulin. If this assumption is true it may be also related to the rate of insulin production in the β-cells. The biosynthetic precursor of insulin is a single chain polypeptide, proinsulin (Steiner *et al.*, 1969). Proinsulin has also been found in cod by Grant & Reid (1968). This substance is cleaved off by proteolytic digestion in the secretory granules of the β-cells to give approximately equimolar amounts of insulin and a smaller peptide-'connecting' or C-peptide. The enzyme responsible for cleavage of the proinsulin in fish β-cells seems to differ somewhat from that of mammals, but generally it has a specificity similar to that of trypsin (Grant *et al.*, 1972). Lukowsky *et al.* (1974) studied the biosynthesis of proinsulin and insulin in isolated Islets of Langerhans from common carp, after incubating them with U-^{14}C-leucine. About 3–4% of the newly synthesized protein during two hours of incubation was proinsulin. However, a conversion of the proinsulin into insulin was not observed. Moule & Yip (1973) carried out a similar study with bullhead (*Ictalurus nebulosus*). They used ^{3}H-leucine as a tracer, and found that after incubation at 22 °C the radioactivity was identified in both proinsulin and insulin. However, at 12 °C ^{3}H-leucine was incorporated only in proinsulin and no radioactivity was found in insulin. Their results suggested that fish proinsulinases are sensitive to temperature and have low activity at 12 °C. In considering the assumption that fish are deficient in the insulin regulation system due to its insufficient production, one must take into account that the levels of insulin normally found in fish blood seem to be of the same order as those of omnivorous mammals. Thus, Thorpe & Ince (1976) found the mean insulin levels in the plasma of various fish species to be 0.26 ng/ml (pike) to 6.35 ng/ml (cod). Insulin levels in plasma of rainbow trout varied between 1.6 and 6.8 ng/ml.

However, not much is known about the increase in insulin production by the β-cells in response to stimulation by dietary glucose. Tashima & Cahill (1968) studied the change in plasma insulin in toadfish (*Opsanus tau*) after a single oral administration of glucose, and failed to notice any such change. Also, Thorpe & Ince (1974) showed that plasma insulin levels in pike (*Esox lucius*) did not increase in response to a single glucose load of 0.5 g/kg, at least for the first two days. In contrast, Yone (1979), who determined the level of insulin in common carp, red sea bream and yellowtail fed a gelatin capsule containing 167 mg glucose/100 g body weight, found that all responded by increasing serum insulin. This response, however, was species related. It was much lower in the carnivorous yellowtail than in the red sea bream and less in the latter than in common carp. Moreover, when glucose was administered orally to these fish, their blood glucose reached maximum levels before or simultaneously with the appearance of maximum plasma insulin level, which occurred only two hours after the glucose administration (Furuichi & Yone, 1981). The activities of the glycolytic enzymes (hexokinase and phosphofructokinase) rose to maximum levels only after the plasma insulin level reached its peak (Furuichi & Yone, 1982a). These authors have suggested that a considerable amount of glucose entered the liver and muscle before the rise in the activity of glycolytic enzymes, and possibly was excreted without being utilized by the fish (see also Malvin *et al.*, 1965; Oguri, 1968). Furuichi & Yone (1982b) suggested that this may be the reason that common carp in their experiment and channel catfish in Dupree's (1966) experiment grew least well on diets containing glucose as the sole carbohydrate, and better as the chain length of carbohydrates increased (maltose, dextrin and α-starch). The longer chain carbohydrates are absorbed (as glucose), according to Furuichi & Yone, at a slower rate, which is more synchronized with the secretion of insulin and the activity of glycolytic enzymes. It follows that anything which slows down the absorption of the shorter chain carbohydrates will increase their utilization. Morita *et al.* (1982) found that the addition of carboxymethylcellulose (CMC) to dextrin containing diet for red sea bream (*Chrysophrys major*) improves the growth rate of the fish and feed efficiency, which could be attributed to the slower absorption of the dextrin. Also, Murai *et al.* (1983a) found that feeding common carp four to six times a day improves the utilization of glucose from the diet, compared with two feedings a day. This can also be attributed to a better synchronization with insulin secretion. Thus, it may be that the problem of deficient glycolysis control lies, at least in part, in the rate of insulin production in response to glucose stimulation, and in synchronization with glucose absorption, rather than in the normal insulin level. Another possible reason for the deficient insulin regulation may be a low sensitivity, in fish, of the target tissues and their receptors to insulin. This possibility, however, has yet to be studied.

Role of carbohydrates in energy supply

In normal mammals carbohydrates incorporated into the body are accumulated as glycogen and any surplus carbohydrates are accumulated as lipids. Under starvation conditions glycogen is utilized prior to lipids. However, the situation is again different in fish. Although Moorthy *et al.* (1980) found a considerable depletion of liver carbohydrates (presumably glycogen) when *Sarotherodon mossambicus* were starved for 30 days, most researchers, working with other species, obtained different results. Dave *et al.* (1975) and Dave (1976) starved European eel (*Anguilla anguilla*) for a very long period (146 days). During the first three months of starvation the eel preferentially used liver and muscle triglycerides for energy. Only after exhaustion of the lipids was a marked depletion of liver glycogen observed. This was found to be true for other fish species (Kiermeir, 1940; Nagai & Ikeda, 1971a; Murat *et al.*, 1978). Also, glucose does not decrease during a long starvation period (Kiermeir, 1940; Chavin & Young, 1970). Sherstneva (1971) found that blood glucose concentration in two-year-old common carp, which is normally about 50 mg%, increased during a 10 day starvation period to about 81 mg%, and then returned to its normal level which did not change even after 100 days of starvation. Also, Bever *et al.* (1977), working on kelp bass (*Paralabrax* sp.), found that plasma glucose levels and glucose turnover are maintained in fish fasted up to 40 days with no apparent increase in carbon recycling. Mean plasma glucose levels remained unchanged in fasted fish and similar to those of fed fish (43 ± 8 mg% and 48 ± 8 mg%). It seems that the energy source of starved fish must be tissue lipid and protein transformed into energy through gluconeogenesis, rather than glucose or glycogen.

Several studies have shown that depressed carbohydrate metabolism is also evident in actively feeding fish. Hayama & Ikeda (1972) and Nagai & Ikeda (1972, 1973) found that in common carp dietary glucose contributes less to energy production than amino acids through an elevated gluconeogenesis. They traced the fate of labelled carbon from glucose, acetate and L-alanine and found that there is a preference for lipogenesis from amino acids and lipids rather than from carbohydrates. When common carp were fed a high carbohydrate–low protein diet, lipogenesis from both acetate and L-alanine was impaired, but it was promoted when the fish were fed high protein–low carbohydrate diets. This was explained by the authors in that, as with diabetic animals, acetyl-CoA production by glycolysis is limited even under the high carbohydrate diet conditions. The amount of acetyl-CoA is sufficient for oxidation but not for synthesis of lipids. That fish behave like diabetic mammals was also evident from a study by Murat & Serfaty (1975). They injected common carp with glucagon, causing hyperglycaemia which lasted for a long time, although it did not appear immediately after injection. In spite of the hyperglycaemia, liver glycogen concentration did not decrease appreciably, which indicated that energy was drawn from another source, probably through gluconeogenesis. In more recent experiments Murat *et al.* (1978) attempted to inhibit gluconeogenesis in common carp by injecting 3-mercaptopiconilic acid (MPA) simultaneously with the injection of glucagon (Figure 13). While a noticeable hyperglycaemia appeared after injecting glucagon alone, the inclusion of MPA caused the hyperglycaemia to be lower, with no difference, after 15 h, from the control fish which did not receive glucagon at all. Only in fish injected MPA or MPA and glucagon was there a noticeable reduction in liver glycogen (Table 17). Since the gluconeogenesis pathway was blocked in this case, energy had to

Figure 13. Effect of glucagon (G) – 0.2 mg/100 g, and/or 3-mercaptopiconilic acid (MPA) – 10 mg/100 g, on common carp blood glucose as compared to control (C) (from Murat *et al.*, 1978).

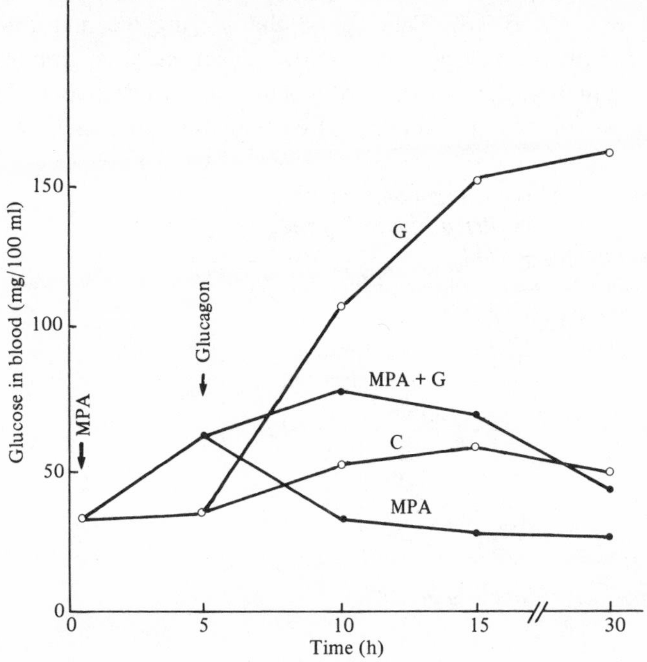

be utilized through the less preferred pathway of glycogenolysis from the glycogen reserves. The importance of gluconeogenesis as a pathway supplying energy from lipid and protein in preference to glycolysis of carbohydrates has also been shown for other fish (Demael-Suard *et al.*, 1974, for tench, *Tinca tinca*; Inui & Yokote, 1974, for Japanese eel, *A. japonica*). By determining the activities of three enzymes regulating gluconeogenesis (D-fructose 1,6-diphosphate, 1-phosphohydrolase; pyruvate carboxylase; and phosphoenolpyruvate carboxykinase) in various organs of rainbow trout, Cowey *et al.* (1977a) localized the major sites of gluconeogenesis in this fish. These were the liver and the kidney.

The relative importance of glycolysis and gluconeogenesis in supplying energy may vary during different physiological states. Kutty (1972) has noted in exercised *Sarotherodon mossambicus*, starved for 36 h before the experiment, that the level of ammonium excretion relative to oxygen consumption is high enough to indicate that the fish were generating all of their aerobic energy from protein catabolism through gluconeogenesis. Onishi *et al.* (1974) studied the activities of several hepatic enzymes participating in the metabolism of carbohydrates (hexokinase, phosphofructokinase, pyruvate kinase, fructose 1,6-diphosphatase, and phosphogluconate dehydrogenase) and amino acids (glutamate dehydrogenase, glutamate pyruvate transaminase, and glutamate oxaloacetate transaminase) of rainbow trout, cherry salmon (*Oncorhynchus masou*) and hime (sockeye) salmon (*O. nerka*). They found that during the maturation process the activity of key enzymes in the glycolysis, and glutamate dehydrogenase, increased, while those of gluconeogenesis decreased. This seems to indicate that the metabolism of carbohydrate increases during

Table 17. *Glycogen content in the liver of common carp at the end of the experimental period after fish have received glucagon and/or 3-mercaptopiconilic acid (MPA)*

Treatment	Liver glycogen (mg/100 g)
Control	4.20 ± 0.28
Glucagon	3.30 ± 0.42
MPA	2.25 ± 0.45
Glucagon + MPA	2.22 ± 0.25

Source: From Murat *et al.* (1978).

the period of maturation. The increase in glutamate dehydrogenase activity also seems to have the same meaning because this enzyme requires an α-ketoglutarate substrate which occurs through glycolysis (see section on metabolism of protein).

Conclusion

Carbohydrate metabolism in fish seems to encounter a number of constraints. Not only is digestibility of carbohydrates in carnivorous fish low, (see Chapter 2), but also, in many fish its intermediate metabolism is depressed. This seems to be due to low activity of the hormones controlling carbohydrate metabolism. The insulin control system, which stimulates glycolysis, seems to be inefficient, making fish similar to diabetic mammals with respect to carbohydrate metabolism. Some studies also show an inefficiency of hormonal control of glycogenesis by phosphorylation (Verjbinskaya & Savina, 1969, quoted from Murat *et al.*, 1978). Red muscle, which is the more active in normal circumstances, is adapted to utilize lipid and protein as sources of energy rather than carbohydrates. Thus it seems that fish possess an intrinsically developed ability to use preferentially lipid and protein for energy through gluconeogenesis.

Although the above is probably true for all fish, a number of studies show that fish, especially omnivorous and herbivorous, adapt to utilization of high carbohydrate diets (Shimeno *et al.*, 1981). Cowey *et al.* (1977a, b) found the highest rates of gluconeogenesis in rainbow trout given a high protein diet and in trout subjected to prolonged starvation. Significantly lower rates of gluconeogenesis were found in trout given a high carbohydrate–low protein diet. Similar results have been obtained by Abel *et al.* (1978, 1979) who found a lower activity of the gluconeogenic enyzme phosphoenolpyruvate carboxykinase and higher activity of glucokinase when fish were fed a diet rich in gelatinized maize starch, though the activity of the former enzyme did not decrease below 75% of its activity when the fish were fed a high-protein diet. Thus, in spite of its ability to adapt, to a certain extent, to high carbohydrate levels in the diet, the amount of carbohydrate that the fish can physiologically cope with is rather limited. Edwards *et al.* (1977) fed rainbow trout on diets similar in protein and energy contents, but differing in the percentage of metabolizable energy present as carbohydrate. Fish growth was best on the diet with the lowest (17%), lower with the intermediate (25%), and worst on the diet with the highest (38%) level of metabolizable energy from carbohydrate.

On the other hand, feeding fish too little carbohydrate may result in inhibited growth (Nagai & Ikeda, 1971a, b; Yone, 1979), which indicates that a certain quantity of carbohydrate must be taken, and an active system consisting of hexoses and hexose-phosphates should be present in order that protein may be efficiently utilized. Dupree & Sneed (1966) found with channel catfish that increasing the level of dextrin in the diet from 2.5 to 10% increased weight gains. (But further increase to a level of 15.2% depressed growth). Similar results were obtained by Likimani & Wilson (1982) with an increase in dextrin content from 0.0 to 11.25 and 31.5%. Tryamkina (1973) fed rainbow trout a diet in which the major source of carbohydrate was wheat flour, comprising 15% of the diet. The exclusion of wheat flour from the diet resulted in a sharp decline in growth rate. Similar results were obtained by Pieper & Pfeffer (1979) for rainbow trout, and by Cowey *et al.* (1975) for plaice (*Pleuronectes platessa*). Weight gains in plaice given diets containing carbohydrate as well as protein and lipid were superior to those given diets lacking carbohydrate. Both *PER* and *NPU* values were greater in the first diet than in the latter ones, even though the total energy content of the diets was approximately the same.

Metabolism of lipids

Triglycerides are a major energy source for metabolism in fish muscle. The calorific value of fatty acids (9.2 kcal/g), derived from the hydrolysis of triglycerides, is more than double that of carbohydrates. The glycerol produced in this way is also easily oxidized, but the energy it contributes is much lower than that of the free fatty acids. After hydrolysis fatty acids are oxidized in the cell mitochondria in a metabolic pathway similar to that of higher vertebrates (Figure 14). Brown & Tappel (1959) studied the ability of an *in vitro* preparation of mitochondria from liver of common carp to oxidize fatty acids. When the medium contained intermediate metabolites of the TCA cycle, such as succinate–oxaloacetate, fumarate, malate, or α-ketoglutarate, as catalytic substances, there was considerably enhanced oxidation of both saturated and unsaturated fatty acids, but oxidation was slow in the absence of these metabolites.

In mammals, fatty acid oxidation is carried out mainly in the liver. In fish, however, a considerable proportion of the fatty acids are oxidized in red muscle. Bilinski (1963) measured the rate of fatty acid oxidation in white and red muscles of rainbow trout with the aid of labelled carbon compounds. Octanoic acid was oxidized in red muscle 50 times as fast as in white muscle. Hexanoic acid was also oxidized at a higher rate in red muscle, but the difference here was not as great as in octanoic acid. Similar results have been obtained by George & Bokadawala (1964) for the oxidation of butyric acid by the red and white muscle of various cyprinids. The ability of the red muscle to oxidize fatty acids and acetate is only

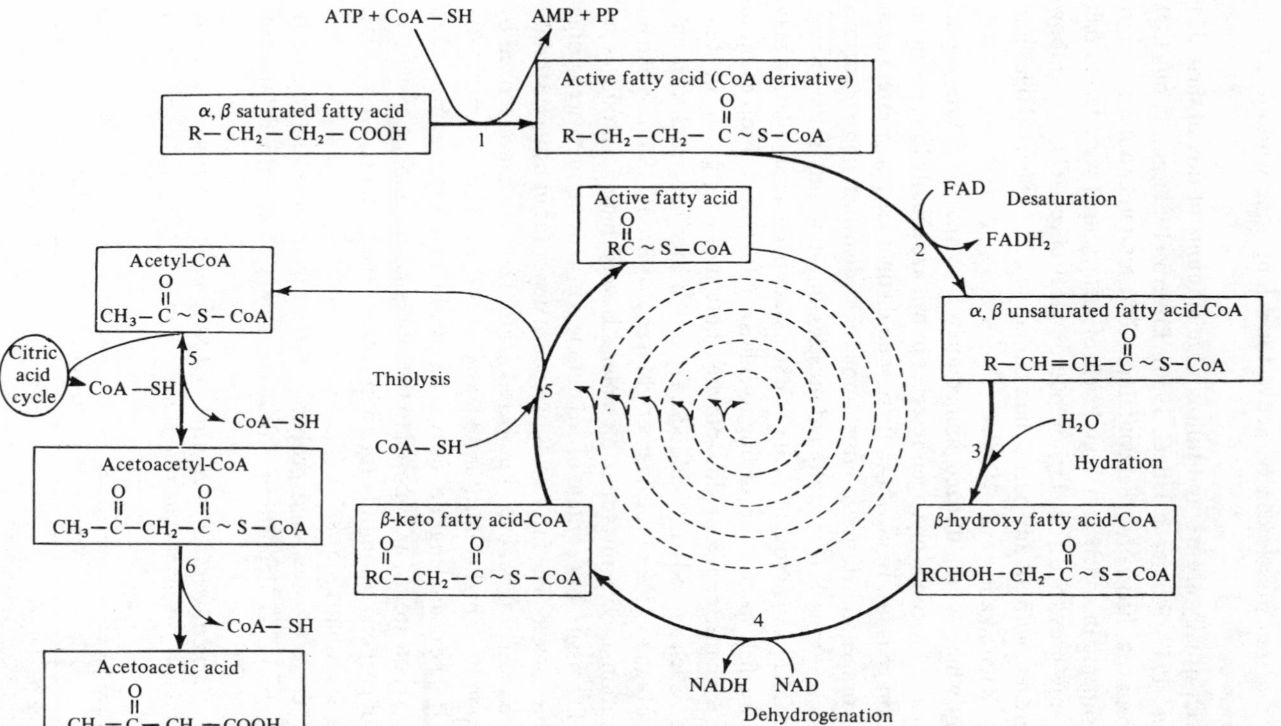

Figure 14. Metabolism of fatty acids. Enzymes: (1) thiokinase and Mg^{2+}; (2) acyl dehydrogenase; (3) enol hydrase (crotonase); (4) β-hydroxyacyl dehydrogenase; (5) thiolase; (6) deacylase (from Hoar, 1975).

somewhat lower than that of cardiac muscle (Table 18), which may also indicate that red muscle can be active for a long time without getting fatigued.

Fish can usually absorb and utilize large amounts of fats in their diet (Yu *et al.*, 1977; Reinitz & Hitzel, 1980). Lovern (1951) points out that brook trout can thrive on a diet containing 57% fat. Other authors report on experiments in which very high levels of dietary lipid were fed to fish without any growth depressing or pathological effects. Thus, rainbow trout were fed, without any harm, diets containing 25–30% fat (Higashi *et al.*, 1964; Kitamikado *et al.*, 1964b).

Energy which is not utilized immediately is stored for future use as glycogen and fat. Since glycogen reserves in fish are usually low, the main energy storage is fat. In most fish there is no distinct fatty (adipose) tissue and fat storage, sometimes in large amounts, is dispersed among various tissues such as muscle (especially red muscle), intestine and its mesentery, and liver. The relative role of each of these tissues in fat accumulation may vary among fish species. According to Bilinski (1969) deposition of fat in the liver is characteristic of slow-moving, bottom-dwelling fish. In some sea fishes, such as cod and haddock, fat in the liver can reach concentrations as high as 75% of the liver weight (see also Brody, 1945). When fat concentration in the intestinal mesentery is very high it produces an adipose-like tissue. Fat content of red muscle is usually about twice as high as in white muscle. The fatty acids present in depot fat in fish are straight chain, monocarboxylic, even-numbered acids. They are characterized by larger amounts of highly unsaturated fatty acids than in higher vertebrates (Table 19). The chain length of fatty acids present in fish ranges from 12 to 26 or 28 carbon atoms, as opposed to the more limited number, usually 16 and 18, of carbons commonly found in higher vertebrates (Lovern, 1951; Tashima & Cahill, 1965).

In higher vertebrates, and probably also in fish, there are two pathways for body fat deposition: (a) from dietary fats, with or without

Table 18. *Oxidation rates of acetate and fatty acids in various tissues of sockeye salmon (in μmol substrate converted into CO_2/mg tissue nitrogen)*

Substrate	White muscle	Red muscle	Heart	Liver	Kidney	Brain
Na acetate-1-^{14}C, 1.0 mM	2.40	23.4	36.60	37.40	120.10	2.80
K octanoate-1-^{14}C, 1.0 mM	0.40	14.6	22.20	24.40	24.00	1.00
K myristate-1-^{14}C, 1.0 mM	0.14	1.4	6.24	7.36	12.66	2.28

Source: From Jonas & Bilinski (1964).

modification (the exogenic pathway); (b) *de novo* synthesis of fat from non-fat nutrients. Acetate is an important intermediate metabolite which serves as a precursor for *de novo* synthesis of higher fatty acids in fish. The conversion of carbohydrates or protein into depot fat is associated with their transformation first into acetate-CoA to form CoA-thioester, through the catalytic effect of the enzyme β-ketothiolase, and through several more steps, phosphatides, fatty acids and fats are formed (Figure 14). Farkas *et al.* (1961) injected fish with acetate-1-^{14}C and found the labelled ^{14}C in the fatty acids of the liver. Most of the labelled carbon was found in saturated acids, although about a fifth to a quarter was associated with unsaturated acids. Highly unsaturated C_{18}, C_{20} and C_{22} fatty acids (HUFA) were labelled only to a small extent in comparison to the saturated C_{16} and C_{18} acids. This was supported by a later work of Farkas (1970) who states that fish, like the lower organisms which serve as fish food, can synthesize *de novo* saturated fatty acids from non-lipid precursors and then desaturate these acids to monoenoic fatty acids. However, long chain polyunsaturated fatty acids are formed mainly in planktonic crustaceans and fish merely redistribute them in the triglycerides and phospholipids of their body. Kanazawa *et al.* (1980c) injected acetate-1-^{14}C into *Tilapia zillii* and found that the fish synthesized palmitic acid (16:0)[1], palmitoleic acid (16:1), stearic acid (18:0) and oleic acid (18:1ω9), but not HUFA. Thus most authors conclude that the main source for HUFA of C_{20} and C_{22} is exogenic (see Chapter 7).

The hepatic tissue seems to be the major site for fatty acid synthesis from carbohydrates. Likimani & Wilson (1982) found that a diet rich in carbohydrate stimulated the lipogenic enzymes mainly in the hepatic

Table 19. *Fatty acid composition (% of weight) of depot fat in fish as compared to some higher vertebrates*

	Saturated			Unsaturated				
	C_{14} & below	C_{16}	C_{18}	C_{14}	C_{16}	C_{18}	C_{20}	C_{22}
Fish								
Cod (*Gadus morhua*)	2–6	6.5–14	0.5–1	0.5–2	10–20	25–31	25–32	10–20
Angler fish (*Lophius piscatorius*)	4.9	9.6	1.3	0.4	12.1	30.9	24.9	15.9
Skate (*Raia maculata*)	4.0	14.0	—	trace	10.5	20.5	32.5	18.5
Hen	1.2	24.0	4.1	—	6.7	63.3	← 0.7 →	
Cat	6.0	29.2	16.6	1.2	4.3	42.7	← trace →	
Man	4.0	19.5	4.2	0.6	6.9	57.7	← 1.9 →	

Source: From Tashima & Cahill (1965).

[1] For method of designation of fatty acids see footnote on p. 217.

tissue and to a lesser extent in the mesentric adipose tissue. However, a diet rich in lipids and poor in carbohydrates depressed the activity of these enzymes, especially that of malic enzyme which has been shown to be the rate-limiting enzyme in the NADH–$NADP^+$ transhydrogenation cycle. In a coupled reaction with citrate cleavage enzyme, malic enzyme provides extramitochondrial NADPH and acetyl-CoA for the fatty acid synthesis.

There seems to be a difference in the utilization of various fats, and especially free fatty acids in comparison to the esters of these acids, among different fish species. Stickney & Andrews (1972) found that at a temperature of 26–30 °C channel catfish grew better on a diet containing beef tallow and fish oil than on a diet containing safflower oil. Hydrogenated corn oil caused a higher growth rate than natural corn oil, and free fatty acids were hardly utilized at all. Gaulitz *et al.* (1979) also found that channel catfish grow faster on diets containing fish oil than on those with corn oil. Reinitz & Yu (1981), however, did not find any difference in the utilization of animal fat, soybean oil or fish oil by rainbow trout, provided that the diet contained the required amounts of essential fatty acids.

Metabolism of proteins

Dietary proteins are hydrolysed in the lumen of the digestive tract under the influence of proteinases and peptidases to free amino acids or short peptide chains. Complete hydrolysis of the latter is effected by intracellular peptidases within the mucosa cells. In any case, dietary proteins eventually appear in the blood in the form of amino acids. Plasma of rainbow trout was examined at different intervals after feeding on protein-rich diets. Immediately after feeding and for the next two hours, most of the free amino acids in the plasma increased in concentration, reaching a peak between 12 and 24 h after feeding and then gradually falling off to fasting levels (Nose, 1972; Schlisio & Nicolai, 1978; Yamada *et al.*, 1981a; Walton & Wilson, 1986). According to Wilson *et al.* (1985), however, in channel catfish the peak in plasma free amino acid concentration occurred earlier, four to eight hours after feeding. Increasing the protein level of the diet delays somewhat the time of the peak, but has little effect on its magnitude. It seems, according to these authors, that the liver strongly regulates the concentration of the free amino acids in the serum.

In addition to digestion and transportation of the amino acids, the mucosa is also engaged, in some cases, in metabolic transformation of amino acids. A proportion of the amino acids may be deaminated and their carbon skeletons catabolized to produce energy. However, the major transformation of amino acids occurs in the liver. The relative proportions of amino acids in the blood after feeding are thus not necessarily identical with the pool of amino acids in the intestinal lumen. In mammals the amino acids of the blood may contain less glutamate and aspartate, but more alanine and glutamine. After feeding most of the essential amino acids in the blood plasma of fish (rainbow trout and channel catfish) reached a concentration of more than twice that found during fasting. These concentrations were positively correlated with their contents in the protein of the diet. However, histidine maintained an almost constant level regardless of its concentration in the feed (Nose, 1972; Yamada *et al.*, 1981a; Wilson *et al.*, 1985; Walton & Wilson, 1986). Zabian & Creach (1979) found a correlation between the essential amino acids in the diet and the same amino acids found in the blood plasma after feeding. They noted, however, that the slope of the regression between the two was higher when the fish were fed a balanced protein–carbohydrate diet than when they were fed a high protein–low carbohydrate diet. This suggests that in the latter case a lower proportion of the dietary amino acids was maintained in the plasma, probably due to accelerated turnover and utilization of free amino acids for energy in lieu of the lacking carbohydrate.

The concentrations of non-essential amino acids in the blood serum behave differently from those of essential amino acids. In the study of Yamada *et al.* (1981a), some of the non-essential amino acids such as alanine, glutamic acid, serine and proline reached a peak only 24–36 h after the ingestion of the protein, proline somewhat later than the others. Other non-essential amino acids such as tyrosine, glutamine, asparagine and aspartic acid showed little change, if any, from the pre-feeding condition. Blood serum concentration of glycine showed a marked decrease after feeding. It seems that the non-essential amino acids undergo major changes in the liver. When Pequin (1967) injected glutamine into blood entering the liver of common carp, some of it was hydrolysed. At the same time, hepatic glutamic acid decreased and blood level of glutamine increased. The study of Wilson *et al.* (1985) on channel catfish showed that there was no correlation between the concentratons of non-essential amino acids in the serum and those in the diet. Serine and alanine were most abundant in the serum, and glutamic acid, glycine and proline remained relatively unchanged. It is obvious that the non-essential amino acid concentration in the blood serum is affected not only by their level in the diet and their absorption rate, but also by interconversion reactions which they undergo, such as absorption into tissues, deamination, etc.

Free amino acids carried with the blood can undergo metabolism in either of two directions: the anabolic direction, which provides for the biosynthesis of new proteins, either functional, such as hormones and enzymes, or structural, as in the formation of new tissues (growth), or the replacement of worn-out tissues; and the catabolic direction, which after deamination of the protein molecules produces carbon skeletons which may be utilized for either energy or lipogenesis.

Protein metabolism takes place in many organs of the body. In addition to the mucosa already mentioned, the most important organs where protein metabolism occurs are the liver and the muscles. So many catabolic pathways for amino acids are localized in the liver that this organ must be considered the major catabolic site for the body (Cowey *et al.*, 1977a). Also, most plasma proteins are synthesized in the liver. Kenyon (1967) hepatectomized European eel (*A. anguilla*) and found that this was followed by a 25% fall in the concentration of the plasma proteins and a 65% increase in the concentration of plasma amino acids. This seems to indicate a reduced protein synthesis due to the removal of the liver. Based on experiments with dogs fed fairly large meals of protein, it has been calculated that the largest fraction of free amino acids reaching the liver through the portal vein after feeding is catabolized. This is probably more so with fish, which, as has been already pointed out, use protein for energy in preference to carbohydrate. Demael-Suard *et al.* (1974) found that labelled glycine injected into tench (*Tinca tinca*) is rapidly incorporated into hepatic glycogen.

Although the rate of protein metabolism in muscle may be slower than in the liver, the mass of muscles so much exceeds that of other tissues that it makes this tissue quantitatively the most important site of protein synthesis in the entire body. Also, much of the degradation and catabolism of amino acids take place in the muscle. Glycine and histidine are two major free amino acids found in fish muscle. In many fish species these two amino acids make up, either individually or in combination, about 50% of the total free amino acid pool (Wood *et al.*, 1960; Creach, 1966; Sakaguchi & Kawai, 1970a; Fontain & Marchelidon, 1971; Siddiqui *et al.*, 1973; Zebian & Creach, 1979). Wood *et al.* (1960) found that during the early stages of spawning migration of sockeye salmon upstream, the free histidine content of the muscle and alimentary tract decreases to a small fraction of its initial value. It can be assumed that the histidine was used as an energy source within the muscle, since the salmon do not feed during the entire period of spawning migration and do not have any alternative source of energy. Zebian & Creach (1979) found that the concentrations of most free amino acids in the blood plasma of common carp starved for 15 days were much lower than those of fed fish. This, however, was not true for glycine and histidine, the concentrations of which were more than double those of the fed fish. These elevated concentrations may be the result of transportation of these amino acids to locations where energy is needed most. The breakdown of histidine for energy is catalysed by histidine deaminase which was found in the red muscle of some fish (Sakaguchi & Kawai, 1968, 1970b, 1971; Sakaguchi *et al.*, 1970). It is probable that other amino acids are also utilized for energy directly in the muscle. Fish muscle contains a large number of transaminating enzymes, with glutamate–pyruvate and glutamate–oxaloacetate transaminases occurring in particularly high concentrations (Siebert *et al.*, 1965).

The actual proportions of proteins anabolized and catabolized depend on the protein requirement of the fish, the content of protein in the diet and the proportions of various amino acids within it, the energy requirement, and the amount of energy available from other sources such as fat and carbohydrates. Amino acid catabolism is favoured by a lack of sufficient energy from dietary carbohydrates and fats. Such diets, even though adequate in proteins, will be utilized for energy. On the other hand, dietary protein provides the raw material for protein anabolism. While non-essential amino acids can by synthesized from carbon skeletons provided by carbohydrates and ammonia, dietary protein is the only source for the essential amino acids. These should, therefore, be supplied in the required proportions for the synthesis of new proteins. Otherwise, synthesis will be governed by the limiting essential amino acid, and those in excess will be catabolized.

In fish, as in mammals, both protein synthesis (particularly in muscle) and translocation of amino acids from liver to muscle are controlled by insulin (Cahill *et al.*, 1972; Thorpe, 1976). According to a number of authors (Thorpe & Ince, 1974; Inui *et al.*, 1975, 1978; Inui & Yokote, 1975a, b) the major role of fish insulin is in protein and lipid metabolism, whereas involvement in glucose homeostasis may be of a secondary significance. The rise in plasma amino acids after feeding acts as a major stimulus for secretion of insulin (Ablett *et al.*, 1983), which enhances the deposition of amino acids in skeletal muscle and their incorporation into muscle proteins. Growth hormones and androgens (primarily testosterone) also promote these effects, probably independently of insulin, although in some test systems the presence of insulin was required for the other hormones to be effective. Glucagon and glucocorticoids, though they may have an anabolic effect on certain proteins such as hepatic enzymes, generally stimulate amino acid catabolism and gluconeogenesis (Brown *et al.*, 1975; Walton & Cowey, 1979). Excessive amounts of thyroid hormones and inadequacy of insulin may also act as stimulants of amino acid catabolism (Cahill *et al.*, 1972). This last factor may be of a special importance in fish where, as was mentioned above, insulin is less efficient than in higher vertebrates. Injecting insulin into fish has been shown to reduce gluconeogenesis (Cowey *et al.*, 1977a, b). Exogenous insulin also reduces plasma and muscle free amino acids (Seshadri, 1959; Thorpe & Ince, 1974; Thorpe & Ince, 1974, 1978; Inui & Yokote, 1975b: Inui *et al.*, 1975), and stimulates their incorporation into skeletal muscle protein (Tashima & Cahill, 1968; Jackim & LaRoch, 1973; Ahmad & Matty, 1975; Castilla & Murat, 1975; Ince & Thorpe, 1976; Inui *et al.*, 1978). Ludwig *et al.* (1977) injected bovine insulin into the peritoneal cavity of coho salmon twice weekly for 70 days. They found that this increased the growth rate of the fish. Following this, Ablett *et al.* (1981) studied the incorporation of ^{14}C-leucine into plasma, liver and skeletal muscle of rainbow trout which had been injected with bovine insulin repeatedly every 48 h for 56 days. In addition to the hypoglycaemic responses, they observed a decrease in ^{14}C-leucine as free amino acid in the liver at the same time as its increase in skeletal muscle. There was a net increase of 22.9% in protein content in skeletal muscle over saline injected controls, as well as an increase in lipid content of the muscle. This resulted in a significantly better growth (by 18.5%) of the insulin injected fish over the saline injected control fish. These results indicate that the major role of insulin in fish metabolism is directed toward proteogenic and lipogenic pathways, and that inefficiency of insulin may be one of the causes for the high protein requirement by fish and its utilization for energy.

Blocking insulin action by alloxan, or increasing the concentration of corticoid hormones, for example by injecting hydrocortisone, result in increased gluconeogenesis. In addition to hyperglycaemia there is a pronounced increase in plasma free amino acids (Inui & Yokote, 1975a, c). Stress may also enhance amino acid catabolism. The adrenocorticotropic hormone (ACTH), which is generally secreted in response to stress causes, in turn, the secretion of glucocorticoid hormones which affect gluconeogenesis (Redgate, 1974; Fryer, 1975).

Catabolism of amino acids usually starts by deamination, in which nitrogen (as well as sulphur and iodine, where they are included in the amino acid composition) is removed. There are a number of deamination pathways in fish. The major one seems to be oxidative deamination coupled with transamination (Forster & Goldstein, 1969). The overall oxidative deamination process, by which amino acids are converted into keto acids and ammonia, is described in Figure 15. It involves two consecutive reactions: first, transamination of amino acid with α-ketoglutarate to form glutamate. This reaction is catalysed by specific transaminases. It is followed by oxidative deamination of glutamate to α-ketoglutaric acid, catalysed by glutamate dehydrogenase (Cohen & Brown, 1960, quoted from Hoar, 1975; Schepartz, 1973). The last reaction is linked with the mitochondrial oxidative chain. It requires NAD^+ and $NADP^+$ which can in their reduced forms NADH and NADPH enter the phosphorylation reaction, producing ATP in the cytochrome system.

The major sites for deamination are the liver, kidney and gill, and to a lesser extent also other tissues. McBean *et al.* (1966) assayed glutamate dehydrogenase in various fish tissues. The occurrence of this enzyme was demonstrated in the livers of five fish species. From the liver ammonia is delivered in the blood to the gills, to be excreted together with ammonia produced in the gill itself. The relative importance of the liver and the gill as sources of ammonia probably varies between species and with the physiological state of the fish (Forster & Goldstein, 1969). Kenyon (1967) hepatectomized European eel (*A. anguilla*) and found that ammonia excretion continued unchanged, but did not increase after injection of alanine. He concluded that the liver in the eel is, therefore, not essential in amino acid deamination. However, excess exogenous amino acids cannot be deaminated in the absence of the liver.

As a result of deamination amino acids produce carbon skeletons of two kinds: (a) Most amino acids produce α-keto acids which can be readily converted to carbohydrate or participate directly in the TCA cycle producing energy. These are 'glucogenic acids'. (b) A few amino acids such as leucine and isoleucine form carbon skeletons which yield intermediate metabolites which are more closely related to the metabolism of fatty acids than to that of carbohydrate, such as acetate, acetoacetate,

Figure 15. Scheme of deamination of amino acids via coupled transaminases and glutamate dehydrogenase (based on Hoar, 1975).

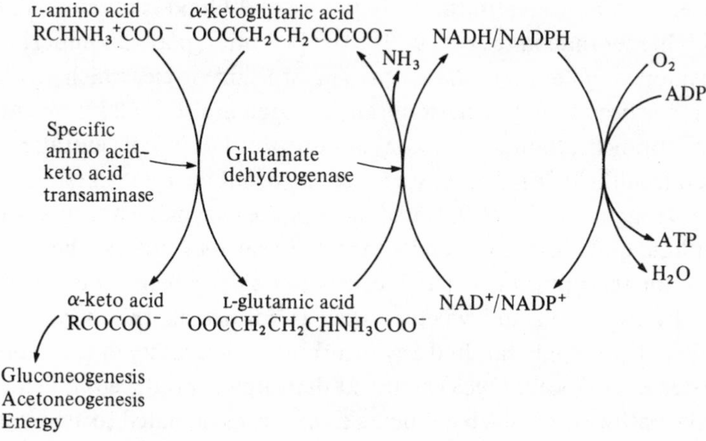

β-hydroxybutyrate, and acetone. Eventually these produce acetyl-CoA which may be either utilized for synthesis of fatty acids or oxidized for production of energy through the TCA cycle. Amino acids yielding such carbon skeletons are said to 'ketogenic'. The glucogenic carbon skeletons of the amino acids (such as pyruvate or α-keto acids), whether provided by deamination or through glycolysis of carbohydrates, serve as precursors to the biosynthesis of new amino acids in a reductive amination which is a reverse reaction to deamination and transamination.

Nitrogen excretion

Ammonia produced by oxidative deamination, if not reused for amination is toxic and therefore cannot accumulate in the blood or body organs. Aquatic animals usually solve this problem by excreting most of the ammonia into the surrounding water. Such animals are said to ammoniotelic. This method of disposal of nitrogenous excretory products is, of course, not possible in terrestrial animals. In these, the toxic ammonia is converted to non-toxic compounds such as urea in mammals (ureotelic) and uric acid in birds (uricotelic). The conversion of ammonia to urea in mammals is performed through the energy expensive ornithine cycle and so affects the total energy requirement and the *SDA*.

The division of animals according to their excretion is very general. In fact, animals excrete a mixture of compounds in which ammonia, urea or uric acid are the major constituents in ammoniotelic, ureotelic and uricotelic animals respectively. Also in fish other compounds beside ammonia are present in the excreta, namely, urea, trimethylamine oxide (TMAO), creatine, creatinine, uric acid, inulin, para-aminohippuric acid, and amino acids (Wood, 1958; Fromm, 1963; Forster & Goldstein, 1969; Masoni & Payan, 1974). Smith (1936) showed that in elasmobranchs the concentration of urea in the body tissues and blood is about 2%. This is much higher than in teleosts (0.01–0.03%), and it plays an important role in osmotic regulation in these fish. The ornithine cycle, which synthesizes urea, has been found in elasmobranchs. Urea is also found in the urine of elasmobranchs, though only at about one third of its concentration in the blood (Smith, 1936). The existence of the ornithine cycle in teleosts is not clear. Huggins *et al.* (1969) found most of the enzymes associated with the ornithine cycle in teleosts and concluded that they possess the structural genes for the ornithine cycle; but it is not clear whether these genes are actually expressed and whether urea is formed in this way. Vellas & Serfaty (1974) could not find any ornithine cycle activity in common carp. Forster & Goldstein (1969) assumed that urea is produced in fish via the purine pathway, in which amino acids are transaminated to aspartate and glutamine and then converted successively into uric acid, allantoin, allantoic acid, and urea.

While in mammals all nitrogenous residues are excreted in urine, in fish only a small proportion of these products is excreted via the urine. Excretion is effected mainly through the gills by diffusion into water. Branchial excretion is usually 5–10 times higher than urinary excretion (Smith, 1929; Wood, 1958; Forster & Goldstein, 1969). According to Fromm (1963) only about 3% of the total waste nitrogen of rainbow trout was excreted through the kidney. The major excretory products in the urine are urea, creatine, creatinine, ammonia, uric acid, amino acids, and trimethylamine oxide (TMAO) (Grollman, 1929; Grafflin & Gould, 1936). TMAO [$(CH_3)_3N=O$], which is related to urea and also has a low toxicity, has a special importance in marine fishes where it is found in appreciable concentrations in the tissues. Harada & Yamada (1973) found this concentration to be 9–134 mg N/100 g in different marine fishes. It also comprises a considerable part of the urine (Kutscher & Ackermann, 1933). In freshwater fishes the concentration of TMAO in tissue is nearly undetectable (Harada & Yamada, 1973), but it does appear in urine, though in much lower concentrations than in marine fishes. TMAO is produced by oxidative methylation of ammonia. Its main role, similar to that of urea in elasmobranchs, is regulation of the osmotic balance. This is especially important in marine fishes which live in a hypertonic medium, though the osmotic pressure developed by TMAO is only a quarter that of urea of an equivalent molecular concentration.

As pointed out above, ammonia, excreted through the gills, is the main excretory product of the fish. During the early developmental stages of the eggs ammonia excretion already amounts to about 80% of the total nitrogen excreted (Kaushik *et al.*, 1982). Excretion of ammonia by fish has some advantages over the excretion of urea and uric acid. It is the simplest compound among the excretory products and has the smallest molecule. Therefore, it easily passes through the gill membrane and the energy cost for its excretion is minimal. Ammonia dissolves easily in water and 99% of it undergoes ionic dissociation when the pH of the medium and blood are almost neutral.

Some data on the percentage of ammonia from the total nitrogenous excretion of fish were gathered from the literature and are presented in Table 20. One can observe a considerable fluctuation in the relative amount of excreted ammonia. Since most of the measurements were made on starved fish the data did not relate to the source of the excreted nitrogen. Brett & Zala (1975), working on nitrogenous excretion by sockeye salmon (*O. nerka*), found that the endogenous nitrogen excretion, measured on starved fish or after a gap of 15 h or more from the last feeding, had a more or less stable ratio of ammonia of about 79%, the rest being mainly urea. Kaushik (1980) has shown that urea excretion of fed rainbow trout is higher than that of starved fish. Daily urea excretion increased with time since the fish were first fed, as also with the amount of protein ingested, but there was no immediate response of urea excretion to feeding. In contrast, immediately after feeding ammonia excretion increased sharply. In Brett & Zalla's (1975) experiment the ammonia reached a peak 4–4$\frac{1}{2}$ h after feeding, when its excretion was four times the normal rate. There was no change in the excretion of urea during this time (Figure 16). This means that urea is a metabolic product of endogenous protein metabolism, while the end product of the dietary protein catabolism is ammonia. With the increase in acclimation temperature of the fish the total amount of nitrogen excreted by starved fish increases. However, it is important to note that the proportion of urea in the total nitrogen increases, while that of ammonia changes only a little (Guerin-Ancey, 1976a). The effect of feeding on the proportion of ammonia in the excreted nitrogen has also been studied by others. Guerin-Ancey (1976b) noticed this effect in sea bass (*Dicentrarchus labrax*) and Rychly & Marina (1977) found that fed rainbow trout excreted four times as much ammonia as trout starved for 12 days. Caulton (1978) stated that the amount of nitrogen excreted as ammonia by *Tilapia rendalli* depends largely on the quantity of protein assimilated during feeding. He assumes that ammonia is the main, if not the only, end product of catabolized dietary protein.

Table 20. *Percentage of ammonia nitrogen from total nitrogen excreted by fish*

Species	% ammonia N	Reference
Abramis brama	56.0	Tatrai (1981)
Cyprinus carpio	54–62	Smith (1929)
(at 10 °C)	41–95	Infante (1974)
(at 20 °C)	66–94	Infante (1974)
Carassius carassius	73.3	Smith (1929)
	77.8	Iwata (1970)
Carassius auratus	75.0	Iwata (1970)
Leptocottus armatus	62.7	Wood (1958)
Platichthys stellatus	83.8	Wood (1958)
Taeniotoca lateralis	48.0	Wood (1958)
Salmo gairdneri	55.1	Fromm (1963)
	80.0	Smith (1971)
(starved)	69.8	Kaushik (1980)
(fed)	84.7–92.5	Kaushik (1980)

This was also supported by an experiment carried out by Zoren (1984) with common carp.

Metabolizable energy of protein

The metabolizable energy value of dietary protein varies with the amino acid composition of the protein, its digestibility, and whether these amino acids are retained by the animal for protein synthesis, or are deaminated and their carbon skeletons utilized for energy. In the latter case, the metabolizable energy of the catabolized protein also depends on the residual energy value of the excreted product, and the digestible energy of the protein must be corrected to 'zero nitrogen balance'.

For zero nitrogen balance in mammals, the calorific value of excreted urea in the aqueous state was calculated as 154.4 kcal/mol, or 5.51 kcal/g urea nitrogen (Elliott & Davison, 1975). Since urea comprises about 90% of the nitrogenous excretory products of mammals, and the rest consists of compounds containing higher residual energy than does urea, the factor used for correction to zero nitrogen balance is higher. In ruminants a factor of 7.45 kcal/g urinary N is used (McDonald *et al.*, 1966). Uric acid, the main excretory product of poultry, has a higher residual calorific value than that of urea, and therefore the correction factor for zero nitrogen balance in poultry is higher and amounts to 8.22 kcal/g urinary N. Since it is assumed that the average nitrogen content of protein is 16%, the above correction factors amount to 7.45/6.25 = 1.19 kcal/g protein in ruminants and 8.22/6.25 = 1.32 kcal/g protein in poultry. This should be subtracted from the digestible energy of the protein utilized for energy. The gross

Figure 16. Pattern of change in rates of nitrogen excretion of sockeye salmon fed a single meal daily and that of starved for 22 days (from Brett & Zala, 1975).

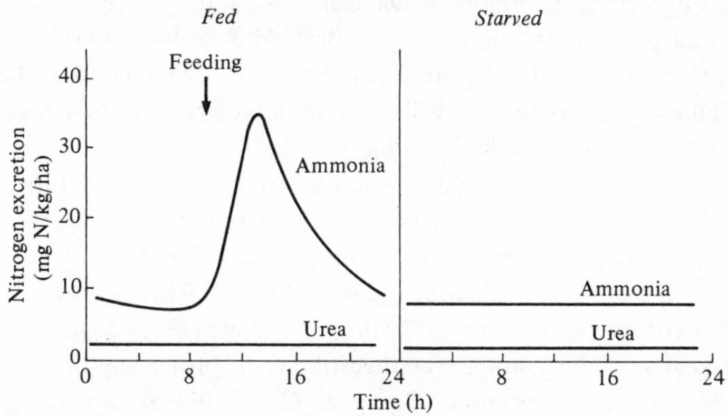

calorific value of protein varies somewhat with its amino acid composition. If, however, a 'standard' protein is considered, its gross calorific value may be taken as 5.65 kcal/g (Brody, 1945; Kleiber, 1961). Taking into account the digestibility of protein, and subtracting the residual calorific value of the nitrogenous excretory products, the metabolizable energy of dietary protein for mammals is generally considered to be about 4.1 kcal/g (Bell *et al.*, 1980). This value has become firmly established in the literature on nutritional physiology, and has often also been used for fish (e.g. Phillips *et al.*, 1966b; Hastings & Dupree, 1969).

There is, however, a serious controversy as to the residual calorific value of ammonia, and thus as to the metabolizable energy value of catabolic protein in ammoniotelic animals. Some authors (e.g. Cowey & Sargent, 1972) maintain that since ammonia has no carbon fragment, it is devoid of residual energy. Thus, if the end product of protein metabolism is ammonia alone, the metabolizable energy of the protein equals the digestible energy, i.e. 5.65 kcal/g digested protein. This approach has been criticized by a number of physiologists. A completely opposed view has been taken by Elliott & Davison (1975). They have calculated the energy value of ammonia, or its 'specific enthalpy of combustion', from its heat of formation. From these calculations they have derived a value of 83.2 kcal/mol or 5.94 kcal/g ammonia N. This value is somewhat higher than that of urea and therefore, if accepted, the metabolizable energy of protein in ammoniotelic animals is not greatly different from that of mammals. A different approach has been taken by Brafield & Solomon (1972). They have back-calculated the calorific value of ammonia from that of urea, subtracting from it the energy required to convert ammonia to urea (13.8 kcal/mol). The resulting calorific value of ammonia was 68.9 kcal/mol, or 4.92 kcal/g ammonia nitrogen. Thus, if ammonia is the sole nitrogenous excretory product of protein catabolism, protein energy losses according to Elliott & Davison (1975) are 5.94/6.25 = 0.95 kcal/g digested protein, and according to Brafield & Solomon (1972) they are 4.92/6.25 = 0.79 kcal/g digestible protein. When deducted from the calorific value of the protein, this will amount, in the first case, to 4.7, and in the second case to 4.9 kcal/g digestible protein.

In view of the preceding discussion it is not surprising that the value of metabolizable energy of protein given by various authors varies so much. Phillips (1969, 1972) gives the 'physiological value' of protein for three species of trout (brook, brown and rainbow) as 3.9 kcal/g, Brett & Groves (1979) considered it to be 4.2 kcal/g, and Brett (1973) gave the value as 4.81 kcal/g. If the average true digestibility of protein by fish is taken as 94–95% (see, for example, Ogino & Chen, 1973a), according to the calculations brought above, the metabolizable energy of protein for fish amounts to about 4.4 kcal/g dietary protein if Elliott & Davison's values are adopted, and if those of Brafield & Solomon are considered this will amount to 4.6 kcal/g. The higher value of metabolizable energy of protein for fish was supported by experimental results of Smith (1971). He fed rainbow trout on protein (casein and gelatin) and measured the digestibility coefficient and the residual calorific value of the excretory products. The average digestibility of the protein was 86.6% and its metabolizable energy 4.54 kcal/g. Smith did not, however, take into account either metabolic faecal nitrogen or endogenous nitrogen excretion, and the subtraction of both would have increased somewhat the metabolizable energy value of the dietary protein. Niimi & Beamish (1974) estimated the energy loss through non-faecal excretion by largemouth bass (*Micropterus salmoides*) at 10% of the intake energy. Considering that the diet, which consisted of emerald shiners (*Notropis atherinoides*), contained about 50% protein, this would make the metabolizable energy of the digested protein equal to about 4.5 kcal/g. A similar value was also suggested by Smith (1971) and Gropp (1979). It should be noted that all these values are higher than those for mammals and fowls, which should be taken into account in nutritional experiments with fish (see p. 187). A more precise value should, of course, be calculated according to the digestibilities of the various feedstuffs by the different fish species.

4

Maintenance

For maintaining the normal processes of life such as respiration, blood circulation, excretion, osmoregulation, digestion and movement, the fish requires energy. These processes and the energy required to maintain them are referred to as 'maintenance metabolism' (Q_M). Those processes most vital for immediate survival, such as respiration and blood circulation etc., constitute the 'basal metabolism' (Q_B). The other processes are vital for a longer duration existence. It should be remembered that maintenance requires not only an energy input, but also a material build-up by biosynthesis. This is required in part to replace worn-out tissues and cells, such as epithelial cells; in part to supply enzymes and hormones which may be 'recycled' rather quickly; and in part to supply continuous secretions such as mucus. From a quantitative point of view these material requirements may be expressed as energy values, but they also have qualitative aspects which must be taken into account when considering the fish diet. These aspects will be discussed later in the book.

When fish are starved and thus do not receive the energy required for maintenance, they utilize their own body tissues for this purpose. Fat depots, if they are present, will be utilized first, but tissue proteins are also utilized. Thus, during fasting fish lose weight. When they are fed they utilize the food first for supplying the necessary energy for vital maintenance processes and replacing tissue catabolism, and only then can the surplus be used for growth. Maintenance requirements have been determined by measuring the calorific value of the tissues utilized during starvation, or the calorific value of the food intake which ensures that the fish will neither lose nor gain weight – this is the maintenance level. The problem involved in determining the maintenance energy requirement through the weight lost during starvation, which does not include *SDA* energy expenditure, has been discussed earlier (Chapter 2). It is, therefore, doubtful whether such determinations can represent true maintenance requirements, unless the *SDA* is known and the ratio between tissue energy and food energy utilized for maintenance, i.e. the partial apparent efficiency of food for maintenance (see p. 9), is determined. For that, more experiments with feeding at maintenance level are required. However, since these experiments are complicated and time-consuming, a few such studies have been attempted. Moreover, when such a study is carried out, it has to refer to a certain given set of conditions with respect to fish weight, water temperature, feed composition, etc., although in many cases it would be of interest to study the effect of these factors on maintenance requirement. This calls for a complicated experimental set-up involving many aquaria or tanks and much work. Maintenance metabolism has, therefore, been studied mainly by means of indirect methods in the laboratory. Obviously, laboratory conditions differ from natural conditions but, since they are more controlled, they can be used to study the various factors affecting metabolism. However, before relating such conclusions to natural conditions, a verifying comparison is necessary, and a correction factor for the natural conditions must be determined for the quantatitive values obtained in the laboratory.

Many of the laboratory studies have been aimed at determining the basal metabolism (Q_M) and how it is affected by temperature, fish body weight and other factors. The fish used in such experiments were usually starved for about 48 h, and during the experiments were held in narrow containers to prevent excessive movement. Such studies have shown that it is difficult to prevent all movement, and there is always some essential movement to enable the fish to remain suspended in the water. This unavoidable movement is therefore included in the metabolism, which is referred to as 'standard metabolism' (Q_S), defined as the nearest attainable approximation to basal metabolism (Winberg, 1956, 1961; Brett, 1962). Determination of basal metabolism is possible by indirect methods. Metabolism is determined for varying degrees of spontaneous activity and the regression of metabolism on activity is found. Extrapolating back to zero activity gives the estimate of basal metabolism. Spontaneous activity was measured in different ways. Spoor (1946) recorded the number of deflections of an aluminium paddle activated by the water currents caused by the activity of fish. Other types of activity recording devices were used by other research workers (e.g. Brett, 1962; Beamish, 1964a; Beamish & Mookherjii, 1964).

When fish are permitted to move more or less at random in their tank during metabolism determination, the resulting value is called 'routine metabolism' (Q_R) (Fry, 1957; Brett, 1962). This is the most commonly measured rate of metabolism. However, it is obvious that in as much as this value approaches the natural metabolism, it cannot allow for energy requirements such as for free movement in search of food, or similar necessary maintenance requirements as they are in nature.

A number of studies have been aimed at determining the energy cost of movement. Various types of apparatus have been devised for such measurements. These include tanks in which fish swim against a water current of known velocity, or revolving systems where fish swim against the direction of the rotation at varying speeds. The metabolism of fish stimulated to maximum sustained activity is determined as 'active metabolism' (Q_A). Some aspects of active metabolism derived from such studies will be discussed below.

Still other metabolic studies have been aimed at determining the extra energy requirements for *SDA*. Here metabolism was measured for starved fish and fish fed known amounts of feed to obtain the extra amount of energy required for the utilization of this feed.

It can be seen that the effects of many factors on metabolism have been studied in the laboratory but, regrettably, only a few have been related to the determination of the overall energetic requirements of growing fish. These include also the energy required for synthesis of body tissues from simple precursors, which Brody (1945) called 'morphogenic work'. Nevertheless, much can be learned from laboratory studies, provided they can be related to natural conditions.

Two major methods have been used in the laboratory to determine the energy requirements of animals: (a) direct calorimetry; (b) indirect calorimetry. In the first the heat produced by the metabolic processes is measured in a calorimeter. This is a thermally insulated chamber (or container) in which the temperature change due to heat production can be accurately determined. Figure 17 describes such a calorimeter adapted to measuring heat production by fish. Though it would seem that fish are ideal subjects for the application of this method, since heat produced by metabolic processes is transferred rapidly into the surrounding water without the losses due to evaporation, convection and radiation which may occur in terrestrial animals, very few studies have been conducted using this method. The reason is that water has a high specific heat value, so that it requires relatively large amounts of heat to produce significant changes in temperature that can be measured accurately (Brett, 1962). The few studies that have been carried out, by Smith *et al.* (1978a), were on various salmonids up to a weight of 60 g. The researchers claim that accuracy of measurement of small quantities of heat could be obtained when the senstivity of the instrument was enhanced by using differential thermometers sensitive to 0.001 °C.

Most of the metabolic studies on fish have, however, been carried out by indirect calorimetry. This method is based on the assumption that energy production in fish is an aerobic process and requires oxygen for oxidizing nutrients either from food or from the fish's own body tissues. According to Blazka (1958), anaerobic metabolism may exist in fish. He found that at a temperature of 5 °C, crucian carp (*Carassius carassius*) can live, even for several months, without oxygen. Also Mourik *et al.* (1982) and Van den Thillart *et al.* (1983) found that goldfish, *C. auratus*, can survive without oxygen for about 16 h at 20 °C and for several days at 4 °C. Van den Thillart *et al.* (1983) suspected that the very long survival period in anoxic condition found by Blazka was due to some aerobic metabolism. According to Hochachka (1961) under anoxic conditions goldfish respire carbon dioxide by means of anaerobic breakdown of organic compounds (metabolic CO_2) rather than from oxidation (acidic CO_2). The resulting respiratory quotient (*RQ* – the ratio between the volume of inspired oxygen and that of expired CO_2) in these conditions is very high. It has been observed that glycogen is mobilized in anoxic goldfish but the quantity of accumulated lactic acid is far too low to account for the decrease in glycogen (Van den Thillart & Kesbeke, 1978). Mourik *et al.* (1982) showed that during anaerobic metabolism pyruvate is converted in muscle mitochondria by a modified pyruvate dehydrogenase into acetaldehyde which, after transport to the cytoplasm, serves as an

Figure 17. A calorimeter for the measurement of fish metabolism (from Smith *et al.*, 1978a).

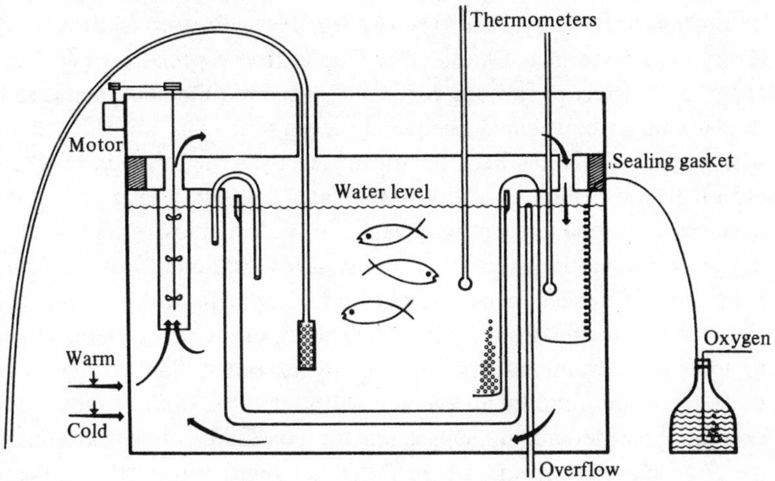

electron acceptor. From the acetaldehyde, due to activity of ethanol dehydrogenase in the cytoplasm, ethanol and carbon dioxide are formed (Shoubridge & Hochachka, 1980). According to Van den Thillart *et al.* (1983) this process should have resulted in a CO_2:ethanol ratio of 1:1. However, since in most of their experiments they obtained a higher ratio, they suggested the existence of other metabolic CO_2-producing pathways. They have concluded (Van den Thillart, 1982; Van den Thillart & Verbeek, 1982; Van den Thillart *et al.*, 1983) that fish have two systems of anaerobic metabolism. A short term exposure to anoxia provokes lactate glycolysis, in which pyruvate is converted into lactic acid, while a long term exposure to anoxia goes with an 'improved' anaerobic metabolism in which part of the pyruvate is converted into ethanol. This pathway is not common to all fish. Eels (*Anguilla anguilla*) can survive in anoxic conditions at 15 °C for almost six hours (Van den Thillart *et al.*, 1983; Van Waarde *et al.*, 1983). During this period muscle creatine phosphate was almost completely exhausted, glycogen decreased and lactate was accumulated, but no conversion of glycogen to ethanol was found. Thus, eels seem to lack the capacity for anaerobic metabolism in this pathway and lactic glycolysis appears to be the main anaerobic energy producing pathway. It is obvious, however, that the conditions mentioned above for anaerobic metabolism are extreme and very rare, so that they may be ignored in any generalization and for normal conditions the assumption of aerobic metabolism may be accepted.

Another assumption made when determining metabolism by indirect calorimetry is that an equilibrium exists between oxygen requirement and its supply, and that no oxygen debt develops during the measurement. However, several studies have shown that a temporary increase RQ over its normal value may occur (Huckabee, 1958; Prosser & Brown, 1961; Kutty, 1968; see also Chapter 3). This is true especially when fish are exercised severely or for long periods. Oxygen requirement then is so high that it cannot be fulfilled adequately, even when the water is saturated with oxygen. This has been demonstrated by Krueger *et al.* (1968) who examined the depletion of lipids and fatty acids in young coho salmon which were forced to swim to exhaustion. Calorific losses estimated from oxygen consumption were only one third of the actual losses calculated from the differences in fish weight and composition. Also Brett (1973) found that calorific loss from body substance of sockeye salmon subjected to long periods of sustained swimming exceeded that estimated from oxygen consumption by an average of about 20%. Both of these authors explain the difference by suggesting the possibility of partial anaerobic metabolism, which means a higher RQ, i.e., higher proportions of carbon dioxide and energy are formed for the amount of oxygen consumed. The fish then accumulates or excretes reduced metabolites such as pyruvic and lactic acids (Blazka, 1958; Brett, 1962; Krueger *et al.*, 1968). Calculations of the oxycaloric value for the oxygen taken up do not hold in this case. In order to avoid such conditions, experiments must be conducted with sufficient oxygen available, the fish should not move excessively in the respirometer, and measurements must continue for sufficient time to make up any oxygen debt which may develop at the beginning of the experiment due to excitement (stress) of the fish[1]. Other factors, such as composition of the ingested food, may affect the *RQ* value. These will be discussed later in connection with the oxycaloric value. However, when the above conditions are observed it may be assumed that the amount of oxygen taken up by respiration will release an equivalent amount of energy which can be calculated from the oxycaloric value.

Winberg (1956) and Beamish & Dickie (1967) describe various metabolic chambers, or 'respirometers' as they are often called, which are used for fish. These are usually of two kinds: (a) sealed respirometers; (b) flow-through respirometers. The first are sealed containers filled completely with water, with no exchange of oxygen or carbon dioxide between water and atmosphere. Fish are introduced into the container and the change in oxygen content during the experiment is recorded. The main drawback of this method is the constant decrease in oxygen concentration and increase in carbon dioxide concentration which may affect fish metabolism. This effect is slight when oxygen is above a critical threshold level and carbon dioxide concentration is low. However, below this level, which is dependent on temperature and carbon dioxide concentration, the metabolism decreases with the decrease in oxygen concentration (Winberg, 1956, 1961). Also, in such conditions the fish may develop an oxygen debt, causing an erroneous energy requirement estimation. This limits the range of variation in oxygen concentration allowed in this kind of respirometer. There is a tendency to reduce the volume of the respirometer in order to limit free space and avoid excessive fish movement. However, the larger the ratio between fish biomass in the respirometer and the volume of water in it, the larger the difference in oxygen concentration for a given period. Therefore the duration of the experiment must be rather short for the permissible change in oxygen concentration, otherwise the oxygen concentration may decrease below the threshold level. On the other hand, if

[1] Most metabolism measurements show higher values at first due to stress caused by the handling of the fish. Ameer Hamsa & Kutty (1972) state that this elevated metabolism corresponds to the highest rate of spontaneous activity.

the water volume is very large relative to fish biomass, differences in oxygen concentrations may be too small to be measured accurately. This is why flow-through respirometers are much more common for the determination of fish metabolism.

The flow-through respirometer is a cylinder, the diameter of which is not much larger than that of the fish which is introduced into it. The respirometer is often immersed in a constant temperature water bath. Water flows into the respirometer towards the fish's head end, and flows out from the tail end. It is possible to regulate the water flow so that the amount of incoming oxygen will be more or less equal to its uptake by the fish, and the gradient in oxygen concentration between inflow and outflow water will be small. However, the current must not be so strong as to cause higher energy consumption by swimming against it. The oxygen requirement of the fish is determined by the difference in oxygen concentration between inflow and outflow and the volume of water passing through the respirometer.

Other, less common methods measure the amount of gaseous oxygen added to the water in the respirometer during the test to maintain it at a certain degree of saturation, or measure the amount of carbon dioxide released to the water through respiration. It should be remembered that in any of the methods described above, whether by direct or indirect calorimetry, the total energy produced by the fish is determined. This includes net energy utilized for maintenance and the heat increment which is lost unutilized (see Chapter 1). If measured by direct calorimetry, both end up as heat; if measured by indirect calorimetry, oxygen is required to produce both. None of these methods can distinguish between the utilized free energy and the unutilized heat increment.

Many factors affect maintenance metabolism. Brett (1962, 1970) discusses many of these factors such as water salinity, pH, photoperiodism, carbon dioxide concentration, ammonium concentration, physiological

Table 21. *Incipient limiting and no excess activity oxygen levels for some fish species*

Fish species	Temperature (°C)	Incipient limiting level (mg/l O_2)	No excess activity level (mg/l O_2)	Source
Salvelinus fontinalis	5	12.3		Job (1955)
	20		2.8	Job (1955)
Oncorhynchus kisutch	20	8.8		Dahlberg *et al.* (1968)
Micopterus salmoides	25	6.0		Dahlberg *et al.* (1968)
Cyprinus carpio			2.3	Blazka (1958)
Carassius auratus	35	1.8	1.1	Fry & Hart (1948b)
	5	1.2	0.3	Fry & Hart (1948b)

state of the fish, stage of gonadal development and many others. The effect of some of these factors on metabolism was studied to a greater depth than the others; these factors include carbon dioxide concentration (Basu, 1959), season (Beamish, 1964c), sex (Beamish, 1964c; Hughes & Knights, 1968) and salinity (Farmer & Beamish, 1969). Ambient oxygen concentration has, obviously, a special importance. When oxygen is present in sufficient amounts there is no limitation on the movement of the fish even though it requires much energy and therefore also much oxygen. There is, however, a threshold level of oxygen below which movement becomes limited and dependent on oxygen concentration. Fry (1947) called this threshold 'the incipient limiting level' and determined it as the level of oxygen tension below which the rate of oxygen uptake at the maximum steady state of activity begins to decrease. Some values for the incipient limiting oxygen level are presented in Table 21. The variations in these values are quite considerable. Coldwater fishes have much higher values than warmwater fishes. Below this threshold level, the lower the oxygen concentration, the lower the activity. The level of oxygen at which the fish cannot satisfy even limited movement, but only its standard requirements for oxygen is the 'level of no excess activity'. This level also varies with fish species and with water temperature as can be seen in Table 21, above. When fish continue their activity at a higher rate than allowed by the ambient oxygen level for too long a duration, they may die due to a lethal oxygen debt (Brett, 1962). However, it should be mentioned here that, although the above is physiologically true, some fish behave differently. Ameer Hamsa & Kutty (1972) studied the metabolism of five species of marine fishes and found, in four of these, that spontaneous activity increased while the ambient oxygen concentration decreased from air saturation to asphyxial values. They concluded that there could apparently be two separate behavioural adaptations to the reduction in ambient oxygen concentration. While the decrease in spontaneous activity at lower oxygen concentrations may help conserve energy, an increase in activity can help the fish escape from the site of unfavourable conditions.

Another method of adaptation to low oxygen concentrations is by changing the ventilation rate, i.e. the rate of respiratory movements and the amount of water passing over the gill filaments. According to Eclancher (1975) the change in the concentration of ambient oxygen leads to change in arterial oxygen tension, which in turn elicits, within 1.5–2.6 sec, cardiac and ventilatory reactions. Rantin & Johansen (1984) studied the response of *Hoplias malabaricus* to reduced oxygen concentrations. They found that lowering of oxygen tension in the water resulted in a two to three-fold increase in gill ventilation, primarily through increased breathing volumes rather than breathing frequency.

Blazka (1958) found that with the decrease in ambient oxygen concentration there was a slight decrease in the metabolism of crucian carp, but when oxygen concentration reached about 1.7–2.0 ml O_2/l there was a marked increase in metabolism, probably due to increased ventilation movement. Then, at lower oxygen concentrations there was a fall in oxygen consumption. Also Beamish (1964d), working on metabolism of brook trout, common carp and goldfish, found that down to an oxygen saturation level of approximately 51% standard oxygen uptake remained quite constant. Below this oxygen level the standard rate first increased to a maximum and then, with further reduction in oxygen level, decreased. Randall & Shelton (1963), who studied the effect of hypoxia on breathing rate and heart rate of the tench (*Tinca tinca*) and Nakanishi & Itazawa (1974), who made a very similar study on several fishes, showed that the increase in oxygen consumption was due to marked increase in ventilation (while there was a decrease in heart rate). They also found that at extreme hypoxia oxygen consumption decreased further.

The 'critical oxygen level' was considered by Blazka (1958) as the lowest oxygen level which does not lower oxygen consumption. Different fish species seem to react differently to oxygen concentrations lower than the critical level (Blazka & Kopecky, 1961). Species adapted to anoxia (e.g. crucian carp, goldfish) go, especially at low temperature, into anaerobic metabolism in which glycogen is converted into ethanol and metabolic carbon dioxide (p. 107). Such metabolism does not create oxygen debt. At higher temperatures these fishes tolerate anoxia for a shorter period. In other species, such as trout, adaptation to hypoxic conditions is limited. Beamish (1964d) suggests that at low levels of oxygen a fraction of the standard metabolism of fish is derived anaerobically. From his studies it appears that acclimation of fish to low level of ambient oxygen enhances this anaerobic fraction of the standard metabolism. This emphasizes the importance of maintaining a relatively high oxygen level during indirect calorimetry metabolic rate measurement.

Assuming that fishponds usually have sufficient oxygen for normal activity of fish, three other factors affecting metabolism are most important from the point of view of fish farming management: (a) fish weight – the larger the fish the more energy it requires for maintenance, and metabolism is higher; (b) temperature – the higher the temperature the higher the rate of metabolic processes; (c) activity – which requires extra energy. Many studies have been conducted to determine the quantitative relationships between these factors and metabolism. Winberg (1956, 1961) has contributed much to the understanding of these relationships by reviewing, summarizing and analysing the available literature on this subject. He determined the regressions between each of these factors and standard metabolism. These will be discussed in greater detail below.

Fish weight and metabolism

Fish weight is the major parameter used by the fish farmer to determine feeding level. Since food must cover the energetic requirements of the fish, and these include maintenance needs, it is important to determine the relationship between fish weight and metabolism. In most animals this relationship can be expressed by an exponential equation of the form:

$$Q = a\,W^{b} \qquad (16)$$

where W is the weight of the animal, and a and b are constants. The regression coefficient b expresses the slope of the regression line describing the log of the rate of metabolism as a function of the log body weight. Thus this regression can also be expressed as:

$$\log_{10} Q = \log_{10} a + b \log_{10} W$$

The constant a is the Y-intercept, and expresses the rate of metabolism of an animal at $\log_{10} W = 0$, i.e. one unit body weight (in fish, usually 1 g), and may be called 'specific metabolism'. According to Brody (1945) the regression expressing the basal metabolism of adult mammals is:

$$Q_B \text{ (kcal/day)} = 70.5\,W^{0.735} \qquad (17)$$

where W is expressed in kg. With more recent data it was found that a regression coefficient of 0.75 ($\frac{3}{4}$) is general to many higher vertebrates of different weights (Kleiber, 1965). This coefficient (weight exponent) has since become generally accepted for higher vertebrates.

Most research workers have accepted the above power function of metabolism–body weight to be also relevant to fish. Glass (1969) compared such functions calculated for data obtained from the literature with non-linear functions derived by iterative least square fitting. He concluded that the latter functions are more accurate, but since the difference is not great and in view of the greater convenience and the accepted function used for higher vertebrates, most researchers use the log–log power function.

From the large amount of data on metabolic rates of fishes of various weights Winberg (1956, 1961) was able to collect from the literature, and after correction for temperature differences, he concluded that the regression coefficient for routine metabolism of fish is similar for many fish species, and is about 0.8. This means that the rate of increase in metabolism with fish weight is higher than that in birds and mammals.

Table 22. *Specific metabolism (a), in mg O_2/h^a, and weight exponent (b) of the metabolism-body weight function ($Q = aW^b$), at standard and routine metabolic levels for various fish species*

Species	Temperature (°C)	*a*	*b*	Source
Freshwater	20	0.56	0.81	Ivlev (1959)
	20	0.438	0.80	Winberg (1961)
Abramis brama	20	0.592	0.93	Kudrinskaya (1969)
Aeguidens latifrons	25	NG[b]	0.7	Heusner *et al.* (1963)
Anguilla anguilla	NG	0.309	0.87	Swierzowski (from Fischer, 1978)
Brevoortia tyrannus	10	0.253	0.782	Hettler (1976)
	15	0.406	0.797	Hettler (1976)
	20	0.721	0.724	Hettler (1976)
	25	0.659	0.816	Hettler (1976)
Carassius auratus	10	0.027	0.882	Beamish & Mookherjii (1964)
	35	0.213	0.887	Beamish & Mookherjii (1964)
Catostomus commersoni	10	0.035	0.994	Beamish (1964b)
	15	0.169	0.828	Beamish (1964b)
	20	0.318	0.770	Beamish (1964b)
Chaenocephalus aceratus	0.5–2	0.038	0.906	Muir & Niimi (1972)
Citharichthys stigmaeus	15	NG	0.905	Hickman (1959)
Clupea harengus	10	0.306	0.978	Muir & Niimi (1972)
Coregonus sp.	10	0.252	0.862	Dabrowski & Kaushik (1984)
Cyprinus carpio				
(fry below 0.7 g)	20	0.52	0.97	Kamler (1976)
(fry below 1.0 g)	20	0.522	0.98	Winberg & Khartova (1953)
(fry 1.2–46 g)	20	0.323	0.802	Kamler (1976)
(red carp – early stages)	20	—	0.84	Huang (1975)
	10	0.018	0.983	Beamish (1964b)
	20	0.072	0.909	Beamish (1964b)
	30	0.180	0.876	Beamish (1964b)
	35	0.280	0.810	Beamish (1964b)
	20	0.46	0.816	Kausch (1968)
	NG	0.8	0.76	Melnichuk (from Kamler, 1976)
	—	—	0.89	Huang (1975)
	23	0.192	0.805	Sondermann *et al.* (1985)
Ctenopharyngodon idella (fed fish)	23	0.662	0.598	Fischer (1970)
Gadus callarias	NG	0.411	0.71	Sundnes (1957)
	5–10	0.315	0.727	Muir & Niimi (1972)
Gadus morhua	3	0.182	0.79	Saunders (1963)
	10	0.150	0.89	Saunders (1963)
	15	0.226	0.85	Saunders (1963)
	12	0.245	0.82	Edwards *et al.* (1972)
(fed fish)	10	0.329	0.83	Saunders (1963)
(fed fish)	15	0.652	0.76	Saunders (1963)
Gadus virens	NG	0.411	0.71	Sundnes (1957)
Hypophthalmichthys molitrix	—	—	0.88	Huang (1975)
Ictalurus nebulosus	10	0.020	0.998	Beamish (1964b)
	20	0.103	0.903	Beamish (1964b)
	30	0.190	0.874	Beamish (1964b)
Kuhlia sandvicensis	23	1.467	0.784	Muir *et al.* (1965)
(freshwater)	23	1.37	0.785	Muir & Niimi (1972)
(saltwater)	23	1.39	0.790	Muir & Niimi (1972)
Lebistes reticulatus	25	NG	0.7	Heusner *et al.* (1963)
Limanda limanda	10	0.597	0.666	Edwards *et al.* (1969)
Lepomis gibbosus	25	0.106	0.710	O'Hara (1968)
	30	0.204	0.791	O'Hara (1968)
Lepomis macrochirus	25	0.108	0.717	O'Hara (1968)
	30	0.164	0.749	O'Hara (1968)
Macrognathus aculeatum	21	0.186	0.709	Ojha & Datta Munshi (1975)
	30	0.292	0.714	Ojha & Datta Munshi (1975)
Micropterus salmoides	10	0.770	0.67	Beamish (1970)
	20	1.220	0.65	Beamish (1970)
	30	1.585	0.66	Beamish (1970)
Notothenis rossi marmorata	1.4	0.152	0.831	Muir & Niimi (1972)
Notothenis corriceps neglecta	−0.5–1.7	0.165	0.782	Muir & Niimi (1972)
Notothenis gibberifrons	−1–2.4	0.068	0.912	Muir & Niimi (1972)
Notothenis ramsayi	10	0.303	0.783	Muir & Niimi (1972)
Oncorhynchus nerka	5.3	0.057	0.914	Brett & Glass (1973)
	15	0.101	0.846	Brett & Glass (1973)
	20	0.199	0.884	Brett & Glass (1973)
	15	0.233	0.78	Brett (1965)
	11.5	0.255	0.65	Staples & Nomura (1976)
(fed to satiation)	11.5	0.487	0.81	Staples & Nomura (1976)
(fed to satiation)	—	0.305	0.79	Yamazaki (1967) (from Staples & Nomura, 1976)
Ophiodon elongatus	10.3–13.3	NG	0.78	Pritchard *et al.* (1958)
Oreochromis mossambicus	15	NG	0.658	Job (1969)
	40	NG	0.821	Job (1969)
	20	0.262	0.74	Mironova (1976)
	31	0.579	0.76	Mironova (1976)

Table 22 (*cont.*)

Species	Temperature (°C)	*a*	*b*	Source
Oreochromis niloticus				
(freshwater)	25	0.620	0.838	Farmer & Beamish (1969)
(salinity–30ppt).	25	0.589	0.814	Farmer & Beamish (1969)
Parophrys vetulus	15	NG	0.842	Hickman (1959)
Perca fluviatilis	20	0.384	0.78	Kudrinskaya (1969)
Platichthys stellatus	15	NG	0.859	Hickman (1959)
Pleuronectes platessa	10	0.285	0.72	Edwards *et al.* (1969)
Rutilus rutilus	20	0.384	0.82	Kudrinskaya (1969)
Salmo gairdneri (direct calorimetry)	15	0.598	0.75	Smith *et al.* (1978a)
Salmo trutta	10	0.142	0.877	Beamish (1964b)
Salvelinus fontinalis	10	0.033	1.107	Beamish (1964b)
	15	0.101	1.014	Beamish (1964b)
	20	0.129	1.036	Beamish (1964b)
	5	NG	0.86	Job (1955) (quoted from Kausch, 1968)
	10	NG	0.85	Job (1955) (quoted from Kausch, 1968)
	15	NG	0.85	Job (1955) (quoted from Kausch, 1968)
	20	NG	0.85	Job (1955) (quoted from Kausch, 1968)
Scophthalmus maximus	7–16	1.969	0.78	Brown *et al.* (1984)
Sprattus fuegensis	10	0.315	0.980	Muir & Niimi (1972)
Squalus suckleyi	12–14	NG	0.74	Pritchard *et al.* (1958)
Stizostedion lucioperca	20	0.418	0.82	Kudrinskaya (1969)
Stizostedion vitreum	20	0.301	0.841	Kelso (1972)
Xiphophorus helleri	25	NG	0.78	Heusner *et al.* (1963)
AVERAGE (n=99)			0.819±0.094 SE	

Notes: [a] All units have been converted to mgO_2/h using conversion units of:
1 litre O_2 = 1.429 g at standard temperature and pressure
1 litre O_2 = 1.33 g at 20 °C and standard pressure
[b] Not given in the paper.

When the body weight doubles (= 100% increase), metabolism in the higher vertebrates increases by 75% and in fish by 80%. Since Winberg's work was published, many papers have appeared presenting new data on metabolism of fish and its relationship to body weight. Some of these gave a lower weight exponent than that given by Winberg. Schmidt-Nielsen (1975) maintained, therefore, that this value does not differ substantially from that of higher vertebrates. Other authors found higher slope values. In Table 22 are presented some of these values, which have been collected from the literature.

The wide range of values found in the literature may indicate that the relationship between metabolism and fish body weight is not simple and is compounded with other factors. The weight exponent *b* may vary with environmental conditions and physiological state of the fish. One of the most important factors interacting with this relationship is water temperature. According to Job (1955) the temperature affects the weight exponent *b*. Brown *et al.* (1984) determined the regression between oxygen consumption and mean body weight of turbot (*Scophthalmus maximus*) in five temperature groups, each of 2 °C spread, from 7–8 °C to 15–16 °C. He found differences not only in the specific metabolism, which increased considerably with temperature, but also in weight exponent, which decreased with increasing temperature from 0.82 at 7–8 °C to 0.71 at 15–16 °C. This interaction led some authors to express the metabolic rate as a multiple regression equation, taking into account both body weight and temperature. A number of similar equations were suggested[1]:

$$Q = aW^b t^c \quad \text{by Liao, 1971;} \quad (18)$$

$$Q = aW^b\, 10^{ct} \quad \text{by Muller-Feuga et al., 1978;} \quad (19)$$

$$Q = aW^b\, e^{ct} \quad \text{by Elliott, 1975b;} \quad (20)$$

where Q is the metabolic rate (in the first two cases this was measured on fed fish in actual culture conditions); t = water temperature; a = specific metabolism; b and c are the regression coefficients of body weight and temperature respectively. Muller-Feuga *et al.* (1978) compared the first two equations using data of rainbow trout metabolism. The difference between them was small and statistically insignificant.

Analyses of these regressions using actual data on metabolism of fish showed a discontinuity along the temperature range. Liao (1971) analysed data on salmon and trout (species not given). Specific metabolism values (a) and temperature exponents (c) for the temperature range below 10 °C differed from those for the range above 10 °C, but no difference was observed in weight exponent. This exponent was 0.806 for salmon and 0.862 for trout. Muller-Feuga *et al.* (1978), working on rainbow trout, also found a certain difference in the weight exponent, which was 0.804 below 10 °C and 0.858 over 12 °C. Elliott (1975b) found the point of deflection for brown trout at 6.6 °C: the weight exponent was 0.716 at the temperature range below 6.6 °C and 0.737 at the range above 6.6 °C.

Other factors also interact with body weight in its effect on metabolism. Brett (1965) contemplates the possibility of changes in metabolic rate during different life stages, similar to the changes which occur in growth rates ('stanzas'). While the weight exponent found by him for standard metabolism of sockeye salmon (*Oncorhynchus nerka*), which takes into account all data treated, was 0.78, when the early freshwater period data were treated separately this value was 0.624. In contrast to this lower than general weight exponent, others have found a higher value for young and small fish. Kamler (1972) found the weight exponent to be 0.8 for common carp of over 1 g but 0.98 for carp smaller than 1 g. Huang (1975) studied

---

[1] The original designations of these regressions have been changed to fit each other and those used in the present work.

the oxygen consumption of the early stages of some fish species and found the weight exponent for young red and regular common carp (*Cyprinus carpio*) to be 0.84 and 0.89 respectively, silver carp 0.88 and Chinese bream (*Parabramis pekinensis*) 0.89. Smith *et al.* (1978a) found an inflection point in the body weight–metabolic rate regression line of young rainbow trout at about 4 g. The weight exponent for the smaller fry was 1.0, while that for fry larger than 4 g was 0.63. Brett (1965) points out the possibility of sex affecting the metabolic rate, especially during maturation.

Of most considerable interest is the effect of activity on the body weight–metabolism relationship. Muir & Niimi (1972) did not find any difference in weight exponent between standard and active metabolism. However, according to Brett (1965) and Brett & Glass (1973) there is in the sockeye salmon (*O. nerka*) a progressive increase in the weight exponent with increasing activity. Starting at 0.78 (Brett, 1965) or 0.88 (Brett & Glass, 1973) for the resting state, the slope rises to near unity (0.97–0.99) for maximum sustained performance. Job (1955) found for *Salvelinus fontinalis* that this difference is temperature dependent. At 5 °C the weight exponent for active metabolism (0.94) was higher than for standard metabolism (0.86); at 15 °C the slopes were the same for both active and standard metabolism; and at 20 °C the relation was reversed and standard metabolism had a slope of 0.8, while that for active metabolism was 0.75. According to Brett (1965) the increase in activity from standard metabolism to one quarter of the active level, which is approximately the routine metabolism for spontaneous activity, caused only a very small difference in weight exponent. Thus, active fish such as salmonids may have in nature a higher weight exponent than warmwater pond fishes which are less active, and their routine metabolism measured in the laboratory may have a similar weight exponent.

It is obvious that in spite of the numerous works on metabolic rate of fish, there is a need to establish the effect of various conditions (species, age, sex, activity, temperature, etc.) on the body weight–metabolism relationship. Only then will an accurate prediction of metabolic rate be possible. However, until this is resolved, it seems that Winberg's value of 0.8 is the nearest available approximation. A number of studies support this choice of weight exponent experimentally by finding a similar weight exponent (Fry, 1957; Basu, 1959; Paloheimo & Dickie, 1966a; Kausch, 1968).

Much has been written on the significance and interpretation of the body weight–metabolism regression and its causative relationship to physiological characteristics, since the 'surface law' developed for endotherms over a century ago. However, as found by Niimi (1975) the regression between body surface and body weight in fish is between 0.64 and 0.74, which obviously does not correspond to the body weight–metabolism regression coefficient. In endotherms it has been found that the regression between the absorption surface of the lung and body weight, and that between oxygen uptake and body weight are similar to that of metabolism–body weight (Schmidt-Nielson, 1975). It is interesting to note that similar relationships exist in fish. A regression coefficient of 0.8 was found for the relationship between the total area of the gill lamellae of the fish *Kuhlia sandvicensis* and its body weight (Muir, 1969). The mean regression coefficient for this relationship found by De Jagers & Dekkers (1975) for 11 fish species was 0.811 ($r = 0.975$). According to Jones & Randall (1978) the slope of this regression is different for the relatively inactive tench (0.72) and the active salmonids (0.91). However, the average weight exponent of this regression is about 0.8, which led some authors (De Jagers & Dekkers, 1975; Pauly, 1981) to state explictly that there is a relationship between the rate of increase in metabolism with fish weight and the increase in gill area and therefore both have the same weight exponent. The ability to absorb oxygen may be one of the causes for the variability of the weight exponent in the metabolism–body weight regression among fish species. This is also suggested from the results of a study by Ojha & Datta Munshi (1975) on the mud-eel (*Macrognathus aculeatum*). This fish has a bimodal oxygen exchange mechanism – through the gills and through the skin. The authors found that the total respiratory area (gills + skin) relationship to body weight is almost identical to that between oxygen uptake for routine metabolism and body weight (0.71). It is possible that the high weight exponent found for active metabolism in salmonids is related to their ability for ram ventilation during their rapid movement or in a swift current. Also, according to Fick's law, under standard conditions, in active fish a considerably lower difference is maintained between oxygen tension in the water and that in the blood, than in slow-moving fish (De Jagers & Dekkers, 1975).

Other physiological variables associated with aerobic metabolism may also be related to body weight in a similar regression form and slope. Somero & Childress (1980) found that the exponent of the regression between the activity, in white muscle, of citrate synthase, an enzyme associated with aerobic metabolism through the TCA cycle, and body weight of various fish species is 0.74, which is near that mentioned above for metabolism–body weight regression.

Specific metabolism can vary among fish species. Generally, coldwater fish such as trout and salmon, which live in oxygen rich flowing water have a higher specific metabolism than warmwater fishes such as common carp and tilapia, living in stagnant water, sometimes low in oxygen. De Jagers & Dekkers (1975) gathered data from the literature on oxygen consumption and activity behaviour of fish. They have corrected these data for a body weight of 200 g and ambient temperature of 20 °C. Their conclusion was that standard metabolism in these conditions is a fairly good quantitative measure of general activity of the fish species in regard to behaviour (see also Chekunova, 1983). Winberg (1956, 1961) gives the following values of specific metabolism at 20 °C of several fishes:

Salmonids	0.498 ml O_2/h
Cyprinids	0.347 ml O_2/h
Common carp	0.343 ml O_2/h
Tench	0.230 ml O_2/h

More recent values for specific metabolism, although for varied environmental conditions, are found in Table 22. Winberg's values as well as those found in the table seem to support De Jagers & Dekkers' (1975) conclusion on the correlation between metabolism and active behaviour of fish.

Since many of the metabolism studies have been performed on common carp, and since common carp is one of the most important pond fishes, special reference will be made to this fish in further discussions on fish metabolism. Values for other fish species can be treated in a similar way. From the foregoing discussion one can already construct the standard metabolism equation for common carp as given by Winberg (1956, 1961):

$$Q_s \text{ (ml } O_2\text{/h, 20 °C)} = 0.343\ W^{0.8} \quad (21)$$

Since the density of oxygen at temperature of 20 °C and at standard atmospheric pressure is 1.33 mg/l, the above equation can be converted into the more common units of mg/l and calculated for one day. It will then be:

$$Q_s \text{ (mg } O_2\text{/day)} = 10.95\ W^{0.8} \quad (22)$$

In order to compare the metabolism of common carp (or any other fish species) as expressed in the above equation with that for mammals as given in equation (17), and especially in order to utilize this information for nutritional purposes, oxygen requirement must be converted to energy units. For that the oxycaloric value must be determined.

Oxycaloric value depends on the characteristics of the oxidized nutrients and the ratio between the end products of their oxidations – carbon dioxide and incomplete oxidation products, if present – and the oxygen used for the process. According to the committee on constants and factors of the European Association for Animal Production (Brouwer, 1964) the

Table 23. *Respiratory quotients (RQ) and oxycaloric values (Q_{ox}) for different nutrients and 'mean' values for mammals and fish*[a]

	Carbohydrates		Lipids		Proteins Mammals		Proteins Fish		'mean' Q_{ox}	
Source	RQ	Q_{ox}	RQ	Q_{ox}	RQ	Q_{ox}	RQ	Q_{ox}	Mammals	Fish
Brody (1945)				3.28		3.37				
Winberg (1956)	1.0	3.53	0.72	3.36						3.75
Kleiber (1961)	1.0		0.71		0.83	3.15				
Brett (1962)				3.22						
Brouwer (1964)	1.0		0.71		0.81					
Petrusewicz & MacFadyen (1970)	1.0	3.57	0.71	3.22	0.79	3.22			3.29	3.38
Dargol'ts (1973) (quoted from Elliot & Davison, 1975)				3.22		3.21		3.07	3.25–3.28	3.24
Brafield & Solomon (1972)		3.53			0.8	3.15	1.0	3.20		
Brett (1973)					0.81	3.35		3.45		3.61
Huisman (1976)										
at hunger										3.53
fed at maintenance level										3.67
growing fish										3.81
Elliot & Davison (1975)										
herbivores		3.53		3.28		3.21	0.95	3.20		3.28
carnivores									3.27	3.24
Windell (1978)										3.61
Brett & Groves (1979)	1.0	3.52	0.71	3.28	0.82	3.37	0.9		3.6–3.76	3.48
Bell *et al.* (1980)	1.0		0.71		0.8					
Sonderman *et al.* (1985)										
at hunger, calculated from RQ										3.41
at hunger, experimentally										3.77

Note: [a] Energy units were converted using the values: 1 cal = 4.181 Joule (20 °C); 1 ml O_2 = 1.33 mg (20 °C).

amount of heat energy produced by mammals by bio-oxidation of nutrients can be determined by the following equation:

$$Q \text{ (kcal)} = 3.866\ O_2 + 1.2\ CO_2 - 0.518\ CH_4 - 1.431\ \text{N in urine} \quad (23)$$

units of the gases being in litres and N in urine in grams. Carbohydrates, lipids and proteins have different ratios between the volume (= mols) of oxygen consumed for oxidation and carbon dioxide respired (*RQ*), and therefore the amount of energy produced by a given volume of oxygen from each of these nutrients is different. *RQ*s and oxycaloric values given by various authors are presented in Table 23. It is commonly accepted that the *RQ* of carbohydrates is 1.0 and the oxycaloric value for oxidizing carbohydrates is 3.53 cal/mg O_2. Since lipids have a more variable ratio between oxygen and carbon, the *RQ* and oxycaloric values are also more variable; the accepted mean *RQ* is 0.71. The oxycaloric value, which is lower than that for carbohydrate, varies between 3.22 and 3.32; the mean of 3.28 is generally accepted by most authors. It is difficult to calculate the *RQ* for proteins, since it is not known which amino acids, if any, are preferentially utilized by the fish for energy and which are used for anabolism. Also, oxidation of protein in the body is not complete and part of the nitrogen is excreted as ammonia, urea, uric acid and other nitrogenous compounds. The *RQ* will, therefore, depend not only on the composition of the protein, but also on the nature of these products. If urea is excreted the *RQ* is about 0.8, whereas, according to Brafield & Solomon (1972) and Brett & Groves (1979) if ammonia is the end product the *RQ* is 0.9–1.0. Thus the nature of the end product also affects the oxycaloric value for oxidizing proteins. In mammals, which excrete mainly urea, it is about 3.2 cal/mg O_2. Elliot & Davison (1975) maintain that since ammonia has somewhat higher residual energy than urea (see Chapter 3), the oxycaloric value for protein in fish is somewhat lower (at about 2%) than that in mammals, and 3.2 cal/mg O_2 may therefore be considered applicable to fish. Also Brafield & Solomon, although they give ammonia a lower residual energy value than urea, give the same oxycaloric value for oxidizing protein in fish – 3.2 cal/mg O_2.

The mean oxycaloric value for an animal depends on the diet of the animal. The *RQ* is an important tool in its determination. For farm animals it has been determined in this way to be about 0.82 and the resulting oxycaloric value as 3.29 cal/mg O_2. However, for fish it is difficult to measure quantitatively the amount of carbon dioxide respired (Kutty, 1968) and, therefore, it is also difficult to determine accurately the *RQ* and the oxycaloric value. The *RQ*s for rainbow trout, goldfish and tilapia (*Oreochromis mossambicus*), starved for at least 36 h, were found to be 0.85, 0.9 and 1.0 respectively (Kutty, 1968, 1972). Kutty (1972) also found, from measurements of ammonia excretion by *O. mossambicus*, that most of the energy is derived by this fish from utilizing protein. The high *RQ*s for the last two species are therefore suggestive of protein having an *RQ* of about 1.0 as suggested by Brafield & Solomon (1972), and/or, as concluded by Kutty, that partial anaerobic metabolism may have occurred even at high ambient oxygen concentrations, possibly due to stress. Either case makes the *RQ* unreliable for oxycaloric determination until more is known on the *RQ* of protein in fish.

Due to these difficulties, various oxycaloric values can be found in the literature (Table 23). Brett (1973) determined the change in energy content of exercised migrating sockeye salmon (*Oncorhynchus nerka*) by bomb calorimetry and compared it with values deduced from oxygen consumption measurements. From this he derived an oxycaloric equivalent of 4.8 kcal/l O_2, which upon conversion are equal to 3.61 cal/mg O_2, but the same authors (Brett & Groves (1979) consider the mean oxycaloric value for fish to be 3.48 cal/mg O_2. Elliot & Davison (1975) argue that high values are perhaps suitable for herbivorous fish, but not for carnivorous ones, for which they suggest an oxycaloric value of 3.25 cal/mg O_2. In view of the high *RQ*s found in pond fishes, and since these fishes are mainly omnivores and herbivores, the relatively higher oxycaloric value of 3.4 cal/mg O_2 suggested by Petrusewicz & MacFadyen (1970) was adopted here for further calculation and discussion. The standard metabolism of common carp at 20 °C will accordingly be:

$$Q_s\ (\text{cal/day}) = 37.2\ W^{0.8} \qquad (24)$$

At this stage the energy costs of maintenance of fish can be compared with those of mammals. A common carp weighing 1 kg requires, according to equation (24), 9.34 kcal/day compared with 70.5 kcal/day for a mammal of the same weight, which is 7.5 times more than for the fish. The difference seems to be due to several energy savings, such as the fact that fish live in water and do not have to support their body weight as terrestrial animals must do; there appears to be no energy expenditure to maintain a body temperature different from the environmental temperature; and the excretion of waste nitrogen requires much less energy in fish than it does in mammals or birds.

One of the important conclusions of the metabolism–body weight regression given above is that since the weight exponent is lower than 1.0, the 'relative metabolism', i.e. metabolism per unit weight (Q_s/W), *decreases* with the increase in body weight. The following example illustrates this:

For a common carp of 100 g, metabolism according to equation (24) is:

$$Q = 37.2 \times 100^{0.8} = 1480 \text{ cal/day or } \frac{1480}{100} = 14.8 \text{ cal/g/day},$$

while for a fish of 500 g it is:

$$Q_s = 37.2 \times 500^{0.8} = 5370 \text{ cal/day or } \frac{5370}{500} = 10.7 \text{ cal/g/day}.$$

Since feeding level is often calculated as a percentage of body weight, the above conclusion becomes relevant. The amount of food which has to cover maintenance costs will increase with body weight, but the percentage of this amount relative to body weight will decrease. This has been realized already in early feeding experiments carried out by Dawes (1930). He fed plaice (*Pleuronectes platessa*) of different weights on oyster meat. An increase in weight of fish from 18 to 130 g entailed an increase in the daily ration required for maintenance from 0.4 to 1.4 g, but the value as a percentage of body weight decreased from 2.4 to 1.1%.

Effect of water temperature

Since fish are ectotherms, their metabolism is affected by water temperature. With increase in temperature, metabolism increases up to a peak reached at a temperature level which is characteristic of the fish species. Above this temperature there is a sharp drop in metabolism, usually associated with the death of the fish. Heath & Hughes (1973) found that with increase in temperature between 16 and 20 °C oxygen consumption by rainbow trout increases. Above the latter temperature the increase in oxygen consumption becomes more marked until it reaches 25 °C, when cardiac disorder causes death. In warmwater fish maximum metabolism is attained at higher temperatures of about 30–35 °C (Fry & Hart, 1948b; Beamish & Dickie, 1967), and in tropical species it may be even higher. Not less important than the peak metabolism temperature is the rate of increase in metabolism up to this peak. A common way to express the effect of temperature on various chemical and physical processes is Van't-Hoff equation:

$$Q_{10} = \left(\frac{L_1}{L_2}\right) \frac{10}{t_1 - t_2} \tag{25}$$

where L_1 = process rate at temperature t_1; L_2 = process rate at temperature t_2. Q_{10} therefore expresses the relative increase in the rate of the process with the increase in temperature by 10 °C. The Q_{10} of many biochemical processes is about 2, which means that with an increase of 10 °C these processes are doubled. Ivleva (1973) examined five species of fish from the Black Sea which were acclimated to various temperatures and stated that in all these fish the dependence of metabolism on temperature could be described by the Van't-Hoff equation. The Q_{10} calculated for these fishes at 5 °C intervals decreased from 2.5 to 2.0 as temperature rose. However, most researchers studying this relationship concluded that Van't-Hoff's equation is insufficient to describe the dependence of biological processes, including metabolism, on temperature. This led Krogh (1914) to present an empirical curve to express this relationship, which he called the 'normal curve' (Figure 18). Fry & Hart (1948b), Winberg (1956) and later others (e.g. Kamler, 1972) found that this curve also fitted the relationship between metabolism and temperature in fish. Beamish & Mookherjii (1964) found for goldfish a metabolism–temperature relationship resembling that of Krogh's 'normal curve' (Figure 18). The log of oxygen consumption at standard metabolism level increased linearly with temperature. Krogh himself did not try to define

Figure 18. Effect of temperature on the metabolism of fish according to the 'normal curve' of Krogh (1914) compared with other curves collected from the literature. For this comparison metabolic rate is expressed as percentage of the metabolism at 20 °C.

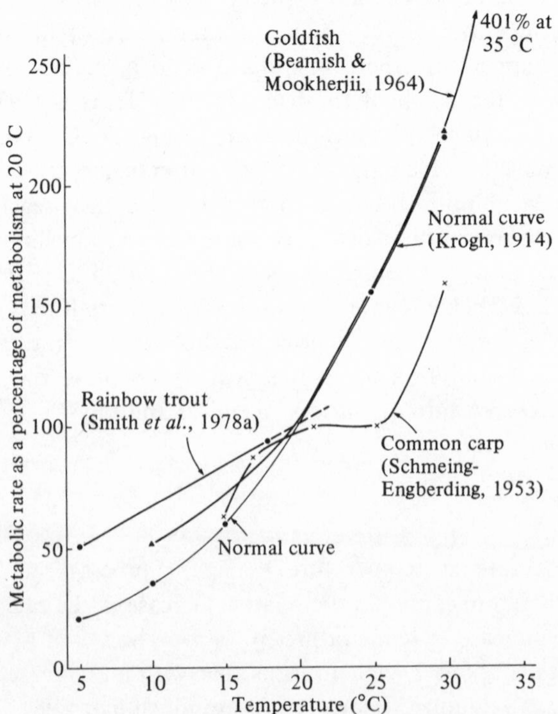

his curve by a mathematical regression, but several such attempts have been made by others. Backiel (1977) gives the following mathematical expression for the normal curve and for conversion of data on metabolism at 20 °C, which is the most common temperature quoted in the literature:

$$Q_t = Q_{20}\, 0.3\, e^{0.071t} - 0.24 \qquad (26)$$

where R_t and R_{20} are the rates of metabolism at temperature t °C and 20 °C respectively. Another way to express Krogh's normal curve has been taken by Krueger (1963, 1966). He tried at first to relate it to Q_{10} values, but found different values for the different temperature ranges (see also Winberg, 1971):

Temperature range (°C)	Q_{10}
0–5	10.9
5–10	3.5
10–15	2.9
15–20	2.5
20–25	2.3
25–30	2.2–1.8

He therefore tried another approach. He suggested a modification of Arrhenius' equation in which, in place of the absolute temperature (−273 °C), he introduced a lower value which he called the 'biological zero temperature'. His equation was then:

$$Y_t = \frac{m}{\frac{1}{n} / t-z} \qquad (26)$$

where Y_t is the biological process rate at temperature t; m = the maximal process rate; z = the biological zero temperature; and n = a very large number. None of these equations has been found to be of much use, which only goes to show that the issue is complex. This is confirmed by the significant differences which occur in the effect of temperature among the various fish species and in varied conditions.

Smith *et al.* (1978a), who studied the metabolism of trout by direct calorimetry, found that when the fish were acclimated to various temperatures for one week prior to the determination of metabolism, the relationship between metabolism and temperature was linear within the temperature range of 8–18 °C (Figure 18). The authors explained this deviation from Krogh's normal curve by possible changes in metabolic pathways in fish due to the increase in temperature, which are related to changes in *RQ*. This results in variation in oxycaloric value with the change in temperature which cannot be detected by indirect calorimetry. The authors suggest that measuring oxygen consumption at different temperatures may result in Krogh's normal curve, but direct calorimetry which measures the actual energy gives a more reliable regression.

When considering results on metabolic rate obtained for fish in warm regions, it is usually found that these rates do not conform with Winberg's values for fish of the same species, weight and temperature. Zaitschek (1960) studied the routine metabolism of common carp of 75–350 g and obtained much lower values than could be expected from calculations by Winberg's equation and Krogh's normal curve. Unpublished studies by the present author have shown similar discrepancies. According to these studies a common carp weighing 100 g, at a temperature of 25 °C, consumed 15.7 ml O_2/h, compared with 30 ml/h expected according to Winberg[1] and the normal curve for the same temperature. Winberg (1961) himself quoted studies on freshwater fish metabolism in tropical regions which showed considerably lower metabolic rates than could be expected from his equations and the normal curve. The rate for a temperature of about 30 °C in the tropics corresponded roughly with that at 20 °C in the temperate region. Other measurements of the effects of temperature on routine metabolism, carried out with different fish species, have also shown considerable deviations from Krogh's normal curve. Usually these showed a flattened middle portion of the temperature–metabolism curve. Increasing evidence is accumulating for the existence of such flattened curves for many fishes. Walsh *et al.*, (1983) found a flattened metabolism curve for American eel *Anguilla rostrata*, acclimated to a temperature range of 10–15 °C. Q_{10} for the increase in metabolism with the increase in temperature from 5 °C to 10 °C was 4.1, but from 10 °C to 15 °C it was only 1.12. Gershanovich (1983) found a flattened portion of the metabolism–temperature curve of young beluga (*Huso huso*) at the range of 16–20 °C. Among the cultured warmwater fishes Korovin (1976) found that with gradual increase in water temperature, at a certain temperature range the metabolism of young common carp is very stable. Caulton (1977) measured oxygen uptake by *Tilapia rendalli* at various temperatures and found a relatively flattened portion of the curve between 28 and 37 °C. Below and above this plateau metabolism increased logarithmically with temperature. A similar plateau, above 30 °C, has been reported to be characteristic of the oxygen uptake by *Sarotherodon mossambicus* (Job, 1969) which, like *T. rendalli*, inhabits shallow littoral areas of tropical lakes and shows a diurnal movement into warm water,

[1] Winberg considered routine metabolism to be twice the standard metabolism.

thus experiencing strong thermal changes during the day. The deviations from Krogh's curve brought above can be explained only by an adaptation of the fish to varying temperatures. Such an adaptation is also evident from results of a study by Moss & Scott (1964) who measured oxygen consumption by channel catfish in a flow-through respirometer. Water temperature in the respirometer was raised suddenly from 25 to 30 °C after the fifth day of the experiment. As a result, metabolism was increased substantially for a brief period, but then the fish showed a gradual and progressive decrease in rates of oxygen consumption, and on the fourteenth day they were only slightly higher than those encountered at 25 °C.

Adaptation may be vital for eurythermal fishes. It should be remembered that such fishes may live in an enviroment subjected to a wide range of temperature variations. Water temperature in ponds or shallow lagoons may vary, in certain geographical regions such as the subtropics, between 3–4 °C in winter and 32–33 °C in summer. According to Krogh's normal curve this would mean, for fish living in these conditions, an increase in metabolism by eight to ten times between winter and summer, which can be a significant physiological burden. Also from an ecological point of view it is doubtful whether, in natural conditions, the fish would be able to find sufficient food in summer to provide for both this high maintenance metabolism and growth. Brett (1965) found that metabolism of salmon (*Oncorhynchus* sp.) at 24 °C was six times as high as at 5 °C, but such differences are unusual for this stenothermal fish which lives in cold water, since 24 °C is near its upper lethal temperature limit. The situation is different with eurythermal fish. Their ability to live in a wide temperature range seems to indicate their ability to adapt their metabolic processes to temperature variation. On these grounds criticism has been made of the concept which Krogh's curve implies, which Wieser (1973) has called 'being at the mercy of the environment'. He argues that for molecular as well as ecological reasons it seems unlikely that such a concept is of general applicability. Adaptation of ectotherms to temperature changes can be achieved in two ways: (a) changes in the behaviour such as reducing activity to conserve energy; (b) adaptation of the internal metabolic processes.

Adaptation to temperature through activity

Most of the studies mentioned above on the effect of temperature on metabolism were related to routine metabolism, in which a certain energy expenditure for spontaneous activity is included. According to Fry (1968) no conclusions can be drawn from these results on the effect of temperature on standard metabolism since temperature also affects activity and as a result the energy expenditure for this purpose. The relationship between activity and temperature may differ from that of metabolism and temperature. Basu (1959) studied the effect of temperature on both standard and active metabolism of four species of fish (speckled trout, *Salvelinus fontinalis*; bullhead, *Ameiurus nebulosus*; common carp and goldfish). The rate of increase of active metabolism with temperature was always less than that of standard metabolism. However, different results were obtained by Brett (1964) who studied the effect of temperature on the metabolism of sockeye salmon (*O. nerka*), at different swimming rates. Determining the standard metabolism by extrapolation to zero movement (see discussion at the beginning of this chapter), he found that the standard metabolism of sockeye salmon did not change significantly between 10 and 15 °C, but active metabolic rates increased sharply between 5 and 15 °C, mainly due to increase in swimming speed. Above 15 °C there was a sharp cut-off in active metabolism rate due to the lack of oxygen required to sustain a higher activity. When, however, more oxygen was supplied (approximately 150% air saturation) there was a further increase in active metabolism with the increase in temperature above 15 °C.

When the nutrition of the fish in ponds is considered, both standard and active metabolism rates are meaningless, and it is the routine metabolism under natural conditions that counts. The food must supply the energy requirement for maintenance which includes normal spontaneous activity rather than that at maximum sustainable swimming speed (active metabolism). Brown (1957) showed that consumption of food by brown trout (*Salmo trutta*) to supply energy for metabolic requirements increased rapidly with an increase in temperature between 9 and 11 °C, but the change in metabolism became more moderate above 11 °C, and at temperatures near 20 °C there was no increase in metabolism. She explained this observation on the basis of reduced activity at the higher temperatures. Brett (1956) and Peterson & Anderson (1969) also noted that with the increase in temperature above a certain level, which according to him is near the upper lethal temperature, there is a decrease in activity of certain fishes such as salmon and trout. Interesting experiments in this connection were carried out by Schmeing-Engberding (1953). He found that when fish are held in a tank where a gradient of temperatures exists, most fish will be found in the section of the tank having a preferred temperature. When he measured routine oxygen consumption at different temperatures after a short acclimation period of 14–24 h, he found that the metabolism–temperature curve could be divided into three sections: (a) below the preferred temperature, where temperature has a considerable effect on metabolism; (b) a middle section which starts at the preferred temperature, where temperature has almost no effect on metabolism and the curve flattens; (c) the upper section, where again temperature affects metabolism. The flattened section of the curve may be very short, such as for the rainbow trout where it was only between 20 and 21 °C. In some fishes he found a wider flattened section, such as with common carp where it was between 21 and 25 °C (Figure 18). When the fish in Schmeing-Engberding's experiment were anaesthetized, the flattened section in the metabolism–temperature curve disappeared, and the curve corresponded, more or less, to Krogh's normal curve. Thus it seems, from the above experiment, that the main cause of flattening of the curve is activity, which was lower at the preferred range of temperature. Beamish & Mookherjii (1964) studied the effect of temperature on oxygen consumption by goldfish at standard and routine metabolism. Their results seem to be in line with the above theory. While they found a linear regression between the log of oxygen consumption at standard metabolism and temperature, oxygen consumption at routine metabolism, which reflects the requirement of energy for spontaneous activity, behaved differently. Up to 20 °C it increased at a higher rate than that of standard metabolism, but between 20 and 30 °C it decreased, to increase again above 30 °C (Figure 19). It should be remembered that the requirement for energy increases exponentially with increase in swimming speed (Fry & Hart, 1948a; Brown, 1957; Brett, 1964, 1970), therefore even a moderate decrease in swimming activity may result in a considerable conservation of energy.

Physiological temperature adaptation

Attributing temperature adaptation to activity cannot, however, explain all deviations from Krogh's curve, since standard metabolism, which does not include energy cost for activity, also showed the discrepancies in metabolism–temperature relationships between fish in the warm and cold geographical regions, as well as between cold and warm acclimated fish. Fry (1958) quoted Shurmann who measured the standard metabolism of common carp in the temperature range of 10–30 °C. Oxygen consumption by fish acclimated to 26 °C was 30% lower than that of fish acclimated to 5 °C, when both were measured at the same temperature. Almost identical results were obtained by Kanungo & Prosser (1959a) for goldfish. It is obvious that variations in activity can, at best, be only a partial explanation of the above differences. As more sophisticated biochemical techniques were developed it became increasingly apparent that more complex metabolic processes are involved in acclimation to changes in temperature (Kanungo & Prosser, 1959b; Hochachka & Somero, 1973).

Some of the intermediate metabolic processes associated with temperature adaptation have already been discussed above (Chapter 3). These are associated mainly with the change of enzymatic activity and membrane permeability. In cold conditions, adaptation can be achieved by an acceleration of enzymatic processes as well as reduction of membrane resistance (the converse applying for warm conditions). Experimental evidence indicates that enzymatic changes associated with metabolic temperature compensation fall into several categories: (a) the quantities of certain enzymes may be increased during cold acclimation; (b) temperature may affect the enzyme–substrate affinity and thus its activity; (c) different enzymes of the same function appear at different temperature ranges (d) changes in isozyme pattern upon acclimation which results in the production of these isozymes exhibiting the minimum *K*m value and higher activities at the approximate acclimation temperature (Hochachka, 1962; Whitmore & Goldberg, 1972b; Hazel & Prosser, 1974; Tsukuda, 1975; Moon, 1975; Somero, 1975; Lapkin *et al.*, 1983).

The role of biomembranes in temperature adaptation was suggested by many investigators who studied the temperature dependent alterations in

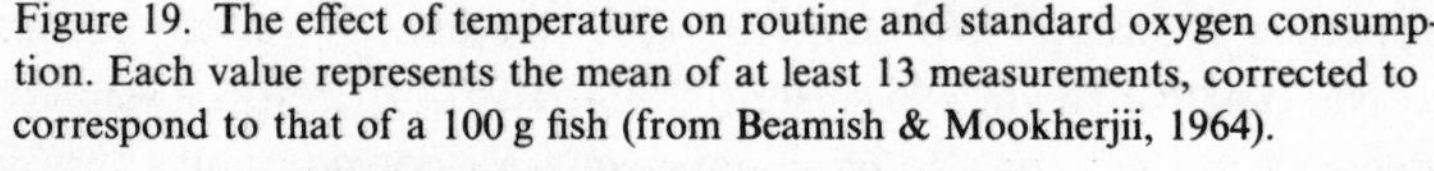

Figure 19. The effect of temperature on routine and standard oxygen consumption. Each value represents the mean of at least 13 measurements, corrected to correspond to that of a 100 g fish (from Beamish & Mookherjii, 1964).

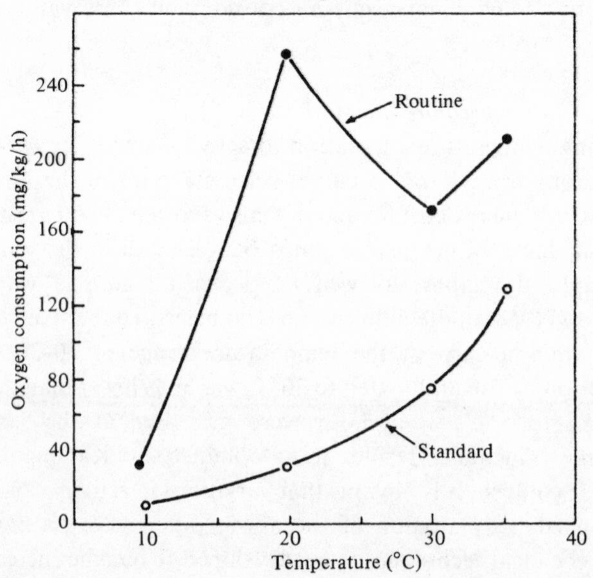

membrane composition. These membranes may involve mitochondrial membranes in fish gills (Caldwell & Vernberg, 1970), membranes in the digestive tract (Smith & Kemp, 1971) and other membranes. It appears that membrane adaptation to cold temperature is involved with the following: (a) increased activity of the liver in lipogenesis and gluconeogenesis; (b) a certain proportion of the lipids produced in the liver are associated with a reorganization of the phospholipids in cell membranes; (c) for this, a larger proportion of highly unsaturated fatty acids of the ω3 group is required, and since this group cannot be synthesized endogenically, the essential linolenic acid is required (see Chapter 7); (d) increased pentose phosphate shunt activity provides for the NADPH required for the above lipogenesis; (e) increased oxygen requirement for all these functions. The last mentioned condition is well demonstrated in experiments by Krebs (1975) with the gibel carp (*Carassius auratus gibelio*), and by Krebs & Brandt (1975) with goldfish (*C. auratus*). Adaptation of oxygen consumption by white muscle and gills to temperature changes was evident only in well oxygenated water, while no adaptation occurred in hypoxic water.

From the previous discussion it seems that membrane phospholipids play an important role in temperature adaptation. Irvine *et al.* (1957) claimed that the addition of phospholipids and cholesterol to the diet increased the resistance of goldfish to cold. No corroboration of this by further experiments has been found.

Results of studies on temperature adaptation could not always be satisfactorily explained. Some researchers found adaptation to temperature only in the intact fish, but not in isolated fish tissues. Ekberg (1958) did not observe any adaptation effect on respiration of isolated tissues of goldfish such as brain, liver and heart. Jankowsky (1964a, 1966), who studied temperature adaptation in the eel *Anguilla vulgaris* and obtained similar results, attributed this to hormonal or nervous effects which could not be manifested in the isolated tissue. Prosser *et al.* (1965) held European eel (*A. anguilla*) in a glass tube and conducted experiments on the adaptive capacity of the head and tail sections, separately, to temperature variations. They came to the conclusion that the metabolism of the intact eel is determined by the temperature of the head, indicative of hormonal or nervous control. They suggested that the spinal cord, by its tonic discharge, causes enzymatic changes in the isolated muscle. Nevertheless, adaptation of various isolated tissues to varying temperatures has been found by a number of researchers. Freeman (1950, quoted by Prosser, 1955) found that goldfish brain acclimated at 20 °C required more oxygen than a brain acclimated at 27 °C, when both were held at the same temperature. Evans *et al.* (1962) found this to be also true for acclimated rainbow trout where the brain showed a complete metabolic compensation to temperature (O_2 consumption at 8 °C was equal to that at 16 °C). Liver even showed an overcompensation (O_2 consumption at 8 °C was higher than at 16 °C), while no compensation occurred in gill respiration. Total respiration of the intact fish showed a partial temperature compensation. The authors suggested that complete compensation in brain would maintain nervous coordinations and motor conduction at optimal levels, thus permitting a large degree of independence of locomotor activity.

Some of the differences discussed above may have been caused by the experimental conditions. Thus, Jankowsky (1964b) showed that muscle tissue of tench (*Tinca tinca*) held in Ringer solution did not show any adaptation to temperature, but when a small amount of tench's blood was added the tissue reacted to the previous adaptation temperature of the blood. With cold adapted blood the metabolism of the tissue was higher. Jankowsky (1964b) attributed this to the change of the nutritional medium by the addition of the blood. Also it seems that some adaptation processes are related to changes in the morphological characteristics of the fish. Houston & De Wilde (1968) found that in common carp red cell number and volume, as well as haemoglobin concentration, increased at low temperature, thus indicating an increase in blood oxygen capacity. This is probably associated with adjustments of the oxidative metabolism to changing temperatures. Wodtke (1974) determined the effects of acclimation to low temperature on oxidative metabolism of eels. He found that in eels acclimated to 7 °C the relative weight of the liver increased about 1.8 times compared with eels acclimated at 25 °C. A comparable increase occurred in the proportion of red muscle to white muscle. The effect of low temperature on liver weight was reversible. Cold acclimation of the eels also increased the mitochondrial protein content in the liver and red muscle by about 1.5 and two times, respectively, compared with eels acclimated to 25 °C. Following the acclimation to the lower temperature the respiration rates of liver mitochondria with endogenous substrates were reduced to 64.5%, but those of red muscle mitochondria were enhanced to 240% compared with fish acclimated to 25 °C. The somewhat contradictory results of these experiments indicate that the processes involved with adaptation to changing temperature still require further study.

According to Prosser (1955) adaptation of ectotherms to temperature changes can be expressed in two forms: (a) a shift in the metabolism–temperature curve, to the left in the case of cold acclimation or to the right in the case of warm acclimation, which he called 'translation'; (b) change in the slope of the metabolism–temperature curve, which flattens in its middle section – 'rotation'. The first kind of adaptation seems to be more common in species having a wide geographical distribution which live in varied climates. Due to the temperature compensation, fish of the same species from different climatic regions will show similar metabolic rates. When under the same temperature, those from colder climates will show a higher metabolism than those from warmer climates. Bullock (1955) in his extensive review on this subject gives many examples of the physiological differentiations of this kind occurring in the same species or genus living in different environments. The second kind of adaptation (rotation) is associated more with the acclimation of the individual fish. This adaptation does not have to be expressed throughout the temperature range in which the fish lives. It is usually restricted to the middle of this range about the preferential temperature, where the curve flattens. Meuwis & Heuts (1957) measured the rate of respiratory movements of the opercula in common carp after acclimation for three to 14 days to various temperatures. Within the range of 3–20 °C there was a sharp increase in the respiratory movements with increase in temperature. Within the range of 20–32 °C the respiratory movements of the small fish scarcely changed at all, and those of large fish increased with temperature only to a small degree. Above 32 °C there was, again, a sharp increase in respiration rate with the increase in temperature. It thus appears that for common carp the flattened portion of the curve where change in temperature will not greatly affect metabolism is between about 20 and 30 °C.

An important question in this respect is, how long does it take the fish to acclimate? Wieser (1973) points out that with the change in temperature two kinds of response occur: some compensatory reactions, possibly associated with modulation in enzyme activities, are immediate, while others set in after days or even weeks. Houston & Rupert (1976) found an immediate response of the haemoglobin system in goldfish, *Carassius auratus*, to temperature changes. The abrupt transfer of fish acclimated at constant 23 °C to 3 °C, and vice versa, led to the appearance or disappearance of an additional haemoglobin component within three hours. The slower adaptive processes are probably associated with complicated structural changes involving RNA and protein synthesis or modifications in membrane structure. Precht (1955) gives a period of one to five days for acclimation of organisms to changes in temperature, which seems to be quite short. Peterson & Anderson (1969) stated that a period of two weeks was required to acclimate the Atlantic salmon (*Salmo salar*). Beamish & Mookherjii (1964) acclimated goldfish by changing the temperature by not more than 1 °C/day and after the desired temperature was reached a further period of at least two weeks was allowed before measuring their oxygen consumption; yet oxygen consumption rate resembled Krogh's 'normal curve', which indicates little or no acclimation in this case. It appears that a longer period than that applied above may be necessary for the acclimation of some fish species to changes in temperature. Would the seasonal changes in temperature be slow enough for such acclimation? Bullock (1955) brings many examples of ectothermic animals having a lower metabolism in summer than in winter (when determined at the same experimental temperature), but only a few of these related to fish. Results of experiments by Roberts (1964) suggest that such acclimation is achieved as a result of an interaction between various seasonal factors, such as photoperiod and temperature, rather than the changes in temperature alone. Hoar (1956), working on the acclimation of goldfish and its resistance to lethal temperatures, showed that when the goldfish were acclimated to different photoperiods but equal temperature, the short day acclimated fish were more resistant to cold and long day acclimated fish were more resistant to heat. He therefore concluded that the seasonal variation in thermal tolerance is also controlled by photoperiodism. However, much more study is needed on this subject. It should be mentioned here that adaptation to temperature may or may not include other living functions besides metabolism, such as growth, reproduction, etc. These may show different relationships with temperature than metabolism.

Energy cost of movement

It is clear that with the increase in movement, metabolism and oxygen consumption also increase. Measurements of standard metabolism carried out in regular respirometers will not reflect this extra energy requirement. Active metabolism can, however, be measured in respirometers where fish are forced to swim and where their swimming speed can be measured. These respirometers can be used for constructing curves describing the relationship between swimming speed and metabolism. Such respirometers are of two kinds: (a) revolving respirometers, where fish move against the direction of rotation, the faster the rotation rate the faster the swimming speed (Figure 20(*a*)), in some of these the fish receive an extra stimulation for movement by electric shocks (Basu, 1959); (b) water tunnel (Figure 20(*b*)), where fish swim against a water current of varying velocity (Rao, 1968; Brett, 1973).

In order to facilitate the comparison between fish of different lengths, swimming speed is usually expressed as bodylengths per second (*L*/sec). Due to the special interest in the amount of energy a salmon expends during the upstream ascent for reproduction, most experiments have been carried out on salmonids (especially sockeye salmon, *Oncorhynchus*

Figure 20. Active metabolism respirometers: (*a*) annular swimming chamber; (*b*) swimming flumes (from Beamish, 1978).

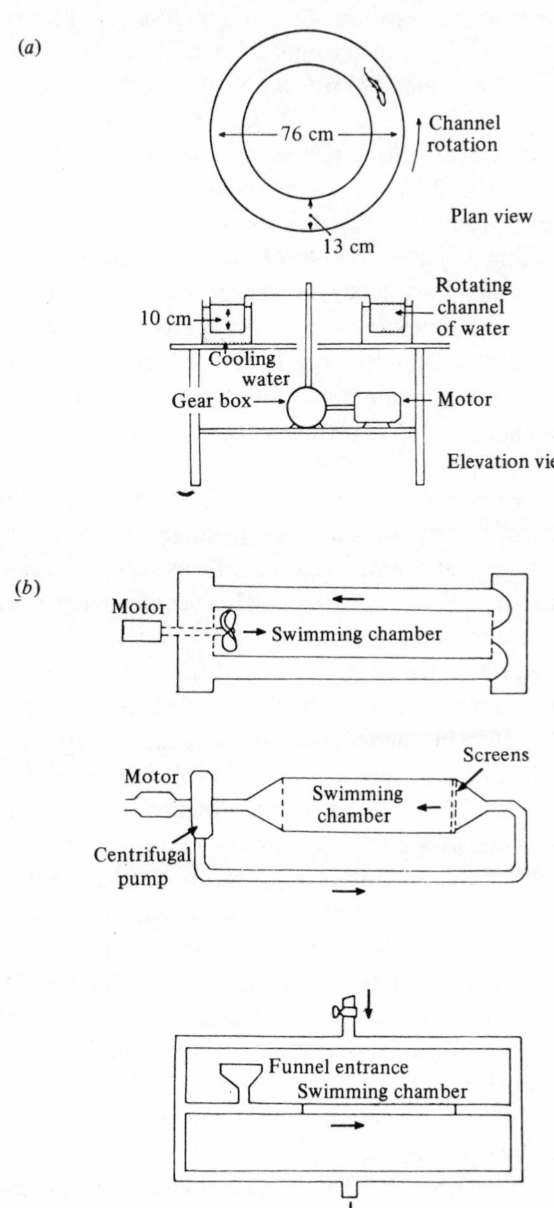

nerka). According to Brett (1964), in given environmental conditions, the relationship between swimming speed and metabolism can be expressed by the following exponential regression:

$$\log Q = a\, b^{x} \qquad (27)$$

where Q = metabolism; x = swimming speed in L/sec; and a and b are constants. a is the metabolism at zero movement, and should, therefore, be equal to the standard metabolism of the fish of the given weight. According to Kausch (1972) the value b is similar for various fish species, but varies with temperature. At a temperature of 20 °C this value was equal to 1.66 for *O. nerka*, 2.04 for *Lepomis gibbosus* (Brett & Sutherland, 1965), and 1.86 for *Cyprinus carpio* (Kausch, 1972).

Very high swimming activities over 4–6 L/sec can be sustained for only short periods, after which the fish becomes fatigued. Active metabolism at the maximum speeds may elevate metabolic rates as much as 15-fold of the standard metabolism (Beamish, 1964b; Brett, 1964). However, these maximum speeds cannot be maintained for any length of time, and the fish becomes fatigued, usually within half an hour of continuous swimming.

Several difficulties arise with the use of equation (27) for the estimation of energy requirement. The main problem is defining normal activity and swimming speeds in natural conditions, especially in standing waters. The metabolic costs associated with activities such as escape from predators, the capture of prey, migration and territorial defence have been studied extensively (see reviews by Beamish, 1978; Brett & Groves, 1979). However, these activities are not regular, and are often exceptional in the routine behaviour of fish in ponds. In this respect they completely differ from migrating fishes such as the salmon and their associated sustained swimming performance. The quantitative measurement of natural routine activity in the water body itself entails enormous practical difficulties.

Various methods have been employed to measure movements of fish such as direct observation, photography, etc. (Beamish, 1978). Bainbridge (1958), working with trout (*Salmo gairdneri*), dace (*Leuciscus leuciscus*) and goldfish (*Carassius auratus*) determined the relationship between tail-beat frequency, tail-beat amplitude, total fish length and swimming speed. The relationship between tail-beat frequency and swimming speed was confirmed later by Puckett & Dill (1984). Feldmeth & Jenkins (1973) tried to utilize this relationship to determine natural swimming speed and energy expenditure by counting the caudal fin beat frequency of trout in a mountain stream. It is obvious that the long chain of relationships involved, metabolism–activity–swimming speed–tail-beat frequency, reduces the reliability of the measurement. Moreover, pond fishes, which do not swim constantly against a current, use their pectoral fins more frequently for locomotion, rather than the tail. Also, fish do not swim at a constant rate, and their activity may, and usually does, vary during the day. Although Feldmeth & Jenkins (1973) state that swimming speeds of trout did not vary statistically with time of day, other researchers did find diurnal cycles in the activity of trout and other fishes. The diurnal cyclic pattern of activity can have one or two peaks, usually at dawn and dusk, or only one peak during the day or night. Swift (1962) showed that brown trout have a low activity during the night and increased activity during the day, with a pronounced increase at dawn. Spencer (1939) studied the diurnal activity rhythms of 10 species of freshwater fishes. While some showed a uniform pattern of activity, others, such as common carp and goldfish, showed a monophasic diurnal pattern. Young common carp were very active by night and less active by day. Goldfish (*C. auratus*) were more active by day than by night. Spencer (1939) also observed long cycles of 4–7 days, leading gradually to a peak of activity followed by a sudden drop in activity. An annual cycle of activity was found in brown trout by Swift (1962) with a maximum during May and June. Temperature, light, food and oxygen concentration may affect the extent and pattern of activity, as would aggressiveness of fish towards each other (Spoor, 1946, Swift, 1962).

According to Fry (1947), activity is determined by the availability of metabolic energy above that already committed to the essential maintenance metabolism. By deducting the amount of energy (or oxygen consumption) directed into maintenance without movement (standard metabolism), the balance, when swimming, can be ascribed to the mechanics of propulsion. This balance of 'available energy' was termed by Fry (1957) 'scope for activity' and is also sometimes called 'metabolic scope'. Thus activity is dependent on the oxygen available to the fish. Brett (1964) and later Nightingale (1975) found that, with increase in temperature, energy expenditure for essential metabolic processes increases and less remains for spontaneous activity. At temperatures up to 15 °C active metabolic rates of sockeye salmon were found by Brett (1964) to be about 10–12 times the standard level. Above 15 °C the ratio fell, and at an acclimation temperature of 24 °C it was four times the standard rate. The cost of movement, as reflected by the ratio of active to standard metabolism, also increases with the increase in fish weight, while the relative ability to maintain sustained speeds decreases as size increases (Brett, 1965).

In view of the numerous factors affecting spontaneous activity, it is very hard to determine an 'average activity' or an 'average swimming speed' and relate it to the metabolism found in the laboratory as Feldmeth & Jenkins (1973) did for trout. It seems more practical to aggregate the extra energy expenses associated with life in natural conditions, such as *SDA*, movement etc, and correct routine metabolism as found in the laboratory for all these expenses together. This has been attempted and will be discussed below.

Utilization of metabolizable energy for maintenance

Standard metabolism as measured in the laboratory interests the practical fish farmer only as a source of information leading to a better understanding of the relationship between metabolism and the various factors affecting it. It cannot be utilized for compiling feeding charts until more information is available on maintenance metabolism in natural conditions, on how this relates to metabolic rate measurements in the laboratory, and how much food is needed to supply this energy requirement, i.e. the partial apparent efficiency of food energy for maintenance (E_{pm}). Generally, maintenance metabolism under natural conditions is higher than standard metabolism due to three major factors: activity, *SDA*, and lower efficiency of utilization of food for energy as compared to body tissue used for energy. The first of these has been discussed above, the other two have been considered briefly in Chapter 1, but in view of their importance they deserve a more detailed discussion.

Specific dynamic action (*SDA*)

Following food ingestion the rate of oxygen consumption by fish increases and then gradually declines back to its resting level. This is associated with the extra energy produced for transportation of food in the alimentary tract, its digestion, absorption and post absorption metabolic processes related to the ingested food. The duration of the increased metabolic rate varies among fish species and has been shown to be dependent on several factors such as fish weight, fish density, meal size, diet composition and temperature (Muir & Niimi, 1972; Beamish, 1974; Caulton, 1978; Smith *et al.*, 1978b; Vahl & Davenport, 1979; Brett & Groves, 1979; Jobling, 1981; Tandler & Beamish, 1981; Medland & Beamish, 1985). It seems to be linked with the digestion rate (Beamish, 1974); Jobling & Davis (1980) found a clear correlation between the two. Post-prandial increase in metabolism was observed for slightly longer than the time it took the food to be transported through the gut.

It may not be easy to distinguish the above *SDA* effect from other variations in metabolic rate. Brett & Zalla (1975) measured the diurnal fluctuations in metabolic rate of sockeye salmon fed daily between 0830 and 0930 h, and compared it with that of starved fish. A similar study was conducted by Sondermann *et al.* (1985) with common carp. Metabolic rate of the fed fish showed a large diurnal fluctuation with a minimum at 0300 h, rising in advance of feeding time to reach a maximum just at the start of the feeding (0830 h) in Brett & Zalla's study, and soon after feeding time in the study of Sondermann *et al.* A similar pulse was observed for the starved fish, diminishing and becoming more diffused with food deprivation time. The increased metabolism was explained by the authors as resulting from increased spontaneous activity of the excited fish anticipating food.

Mann (1968) found that the post-prandial oxygen requirement of fed trout was 15–30% higher than of starved fish. Most observers, however, show the peak in metabolic rate to be approximately double that of fasting level (e.g. Jobling & David, 1980). Tandler & Beamish (1981), working with largemouth bass (*Micropterus salmoides*) found that maximum oxygen uptake by fish of about 100 g fed to satiation equalled that of active metabolism, but maximum rate of oxygen uptake at satiation of a fish of 250 g reached only about 50% of their active metabolism.

Krayukhin (1962) measured oxygen consumption by common carp fingerlings fed different diets. He showed that large variations in oxygen uptake can occur at different diets and temperatures. When fed rapeseed oil meal, oxygen uptake at 18–20, 5–7, 4–5, and 3–4 °C was, respectively, 2.58–2.67, 1.43–1.48, 1.25–1.27 and 1.05–1.12 times higher than that of starved control fish. When, however, the fish were fed *Tubifex* sp., the increases in oxygen uptake at the corresponding temperatures were 2.16–2.26, 1.37–1.38, 1.13–1.38, and 1.05–1.06 times as great as that of starved fish, respectively. At 1–2 °C the control did not differ from the experimental fish in oxygen utilization since the fish did not eat. Also Tandler & Beamish (1980) found differences in the post-prandial maximum oxygen uptake with different diets. Maximum oxygen uptake was always positively correlated with energy intake. For a given energy intake the peak of oxygen uptake was highest for a diet containing 25 or 50% protein and lower for all protein or all carbohydrate diet.

The magnitude of *SDA* can be determined by plotting the curve of increased oxygen consumption after feeding against time until it subsides to the pre-feeding rates, then integrating the area beneath the curve. Many workers have found that the magnitude of *SDA* increased linearly in relation to meal size (Hamada & Ida, 1973; Schalles & Wissing, 1976; Caulton, 1978; Vahl & Davenport, 1979; Jobling & Davis, 1980). Others, however, have reported an exponential relationship (Averett, 1969,

Table 24. *SDA energy consumption relative to ingested food energy (in %)*

Species	Food type	*SDA*	Remarks	Source
Blennius pholis	Mussel flesh	7.5–11.9 (Av. 9.7)	Estimated energy of food	Vahl & Davenport (1979)
Coregonus schinizi	*Artemia salina*	28.7	Fish at larval stage	Dabrowski & Kaushik (1984)
	Compounded diet	4.9–12.7	Fish at larval stage	
Cyprinus carpio	Pelleted feed	16		Yarzhombek *et al.* (1983)
Histrio histrio		15.2–36.2 (Av. 23.3)		Smith (1973)
Kuhlia sandvicensis	Tuna flesh	16–19	Estimated energy of food	Muir & Niimi (1972)
Lepomis macrochirus	Mayfly larvae	4.8–24.4 (Av. 12.7)	No significant differences at temperatures of 15, 20 and 25 °C	Pierce & Wissing (1974)
	Pelleted feed	7.5–32.3 (Av. 14.9)	No significant differences for diets of varying composition	Schalles & Wissing (1976)
Micropterus salmoides	Emerald shiners	14.2 ± 4.2	Values from various sizes at swimming speeds; fed 2–8% body weight/day	Beamish (1974)
	Pelleted feed	11.3 ± 1.3	Average of various fish weights at 25 °C	Tandler & Beamish (1981)
Oncorhynchus rhodorus	Trout and ayu flesh	9.5–25.9 (Av. 16.9)	Temperature varies from 9 to 13 °C	Miura *et al.* (1976)
Pleuronectes platessa	Moist feed	10.0–18.0 (Av. 13.1)	Varies with dietary composition but indifferent to temperature	Jobling & Davis (1980)
Salmo gairdneri	Pelleted feed	8–12 (Av. 9.7)	Influenced by dietary composition	Cho *et al.* (1976b)
	Pelleted feed	9.7		Yarzhombek *et al.* (1983)
Tilapia rendalli	Aquatic plants	4.9–13.8 (Av. 9.5)	Values increase with increasing temperature	Caulton (1978)

quoted from Tandler & Beamish, 1979; Tandler & Beamish, 1979). In the first case it is possible to express *SDA* in terms of proportion of the energy ingested. Data on *SDA* expressed in this way were collected from the literature and are presented in Table 24. The data show quite a large range of variation, which indicates that *SDA* is a compounded effect. Tandler & Beamish (1981) have shown that in largemouth bass (*Micropterus salmoides*) *SDA* is negatively correlated with fish weight. They also have shown that not only the maximum oxygen uptake but also the magnitude of apparent *SDA* is affected by diet composition (Tandler & Beamish, 1980). *SDA* was positively correlated with the proportion of protein in the diet. When largemouth bass received 0.75 kcal gross energy in food containing 100% protein, *SDA* was 40 mg O_2/meal, but when it had ingested an equivalent amount of energy as carbohydrate, *SDA* was only 9.24 mg O_2/meal.

Tandler & Beamish (1979) attempted at differentiating the mechanical and biochemical components of *SDA*. The mechanical *SDA*, which is the energy cost for transport of a meal through the alimentary canal, was measured as post-prandial energy requirement in largemouth bass (*M. salmoides*), fed a non-digestible (high cellulose) diet. This was compared with the *SDA* of fish fed a standard diet and the difference was taken as biochemical *SDA*, assumed to be associated with the post-absorptive catabolic and anabolic processes. Mechanical *SDA* increased asymptotically approaching a plateau already with quite small meal sizes. Biochemical *SDA* rose exponentially with the increase in ingested energy. Mechanical *SDA* was only 10–30% of the total *SDA*. According to Tandler & Beamish (1980), the higher the protein content of the diet the higher the proportion of the biochemical to the mechanical *SDA* in the total apparent *SDA*. Also Jobling & David (1980), working with plaice (*Pleuronectes platessa*) fed kaolin, concluded that the contribution of mechanical *SDA* to the post-prandial rise in metabolic rate is small. This led Jobling (1981) to the obvious conclusion that *SDA* should be considered in terms of absorbed rather than ingested energy. When he considered previous results of Jobling & Davis (1980) on *SDA* of plaice in respect to the amount of digestible nutrients, it was apparent that the relative *SDA* tends to increase with increasing levels of digestible energy. Tandler & Beamish (1979) and Jobling (1981) interpreted this as indicating the higher energy expenditure for the synthesis of body tissues as compared to catabolism of food for energy.

Efficiency of food utilization

None of the studies mentioned above differentiated between the cost of food utilization for maintenance and that for growth. They also did not differentiate between the cost of energy for feed consumption (*SDA*) and that for the activity associated with it. Kausch (1968) found that the activity increases with feeding. The metabolic rate of fed common carp, at 20 °C, was 73% higher than that of starved fish. An identical difference has been found by the present author (unpublished data). Oxygen consumption at routine level by starved common carp, at 25 °C, was 20.9 mg O_2/h as compared to 36.2 mg O_2/h by fed fish. In both cases the difference is larger than could have been expected from *SDA* alone, according to the discussion in the previous section. The higher oxygen consumption by fed fish can also be attributed to the extra movement of the fed fish, but it is very difficult to separate this energy expenditure from the *SDA*. It seems, therefore, more rewarding to aggregate, as was already suggested above, all extra expenditure of energy in natural condition and to express it as a ratio of the standard metabolism, information on which is more common in the literature.

An interesting study on this subject was conducted by Huisman (1976) with common carp and with rainbow trout. Each of the treatments in his experiments involved 19 fish held in large tanks (275 l) with a constant flow of water through them, so that the fish could move about freely. The fish were fed different levels of feed of known caloric value. Oxygen consumption, changes in body weight and changes in the energy content of the fish were also measured. Data calculated from Huisman's results, which enable the determination of the 'correction factor' of the standard metabolism for conditions approaching those in nature, are presented in Table 25.

Table 25. *Maintenance metabolism of common carp at 23 °C as determined by Huisman (1976) from the change in body weight and composition and from oxygen consumption of starved fish and fish fed 1% their body weight/day*

Feeding level (%) body wt/day)	Fish wt at start of expt (g)	Fish wt at end of expt (g)	Average fish wt (g)	Daily loss in calories (cal/day)	Caloric value of O_2 consumed (cal/day)[a]	Gross energy of food (cal/day)
0	42.1	36.1	39.1	−828.6	774.2	0
1	40.8	48.4	44.6	−114.3	1203.8	1785.7

Note: [a] Calculated from Huisman's data using an oxycaloric value of 3.4 cal/mg O_2 (see discussion above).

In order to be able to compare the metabolism of fish of different body weights, the specific metabolism should be calculated. From equation (16) it follows that the specific metabolism (energy requirement by a fish of 1 g/day) is:

$$a = \frac{Q}{W^b} \tag{28}$$

or, to put it in words, for the specific metabolism of a fish, the actual energy expenditure should be divided by $W^{0.8}$, which some have also called the 'metabolic weight'.

Specific metabolism of the starved fish in Huisman's experiment (Table 25), when calculated from oxygen consumption, was $774.2/39.1^{0.8}$ = 41.2 cal/day, while that calculated from the loss of body energy was $828.6/39.1^{0.8}$ = 44.1 cal/day. These two values are very close, indicating that the two methods agree quite well. The average value between the two – 42.7 cal/day – is about 15% higher than the standard specific metabolism calculated from Winberg's data and given in equation (24). This difference may be attributed to activity, since in both cases the fish were not fed.

Also, the partial efficiency of the food for maintenance (E_{pm}) can be calculated from the data in Table 25, using equation (5). Due to feeding at a rate of 1% of body weight/day, the fish lost less of their body energy than starved fish. In order to compare the difference in energy loss with the food energy ingested by the fish, the two groups of fish must first be brought to the same weight basis, i.e. to their specific metabolism. 'Specific' energy loss of the fed fish was calculated as: $114.3/44.6^{0.8}$ = 5.5 cal/day. The difference in the calorific value of the lost tissue between fed and starved fish is thus 44.1 − 5.5 = 38.6 cal/day. Huisman gave the gross calorific value of the feed consumed by the fish (trout pellets) which caused the reduction in the loss of tissue energy 1785.7 cal/day. However, not all this energy was absorbed by the fish. Some was egested in faeces and some was excreted as catabolites. Taking the composition of the pellets given by Huisman, using the digestibility coefficients for each of the nutrients from Appendix I, and assuming metabolic excretion losses of nitrogenous compounds from protein to be 10% of the digestible protein energy (see Chapter 3), the proportional contribution of energy from each of the nutrients was calculated. The metabolizable energy of the whole diet was thus found to be 1393 cal/day, or 78% of the gross dietary energy. It may be assumed that energy intake is proportional to energy requirement. This is supported by the study of De Silva & Balbontin (1974) who found that the weight exponent between food intake and body weight of young herring (*Clupea harengus*), at 14.5 °C, was 0.744, which is near the 0.8 weight exponent used here. The metabolizable energy of the ingested feed was treated, therefore, in the same way as was the metabolic rate, and was likewise divided by the metabolic weight. This gave: $1393/44.6^{0.8}$ = 66.8 cal/day (for a fish of 1 g weight). The efficiency of food utilization for maintenance can now be calculated as:

$$E_{pm} = \frac{38.6}{66.8} = 0.58$$

'Specific' dietary energy requirement for maintenance can be calculated from Huisman's data:

$$[M] = \frac{\text{Specific metabolism of starved fish}}{E_{pm}} = \frac{42.7}{0.58} = 73.6 \text{ cal/day}$$

This value includes losses for activity, *SDA*, or any other loss associated with life under near natural conditions. The above value may be compared with the value of 37.2 cal/day given as specific metabolism at the standard level in equation (24). It follows that the 'correction factor' of standard metabolism to routine metabolism under natural conditions is, in this case: 73.6/37.2 = 1.98.

A similar study was carried out by Huisman (1976) on rainbow trout. Following the same procedure outlined above for common carp, it could be found that the specific metabolism of starved fish (average between that calculated from weight loss and that found by oxygen consumption) was 32.3 cal/day. This value is lower than that for common carp as also from that of Winberg for trout (54.0 cal/day, see p. 119). The last can be attributed to the lower temperature (15 °C) in Huisman's study as compared to Winberg's reference temperature of 20 °C. In spite of these differences, the E_{pm} was 0.59, very near that calculated for common carp.

The above calculations of the 'correction factor' for common carp are supported by a number of studies carried out with this fish. Schaeperclaus (1966a) studied the metabolism of common carp of an average weight of 300 g held in aquaria and fed at maintenance level. The energy content of the daily food ration was 6.3 kcal, compared with 3.57 kcal/day according to equation (24), or 1.76 times higher. Kevern (1966) determined the maintenance feeding level of common carp. He used fish weighing about 35–2663 g and referred to their arithmetic mean, which may be questionable. However, since most of the fish were within the weight range of 100–200 g, the error should not be too large. The daily food requirement for maintenance found by him for a fish of an average weight of 137 g was 3.7 kcal, compared with 1.9 according to equation (24), or 1.95 times higher. Similarly, Kausch (1968) found the metabolism of common carp, when fed and swimming freely in the respirometer, to be 1.9 times higher than Winberg's standard metabolism. The data given above and those found by a number of authors for various other fish species are tabulated and presented in Table 26. The average E_{pm} resulting from the data in the table is about 0.55, and the resulting average correction factor for converting standard metabolism as determined in the laboratory to maintenance metabolism in natural conditions is 1.90. Winberg (1956) himself suggested doubling the standard metabolism to obtain maintenance metabolism under natural conditions, which is quite near the value given above, although Winberg's suggestion was based on only a few, relatively incomplete, experimental data.

Brett (1970) shows a much larger difference between the standard and maintenance metabolism than suggested above, for sockeye salmon (*Oncorhynchus nerka*) during its homeward migration, when the velocity of its swimming is high. In such conditions maintenance metabolism was four to five-fold higher than standard metabolism of fasting fish. However, such conditions are exceptional for warmwater pond fishes. Even with the salmon, when swimming speed was limited to 2 *L*/sec and with an intermediate feeding level (3% of body weight/day), the requirement of energy for maintenance in Brett's (1970) experiments was only somewhat higher than double the standard metabolism.

Table 26. *Partial utilization efficiency of food for maintenance (E_{pm}) and correction factor for converting routine metabolism to maintenance metabolism under natural conditions, as calculated from experimental data in the literature*

Fish species	E_{pm}	Correction factor[a]	Source
Cottus perplexus	0.5	2.0	Warren & Davis (1967)
Cyprinus carpio	—	1.76	Schaeperclaus (1966b)
	—	1.9	Kevern (1966)
	0.58	1.98	Huisman (1976)
	—	1.9	Kausch (1968)
	—	2.08	Yarzhombek *et al.* (1983)
Micropterus salmoides	0.5	2.0	Niimi & Beamish (1974)
Salmo clarki	0.6	1.67	Warren & Davis (1967)
Salmo gairdneri	0.59	1.69	Huisman (1976)
	—	1.98	Yarzhombek *et al.* (1983)
Red tilapia	0.5	2.0	Hepher *et al.* (1983)
Average		1.9	

Note: [a] Where only the maintenance ration was given, correction factor was calculated relative to routine metabolism (Equation 29). Where E_{pm} was determined, the correction factor is the reciprocal of the E_{pm}.

The corrected equation expressing metabolic energy expenses for maintenance of common carp under natural conditions, which takes into account natural movement, *SDA* and other losses which may occur in these conditions will thus be:

$$M \text{ (cal/day)} = 37.2 \times 1.90W^{0.8} \quad (29)$$

For the above equation it is assumed that the weight exponent (0.8) for standard or routine metabolism is equally suitable for fed fish under natural conditions. Brett & Groves (1979) argue that there is no convincing proof for this, although they did not discredit such an application since the results of some experiments (e.g. Saunders, 1963) indicate that the difference in the weight exponent should not be large.

Metabolic energy for maintenance of other fish species can be similarly calculated using the respective specific routine metabolism.

5

Growth

Growth depends on a number of various factors amongst which food ration and the weight of the fish are of special importance. When food is insufficient for both maintenance and growth, growth will be inhibited or will cease entirely. In order to determine the amount of food required for both maintenance and growth, it is necessary to know the maximum rate of growth possible when food is not limiting. This does not mean that at this growth rate food is utilized efficiently. It is often possible to achieve a high rate of growth at the expense of excessive food and low utilization so as to make this gain uneconomical. Therefore, the relationship between rate of growth and food utilization for growth must also be studied, which will be done in later sections (pp. 169, 301).

Growth is also a function of body size (weight). Body weight or length are the main parameters by which fish farmers determine the feeding level. If feed ration is to produce optimum growth, it is essential to learn the relationships between body weight (or length) and growth rate. This will also, therefore, be discussed in detail in a later section (p. 156).

Except food and weight, growth also depends on a number of other factors, which often interact with food ration and body weight. These were thoroughly reviewed by Brett (1979), therefore they will be discussed here only briefly. Factors affecting growth can be divided into two main groups: (a) those related to the fish itself (internal factors); (b) environmental (external) factors.

Growth factors related to the fish

Different fish species may vary in their potential growth rate. Some species are capable of growing faster than others. However, considerable variations in growth also exist among individuals or groups within the same species. Some of the factors affecting growth, which are responsible for these variations, are discussed below.

Sex In some fish there is a distinct difference in growth rate between sexes. In some of the species, such as the tilapias, the male grows faster than the female (Mabaye, 1971; Fryer & Iles, 1972; Hepher & Pruginin, 1981), while in others, such as common carp and eels, the female grows faster than the male (common carp, Kempinska, 1968; eels, Tesch, 1977).

Genetic characteristics These may be associated with a number of traits affecting growth, such as the capability of the fish to search for and utilize food; its ability to compete for food with other fish; the physiological utilization of food; or any other physiological attribute. All these are ultimately expressed in the growth rate of the fish (Brown, 1957; Steffens, 1964; Laarman, 1969). Moav & Wohlfarth (1968) selected some genotypes of common carp which showed a noticeably better growth rate than others under the same conditions. However, these genotypes did not always retain their rank order for growth when conditions were changed. Thus, Wohlfarth *et al.* (1983) measured growth rates of two different races of common carp (*Cyprinus carpio*): the Chinese race ('big belly carp') and a European race. Both these races and their hybrids were stocked, together with other fish species (tilapia, silver carp and grass carp), in a polyculture, into ponds having different environments. The quality of the environments was expressed by the mean growth rate of all fish in the pond. In the poorest environment, where fish were stocked at high density and poultry manure was the only nutrient input (mean growth rate of 2.0 g/fish/day), the Chinese race grew best, while in the best environment, where stocking density was relatively low and in addition to poultry manure high-protein pellets were also added (mean growth rate, 7.1 g/fish/day), the European race grew best. Suzuki *et al.* (1976) also found differences in growth among different races of common carp, mirror and German scaly carp growing faster than Japanese races of carp.

Inbreeding can cause morphological degeneration and decreased growth rate, while cross breeding between different genotypes results in many cases in heterosis, i.e. a better growth of the offspring than any of their parents (Moav & Wohlfarth, 1968; Bakos, 1979). In common carp a linkage was found between growth rate and the scale coverage. The genotype having no scales (leather carp), or that having one row of scales on each side over the lateral lines, showed an inferior growth rate in comparison with genotypes covered entirely by scales or having a limited scale coverage under the dorsal fin and a few scales on the ventral side ('mirror carp'). The mode of inheritance of these four scale patterns of common carp is well known from Europe (Wunder, 1955) and was also studied in Israel by Wohlfarth *et al.* (1962).

Physiological state The physiological state of the fish affects growth considerably. Fish suffering from diseases or parasites have a retarded growth or cease growing entirely. Generally stress, resulting from any cause such as anoxia or poisoning, results in retarded growth. Stress may also be caused by social factors such as the presence of fish of other species or a territorial struggle among fish of the same species. In tilapia, for instance, the loser in such a struggle becomes darker in colour due to physiological stress and may even die (Rothbard, 1979; Santiago *et al.*, 1981).

During the ripening of the gonads, many fish show temporary inhibition or even cessation of growth. Gonads when fully developed may constitute a considerable part of the total body weight. Bagenal (1967) stated that in brown trout (*Salmo trutta*) the gonads can constitute up to 20% of the total fish weight, and in other fish they can exceed 30% of the total weight. Material for the gonads is drawn from food and/or somatic tissues. Bagenal (1973) showed that the gonads of pike (*Esox lucius*) increased five-fold in weight during the period October–March, but this increase corresponded to a similar decrease in weight of the other body tissues. Even when the increase in gonad weight is derived mainly from the food, the latter is usually not sufficient to provide for both gonad development and maximum growth rate and growth is therefore inhibited. Schaeperclaus (1961) working with common carp, and Kadmon *et al.* (1985), with sea bream (*Sparus aurata*), state that during gonad development the fish showed temporary inhibition of growth. It is clear that on spawning much of the gonad weight is lost. Thus, early and frequent development of gonads, as for instance in tilapia, is a great disadvantage in the culture of these fishes and should be avoided. Further discussion on metabolism and nutrition of pond fishes therefore will be limited to fish which are not sexually mature and whose gonads do not develop during the period of culture.

Physiological effects on growth are not always associated with traumatic conditions. It seems that some latent physiological facts cause unexplained cycles in growth rate during normal life conditions. Brown (1946b) found that in spite of constant environmental conditions, brown trout (*Salmo trutta*) had an annual growth cycle, with an autumn check, a spring maximum, rapid summer growth and another autumn check, which coincided with maturation of gonads when the fish became a three-year-old. Individual specific growth rate fluctuations over a period of four to six weeks were also observed by Brown (1946b), during which time rapid growth in length alternated with rapid growth in weight.

The latent physiological factors affecting growth may be associated with 'social' conditions. Brown (1946a) found that when larger fry were removed from a group of brown trout, the smaller ones grew at an increased specific rate, and when larger fry were added the smaller ones grew more slowly. Brown suggested that a 'size hierarchy' exists within each group, and that the individual's specific growth rate depends on its position in the order of decreasing weight.

Environmental conditions affecting growth

The main environmental factors affecting growth of fish are temperature, light, permanent chemical constituents in water such as salts and organic compounds (water quality), oxygen concentration and the presence of fish catabolites which inhibit growth. While the first three factors are not dependent on fish density in the pond, the latter two are density dependent. The higher the fish biomass in the pond (due to increased density or increased individual weight) the higher the requirement for oxygen. When oxygen is limited, this can result in a reduction in oxygen concentration in water to a harmful level. Also, with higher fish biomass excreted catabolites, especially ammonia, may accumulate, poison the fish and inhibit their growth.

Temperature Like all other life processes in ectotherms growth is affected by temperature. It increases with increase in temperature up to an optimum above which growth decreases, until the upper lethal temperature is reached. It seems that the optimal temperature range for growth is narrower than that for maintenance metabolism.

The effect of temperature (or light) on growth may be mediated by hormones such as somatotropin produced in the pituitary, or thyroxine produced in the thyroid, which are involved in the metabolism and, according to some authors, with growth (Gorbman, 1969). Swift (1955) has shown that seasonal variations in activity of the thyroid gland of three-year-old brown trout (*S. trutta*) can be related to the daylength. Peak activity at midsummer coincided with temperatures of 12–15 °C.

Adaptation of growth processes to temperature seems to exist, although it is slower than adaptation of metabolism to temperature. Pessah & Powles (1974) exposed pumpkinseed sunfish (*Lepomis gibbosus*) to longer periods at different constant temperatures. While during the first six weeks there was a marked effect of temperature on growth rate, which was highest at the highest temperature tested, after six weeks the growth rate declined, becoming uniform between 15 and 30 °C. The initial high, followed by a low uniform growth stanza suggests, according to these authors, a 'growth acclimation'. It should be borne in mind, however, that there is an interaction between temperature and a number of other factors such as photoperiod and feeding rate in affecting growth. A significant statistical interaction was found by Huh *et al.* (1976) between the effects of temperature and photoperiod on the growth of walleye (*Stizostedion vitreum*). An interaction much more studied is that between temperature and feeding rate. An increase in temperature will increase growth rate only if it is accompanied by an increase in food consumption to meet increased requirements for both maintenance and growth. Elliott (1975b) studied the growth rate of brown trout fed various ration sizes, ranging from no feed to maximum ration, at 12 different water temperatures (3.8–19.5 °C). He found that the optimum temperature for growth decreased with decreasing ration from about 13 °C at maximum ration to about 4 °C at a ration just above maintenance. This seems to be due to the increased energy requirements for maintenance at the higher temperatures. When fed reduced rations at these temperatures less food is available for growth, reducing what the author called 'scope for growth'. The difference in response of fish of different ages to temperature may also be the result of the same phenomenon, since the older (and usually heavier) fish require more energy for maintenance than the younger (and usually lighter) fish. Indeed, Elliott (1975b) found that as the weight of the trout increased, the effect of temperature on growth, at each ration size, was progressively reduced.

The lowest temperature limit for growth of coldwater fishes is lower than that of warmwater fishes, and likewise the optimal temperatures for growth. Experiments by Elliott (1975a) showed that the specific relative growth (expressed as per cent of body weight per day) of brown trout fed maximum rations increased with increasing temperature from 3.8 to 12.8 °C. Maximum growth was achieved between 12.8 and 13.6 °C, above which growth decreased with increasing temperatures up to 19.5 °C. A similar optimum temperature for growth has also been found for other salmonids (e.g. Cooper, 1953; Shelbourn *et al.*, 1973). In most warmwater fishes, however, fish growth starts when the temperature reaches 17–18 °C and attains its maximum rate at 28–30 °C (Kinne, 1960; Niimi & Beamish, 1974; Stickney & Andrews, 1971; Andrews & Stickney, 1972). Jauncey (1979, quoted from Jauncey 1982b) found a certain interaction between temperature and ration size on the growth rate of carp (Figure 21), although not as strong as for brown trout. Optimum temperature remained approximately the same (about 28 °C) at feeding levels of 3, 6, and 9% of body weight per day. This may be due to insufficient data at lower feeding rates.

The physiological pathway by which temperature affects growth is not clear enough. Jackim & LaRoch (1973) studied the kinetics of labelled leucine incorporation into the muscle of *Fundulus heteroclitus* as a function of environmental factors. They found that protein synthesis increased with increase in temperature up to a critical point of 26–29 °C, beyond which it decreased sharply. Fry (1971) has suggested that it may be the respiratory system which limits growth above the optimum temperature, and likewise Swift (1961) suggested that above 12 °C the incapability of the respiratory system of brown trout (*Salmo trutta*) to meet respiratory needs caused the decrease in growth. However, in a later study Swift (1963) found that hypoxia affects growth only when oxygen concentration is reduced to below 5 mg/l. It is not likely, therefore, that it is the restriction of respiratory capacity which causes reduction of growth rate at higher temperatures.

Photoperiod The cyclic and seasonal variations in growth are undoubtedly regulated, in addition to temperature, also by the light period, which often interacts with temperature. Some experiments have shown that fish

Figure 21. Growth response of common carp to four temperatures at three feeding levels (as dry food percentage of wet body weight per day) (from Jauncey, 1979, quoted from Jauncey, 1982b).

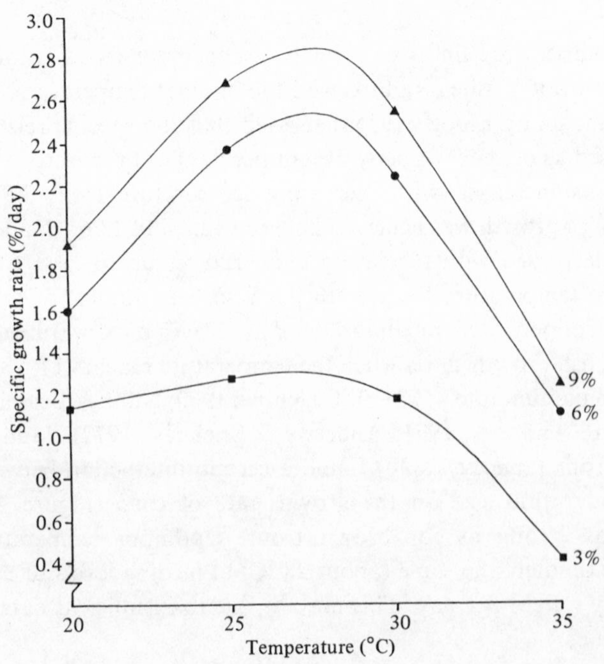

growth is affected by the length of the daily period of light more than by temperature. Huh *et al.* (1976) studied the growth of yellow perch (*Perca flavescens*) at two temperatures (16 and 22 °C) and two light periods (8 and 16 h), and found the effect of the photoperiod to be more significant than the temperature. Gross *et al.* (1965) working with green sunfish (*Lepomis cyanellus*) and Kadmon *et al.* (1985) with sea bream (*Sparus aurata*), found that growth was significantly higher when the fish were exposed to a long photoperiod. This seems to be due to increased feed consumption and possibly to a better feed conversion associated with the longer day. The opposite was found by Brown (1946b) for brown trout (*S. trutta*). At a constant temperature of 11.5 °C specific growth rates were significantly lower with 12 or 18 h of light per day than with only 6 h. Varying daylengths exerted a greater influence on fish growth than a constant daylength.

Water quality Water quality affects growth of fish to a large extent. Excessively low or high pH beyond the range of 6.5–9.0 can lower the growth rate of most pond fishes, and extreme pH values can be detrimental (Swingle, 1961; Alabaster & Lloyd, 1980). Variability in the effect of pH on fish depends on species, size of fish, temperature, the concentration of carbon dioxide and the presence of heavy metals such as iron.

Many species of fish can tolerate a wide range of salinity ('euryhaline fish'), but others can tolerate only a narrow range of salinity ('stenohaline fish'). However, the change of salinity within the tolerance range of both groups may affect growth to a large extent. For example, the lethal concentration of salts for common carp is about 11.5‰ (about 7000 ppm Cl^-), and it can, therefore, be considered as a stenohaline fish. Up to about 5‰ no appreciable effect on its growth rate was noticed, but above this level growth is affected (Soller *et al.*, 1965). Tilapia, which is euryhaline, grows best in brackish water, and its growth rate decreases with a decrease or increase in salinity. However, some tilapia species (e.g. *Oreochromis mossambicus* and *O. niloticus*) still show a fair growth in sea water (35–40‰ salinity). Similar growth curves, although at higher salinities, have been obtained by Kinne (1960) for the desert pupfish (*Cyprinodon macularius*). Best growth was obtained at a salinity of 35‰, but growth was reduced when salinity had been decreased to 15‰, or increased to 55‰. The lowest growth rate was observed in freshwater. Rainbow trout also grow better in seawater than in freshwater (Lall & Bishop, 1979).

Oxygen concentration The concentration of oxygen in water may affect growth of fish either through their appetite and food consumption or their food utilization. Herrmann *et al.* (1962) studied the effect of low oxygen concentrations on the growth of coho salmon (*Oncorhynchus kisutch*). Similar studies have been conducted by Swift (1963) on brown trout (*S. trutta*), Stewart *et al.* (1967) on largemouth bass (*Micropterus salmoides*), Adelman & Smith (1970) on northern pike (*Esox lucius*), and Brungs (1971) on fathead minnow (*Pimephales promelas*). In all these studies the fish received *ad libitum* feeding, but in spite of this, growth rate decreased with the decrease in oxygen concentration below saturation level. At first, the effect of reduced oxygen concentration on growth rate was slight, but below about 4 mg/l, the decrease in growth rate became pronounced. According to the above authors the decrease in growth rate with the decrease in oxygen concentration down to 4 mg/l is mainly due to decrease in appetite and food consumption (Figure 22), since feed utilization efficiency was not affected at these concentrations. Growth decreased more or less in parallel with the decrease in feed consumption. However, when oxygen concentration was reduced to below about 4 mg/l, feed utilization efficiency was also reduced.

According to Stewart *et al.* (1967) the growth of largemouth bass was also inhibited at oxygen concentrations above saturation level. This may be specific to some fish species living in a fairly constant oxygen environment such as coldwater fishes living in running water, where oxygen concentration is usually at or near saturation. For most warmwater pond fishes exposure to supersaturation of oxygen is quite common and it is doubtful whether it affects growth. Thiel (1977, quoted from Pauly, 1981) kept common carp in pressurized tanks so that the oxygen concentration

Figure 22. The effect of oxygen concentration on food consumption and growth rate of northern pike (*Esox lucius*) (according to data from Adelman & Smith, 1970).

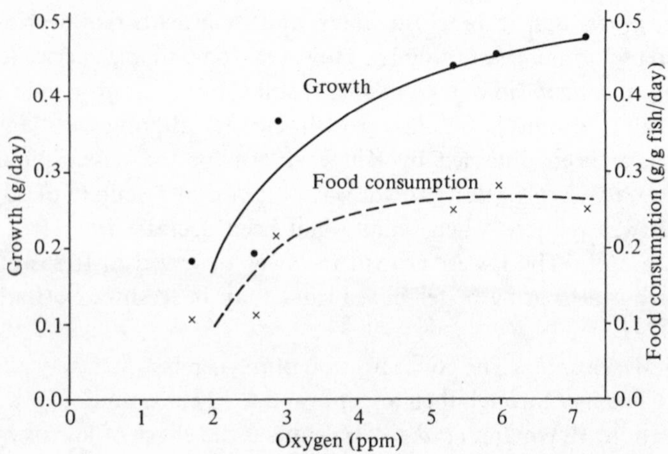

of the water could be increased well above the normal levels, without unduly increasing oxygen saturation. He found a significant increase in growth with increased oxygen concentration.

The effect of low oxygen concentrations on growth of warmwater fishes occurs at lower oxygen level than with coldwater fishes. Chiba (1965) found that the lowest oxygen concentration below which food consumption and growth of common carp is affected was about 3 mg/l. Andrews *et al.* (1973) studied the effect of various oxygen concentrations on the growth of fed channel catfish. The fish were held at three oxygen levels: 100, 60 and 36% saturation (8.0, 4.8 and 2.9 ppm, respectively). At these concentrations the average gain was 159, 124 and 65 g respectively. Food consumption and utilization decreased considerably at 36% oxygen saturation. Since survival rate in all treatments was 100%, it is obvious that this reduced growth rate was not due to any disease but rather to the oxygen concentration.

Toxicants and catabolites The presence of toxicants, usually originating from outside sources, and the concentration of catabolites excreted by the fish, may also affect growth. The literature contains considerable data on the effect of sublethal concentrations of toxicants, such as pesticides, herbicides and detergents, on the growth of fish. However, since such a situation is not regular and the number of toxicants is extremely large, this topic will not be discussed here.

The situation is, however, different with regard to catabolites, since these are associated with the biomass of fish in the pond. The higher the biomass, the higher the excretion of catabolites which may accumulate and reach high concentrations, poison the fish, and affect their growth. Little is known about the nature of these catabolites. Since ammonia is the major catabolite excreted by fish, some have related the toxic effect to the increase in ammonia concentration alone. Though no doubt ammonia in its non-ionized form (NH_3) has a toxic effect on fish, it has not been proved that it is the only catabolic compound which affects the fish. Pfuderer *et al.* (1974) partly purified the 'crowding factor' from *Carassius auratus* and *Cyprinus carpio* and showed the partly purified eluted substances to depress the heart-beat rate of the fish. It is obvious that under natural conditions the effect of such catabolites is compounded by that of ammonia.

Two factors affect the concentration of catabolites in pond water: the biomass of fish producing the catabolites on the one hand the period of their accumulation on the other. The higher the biomass, the faster the concentration affecting growth may be reached. From experiments at the Fish Culture Research Station at Dor, Israel (unpublished data) it was found that when pellets containing 25% crude protein were fed to common carp, growth proceeded at the maximum potential rate up to a standing crop of 2.4 t/ha (Figure 23). Thus, it seems that at least up to this standing crop the effect of catabolites was negligible. However, Marek & Sarig (1971) have shown that under a more intensive culture method, when a pond was stocked with common carp each weighing 250–300 g, at a rate of 10 000/ha, without exchanging the water, the fish grew to a market size of 600 g in about 60 days, after which growth ceased. Since such growth inhibition is not observed in ponds with a strong exchange of water which removes the wastes, it seems that growth inhibition was a result of accumulation of catabolites. Thus, the maximum standing crop of common carp, in monoculture in stagnant water ponds, beyond which growth is affected by catabolites is between 3 and 6 t/ha. This, however, may vary with environmental conditions.

Maximum potential growth rate

From a bioenergetic point of view, there is much interest in the relationship between growth rate and food intake, so that one can determine feeding ration as a function of growth rate. In this respect it

Figure 23. The relationship between growth rate (g/day) and the average weight (g) of common carp, as determined by two week interval sample weighings for four treatments: (1) no fertilization and no feeding (empty triangles); (2) fertilization but no feeding (black triangles); (3) feeding on sorghum (black circles); (4) feeding on protein-rich diet (empty circles). Each point is an average value determined from four replicated ponds.

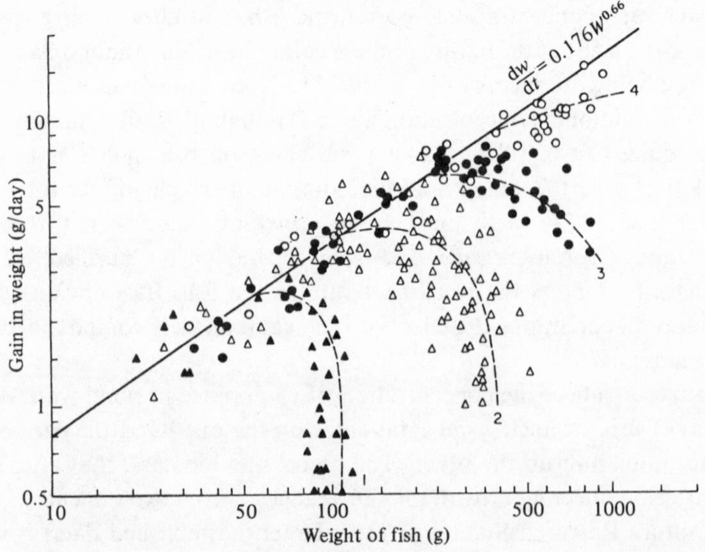

would be important to know the maximum potential growth rate when food does not limit growth. As was already mentioned above, at a given environmental set of conditions this potential growth rate depends on the weight of the fish. Growth of fish, relative to their body weight, is very high during the larval and juvenile stages of development. It can reach 40% of the fry weight/day and even more. Fish growth decreases with increasing weight, and a fish of 1 kg usually grows less than 1% per day.

Many attempts have been made to express the rate of fish growth in relation to its weight in a mathematical form (for extensive reviews see Ghosh, 1974; Ricker, 1979; Pauly 1981). One of the most widely accepted models which describe this relationship is that of Bertalanffy (1938). His equation is based on the assumption that growth is a resultant of two opposing processes – the anabolic and catabolic processes – and that both are related allometrically to the size of the organism. The basic equation of Bertalanffy is:

$$\frac{\mathrm{d}w}{\mathrm{d}t} = HW^n - kW^m \tag{30}$$

where H and k are the quotients of anabolism and catabolism, respectively, and W is the weight of the organism. Bertalanffy assumed that anabolism is affected by the surface area of the organism while catabolism is linearly related to weight, and that growth is isometric. Since the surface area of a regular body increases at about the power of two-thirds of its volume, $n = \frac{2}{3}$ and $m = 1$. The equation thus takes the form:

$$\frac{\mathrm{d}w}{\mathrm{d}t} = HW^{\frac{2}{3}} - kW \tag{31}$$

Bertalanffy continued to develop his equation by integration so as to find the weight of the fish at age t and received:

$$W_t = W\infty\ (1 - \mathrm{e}^{-k(t-t_0)})^3 \tag{32}$$

where $W\infty$ is the asymptotic maximum weight; t_0 = the time when the organism was theoretically at zero weight; and $k = \frac{k}{3}$ of the basic equation.

As pointed out by Bertalanffy (1964) himself, his assumptions were not based on experimental results. While he was quite confident about the linear relationship between the catabolic processes and the weight of the organism, he was less confident about the relationship between anabolism and body weight and pointed to the possibility of other exponents expressing this relationship. In choosing the exponent of $\frac{2}{3}$ Bertalanffy was influenced by the studies on metabolism which also related, at first, the metabolic rate to the surface of the body and used the exponent of $\frac{2}{3}$, though later it was found that the true exponent for endotherms is nearer

$\frac{3}{4}$ and for fish about $\frac{4}{5}$. He reviewed other possible values for *n*, as found from the studies on metabolism and oxygen consumption, but in view of the good fit which was generally found between data on fish growth and his original equation, he came to the conclusion that, at least for fish, the value of $\frac{2}{3}$ is most appropriate. Many attempts have been made to verify Bertalanffy's equation (e.g. Allen, 1969; Kerr, 1971a, b; Nielson, 1973; Krueger, 1973), not always with success. A major problem was that most studies were based on data from natural waters such as lakes and seas. Fish caught in such water bodies may be a mixture of different populations and age groups. The samples were taken, very often, at long intervals, during which the population may have changed.

Some of the empirical works have assumed a growth equation which is equivalent only to the first half of Bertalanffy's equation (29). This equation was also used by Parker & Larkin (1959):

$$\frac{\mathrm{d}w}{\mathrm{d}t} = a\,W^{b} \tag{33}$$

Corey *et al.* (1983) pointed out that this equation is implicit in these works because of the assumption usually made that time rate change of fish length (l) is constant, i.e.

$$\frac{\mathrm{d}l}{\mathrm{d}t} = \text{constant} \tag{34}$$

It can be shown, using calculus, that equation (34) is equal to equation (33) if l is proportional to $W^{\frac{1}{3}}$, which has been shown in many empirical studies, and if $b = \frac{2}{3}$. Experiments by Haskell (1959) demonstrated that maximum growth rate of brook trout (*Salvelinus fontinalis*) fits equation (33) when the exponent (b) equals $\frac{2}{3}$. This equation then takes the form:

$$\dot{W}_{\max} = a\,W^{\frac{2}{3}} \tag{35}$$

where $\dot{W}_{\max}$ is equal to $\frac{\mathrm{d}w}{\mathrm{d}t}_{\max}$. The study of fish growth in ponds may have a good chance to verify Haskell's (1959) equation since here samples of the same fish populations can be weighed frequently. Analyses of maximum growth of fish in ponds have been made, utilizing published data of a number of series of sample weighings made at regular intervals, as well as original data. In each case it was first verified that the fish had enough food for maximum growth. It must be remembered that, due to the intervention of the fish farmer, the conditions in the fish pond are different from those of natural waters. In order to maximize growth rate of the fish the fish farmer controls, to the best of his ability, the environmental conditions in the pond (temperature, oxygen, etc.); in order to avoid competition between fish the fish farmer tries to prevent uncontrolled breeding in the grow-out ponds, usually by culturing young fish, prior to their sexual maturity; and lastly, in a well managed pond the fish farmer feeds his fish adequately with supplementary feed. In these conditions growth rate approaches its maximum physiological potential. The relative importance of catabolism then becomes negligible in comparison to anabolism. This may be different for fish in the wild. Growth curves for fish in natural populations are usually fitted to data on size at yearly intervals. During this relatively long period the growth rate may be differentiated into various stanzas (Ricker, 1979) which occur at certain seasons or at certain stages of development. Growth rate may decrease due to unfavourable conditions, especially lack of food. Ricker (1979) also points to the bias which may occur in the growth line in natural waters due to selective mortality or fishing of either the larger or smaller individuals. Thus, it seems that in natural waters catabolism may be of relatively greater importance than in ponds, and growth estimation less reliable than in ponds.

For each of the series of weighings of pond fish, a regression was determined where weight gain per day ($\frac{dw}{dt}$) was calculated for the period between two successive sample weighings ($W_t - W_0/t$), and body weight ($\bar{W}$) was the average weight of fish during this period ($W_t + W_0/2$). Yashouv (1969) gave data on growth rate of common carp in very thinly populated ponds – 400 fish/ha. The results of the frequent sample weighings were analysed by the method described above and the resulting regression was:

$$\dot{W}_{max} \text{ (g/day)} = 0.2\ W^{0.65}\ (r = 0.924) \tag{36}$$

In another experiment, which was carried out at Dor, Israel (Hepher & Chervinski, 1965), two treatments were tested. In one fish were fed sorghum, while in the second a protein-rich diet was given. The results showed that, at a density below 2000 fish/ha and up to an individual weight of 300 g, there was no difference in growth between treatments. It was concluded, therefore, that the fish in both treatments did not lack food or protein. The data on periodic sample weighings from this experiment were combined with those of two other treatments of an experiment performed previously in the same ponds (Hepher, 1962a). In one treatment, no feed or fertilizers were added, while in the other the ponds were fertilized but no feed was added. All four treatments were analysed as above and the results are given in Figure 23. It can be seen that up to a certain individual fish weight (which is determined by the standing crop of fish in the pond), where natural food was sufficient to sustain maximum growth even in the poorest pond, there was no difference in growth rate among the four treatments. When, however, the standing crop in any of these treatments increased to a level where fish food was not sufficient for both maintenance and maximum growth, the latter decreased. With the increasing supply of food or improvement in its quality in the different treatments (fertilization, sorghum, and protein-rich pellets), the decrease in growth rate occurred at a higher standing crop. When densities were similar, the deflection from the maximum growth rate occurred at a higher individual weight. The regression of the line combining the maximum growth rates of all treatments, when food did not limit growth was;

$$\dot{W}_{max}\ (\text{g/day}) = 0.176\ W^{0.66}\ (r = 0.88) \tag{37}$$

Regressions of this kind were also calculated for other series of data collected from the literature. The results are summarized in Table 27. From the information collected in this Table it appears that the relationship between weight gain and body weight of fish given in equation (35), which equals the first half of Bertalanffy's equation (31) is confirmed experimentally. The regression coefficient 0.66 ($= \frac{2}{3}$) seems to be determined by intrinsic physiological factors, since it was found to be common to different fish species and holds true for varied environmental conditions. The numerous, frequent and short interval data obtained from sample weighings of fish ponds seem to be more reliable than those obtained from larger water bodies. The results also show that for young pond fish, the last part of Bertalanffy's equation is so small that it can be

Table 27. *Regression equations for maximum growth rates, as functions of average body weight, when food did not limit growth, calculated from data given in the literature*

Fish species	$\dot{W}_{max}$ regression (g/day)	Remarks	Source
Channel catfish	$0.086W^{0.674}$	Cultured in favouarable conditions	Andrews (1972)
Common carp	$0.09W^{0.667}$	Cultured in a running water system and fed protein-rich pellets	Luer (1967)
Common carp	$0.2W^{0.65}$	Density: 400 fish/ha	Yashouv (1969)
Common carp	$0.176W^{0.66}$	From about 900 sample weighings carried in ponds during 15 years (see Figure 23)	Hepher (1978)
Common carp	$0.188W^{0.66}$	Reared in ponds	Hamada *et al.* (1975)
Salmonids	NG[a] $W^{0.5-0.83}$	(average exponent 0.66)	Parker & Larkin (1959)
Rainbow trout	$0.079W^{0.662}$	Cultured at 12 °C	Weatherley & Gill (1983)
Brook trout	NG[a] $W^{0.667}$		Haskell (1959)
Brown trout	NG[a] $W^{0.688}$	Fed maximum ration	Elliott (1975a)

Note: [a] Y-intercept not given.

ignored. This, however, may be not true for large, less productive, water bodies and older fish.

The coefficient a in equation (35) – the Y-intercept – expresses the growth rate of a fish of a unit weight (usually 1 g) and, therefore, may be called 'specific growth coefficient'[1]. It is affected by the fish species, the genetic characteristics within that species (the genotype), and the environmental conditions such as temperature or water quality prevalent during growth. It can, therefore, be used for comparing growth rates among different fish groups or, using the same fish, the effect of environmental factors.

The difficulty in comparing growth rates between groups of fish having different average weights is obvious in view of the fact that large fish have a higher potential growth rate than small ones. Some of the early studies involved in comparison of growth rates tried to solve this problem by comparing the relative growth rates, expressed as percentages of initial fish weight, rather than comparing the absolute growth rates (g/fish). This, however, leads to another pitfall. When fish are in their early stages of development, the percentage gain usually amounts during a period of few weeks to hundreds or thousands, and a slight difference in the initial weight results in a considerable difference in percentage gain. Moreover, this method did not really solve the major problem, since the relative growth rate also varies considerably with body weight. To correct this Brown (1946b) suggested the use of 'specific growth rate', as it is commonly called today, which Winberg (1956, 1971) has called 'average daily percentage gain' (see also Cooper, 1961). As a basis for calculating this value Winberg (1956) assumed an exponential relationship between the weight of the fish and its age. For this relationship he suggested the equation:

$$W_t = W_0 \, e^{k(t-t_0)} = W \, 10^{K(t-t_0)} \tag{38}$$

where k and K are constants which characterize the rate of growth; W_t is the weight at the time of observation; W_0 is the initial weight; and $t-t_0$ is the period under study expressed in days. From equation (38) it is obvious that:

---

[1] The term 'specific growth rate' is commonly used today for the average daily growth rate expressed as a percentage (G_w) (see following discussion). It is argued that the term is more appropriate for the potential growth rate of a fish of a unit of weight, since like 'specific heat' or 'specific weight' it refers to a *unit* weight, it does not vary with fish weight, and it is specific to the fish whose growth is measured.

$$k = \frac{\ln W_t - \ln W_0}{t - t_0} \text{ and } K = \frac{\log W_t - \log W_0}{t - t_0} \qquad (39)$$

Winberg further assumed that the value K (or k) is constant with age (or at least for the duration of any experiment) and thus the 'average daily percentage gain' (G_w)[1] could be calculated as:

$$G_w = \left[\frac{\ln W_t - \ln W_0}{t - t_0}\right] \times 100 = \left[10^{\frac{\log W_t - \log W_0}{t - t_0}} - 1\right] \times 100 \qquad (40)$$

While this equation is an improvement over the relative growth rate, a decrease in G_w with increasing body weight was apparent in many studies (Cooper, 1961; Laarman, 1969; Winberg, 1971; Elliott, 1975a). It has been noted (Elliott, 1975a; Brett, 1979) that the rate of decrease in the daily percentage gain ('specific growth') is gradual and that its relationship to body weight can be described by an equation of the form:

$$\ln G_w = a + b \ln W \qquad (41)$$

Jobling (1983a) confirmed this relationship and suggested that the weight exponent b is common to all fish species and can be taken to be -0.4. Jobling (1983a) further suggested using the intercept a which represents the $\ln G_w$ of a fish of a unit weight for comparing growth of fish at variable environmental conditions. It is obvious that the value $\ln G_w$ is far from expressing growth rate in a meaningful unit. In these circumstances the use of the 'specific growth coefficient' (*SG*), as suggested above for comparing growth rates of fish having different body weights should be considered. Where the *SG* is not found from the regression of growth rate on average body weight, it can be calculated by the following equation:[2]

$$SG \text{ (g/day)} = \frac{3(W_t^{0.33} - W_0^{0.33})}{t - t_0} \qquad (42)$$

Multiplying *SG* by 100 gives the instantaneous percentage growth of a fish of a unit weight (usually 1 g).

The fish farmer usually calculates the amount of feed to be given to the fish during a certain period on the basis of the weight of the fish sampled at the beginning of this period. It is obvious that the growth rate calculated by equation (35), which is an instantaneous value, cannot serve this purpose well. The projected average weight of the fish during the period in question should be applied instead. This can be done by calculus, assuming that growth continues at the maximum rate for the entire

---

[1] The original designation (*C*) has been changed for the sake of conformity with those adopted here.

[2] Thanks are due to Dr G. L. Schroeder for developing this equation

period. Thus, based on equation (37) for the growth of common carp at Dor, the following equation was developed to find the projected weight of the fish at time t_f

$$W_{tf} = [0.176 \times 0.33\,(t_f - t) + W_t^{0.33}]^3 \qquad (43)$$

where W_{tf} is the weight of the fish at a future time t_f; W_t is the average weight of the fish at time of sampling at the beginning of the period; $t_f - t$ is the length of the period, in days, for which the feeding level is calculated. The first term in equation (43) is taken from equation (37) and refers to conditions at the Fish and Aquaculture Research Station, Dor, Israel. This term may, of course, be somewhat different with different values of '*a*' in equation (35). The average fish weight during the period would be $(W_t + W_{tf})/2$. Both feeding rate and growth rate can be calculated based on this average weight, but the period should not be longer than 10–14 days.

Growth on reduced food rations

The range of dietary energy ingested between maintenance ration, when no growth occurs, and maximum ration, when maximum growth occurs, is the 'scope for growth'. When food is limited the scope for growth is that portion of it available for growth after satisfying the maintenance requirement. Since both maintenance requirement and growth potential are dependent, among other factors, on body weight and temperature, these will also affect the scope for growth, which means a strong interaction between body weight, water temperature, the food ration and the rate of growth. Elliott (1975b), working with brown trout (*Salmo trutta*), showed that at a constant limited food ration growth rate decreased markedly with increasing body weight and water temperature. This is no doubt due to the fact that large fish require more food for maintenance than small fish, which leaves a smaller scope for growth. The same is true for the increase in temperature, which also increases maintenance metabolic rate.

Within the scope for growth, the higher the ration the higher the growth rate. The relationship between the two has been described by Ivlev (1945) and termed by him 'the growth coefficient of the first order (K_1)', and when the assimilated part of the food was considered 'the growth coefficient of the second order (K_2)'.[1]

[1] To be distinguished from K_2 – net conversion efficiency in Brett & Groves (1979) which is equal to E_{pg} in the present book.

$$K_1 = \frac{\Delta W}{R\Delta t} \text{ and } K_2 = \frac{\Delta W}{pR\Delta t} \tag{44}$$

where $R\Delta t$ = total ration and $pR\Delta t$ = assimilated ration.

Paloheimo & Dickie (1965, 1966b) reviewed results of various feeding experiments and found that for a given type of food

$$\log K_1 = \log \frac{\Delta W}{R\Delta t} = -a - bR \tag{45}$$

or $$K_1 = \frac{\Delta W}{R\Delta t} = \mathrm{e}^{-a-bR}$$

or $$\frac{\Delta W}{\Delta t} = R\,\mathrm{e}^{-a-bR}$$

where a and b are constants. This means that, for a given type of food, the logarithm of the growth efficiency K_1 is linearly related to the ration. This relationship was termed by Paloheimo & Dickie (1965) the K line. It implies that, beginning with a maximum value at low feeding level, when $K_1 = \mathrm{e}^{-a}$, growth efficiency decreases by a constant fraction (e^{-b}) for each unit increase in the amount of feed consumed during the relevant period, irrespective of the weight of the fish. Many studies have confirmed this form of relationship between growth and ration (e.g. Huisman, 1976). Others, however, suggested a straight line for this relationship. Brett (1979) attributes this conclusion to the scatter of the points of growth: ration and to a not well defined growth rate at maximum feeding level.

Brett (1979) described the relation between growth rate of sockeye salmon fingerlings (as G_w – per cent of body weight per day) and food ration (per cent of body weight per day), which is presented in Figure 24. According to Brett (1979) the tangent from the origin to the G_w:R line touches it at a point where the ratio is maximal. Brett (1979) considers food ration at this point as the optimal one. This may be so from a biological point of view but not necessarily from a practical one. The increase in food ration above the 'optimal level' does increase growth rate, although at a lower efficiency. As long as the cost of the marginal gain in fish weight, i.e. that added over the gain on lower feeding rate, is higher than that of marginal feeding ration, added to achieve the extra gain, fish farmers will consider it economically worthwhile. Maximum economical ration will be attained when the two balance (see also p. 301).

The calorific value of growth

The calorific value of weight gain depends on the composition of tissues laid down, especially the content of lipid in them. The ability to accumulate fat reserves may vary among fish species. Coldwater fishes generally accumulate less fat than warmwater fishes. Fat accumulation may be especially high in some fish species, such as common carp fed a carbohydrate-rich diet. Analyses carried out by the author (unpublished data) showed that common carp fed on sorghum accumulated as much as 34% fat (wet weight basis).

The composition of the diet determines, to a large extent, the fat content of the fish (Amlacher, 1960; Janecek, 1968; Schaeperclaus, 1968; Brett *et al.*, 1969). Hepher *et al.* (1971) showed that, when fed at similar rates, more fat was accumulated by common carp fed on sorghum than by those fed on wheat, and the latter accumulated more fat than carp fed on protein-rich pellets (Figure 25). The higher the ratio between the metabolizable energy and protein in the diet, the higher is the fat content in the tissues. Winfree (1979) found that the tissues of tilapia (*Oreochromis aureus*) which were fed a diet having 7.8 kcal/g protein, contained 2.2% fat, while those of fish fed a diet with 12.2 kcal/g protein contained 11.9% fat. Similar trends have been found in experiments with yellowtail (Takeda *et al.*, 1975), brook trout (Ringrose, 1971), channel catfish

Figure 24. Specific growth rate (*G*) of sockeye salmon fingerlings (mean weight 13 g) in relation to ration (*R*), at 10 °C (from Brett & Groves, 1979).

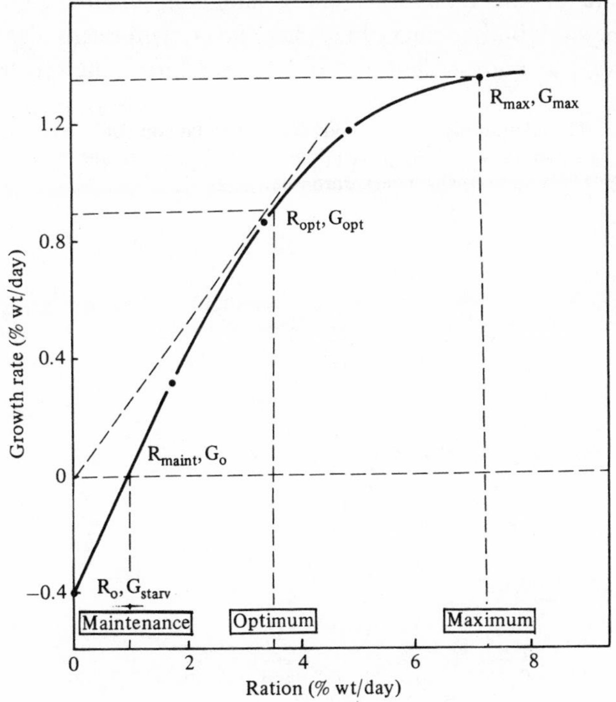

(Prather & Lovell, 1973; Garling & Wilson, 1976) and common carp (Zeitler *et al.*, 1984).

Fat accumulation also depends on the amount of food absorbed. Experiments in which common carp were fed three diets (sorghum, wheat, and a protein-rich diet) showed that fat content was higher when any of these diets was fed *ad libitum* than when fed a restricted ration of 4% of their body weight per day (Hepher *et al.*, 1971). Also in rainbow trout, fat accumulation and calorific value of the tissue increased linearly with increasing feeding level (Staples & Nomura, 1976). Emphasis in studies of body composition has been also given to the effects of season, water temperature, and sexual maturation (Bottesch, 1958, on common carp; Craig, 1977, on *Perca fluviatilis*; Idler & Clemens, 1959; Change & Idler, 1960; Brett *et al.*, 1969, on *Oncorhynchus nerka*; Dawson & Grimm, 1980, on female plaice, *Pleuronectes platessa*; Niimi & Beamish, 1974, on largemouth bass, *Micropterus salmoides*; Medford & Mackay, 1978, on pike, *Esox lucius*; Stirling, 1972, on *Dicentrarchus labrax*; Penczak *et al.*, 1976, on *Rutilus rutilus*).

An increase in the weight of an animal is very often associated with an increase in the relative content of lipid in the tissues. Table 28 shows such a relationship between fat content and body weight for a number of farm animals. In spite of the many interacting and confounding factors affecting fat content in body tissues of fish, here fat content seems also to be strongly related to body weight and to increase during the growing

Figure 25. The relationship between body weight and fat content in common carp fed on sorghum (●) and on protein-rich diet (x) at the Fish and Aquaculture Research Station, Dor Israel, during 1963–72.

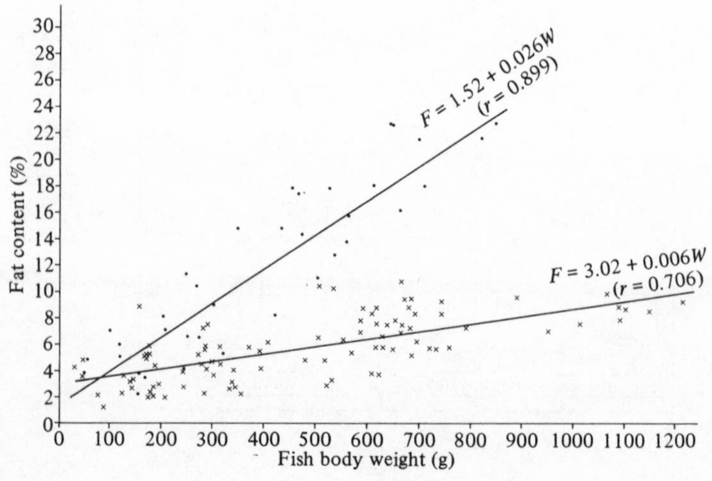

season (e.g. Suppes *et al.*, 1968, for channel catfish). Love (1970) quoted Hornell & Nayudu (1923) who considered a higher fat accumulation rate possible in larger fish since the relative growth rate of the larger fish is lower, allowing for the excess food, if available, to be converted into fat. Later studies, however, show that fat accumulation is not merely a function of available food. Love (1970) considered this phenomenon a natural process of ageing, by which the storage of energy in the body increases toward spawning, enabling an easier recuperation after spawning.

When considering the relationship between body weight and fat content of the tissues, or any change in the body fat content, it should be borne in mind that when fat content is expressed as a percentage of dry matter these changes may be masked. As in higher animals, there is a strong inverse linear relationship between fat and moisture contents of fish tissues (Bottesch, 1958; Brett *et al.*, 1969; Love, 1970; Vavruska & Janecek, 1973; Niimi & Beamish, 1974; Ionas, 1974; Stirling, 1976). Variations in fat content of the fish will, therefore, be parallel to their dry matter content (reciprocal of moisture) and will not be apparent.

Fluctuations in protein content of fish tissue with change in body weight are much less marked than those in fat (Brett *et al.*, 1969; Kausch & Ballion-Cusmano, 1976). It usually remains between 14.5 and 17.5%. Since fat and protein are the major energetic constituents of the fish body, the change in the more variable fat content becomes the main factor which

Table 28. *Percentage composition and energy content of weight gain of various farm animals of different body weights*

Animal	Live body weight (kg)	Moisture (%)	Protein (%)	Fat (%)	Ash (%)	Calorific value (kcal/g)
Chicken	0.225	69.5	22.2	5.6	3.9	1.49
	0.675	61.9	23.3	8.6	3.7	2.39
	1.350	56.5	14.4	25.1	2.2	3.07
Sheep	8.5	57.9	15.3	24.8	2.2	3.32
	33.0	48.0	16.3	32.4	3.1	3.94
	59.0	25.1	15.8	52.8	6.3	4.97
Pig	22.5	39.0	12.7	46.0	2.9	5.03
	95.0	38.0	12.4	47.0	2.8	5.11
	112.0	34.0	11.0	52.0	2.4	5.58
Cow	67.5	67.1	19.0	8.4	—	1.87
	225.0	59.4	16.5	18.9	—	2.72
	450.0	55.2	20.9	18.7	—	2.95

Source: From McDonald *et al.* (1966).

determines the calorific value of the tissue. If fat content is known, the calorific value of the tissues can be estimated with reasonable accuracy.

In view of the above, it can be taken that the content of fat in cultured fish can be regulated by the composition and amount of the supplementary feed. However, there are only few data on the regression of fat content and body weight, according which feeding level could be determined, for any specific diet. Mishvelov (1983) found that the regression between body weight and moisture content in common carp is a hyperbolic function:

$$Y = \frac{a}{W} + b \qquad (46)$$

where Y is the moisture content (%). Since fat and moisture are linearly negatively correlated, it may be expected that fat will also be related to body weight in the same way. However, most studies have shown a linear relationship between fat content and body weight. Wolny *et al.* (1965) studied the changes in fat content of common carp during growth prior to maturation when stocked in ponds at varying densities (2500, 7500, and 15 000/ha). Analysis of their results shows very highly significant linear regressions between body weight and fat content (as percentage of wet weight) in each of these densities. These were:

for 2500 fish/ha $\quad F = 4.339 + 0.011\ W\ (r = 0.755)$
for 7500 fish/ha $\quad F = 2.845 + 0.035\ W\ (r = 0.802)$
for 15 000 fish/ha $\quad F = 2.110 + 0.051\ W\ (r = 0.787)$

The relationship between body weight of common carp and its fat content was studied by the author (unpublished) for fish fed two diets, sorghum and pellets containing 25% protein, at 4% of their body weight/day. The resulting regressions are presented in Figure 25. Both these linear regressions are statistically significant. Likewise, Staple & Nomura (1976) and Weatherley & Gill (1983) found in rainbow trout, and Hogendoorn (1983) in African catfish (*Clarias gariepinus*), high correlation coefficients between fat content, and/or calorific value, and body weight. They have concluded that estimates of body constituents can be made from body weight. This also means that the calorific value of the fish tissues can be estimated from its weight. However, in doing so three points should be considered:

(a) Since the fat content can be regulated by feeding, it is important to set the target fat content of the market fish. For example, according to Janecek (1972) the desirable fat content of a 1.5 kg common carp, which is the marketable weight of this fish in Europe, is about 15%. Janecek (1972) claims that this fat content improves the taste of the fish. The target fat contents will depend on the fish species and market preferences.

(b) As the main purpose of estimating the calorific value of the tissues is the determination of the feeding level, and since the increase in fat content seems to be linearly correlated with body weight, the calorific value of any of the intermediate body weights can also be calculated. Thus, if the above example is taken, an increase by 1% of fat/100 g increase in body weight can be expected. The calculated calorific value of the tissues of a common carp of 100 g (1% fat, 16% protein) will then be 1.0 kcal/g (or about 5.3 kcal/g dry matter), while for a fish of 1000 g (10% fat, 16% protein) it will be 1.85 kcal/g (or 6.67 kcal/g dry matter). This range is similar to that found by Ivlev (1939a) in his experiments (4.3 kcal/g for a 13.8 g fish to 5.6 kcal/g for a 421 g fish).

(c) It is obvious that for determining feeding level, it is the calorific value of the *gained* tissues which is important rather than that of the entire fish. Since the calorific value of the tissues increases with body weight, the calorific value of the gain must be higher than that of the fish tissues. Calculation for the example given above will show that it amounts to 1.09 and 2.79 kcal/g for carp of 100 g and 1000 g, respectively. The relationship between the calorific value of weight gain (C_g, kcal/g) and body weight (W, g) can be expressed in a form of a regression, which for the example given above is:

$$C_g = 0.905 + 0.00189W \tag{47}$$

If, in view of the above discussion, the regression of growth rate on body weight is calculated in units of cal/day rather than g/day, the regression coefficient will be higher than the 0.66 of equation (35). This regression coefficient will probably vary with different species and diets will be higher for species which tend to accumulate fat easily, and for diets of high *ME*/protein ratios.

Utilization of food for growth

The partial efficiency of food utilization for growth (E_{pg}) depends on many factors. The composition of the diet and its compatibility with the requirements for growth constitute one of the major factors among these. When the diet is deficient in any nutrient essential for growth, such as an essential amino acid, fatty acid, vitamin or mineral, a larger amount of food is required to supply this deficient element and the efficiency of the food utilization decreases. Also the fate of the food in the body is important, whether it is converted into proteinaceous tissue or accumulated as lipid. The maximum theoretical efficiency of conversion of glucose to fat is about 70% of the metabolizable energy. For the conversion of amino acids to protein, if the former are present in the right proportions, the maximum theoretical efficiency is about 80% of the metabolizable energy. However, the actual efficiencies of utilization of food for growth are much lower. The highest efficiencies of utilization of nutrients for growth seem to be those of egg protein by the developing embryo. This protein is of the highest biological value (see Chapter 6) and is especially suitable for the development of the embryo. According to Winberg (1956), the utilization of egg protein by the fish embryo amounts to 67%. Kamler (1976) studied the early developmental stages of the common carp and found that the calorific efficiency of yolk transformation into body tissues was 50–67%. Ando (1968) studied the composition of rainbow trout eggs at different stages of development and of fry hatched from them. Some of his results and the calculated efficiency of utilization for growth are presented in Table 29. Assuming that the requirement of energy for maintenance by the embryo is very small, and can therefore be neglected, the efficiency found from these data was 54.4%.

Few studies have been made on the actual efficiency of food utilization for growth by actively feeding fish, and on the effect of diet composition, feeding level or environmental conditions on it. Ivlev (1939a) found the utilization of carbohydrate by common carp for production of body fat to be 30% of the gross energy ingested, or if metabolizable energy is considered to be about 75% of the gross energy (see Chapter 1), this

Table 29. *Composition of rainbow trout eggs and of the fry after hatching*[a] *and the calculated efficiency of energy utilization for growth*

	Unfertilized eggs	Hatched fry
Average weight (mg)	91	113
Composition (mg)		
Moisture	53.6	90.1
Ash	1.4	1.4
Protein	26.25	15.65
Lipid	10.7	5.0
Calculated energy (cal)[b]	249	136
Efficiency of utilization (%)		54.5

Notes: [a] From Ando (1968).
[b] Energy equivalents used were: 5.65 kcal/g protein and 9.45 kcal/g lipid.

amounts to 40% of the metabolizable energy. An experiment with common carp was also performed by the author (unpublished data) in ponds each of 0.1 ha. The ponds were stocked with 500 fish of an average weight of 65 g which were fed on sorghum at two feeding levels: 4% of their body weight per day, and *ad libitum* (three ponds for each treatment). Both groups were analysed for the major body constituents at the beginning and end of the experiment. The results of these analyses served to calculate the calorific value of the gain. Since 4% of the feeding level is more than sufficient for maintenance, any additional energy consumed by the fish fed *ad libitum* was used for growth. Most of it was converted to body fat. *Ad libitum* feeding supplied an additional 2008 kcal/fish, which resulted in an extra gain of 605 kcal/fish. This is equivalent to a utilization efficiency of 30.2% of the gross energy, or about 40% of the metabolizable energy, if the same factor of 75% is applied to convert gross energy to metabolizable energy. A very similar experiment was carried out by Mironova (1976) with *Oreochromis niloticus*. She found that the efficiency of food utilization for growth decreased with increasing weight of the fish. It was 27% of the gross energy (mean of three temperatures tested) for fish of 3–6 g; 22% for fish of 10–33 g; and only 13.5% for fish of 13–40 g. Hepher *et al.* (1983) studied the efficiency of utilization of dietary energy for growth of red tilapia (a hybrid of obscure origin) when fed three diets, containing different levels of protein and carbohydrate, at a rate of 3% of their body weight per day. At a temperature of 24.3 °C, and after deducting the energy requirement for maintenance, a high protein–low carbohydrate diet (43.9 and 23.3% respectively) was utilized at 33% of the gross dietary energy ingested. If the above conversion factor of 75% is used, this amounts to about 45% of the metabolizable energy. A medium protein and carbohydrate diet (28.7 and 43.6% respectively) was utilized at 22.5% of the gross, or 30% of the metabolizable energy. The efficiency of utilization for low protein–high carbohydrate diet (13.1 and 69.2% respectively), which favoured a higher deposition of body fat, was only 9% of the gross and 12% of the metabolizable energy.

The efficiency of utilization of dietary energy for growth is affected by the amount of feed consumed. De Silva & Balbontin (1974) found that the mean feed conversion efficiency of young herring (*Clupea harengus*) fed a ration of 1.3% of its body weight/day was higher than that of fish fed to satiation. However, this study did not differentiate between the effect of ration size on requirement for maintenance and that for growth, nor was it based on energy units. Among the few studies on the efficiency of food utilization for growth in cultured fishes are those by Huisman (1976) and Huisman *et al.* (1979). They fed common carp and rainbow trout with increasing amounts of feed and determined the resulting gain in body energy. The net efficiency of energy utilization for growth, after deducting the requirement for maintenance, calculated from the results of their studies, is presented in Figure 26. It can be seen that the efficiency of utilization decreased rather sharply with the increase in the feeding level. This is more pronounced when the marginal efficiencies are considered, i.e. the amount of extra energy gained as a result of the additional amount of dietary energy consumed over that of the lower level. Increasing the feeding level of common carp from 5 to 7% of body weight/day resulted in lowering the marginal utilization to about 10%, and further increase in the feeding level to 9% even resulted in a reduced growth. Similar results were obtained by Warren & Davis (1967) for the sculpin (*Cottus perplexus*). Here the net efficiency of energy utilization for growth decreased with increase in food consumption from about 50% when 7 kcal/day were consumed to less than 40% at a consumption of about 30 kcal/day.

Figure 26. The relationship between the metabolizable energy in the diet (in multiples of requirement for maintenance) and the net efficiency of its utilization for growth by rainbow trout (x) and common carp (●). Data are taken from Huisman (1976). The broken line designates marginal efficiency by common carp.

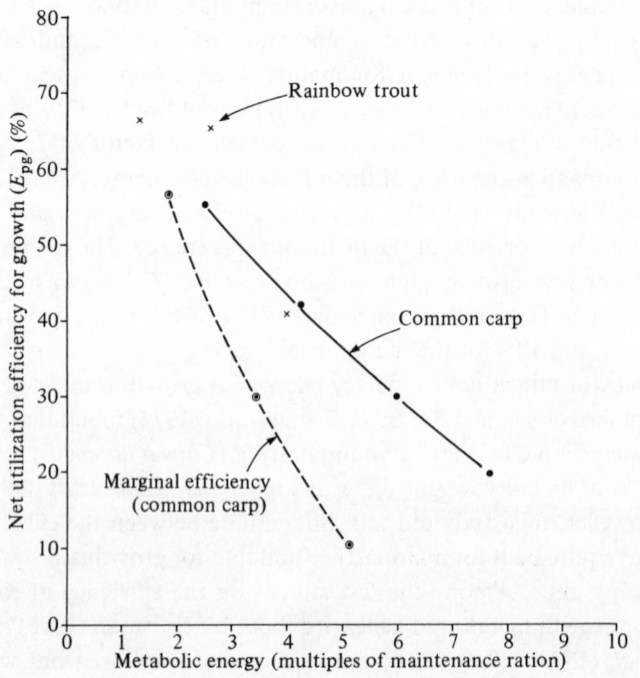

Temperature may have a noticeable effect on energy utilization for growth. Warren & Davis (1967) showed that when the temperature decreased in winter to 6.9 °C the energy utilization for growth of sculpin (*C. perplexus*) decreased to 35% compared with 44% in autumn, when the temperature was 11.6 °C. Also, Hepher *et al.* (1983) found that when the average temperature was reduced from 24.3 to 20.9 °C the utilization of protein-rich diet by red tilapia was reduced from 45% to 27% of the metabolizable energy. However, there was an increase in the utilization efficiency of carbohydrate-rich diet from 12% to 25% with the same decrease in temperature, probably due to a higher fat deposition at the lower temperature. According to Mironova (1976) the highest efficiency of energy utilization for growth is attained at an optimum temperature for the species. Efficiency decreases with the deviation of the temperature from this optimum.

It can thus be concluded that when the diet composition is compatible with the requirements for growth, and when environmental conditions are favourable for growth, the net efficiency of utilization of metabolizable energy for growth is about 40–50%, which is somewhat lower than the efficiency of utilization of metabolizable energy for maintenance. However, when the food is not suitable, such as with diets low in protein and high in carbohydrate, when fish are fed in excess, or when environmental conditions are unfavourable, efficiency decreases rapidly. Carbohydrate-rich diets may still have a relatively high efficiency with species that tend to accumulate fat, such as common carp.

Summary of dietary energy requirements

At this stage, the energy requirement and feeding level can be determined for any fish species, provided the appropriate coefficients of specific metabolism, specific growth rate, and the rate of fat accumulation are known for the particular fish species and for the environmental conditions in which it lives. If such determinations, made for a certain period, are based on average weight of the fish at the beginning of this period, the projected average weight during the period should be calculated first. If the period is not too long, the error caused by taking the average weight rather than the integrated weight is not too large. This average weight can serve for calculating both the maintenance requirement and the growth rate of the fish. The calorific value of the gain and the efficiency of energy utilization for growth should also be taken into account.

Since in the foregoing discussions sufficient data were given on the growth of common carp at the Fish and Aquaculture Research Station, Dor, Israel, an attempt is made here to develop an equation which can serve for calculating the feeding level of fish in the above conditions. This equation is a combination of equations (29), (37) and (47) given above, and takes into account an efficiency of metabolizable energy utilization for growth of 45%. It is assumed that the diet provided is the most suitable for growth, and that the temperature is within the optimal range for growth. Under these conditions the metabolizable energy requirement (kcal/day) is:

$$ME \text{ req.} = 0.070\ W^{0.8} + \frac{0.176\ W^{0.66}\,(0.905 + 0.00189\mathrm{W})}{0.45} \qquad (48)$$

A similar equation can be constructed for other fish species and for different environmental conditions.

It should be noted that while larger fish have a greater absolute requirement of food for maintenance and growth, the situation is reversed when the amount of food is considered relative to the body weight of the fish. Since the relative requirement for maintenance ($\frac{M}{W}$) and the relative requirement for growth ($\frac{G}{W}$) decrease with the increase in body weight, it can be expected that also the total food requirement ($\frac{R}{W}$) will decrease with increasing body weight. Elliott (1975a) has shown that the relation between fish weight and food ration (*Gammarus*) eaten in one meal by brown trout (*S. trutta*) can be described in the form:

$$R = a\ W^{b}$$

where b was 0.75. The value of a depended on temperature and was highest at the optimum temperature for growth (14–18 °C). Thus, if feeding level is expressed as percentage of body weight, this percentage decreases with body weight. This is noticed quite clearly in feeding charts for fish which do not accumulate much fat, such as salmonids. In the feeding chart for rainbow trout given in Appendix IV, feeding rate for fish weighing 0.2–1.5 g at 15 °C is 6.0%, while fish weighing over 167 g receive only 1.2% of their body weight per day. Most feeding charts for salmonids are based on expected gain and food conversion ratios (p. 295) and usually do not distinguish between requirement for maintenance and growth as was attempted above (Freeman *et al.*, 1967; Buterbaugh & Willoughby, 1967). Nevertheless, the rather sharp decrease in percentage of feed per body weight is noticed in all these charts. When, however, fish accumulate much fat during growth, and require increasing amounts of feed for this purpose, as often is the case for common carp, the decrease in the relative feeding level with increasing body weight is checked to a large extent. Feeding at a fixed percentage of body weight, as often practised for these fishes, is therefore not wide of the mark, at least at a medium range of fish weight (100–500 g).

6

Requirement for protein

Protein is the basic component of animal tissues and is, therefore, an essential nutrient for both maintenance and growth. At maintenance level the fish requires protein for replacement of worn-out tissues and proteinaceous products such as intestinal epithelial cells, enzymes and hormones, which are vital for the proper function of the body, and are recycled quite rapidly. The requirement of protein for the synthesis of new tissues is obvious, since protein constitutes 45–75% of the tissue dry matter. The capacity of the fish to synthesize protein *de novo* from carbon skeletons is limited, and most of the protein must, therefore, be supplied through the diet. Thus, the content of protein in the diet and its ratio to the metabolizable energy becomes of prime importance.

Protein for maintenance

As with energy, the amount of protein required for maintenance can be measured by feeding fish a diet containing just enough protein to balance the loss due to the recycling of tissues, enyzmes, etc., so that the protein content of the body will remain unchanged (zero nitrogen balance). However, adjustment of this amount is rather difficult. Since loss of endogenous nitrogen in faeces and metabolic excretion continues whether the fish consumes protein or not, very often protein requirement for maintenance is determined by measuring this nitrogen loss on fish fed a protein-free diet. The rate of protein (nitrogen) loss can be measured either by determining the content of body nitrogen on successive samples, or by determining the content of nitrogen in the excretions. Kaushik (1980) and Ogino *et al.* (1980) compared these two methods and found a small difference in the results. The amount of excreted nitrogen in the study of Ogino *et al.* (1980) was higher by 5.8–12% than the loss of body nitrogen.

Some fishes, such as salmonids, are very difficult to feed a diet without protein, and therefore the endogenous nitrogen excretion is sometimes measured on fasting fish. This, however, introduces a certain degree of error. Since, with the consumption of food, even if it does not contain protein, there is an increased secretion of digestive enzymes and loss of intestinal epithelial cells, excreted endogenous nitrogen measured on fasting fish will be lower than that measured on fed fish. Nose (1961) found that the endogenous nitrogen measured on fasting rainbow trout was 14 mg/100 g body weight/day, compared with 25 mg/100 g/day in similar fish fed a protein-free diet. Also Savitz (1969, 1971) reported that fasting bluegill sunfish (*Lepomis macrochirus*) excreted less nitrogen than those fed glucose.

Dietary protein is not fully utilized for the synthesis of tissues. The unutilized fraction of dietary nitrogen is excreted along with that originating from the recycled tissues. Endogenous nitrogen excreted by fish fed a protein-free diet is therefore lower than that excreted by fish fed protein at maintenance level. Gerking (1955a, b) measured the nitrogen content of bluegill sunfish fed various amounts of mealworms, and found a relationship between the retained nitrogen, the excreted endogenous nitrogen and the ingested dietary nitrogen. The amount of protein nitrogen required for maintenance (zero nitrogen balance), for a fish of about 30 g at 25 °C, was 7.2 mg/day, as compared to an excretion of 5.8 mg N/day by a similar fish fed a diet without protein. The loss of the unutilized portion of the dietary protein is unavoidable, and must therefore also be taken into account when calculating protein requirement for maintenance. It is obvious that endogenous nitrogen excretion, when measured on fish fed a protein-free diet is not sufficient for such calculation, and the utilization efficiency of the protein must also be known. Different sources of dietary protein may vary in their utilization rate, which may alter the requirement for maintenance of this particular protein. Thus, for instance, Nose (1971) fed rainbow trout various levels of casein, fish meal and soybean meal. When fed on casein or fish meal, zero nitrogen balance was achieved at 39 mg N/100 g body weight/day. However, on feeding soybean meal this balance was achieved only at 55 mg/100 g/day.

The requirement of protein for maintenance depends to a considerable extent on water temperature. The effects of temperature on endogenous nitrogen excretion by bluegill sunfish (Savitz, 1969) and common carp (Ogino *et al.*, 1973) are given in Table 30, from which it can be seen that as expected the excretion increases with the increase in temperature.

Another important factor affecting protein requirement for maintenance is the body weight of the fish. In higher animals it has been found that the requirement of protein for maintenance more or less parallels the requirement of energy for basal metabolism (Smuts, 1935). According to McDonald *et al.* (1966) the ratio between the two is about 2 mg nitrogen (= 12.5 mg protein)/kcal of basal metabolism. If this ratio is also true for fish, the regression of protein requirement for maintenance (or endogenous nitrogen excretion) on body weight should have the same weight exponent as that of metabolism, i.e. 0.8. However, Gerking (1955b), who studied the nitrogen balance of bluegill sunfish, found the following regression for the above variables:

$$N_E = 0.937W^{0.54} \quad (49)$$

where N_E = endogenous nitrogen excretion in mg/day by a fish fed a diet devoid of protein. From this equation it appears that endogenous nitrogen excretion increases with body weight to a much lesser degree than metabolism. The 'specific endogenous nitrogen excretion' by a fish of 1 g, as expressed in Gerking's equation by the factor of 0.937 (Y-intercept), is much higher than quoted for higher animals. A weight exponent of 0.9, obviously very different from the one given above, was found by Iwata (1970) and by Infante (1974) for crucian and common carp. Even higher weight exponents were found by Savitz (1969) for bluegill sunfish. The regressions differed somewhat with temperature:

At a temperature of 23.9 °C $$N_E = 0.129W^{0.977} \quad (50)$$

Table 30. Average endogenous nitrogen excretion by bluegill sunfish and common carp as related to temperature

Temperature (°C)	mg N/100 g/day
Bluegill sunfish[a]	
7.2	5.55
15.6	4.85
23.9	11.54
29.4–32.2	28.56
Common carp[b]	
20	10.5
22	11.2
24	11.9
26	12.7
27	13.0

Notes: [a] Average fish weight about 100 g; from Savitz (1969).
[b] Average fish weight about 200 g; from Ogino *et al.* (1973).

At a temperature of 29.4–32.2 °C $N_E = 0.298 W^{0.99}$ (51)

Here the regression coefficients approached 1.0 and the relationship between endogenous nitrogen excretion and body weight was almost linear. In a later study Gerking (1971) received similar results to those of Savitz. Ogino *et al.* (1980) also considered the relationship between endogenous nitrogen excretion and body weight to be linear. If this is true, the endogenous nitrogen excretion can be expressed in relation to a unit body weight, as has been done by a number of authors. Values for the endogenous nitrogen excretion given by these authors are presented in Table 31. The high values of endogenous nitrogen excretion found by Dabrowski (1977) and Ehrlich (1974) given in this Table are explained by Dabrowski as due to the small average weight of the fish. If he is right, it would indicate that the assumption that endogenous nitrogen excretion is linearly related to body weight is erroneous. It should be remembered that most studies quoted above as giving a linear relationship between endogenous nitrogen excretion and body weight were carried out on fish weighing 2–10 g. It is doubtful whether this range is wide enough to give a reliable regression. Thus, from the above discussion it is evident that further studies are necessary to establish the relationship between endogenous nitrogen excretion and body weight, and protein requirement for maintenance.

Protein level

Many experiments have been conducted to determine the optimal level of protein in the diet for various fish species. However, the inter-

Table 31. *Endogenous nitrogen excretion (mg N/100 g body wt/day) by various fish species*

Fish species	Fish weight (g)	Temperature (°C)	N_E	Source
Carassius auratus (goldfish)	1.8–1.95	17	10.8	Nose (1961)
	3.5–3.8	25	8.78	Nose (1961)
	NG[a]	NG	9.0	Migita *et al.* (1958) (quoted from Nose, 1961)
Cyprinus carpio (common carp)	78–370	20–27	10.5–13.0	Ogino *et al.* (1973)
	NG	NG	12.5–14.5	Creach (1972) (quoted from Dabrowski, 1977)
	NG	NG	14.5	Dabrowski (1976) (quoted from Dabrowski, 1977)
(fed glucose)	19–155	20	38.1	Infante (1974)
Ctenopharyngodon idella (grass carp)	0.15–0.2	22–23	54.9	Dabrowski (1977)
Lepomis macrochirus (bluegill)	13.9–85.2	20.5	3.2–39.7	Gerking (1971)
Salmo gairdneri (rainbow trout)	5.3	17	13.9	Nose (1961)
Oreochromis mossambicus (tilapia)	1.6–1.8	27	10.5	Jauncey (1982)

Note: [a] Not given.

pretation of the results of these experiments is not always easy. Since protein requirement is affected by, and interacts with, many factors such as environmental conditions, specific physiology and feeding habits, as well as age and developmental stage of the fish, the results of one experiment, under a certain set of experimental conditions, are not necessarily true for a different set of conditions. Without understanding more about the relationships between protein requirement and factors affecting it, the results of the experiments cannot be generally applied. There is also the problem of how to evaluate the effect of the protein.

Evaluation criteria

A number of criteria are usually used for this evaluation, but these do not always agree. A better understanding of the nature of these criteria and how they are affected by the protein level and other dietary and environmental factors is important for a better evaluation of the diets.

Growth rate This is one of the most common criteria by which the diet and its protein level are evaluated. Since fish farmers are interested in a higher growth rate of the fish, this criterion is the most practical and therefore more often used. However, growth is not always the result of protein synthesis. Quite often growth can result from increase in the fat contents of the tissues. Though protein may cause fat deposition and thus 'growth', this may also be achieved by carbohydrate-rich diets.

Feed conversion ratio (*FCR*) The ratio between the weight of the food consumed and the weight gain of the fish often serves as a measure of efficiency of the diet. The more suitable the diet for growth, the less food is required to produce a unit weight gain, i.e. a lower *FCR*. The reciprocal of *FCR* is the food utilization ratio (often called food conversion efficiency), which is the weight gain obtained from one unit weight of food. This is often expressed as a percentage of the latter. The main problem here is that *FCR* is usually given in wet weights, of both food and weight gain. Some foods, such as plants or natural food, contain much more moisture than others, such as grains or dry pellets. This may cause a bias not necessarily related to the protein content. Also, the weight gain in the fish may comprise varied amounts of moisture or fat, which further complicate the evaluation of the diets with regard to their protein level.

Protein efficiency ratio (*PER*) The *PER* is defined as the ratio between the weight gain of fish and the amount of protein consumed:

$$PER = \frac{\text{Fish wet weight gain (g)}}{\text{Crude protein fed (g)}} \qquad (52)$$

This corrects the error due to variability in moisture contents in the diets which was pointed out above. However, this method of evaluation has three main drawbacks: (a) It evaluates the protein in the diet rather than the diet itself. The higher the *PER* the more efficient the dietary protein, but not necessarily the level of protein in the diet. (b) Like *FCR* it does not consider differences in the composition of the weight gained and the accumulation of fat in fish tissues. Therefore, *PER* may increase due to increase in feeding level. (c) It is assumed that all protein in the diet is utilized for the synthesis of new tissues, whereas, in fact, some of it is utilized for maintenance. This last drawback is sometimes corrected by adding to the weight gain the body weight lost when the fish are fed a diet with no protein (Nose, 1971). The evaluating criterion is then called 'net protein ratio'.

Productive protein value (*PPV*) This criterion, which is sometimes also called 'efficiency of protein utilization' (Gerking, 1971), evaluates the protein in the diet by the ratio between the protein retained in fish tissues and the dietary protein consumed. The former is usually determined by carcass analyses of samples of fish taken before and after feeding with the evaluated protein, and generally expressed as a percentage of the protein consumed.

$$PPV\,(\%) = \frac{\text{Protein retained in tissues} \times 100}{\text{Dietary protein consumed}} \tag{53}$$

It is obvious that *PPV* is a more refined criterion for the evaluation of dietary protein since it takes into account the transformation of the dietary protein into body protein rather than the overall increase in body weight.

It has been shown for a number of fish species, fed a variety of live foods over a wide range of protein consumption, that the relationship between nitrogen retention (N_r) and nitrogen consumption (N_I), the two components of *PPV*, is linear or nearly linear (Birkett, 1969, for plaice (*Pleuronectes platessa*), sole (*Solea vulgaris*) and perch (*Perca fluviatilis*); Davis, 1967, quoted from Gerking, 1971, for cut-throat trout, *Salmo clarki*; Gerking, 1971, for bluegill sunfish, *Lepomis macrochirus*). The more protein is consumed, the more is retained.

$$N_r\,(\text{mg/day}) = a + bN_I \tag{54}$$

a is the Y-intercept, a negative value giving the amount of nitrogen lost (endogenous nitrogen excreted) when the fish do not consume protein at all. It follows that the basic ratio in the *PPV* equation $\frac{N_r}{N_I} = (a + bN_I)/N_I = a/N_I + b$. From the above it is obvious that this ratio, and thus *PPV*, will be affected by the amount of nitrogen consumed. When the amount of nitrogen consumed is small the *PPV* will be small, and only when the amount of nitrogen consumed is large *PPV* approaches *b*, which is the coefficient of regression giving the slope of the line describing the relationship between nitrogen retained and nitrogen consumed. According to Gerking (1971) the regression coefficient *b*, and thus *PPV*, is also affected by the body weight of the fish and decreases with increase in body weight.

Net protein utilization (*NPU*) Like the previous case, *NPU* is the ratio of protein retained to protein consumed expressed as a percentage of the latter. However, it is an improved criterion since it also takes into account the protein utilized for maintenance, which is subtracted from the retained protein. In order to determine *NPU* two parallel treatments must be compared. In the one fish are fed on the tested protein, while in the other they are fed on a control diet devoid of protein. Since in some fishes a protein-free diet is not accepted by the fish, a low protein diet can be used instead (Cowey *et al.*, 1974). The assimilated nitrogen retained by the fish is calculated either by subtracting losses of metabolic faecal nitrogen and endogenous excreted nitrogen from the ingested nitrogen, or by difference in carcass analyses (Miller & Bender, 1955):

$$NPU = \frac{N_I - (N_f - N_{fk}) - N_u - N_{uk})}{N_I - N_{Ik}} \times 100 \tag{55}$$

$$NPU = \frac{N_B - N_{Bk}}{NI - N_{Ik}} \times 100 \tag{56}$$

where N_f and N_u are, respectively, nitrogen faecal and urinary losses from fish fed the test diet; N_{fk} and N_{uk} are, respectively, nitrogen faecal and urinary losses from fish fed the same amount of low protein or protein-free diet; N_B and N_{Bk} are body nitrogen of fish fed the test and control diets, respectively; N_I and N_{Ik} are the nitrogen intake of fish fed the test diet and low protein or protein-free diets, respectively.

Biological value (*BV*) This evaluating criterion of the proteins is based on the ratio between the nitrogen retained by the fish and the absorbed nitrogen, and it is expressed as a percentage of the latter. Thus *BV* is *NPU* corrected for digestibility:

$$BV = \frac{NPU}{\text{True digestibility}} \tag{57}$$

As the previous criteria, *BV* is also affected by the level of the dietary protein. At low protein levels the endogenous nitrogen assumes a relatively high proportion, which causes a bias in the *BV* to the lower side. On the other hand, above a certain level of dietary protein an appreciable portion of it is deaminated and utilized for energy, leaving a smaller portion for protein synthesis. A decrease in *BV* at high protein levels was found to be the case in fish as in mammals. Tunison *et al.* (1942) found it in brook trout. Also, Ogino & Chen (1973b) and Ogino *et al.* (1980), who studied the *BV* of proteins at various levels of inclusion in the diet, found that it decreased considerably with the increase in protein level (Table 32). It is worth noticing that in the earlier study by Ogino & Chen (1973b) relatively high *BV* values were obtained even at high protein levels.

Dietary protein requirement

Growth rate was the evaluating criterion used in most studies on dietary protein level required by fish. Table 33 presents data collected from the literature on protein levels resulting in highest growth rates of various fishes and the ratios between the protein levels and the energy contained in the diets. The most striking fact in this table is the high protein level required by fish for maximum growth. Most fishes require 35–50% protein in their diets. Hastings (1973) explained the lower protein requirement found for channel catfish by Tiemeier & Deyoe (1980) by the low water temperature in the ponds in Kansas, where the experiments

Table 32. *Biological value (BV) of various proteins fed to common carp and rainbow trout at various dietary levels*

Protein source	Protein level (%)	*BV* (%)
(a) Common carp		
Casein	11.1	80
	21.8	74
	33.9	69
	35.0	65
	42.8	63
White fish meal	5.5	70
	10.7	76
	21.3	72
	43.4	64
Dried egg yolk	20.7	83
	41.5	70
(b) Rainbow trout		
Whole egg protein	5.5	89
	10.9	80
	15.7	77
	30.8	58
	41.6	47

Source: (a) From Ogino & Chen (1973b); (b) From Ogino *et al.* (1980).

Table 33. *Dietary protein and energy levels resulting in highest growth rates in various fish species*

Fish species	*Crude dietary protein (%)*	*Gross[a] dietary energy (kcal/kg)*	*Protein energy ratio (mg/kcal)*	*Source*
Anguilla japonica	44.5	—	—	Nose & Arai (1973)
Chanos chanos	40.0	3650[b]	110	Lim *et al.* (1979)
Chrysophrys auratus	40.0	5370[c]	74	Luquet & Sabaut (1973)
Clarias gariepinus	40.0	4441[c]	90	Machiels & Henken (1985)
Ctenopharyngodon idella	45.6	—	—	Dabrowski (1977)
Cyprinus carpio	31.0	3100	100	Takeuchi *et al.* (1979)
	33.0	3060	108	Sin (1973b)
	35.0	5430[c]	64	Jauncey (1981)
	38.4	3860[c]	99	Sin (1973a)
	38,5	3700	104	Ogino & Saito (1970)
	40.6	3265	124	Ogino *et al.* (1976)
	42.0	—	—	Meske & Becker (1981)
	42.0	4883	86	Eckhardt *et al.* (1982)
	45.0	4920	91	Sen *et al.* (1978)
	45.9	—	—	Mann (1974)
	45.9	5330[b]	86	Schwarz *et al.* (1983)
Dicentrarchus labrax	40.0	3000	133	Alliot *et al.* (1979)
Epinephelus salmoides	50.0	3410	147	Teng *et al.* (1978)
Fugu rubripes	50.0	—	—	Kanazawa *et al.* (1980b)
Ictalurus punctatus	25.0	2670[c]	94	Tiemeir & Deyoe (1980)
	25.3	—	—	Nail (1962)
	35.0	—	—	Murray *et al.* (1977)
	35.0	—	—	Lovell (1972)
	35–40	4040[c]	93	Dupree & Sneed (1966)
	36–40	2750–3410	85	Garling & Wilson (1976)

Table 33 (*cont.*)

Fish species	*Crude dietary protein (%)*	*Gross[a] dietary energy (kcal/kg)*	*Protein energy ratio (mg/kcal)*	*Source*
	40–44	—	100	Winfree & Stickney (1984)
	42.0	4090[c]	103	Prather & Lovell (1973)
Labeo rohita	45.0	4920	91	Sen *et al.* (1978)
Micropterus dolomieiu	45.0	4400	102	Anderson *et al.* (1981)
Micropterus salmoides	40.0	4360	92	Anderson *et al.* (1981)
Morone saxatilis	55.0	5930[c]	93	Millikin (1982)
Oncorhynchus kisutch	40.0	—	—	Zeitoun *et al.* (1974)
Oncorhynchus nerka	45.0	—	—	Halver *et al.* (1964)
Oncorhynchus tshawytscha	55.0	—	—	DeLong *et al.* (1957, 1958)
Oreochromis aureus	34–56	3200–4600	115	Winfree (1979)
	36.0	4200	86	Davis & Stickney (1978)
Oreochromis niloticus	52.7	3330	158	Santiago *et al.* (1981)
Oreochromis sp. (red tilapia)	34.4	—	—	Cismeros-Moreno (1981)
Oreochromis hybrid (*O. niloticus* × *O. aureus*)	30–35	4130	127	Viola & Zohar (1984)
Pleuronectes platessa	50.0	—	—	Cowey *et al.* (1972)
	52.0	—	—	Anonymous (1973)
Salmo gairdneri	35.0	7260[b]	48	Watanabe *et al.* (1979)
	36–49	—	—	Takeuchi *et al.* (1978)
	39.2	4570[c]	86	Ogino *et al.* (1976)
	40.0	—	—	Cho *et al.* (1976a)
	40–50	—	—	Satia (1974)
	42–51.2	4969–4993	94	Austreng & Refstie (1979)
	45.0	—	—	Halver *et al.* (1964)
	50.0	—	—	Lall & Bishop (1979)
	50.0	—	—	Zeitoun *et al.* (1976)
	60.0	—	—	Nose (1963)

Fish species	*Crude dietary protein (%)*	*Gross[a] dietary energy (kcal/kg)*	*Protein energy ratio (mg/kcal)*	*Source*
Seriola quinqueradiata	42.0	3830	110	Jauncey (1982)
Tilapia zillii	35.0	5200	67	Mazid *et al.* (1979)

Notes: [a] Original values of energy were given in various forms and were all converted to gross energy using the following somewhat arbitrary conversion factors:

$$\text{Gross energy} = \frac{\text{Digestibile energy}}{0.75} \text{ (see Appendix I)}$$

$$\text{Gross energy} = \frac{\text{Metabolizable energy}}{0.7} \text{ (see p. 13).}$$

[b] Original energy values were given as digestible energy.
[c] Original energy values were given as metabolizable energy.

were conducted. He assumed that this temperature (about 23 °C) allows less than optimum growth and hence accounts for the lower protein requirement. Hastings found that at temperatures below 24 °C channel catfish grew no better on 35% protein than on 25%, but when water temperature exceeded 24 °C the fish gained more on 30 and 35% protein diet. This supported results obtained by DeLong *et al.* (1957, 1958) with chinook salmon (*Oncorhnychus tshawytscha*) fed increasing levels of dietary protein. At 6–8 °C maximum gains were observed at 39–40% protein level. However, at a water temperature of 14.5 °C maximum gains were observed at 55–65% protein. Average protein requirement for maximum growth for the 50 cases cited in Table 33 was 41.9 ± 7.2(SD). This is two to three times higher than protein requirement of growing terrestrial animals. Some authors have explained this high protein requirement by relating it to the feeding habits of fish, pointing out that most fishes are carnivorous (Cowey, 1975; Watanabe *et al.*, 1979). However, high protein levels are also required by omnivorous and even herbivorous fish such as common carp, tilapia and grass carp, and no differences could be noted in Table 33 between the protein requirement of these fishes and that of carnivorous fishes. Mertz (1968) relates the high protein requirement to the high amino acid level in the plasma of the fish, which is three to six times higher than that in mammals. He postulated that this is caused by a deficiency in deamination enzymes, and high recycling of amino acids. However, there is no experimental evidence to support this assumption. In any case, since minimal amounts of amino acids are excreted, they must eventually be deaminated and utilized for energy. A more plausible explanation is that a portion of the consumed protein is immediately catabolized for energy. From the discussion in Chapter 3 it appears that this source of energy is preferred over carbohydrate (Gerking, 1955a). Kutty & Peer Mohamed (1975) estimated from experiments on the mullet *Rhinomugil corsula* that protein contributed 14–15% of the total energy expended in routine metabolism.

The average ratio between protein requirement (mg) and energy requirement (kcal) as calculated from the 30 values in Table 33 is 99.0 ± 22.8(SD), while it is much lower for poultry (60–70) and mammals (about 40–65) (National Research Council, 1977b, 1979). The higher protein:-energy ratio in fish may, in part, result from a lower energy requirement, which has already been discussed above (p. 122). In part, however, it may be due to the contribution of energy by protein. A number of questions remain to be answered such as: Is the use of protein for energy facultative or obligatory? Is it related to the use of protein in the red muscle? If it is, would fish having different proportions of red muscle also have a different requirement for protein? If utilization of protein for energy is obligatory, and carbohydrate can supply only a part of the energy requirement, can lipid replace some of the protein and to what extent? The answers to these questions, which are important in applied fish farming, can only be found in further basic studies on the intermediate metabolism of the fish.

Many of the experiments, the results of which are quoted in Table 32, showed that, with the increase in protein content in the diet, there was an increase in growth rate of the fish. In many cases this increase was linear, at least for part of the range of protein level tested (Figures 27, 28). In most of these experiments an attempt was made to keep the compared diets isocaloric. This was generally done by calculating the calorific values of the diets, using calorific equivalents commonly used for endotherms, mainly chicken. The problems involved in using isocaloric diets based on

Figure 27. The effect of dietary protein level on the growth rate of rainbow trout.

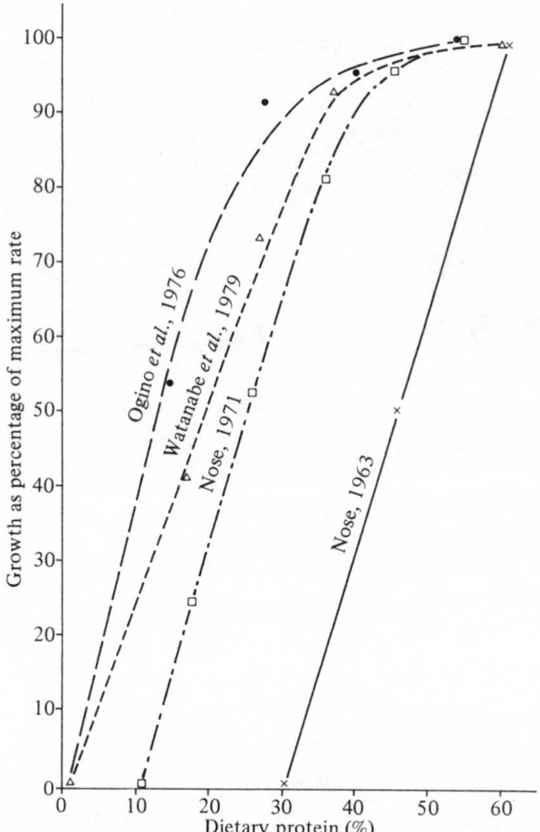

calculation rather than on calorimetric analysis are well discussed by Jobling (1983b). Moreover, since endotherms are ureotelic or uricotelic, the metabolizable energy of the protein is more or less equal to that of carbohydrate (about 4.1 kcal/g). The replacement of protein by carbohydrate does not greatly change the calorific value of the diet. However, from the discussion on this subject in Chapter 3, it appears that this is different for the ammoniotelic fish, where the metabolizable energy of the protein is higher than that of carbohydrate. Therefore, contrary to the assumption of many authors, the metabolizable energy of the diet increases with the increase in protein content. In some cases, such as that given by Dupree & Sneed (1966), the increase in energy with the increase in protein content was intentional. Excess of energy, including that from protein, may result in an accumulation of body fat, as indeed was found by Lee & Putnam

Figure 28. The effect of dietary protein level on the growth rate of common carp.

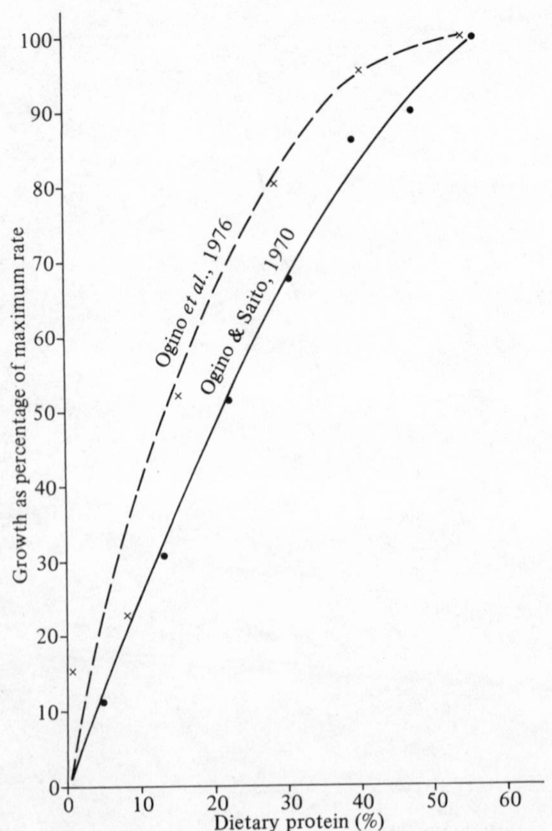

(1973), Takeuchi *et al.* (1978) and Watanabe *et al.* (1979). Thus, the increase in 'growth rate' with the increase in protein content of the diet may have been due to increase in fat content rather than protein anabolism. Ogino & Saito (1970) studied the effect of diets containing increasing levels of casein protein, from 0.4 to 55%, on the growth of young common carp. The higher the protein level in the diet, the higher the growth rate they observed. However, protein gain reached a peak at a protein level of 38%; further increase in the dietary protein level only resulted in fat deposition. In some cases, a peak in growth rate was observed at a certain protein level, beyond which there was no additional growth, and sometimes there was even a decrease in growth rate (Dupree & Sneed, 1966; Ogino *et al.*, 1976). It seems that this occurs when conditions are not suitable for fat deposition, such as at a very high energy level, when fat deposition is anyway at its maximum (Lee & Putnam, 1973) or when the temperatures are low, limiting fat deposition (Stickney & Andrews, 1971). Andrews & Stickney (1972) found that fat accumulation by channel catfish fed on lipid-rich diets increased with increase in temperature. It can be assumed that the effect of temperature on fat deposition from excess protein will be similar.

Though fat can be deposited as a result of excess of energy from proteinaceous sources, the conversion of carbohydrates and lipids to fat seems to be easier and faster. Fish fed on diets low in protein and rich in carbohydrate are fatter than those fed a protein-rich diet (Lee & Putnam, 1973; Ogino *et al.*, 1976; Reinitz & Hitzel, 1980; Hepher *et al.*, 1983).

Protein efficiency ratio (*PER*) is usually used for evaluating proteins of different sources for growing animals, but sometimes it is also used to evaluate the level of protein in the diet. As with higher animals, *PER* changes with the level of protein in the diet. *PER* increases with protein up to an optimum amount of consumed protein when its utilization is optimal, decreasing thereafter. These changes in *PER* are related both to the utilization of a portion of the protein for maintenance and to the overall energy level of the ration. Such curves for *PER* with the change in protein have been found by Nose (1971) for rainbow trout (Figure 29), and by Cowey *et al.* (1972) for plaice (*Pleuronectes platessa*). In the last case *PER* increased from 1.0 at 20% protein to 1.7 at 40% protein and then dropped again to 1.0 at 70% protein. Cowey *et al.* (1972) point out that the maximum *PER* in fish occurs at a much higher level of protein than in rats (where it occurs at about 10% protein). Ogino *et al.* (1976), however, have found a different relationship between *PER* and protein level. *PER* was already high at a relatively low protein level and decreased with increasing protein content. Similar curves have been found by many other authors (Ogino & Saito, 1970, for common carp; Dabrowski, 1977, for grass carp; Takeuchi *et al.*, 1978, and Watanabe *et al.*, 1979, for rainbow trout; Mazid *et al.*, 1979, for *Tilapia zillii*; Jauncey, 1982, for *Sarotherodon mossambicus*; Papaparaskeva-Papoutsoglou & Alexis, 1986, for *Mugil capito*). The difference between the two curves may be due to the level of dietary energy. With the lower dietary energy level, part of the protein may have been diverted to catabolic metabolism rather than to growth, hence the low *PER*. Only when the energy requirement was satisfied, was most of the protein utilized for growth, although with decreasing efficiency accompanying the increase in the level of protein. This is supported by the results of another treatment by Ogino *et al.* (1976) (Figure 29), where the level of dietary energy was lower, and so was *PER*.

PPV and *NPU* are used occasionally to evaluate protein levels for fish. At low levels of dietary protein the efficiency of certain amino acids may limit protein utilization, while with the increase in the level of protein a decrease in *PPV* or *NPU* usually occurs (Cho *et al.*, 1976a). Cowey *et al.* (1972) fed plaice (*Pleuronectes platessa*) six diets containing protein at

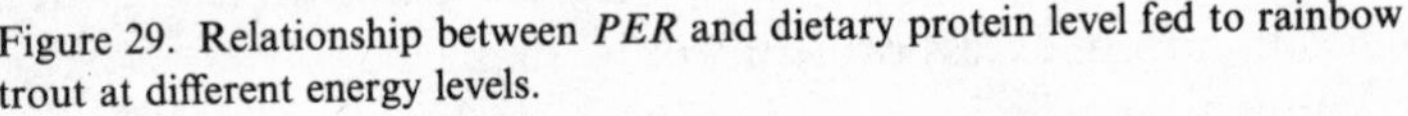
Figure 29. Relationship between *PER* and dietary protein level fed to rainbow trout at different energy levels.

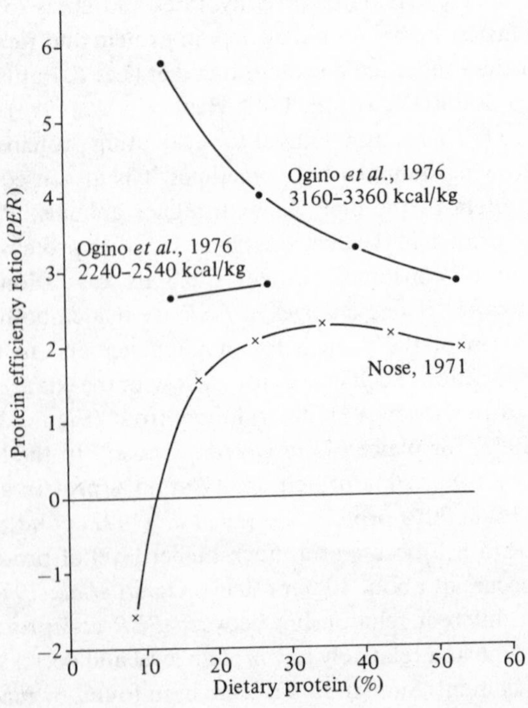

equally spaced concentrations between 20 and 70% of dry diet. *NPU* decreased linearly with the increase in dietary protein content. It should be kept in mind, however, that when the experimental conditions are constant, i.e. at equal fish weights, dietary protein and energy levels, both *PER* and *NPU* can serve as very good criteria to compare proteins from different sources. This will be discussed in greater detail in a later section.

The utilization of part of the protein through catabolism for the supply of energy for metabolism is well recognized in mammals, when the diet is poor in energy from other sources. Since the energetic value of protein for fish is higher than that of carbohydrate, such utilization in fish may also be more efficient biologically. However, since the cost of protein is higher than carbohydrates or lipids, this is a wasteful practice from an economic point of view. It is worthwhile to supply as much as possible of the required energy as carbohydrates and lipids rather than protein. Many experiments show that energy supplied from both carbohydrate and lipids has a sparing effect on the requirement of protein. Phillips *et al.* (1966a) fed brook trout (*Salvelinus fontinalis*) diets having two protein levels – 19 and 27%. At equal energy levels the growth rate was better on the higher protein level, but when lipid was added to the lower protein diet, the fish grew as well as on the higher protein level or better. Nail (1962) attempted at quantifying the protein sparing effect. He found that each addition of 0.23 g of dextrin to catfish diet saved 0.05 g protein/100 g weight of fish.

As long as the amount of energy added is not too high, both carbohydrate and lipid can be equally used to spare protein. Garling & Wilson (1977) fed channel catfish fingerlings with diets containing 2750 kcal/kg, but with varying proportions of carbohydrate (dextrin) to lipid. They found no differences in growth rate or *FCR* among these diets. Anderson *et al.* (1984) added carbohydrates of different molecular weight to diets of *Oreochromis niloticus*. Both growth rate and *PPV* increased with increasing proportion of carbohydrate energy in the diet. Glucose spared more protein than dextrin and starch at low inclusion levels, but dextrin and sucrose were more efficient than glucose at high inclusion level (40% of dietary energy).

Some fishes, however, do not digest and utilize well carbohydrates of high molecular weight, differences in the sources of energy are more apparent. Glucose seems to be more effective than starch in sparing dietary protein in rainbow trout (Pieper & Peffer, 1979; Bergot, 1979b). The sparing effect is more pronounced when lipids are added to the diet than with carbohydrate. This is why the use of lipid as a protein sparing energy source has received much more attention than carbohydrate, for rainbow trout (Lee & Putnam, 1973; Steffens & Albrecht, 1973; Higuera *et al.*, 1977; Reinitz *et al.*, 1978; Takeuchi *et al.*, 1978, 1981).

When large amounts of energy must be supplied, it is easier to add energy-rich lipid rather than the more bulky carbohydrate. Therefore, even when the fish can utilize carbohydrates well, lipid is used as the energy source. This has been done with channel catfish diets (Tiemier *et al.*, 1965; Stickney & Andrews, 1972; Gaulitz *et al.*, 1979; Dupree *et al.*, 1979) and common carp diets (Viola & Rappaport, 1979b). Of course there is a possibility that the addition of lipid to the diet stimulated fat deposition and did not increase the synthesis of proteinaceous tissue, thus truly sparing protein. However, according to Watanabe *et al.* (1979) this occurs only when lipid is added to a diet already high in protein. A true protein sparing can be achieved, according to these authors, at somewhat reduced protein levels. The effect on protein utilization can be seen then by calculating the net protein utilization (*NPU*). Watanabe *et al.* (1979) reduced the protein level, in a diet fed to rainbow trout, from 48 to 35% with the addition of 15% lipid and found that body protein production increased. At this level of protein, the higher the lipid content of the diet, the higher were *PER* and *NPU* (Figure 30), i.e. both growth rate and protein production per unit of dietary protein increased. Also Ufodike & Matty (1983) found that with the increase in dietary energy, added as rice meal or cassava meal, fed to common carp, *NPU* increased. The increase

Figure 30. Variations in *PER* and *NPU* in rainbow trout with increasing content of lipid in the diet, when fish were fed a diet of 35% protein (after Watanabe *et al.*, 1979).

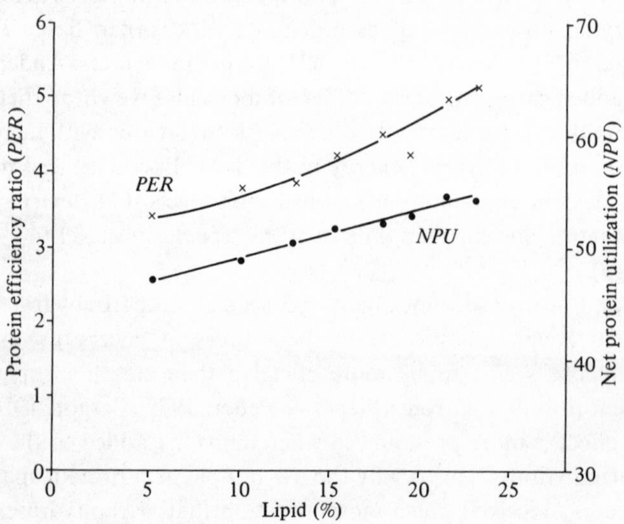

in the content of rice meal from 15% to 45% increased *NPU* from 14.6 to 38.5%, and increasing the content of cassava meal at the same rate increased *NPU* from 11.3 to 27.1% (see also similar results for turbot, *Scophthalmus maximus*, by Adron *et al.*, 1976).

The increased utilization of dietary protein may explain the 'extra-caloric value' of the lipid as pointed out by Viola & Rappaport (1979a) and Viola *et al.* (1981). The gain in body energy due to the addition of lipid may be higher than that of the added lipid. In a study by Watanabe *et al.* (1979), by adding 46.9 kcal to the diet as lipid, the fish gained 201 kcal in its tissues. An extracaloric value of lipid has also been observed by Eckhardt *et al.* (1982) with common carp. The extracaloric value decreases with increasing addition of lipid or protein until it finally disappears. In mammals a protein sparing effect by carbohydrate in protein inadequate diets was explained, in part, in the ability to provide carbon skeletons (α-keto acids) for synthesis of non-essential amino acids (Schepartz, 1973). However, no such studies have been performed on fish.

From the above discussion it can be seen that the content of energy in the diet, especially as lipid, and the ratio *P*:*ME*, is important for achieving best growth rate and protein utilization. In most cases quoted in Table 33 *P*:*ME* ratios are within the range of 80–120 mg/kcal. Ringrose (1971) studied the effect of increasing the dietary energy level on growth of brook trout by adding glucose to a diet relatively low in protein (32%). The best results were obtained at a *P*:*ME* ratio of 127–152 mg/kcal, which is somewhat higher than the range mentioned above. The narrow range observed among the different species and at different environmental conditions may be due to the dominant high protein content which, even when enough energy from carbohydrate and lipid is present, cannot be reduced below 32–35%.

Environmental conditions may affect protein requirement by fish. The effect of temperature on protein requirement has already been discussed above (p. 182). Zeitoun *et al.* (1973) found that water salinity also affects protein requirement by young rainbow trout. At a salinity of 10‰ the optimal dietary protein level was 40%, while at a salinity of 20‰ it was 45%.

Other factors affecting protein requirement are related to the fish itself. According to Page & Andrews (1973), protein requirement varies with body weight. They found that channel catfish of 14–100 g required a diet containing 35% protein, while those of 114–500 g required only 25%, when enough energy was provided in both cases. A reduction in protein requirement with increasing body weight was also noted by Winfree (1979) for *Oreochromis aureus*. A diet providing roughly 56% protein and

4600 kcal/kg produced highest gains in fry of 2.5 g. By the time the fish reached 5 g, 46% protein and 4000 kcal/kg was best, and at 7.5 g a 34% protein and 3200 kcal/kg gave highest gains.

Dietary protein composition

The amount of dietary protein is determined not only by the requirements for maintenance and growth, but also by its utilization rate for these purposes. This, in turn, depends to a large extent on the composition of the protein with respect to the essential amino acids and the presence in the diet of sufficient vitamins and minerals necessary for anabolic processes.

As was mentioned above, *PER*, which indicates the gain in body weight per unit of protein consumed, and *PPV* or *NPU*, both indicating the gain in body protein per unit of protein consumed, can serve as criteria for

Table 34. *A comparison between PER, NPU and BV of various proteins and their mixtures for the nutrition of rainbow trout, as found in five different studies*

	PER			*NPU*				*BV*			
Protein	(1)	(2)	(3)	(1)	(2)	(3)	(4)	(1)	(3)	(4)	(5)
Whole egg		3.8			61						
Egg yolk		3.8			60						89
Egg albumin		3.9			62						
Casein	2.3		2.0	52		40	38	53	41	38	80
Casein + gelatin		3.6			57						
Casein + methionine + tryptophan							45			45	
Casein + squid muscle		3.9			61						
Casein + whole egg		3.8			63						
White fish meal	2.6			45				50			76
Herring meal			1.9			38			41		
Fish muscle		3.8			62						
Squid muscle		3.9			62						
Pelleted trout feed											73
Gelatin											23
Soybean meal	−0.5			33				37			74
Extracted soybean						18			41		
Wheat germ											78
Zein + gelatin							5			7	
Zein + gelatin + lysine							26			39	
Petroleum yeast			2.0			42			46		73–79
Brewers yeast			1.2			30			38		
Methanophilic bacteria			1.6			37			40		
Algae (*Spirulina maxima*)			1.3			32			38		
Corn gluten meal							35			36	55

Sources: (1) Nose (1971), dietary protein content about 30%; (2) Ogino & Nanri (1980), dietary protein content 30%; (3) Atack & Matty (1979), dietary protein content 30%; (4) Nose (1963), dietary protein content 70%; (5) Ogino & Chen (1973b), calculated for 10% dietary protein content.

evaluating dietary proteins of different composition. However, these criteria relate to gross and not to metabolized protein on the one hand, and to body weight gain or protein gain alone and not to that required for maintenance, on the other. They are therefore more sensitive to environmental and other factors affecting growth. Such a comparison between proteins is possible as long as these factors are constant, usually within the same experiment, but does not hold under varying conditions. Results of *PER* and *NPU* determinations for different proteins and their mixtures as found in five different studies are given in Table 34. Due to the different conditions, no comparison can be made between them. Thus, for example, *PER* and *NPU* of casein were much higher in the study of Ogino & Nanri (1980) than that found by Nose (1971). However, these experiments enable the determination of the relative value of these proteins under the conditions of each experiment separately. It can be seen that, as with the higher animals, egg protein has a high value, and that the value of certain proteins can be improved by mixing them together or by supplementing them with amino acids.

A more accurate protein evaluating criterion for this purpose is the 'biological value' (*BV*). Here again, it should be remembered that dietary protein level may affect *BV* and thus cause differences among proteins tested at different dietary levels. At high protein levels, when the requirement for essential amino acids is satisfied, even though some proteins have low concentrations of these amino acids, the differences between the proteins tend to become less distinct. Cowey (1975), however, points out that differences in *BV* may be apparent even at high protein levels. He explains it by the higher essential amino acid requirement of fish compared with terrestrial mammals and birds.

PER, *NPU* and *BV* can differ for the same proteins utilized by different species. Thus proteins used for rainbow trout diets have lower *PER*, *NPU* and *BV* than the same proteins used for common carp. This difference was attributed by Nose (1971) to the ability of common carp to utilize carbohydrate as a source of energy more efficiently than the trout, which have to utilize a larger proportion of protein for this purpose. Atack *et al.* (1979) determined *PER*, *NPU* and *BV* for both species (Table 35), and also attributed the difference to the utilization of dietary energy.

As could have been surmised from the preceding discussion, the determination of the biological value of proteins is a long and tedious process. Attempts have been made, therefore, to correlate the biological value with other parameters, easier to determine, such as the relative proportion of the essential amino acids of the proteins used. Quotients calculated from such proportions are used to predict the biological values of these proteins. One of these quotients is the 'chemical score', which is based on the assumption that whole egg protein is of the highest *BV* and therefore the most suitable for growth, and that growth will be limited by that essential amino acid in the diet whose ratio to its content in the whole egg protein is the lowest. In order to find the chemical score of a protein, the percentage of each essential amino acid in the tested protein is divided by the percentage of the corresponding amino acid in egg protein. The one that shows the lowest ratio is considered to be the limiting amino acid. Then:

$$\text{Chemical score} = \frac{\text{Percentage of limiting EAA in protein} \times 100}{\text{Percentage of corresponding EAA in egg protein}} \tag{58}$$

In time it was found that although the first, limiting amino acid has an important role in determining the relative value of the dietary protein, other essential amino acids may also have some effect on it. This resulted in the development of the 'essential amino acid index' (*EAAI*). It is the geometrical mean of the ratio of all essential amino acids in the evaluated protein relative to their content in a highly nutritive reference protein, viz., whole egg (Oser, 1959)

$$EAAI = \sqrt[n]{\left(\frac{100a}{a_e} \times \frac{100b}{b_e} \times \frac{100c}{c_e} \times \ldots \frac{100j}{j_e}\right)} \tag{59}$$

where: $a, b, c \ldots j$ are the percentages of essential amino acids in the evaluated protein; $a_e, b_e, c_e \ldots j_e$ are the percentages of the corresponding amino acids in egg protein; n = number of amino acids taken into account. Close agreement has been found between *EAAI* and the biological value of dietary proteins.

Table 35. *A comparison between PER, NPU and BV of proteins used in diets of common carp and those of the same proteins used in diets of rainbow trout*

	PER		*NPU*		*BV*	
Protein	carp	trout	carp	trout	carp	trout
Herring meal	2.82	1.91	64	38	79	41
Methanophilic bacteria	2.54	1.62	49	37	52	40
Casein	2.48	1.97	49	40	52	41
Petroleum yeast	2.08	2.01	47	42	49	46
Soybean extract	1.35	—	42	18	51	41
Algae (*Spirulina maxima*)	1.15	1.33	36	32	51	38

Source: From Atack *et al.* (1979).

Chemical score and *EAAI* were developed mainly for evaluating proteins in diets of farm animals. In using those quotients for evaluating proteins in fish diets it is assumed that chicken egg protein is also the best for fish. Murai *et al.* (1984a) fed common carp fingerlings on purified diets containing variable ratios between casein and gelatin, thus changing the amino acid pattern of the ingested protein. He found highly significant correlations ($P < 0.01$) between weight gain, feed efficiency and *NPU*, and the *EAAI* calculated on the basis of chicken whole egg protein. Nose (1963) evaluated various proteins in trout diets by three parameters: the effect on growth rate, biological value, and *EAAI*. The *EAAI* was calculated in two ways – with reference to either chicken whole egg protein, or rainbow trout muscle protein. The results are given in Table 36. The analyses have shown a good correlation between both *EAAI*s and growth rate, but lesser ones (coefficients of 0.69 and 0.64 respectively) between *EAAI* and biological value. The difference between the two *EAAI*s was small and the ranking among the various proteins was the same in both. A similar analysis was made by Atack *et al.* (1979) for proteins used in common carp diets (Table 36). The results show that in absolute terms none of the chemical methods accurately predicted the biological value. According to the authors, this is probably due to the fact that the

Table 36. *Comparison of biological value of various proteins with their chemical quotients derived from the amino acid composition*

Protein	*BV*	*EEAI* (1)	*EAAI* (2)	Chem. score (1)	Chem. score (3)	Chem. score (4)
(a) Rainbow trout diets						
Casein + gelatin + methionine + tryptophan	45.4	82.8	85.8			
Zein + gelatin + lysine	38.7	50.9	55.1			
Casein + gelatin	38.3	79	84.5			
Gluten + gelatin	35.9	60.6	67.9			
Zein + gelatin	7.1	39.5	42.3			
(b) Common carp diets						
Herring meal	79	74		58	110	93
Methanophilic bacteria	52	66		47	87	75
Casein	52	61		46	74	53
Soybean extract	51	71		40	82	63
Petroleum yeast	49	68		40	80	63
Algae (*Spirulina maxima*)	41	68		42	70	67

Notes: (1) Using whole egg protein as reference.
(2) Using rainbow trout muscle protein as reference.
(3) Using amino acid requirement of carp (Nose, 1979) as reference.
(4) Using amino acid requirement of chinook salmon (Mertz, 1968) as reference.

Source: (a) From Nose (1963); (b) from Atack *et al.* (1979).

biological value of a protein varies with the level of that protein in the diet and also with variations in environmental conditions. The chemical scores were more consistent with the biological value than the *EAAI*. Each of the calculated chemical scores misplaced one of the evaluated proteins, but on the whole they could be used to determine the relative value of dietary proteins. Lieder (1965a) calculated the chemical scores of protein in natural food organisms and that in supplementary feed ingredients, using both whole egg protein and common carp flesh protein as references. Differences in the scores calculated for these two references were small. Proteins of natural food organisms showed a high chemical score whilst those of various cereal grains and legume seeds were much lower, showing either lysine or methionine to be the limiting amino acid.

The high biological value of chicken egg protein is no doubt associated with the fact that this is the only protein supplied to the developing embryo and must be efficiently utilized for its growth. By the same token it may be argued that fish egg protein would be better adjusted for fish. However, no direct comparisons of the biological value of these two proteins for fish have been made. The few indirect studies made gave conflicting results. Halver (1957b) had developed a test diet for trout composed of crystalline L-amino acids. Such a diet with the amino acid composition approximating that of fish fry yolk failed to yield acceptable growth, while a test diet with an amino acids pattern approximating that of casein gave favourable results. Also, Ogata *et al.* (1983) obtained a better growth of cherry salmon (*Oncorhynchus masou*) and amago salmon (*O. rhodurus*) when fed with purified casein diets supplemented with crystalline amino acids to simulate the amino acid pattern of these fish carcasses, than those fed a similar diet which simulated the amino acid pattern of cherry salmon eyed eggs. On the other hand, Rumsey & Ketola (1975) showed that the fish egg amino acid pattern served as a useful guide for formulating feeds. They supplemented a rainbow trout diet, containing commercial soybean meal as the sole source of protein, with amino acids to correct their concentrations quantitatively to the level of trout carcass or trout eggs. No significant improvement was achieved with the first, but growth was significantly enhanced – by 43% – with the latter.

These conflicting results may be due to variations which may occur in the amino acid composition of fish eggs as they become mature and after fertilization (Maslennikova, 1974). It has been demonstrated (Petrenko & Karasikova, 1960) that as Baltic herring eggs mature their protein is enriched with arginine and histidine and depleted of methionine, tryptophan and tyrosine. This indicates the need for further study on this topic.

However, until more information is available, it seems that, except for direct growth experiments, the chemical score and *EAAI* may serve as practical means for evaluating protein in fish diets.

In Appendix II information is given on the amino acid composition of fish muscle and fish eggs as well as of some proteins used in fish diets as gathered from the literature. The chemical scores and *EAAI*s of these proteins were calculated using both chicken and fish egg proteins as references. Maslennikova (1974) who reviewed the amino acid composition of fish muscle and inner organs points out of the small variation in the amino acid composition of muscle among different fish species. The average, which is also given in Appendix II, can therefore serve for calculating the chemical score and *EAAI*.

The comparison of the amino acid composition of the evaluated protein to any reference of 'best' protein such as egg protein is relevant only as long as the actual requirements for amino acids by the fish are not known. The requirements for amino acids have been established for a number of species. These requirements may be better references for evaluating dietary proteins than egg protein. The first study to determine the qualitative and quantitative requirements of essential amino acids was made for salmon (*Oncorhynchus tshawytscha*) by Halver (1957b), by techniques similar to those employed for higher vertebtates. Halver used a test diet containing 18 amino acids in proportions similar to those found in casein. The amino acids constituted 70% of the diet, while the remaining 30% were energetic constituents, such as white dextrin (6%), corn oil (5%), cod liver oil (2%), mineral and vitamin mixtures and a binder (carboxymethylcellulose – 10%). The essentiality of any of the amino acids was determined by omitting it from the test diet and comparing the performance of the deficient diet with that of the full test diet. In his experiments he found that the same ten amino acids essential for normal development of mammals are also essential to salmon and trout. These acids are identified in Table 37.

Halver's test diet for salmon and trout could be used, without any modification, for some other fish species, as was done by Mazid *et al.* (1978) for *Tilapia zillii*. However, in other cases the test diet was found to be unsuitable, since it either failed to sustain any growth or was not accepted by the fish. In these cases Halver's test diet has to be modified. In channel catfish it has been observed that the test diet was emptied from the intestine much faster than the commercial diet and therefore did not sustain growth. This was remedied by reducing the concentration of magnesium and sulphates in the mineral mixture, which were suspected to have cathartic effect (Dupree & Halver, 1970). Maximum growth and feed

Table 37. *Essential amino acid requirements by various fish species (as per cent of the dietary protein), as compared to that of chicken and young pig*

Fish species	Arg	His	Iso	Leu	Lys	Met	(Cys[a])	Phe	(Tyr[a])	Thr	Try	Val	Source
Anguilla japonica	4.5	2.1	4.0	5.3	5.3	3.2	(0.0)	5.8	(0.0)	4.0	1.0	4.0	(1)
						2.4	(2.7)	3.2	(5.3)				(1)
Cyprinus carpio	4.3	2.1	2.5	3.3	5.7	3.1	(0.0)	6.5	(0.0)	3.9	0.8	3.6	(2)
						2.1	(5.2)	3.4	(2.6)				(2)
	3.3	1.2	2.0	3.7	4.7	1.4	(0.7)	2.6	(1.8)	2.9	0.6	2.6	(3)
Ictalurus punctatus	4.3	1.5	2.6	3.5	5.1	2.3	(0.0)	5.0	(0.0)	2.3	0.5	3.0	(4)
		1.5	2.6	3.5								3.0	(5)
					5.1								(6)
						← 2.3 →							(7)
								2.0	(0.6)				(8)
										2.2	0.5		(9)
Oncorhynchus tshawytscha	6.0	1.8	2.2	3.9	5.0	4.0	(0.0)	5.1	(0.0)	2.2	0.5	3.2	(10)
						1.5	(2.5)	4.1	(1.0)				(10)
			2.2	3.9				5.1	(1.0)			3.1	(11)
					5.0					2.2			(12)
						1.5	(2.5)						(13)
	6.0	1.8				← 3.8 →				2.2			(14)
O. nerka											0.25		(15)
O. kisutch	6.0	1.8									0.25		(16)
Salmo gairdneri	3.1	1.4	2.1	3.9	4.7	1.6	(0.8)	2.8	(1.9)	3.0	0.4	2.8	(17)
						← 3.0 →							(14)
Sarotherodon mossambicus	4.0				4.0	1.3	(1.9)						(18)
Chicken	6.1	1.7	4.4	6.7	6.1	4.4	(0.0)	7.2	(0.0)	3.3	1.1	4.4	(19)
Young pig	1.5	1.5	4.6	4.6	4.7	3.0	(0.0)	3.6	(0.0)	3.0	0.8	3.1	(20)

Note: [a] Cystine and tyrosine present are given in brackets.

Sources: (1) Nose (1979), dietary protein level 37.7%; (2) Nose (1979), dietary protein level 38.5%; (3) Ogino (1980), dietary protein level 40%; assumed protein digestibility 90%; feeding rate 3% of body weight per day; (4) Robinson *et al.* (1980b), dietary protein level 24%; (5) Wilson *et al.* (1980); (6) Wilson *et al.* (1977); (7) Harding *et al.* (1977), dietary protein level 24%; (8) Robinson *et al.* (1980a); (9) Wilson *et al.* (1978); (10) Mertz (1968), dietary protein level 40%; (11) Chance *et al.* (1964), dietary protein level 40%; (12) Halver *et al.* (1958), dietary protein level 40%; (13) Halver *et al.* (1969), dietary protein level 40%; (14) Ketola (1982); (15) Halver (1965); (16) Klein & Halver (1970); (17) Ogino (1980), dietary protein level 40%; assumed protein digestibility 90%; feeding rate 3% of body weight per day; (18) Jackson & Capper (1982); (19) Mertz (1968), dietary protein level 18%; (20) Mertz (1968), dietary protein level 13%.

conversion was observed when the test diet was adjusted to pH 7 (Wilson *et al.*, 1977). Greater difficulties were encountered in developing a test diet for common carp. Aoe *et al.* (1970) tried feeding young common carp a test diet composed of crystalline amino acids as the only protein precursors. Though the fish ate this diet, they lost considerable weight. No improvement was noticed when magnesium and sulphate concentrations were reduced as with the channel catfish test diet. While casein was utilized well by common carp for growth, casein hydrolysates obtained with trypsin and pepsin, which sustained the growth of rainbow trout, failed to do so with common carp (Aoe *et al.*, 1974). It was therefore concluded by the authors that common carp cannot utilize free amino acids as well as peptides. In a later study Nose *et al.* (1974) succeeded in developing a test diet for common carp, composed of crystalline L-amino acids, by deleting the serine from Halver's test diet and, what seems to be more important, neutralizing the diet acidity through mixing the dry ingredients with water containing suitable amounts of sodium hydroxide to bring the pH to 6.5–6.7. This seems to have improved absorption of the amino acids in the intestine of this stomachless fish, which is normally neutral or slightly alkaline. The effect of pH on absorption may be compounded with the effect of the electrolyte level. It has been shown that the metabolism of amino acids is markedly influenced by the level of dietary electrolytes (Austic & Calvert, 1981), and that the transport of amino acids through the intestinal wall is also often closely associated with Na^+ and K^+ present (Christenson, 1977; see also p. 42). Murai *et al.* (1983b) tried to separate the effect of pH from that of the electrolytes. They found that both play a role in improving amino acid utilization. Adding K^+ to the diet to equimolar level of the Cl^- from supplemented amino acid–HCl improved utilization, but not excess supplements of K^+ and Na^+. Both Nose *et al.* (1974) and Murai *et al.* (1983b) showed that even neutralized test diets resulted in growth rates less than 60% of those of comparable casein controls. Murai *et al.* (1981a) suggested, on the basis of studies made with mammals (Silk *et al.*, 1973) that this may be due to differences in absorption rates among the free amino acids. Some amino acids are absorbed very rapidly, others more slowly. This causes an imbalance in the availability of the amino acids in the tissue during protein synthesis and thus hampers growth. Melnick *et al.* (1946) stated that 'for optimum utilization of food protein all essential amino acids must not only be available for absorption but must also be liberated during digestion at rates permitting mutual supplementation'. Tanaka *et al.* (1977a) explain the low growth rate of carp fed a free amino acid test diet by the short retention time of some of the amino acids in the intestine. These acids are absorbed into the blood and seem to be excreted at a higher rate than they are metabolized, as was found by Murai *et al.* (1984b). According to their study, when carp were fed a free amino acid mixture, the amount of free amino acid excreted reached 36% of the total nitrogenous substances excreted by the fish as compared to only 12.8% excreted by carp fed a casein diet. The rate of absorption of the free amino acids must be synchronized with their requirement for the biosynthesis of body proteins. It has been shown in studies with protein hydrolysates, involving the omission of a single essential amino acid from an otherwise adequate mixture of amino acids, and the injection of this missing amino acid eight hours after the basic mixture was given, that the utilization of the absorbed amino acids in the body remained as poor as if this amino acid had not been added at all (Melnick *et al.*, 1946).

In nature, common carp seem to absorb small peptides more efficiently and with less metabolic energy than required for the absorption of free amino acids (see p. 44). Murai *et al.* (1981a) fed common carp a gelatin diet supplemented with 10 amino acids coated with casein. The growth of these fish was comparable to that of fish fed a complete control diet, and four times faster than that of fish fed the gelatin diet supplemented with the same 10 amino acids but uncoated. They attributed this effect to the reduction of the absorption rate of coated amino acids, thus minimizing variations in the absorption rates among them. Yamada *et al.* (1981b) tried to improve the efficiency of test diets containing crystalline amino acids for common carp by increasing the frequency of feeding. When the fish were fed the test diet three times a day, growth and feed efficiency were very low in comparison to a casein diet. However, as feeding frequency increased (with the same total amount of feed) weight gain and feed efficiency increased proportionally. When fed 18 times a day, weight gain was about 72% that obtained with casein diet. The more frequent feeding seems to have synchronized the availability of the amino acids in the tissues. Halver's test diet was also found to be unacceptable by the red sea bream (*Chrysophrys major*). Yone *et al.* (1974) modified it by adding L-phenylalanine and L-aspartic acid, which made it more palatable to these fish.

A different way of determining the essential amino acid requirements is to administer ^{14}C-glucose and follow its incorporation into amino acids within the fish. Those which accept the labelled carbon are non-essential, while those which do not accept it are essential. This method was employed for sea bass (*Dicentrarchus labrax*) by Metailler *et al.* (1973).

Results of all studies mentioned above, as well as those for other fish species, such as the Japanese eel (Arai *et al.*, 1971, 1972b), and *Tilapia zillii*

(Mazid *et al.*, 1978), showed the essentiality of the same 10 amino acids as those for trout and higher vertebrates. Signs of deficiency of any of these generally include a decrease in growth rate and poor food conversion. Some amino acid deficiencies also lead to specific signs of anatomical abnormalities. Deficiency in tryptophan in rainbow trout and sockeye salmon causes scoliosis due to histological deformation of the myomeres and the notochord (Halver & Shanks, 1960; Kloppel & Post, 1975); deficiency in methionine causes lake trout and rainbow trout to develop bilateral lens cataracts (Poston *et al.*, 1977; Page *et al.*, 1978); lysine deficiency in rainbow trout causes caudal fin rot (Ketola, 1979b).

Quantitative requirements of essential amino acids by fish were studied by gradual increase in the amount of each of these acids to a test diet deficient in the respective amino acid, following the effect of the added amount on growth and well-being of the fish. Such experiments were carried out for salmonids by Halver *et al.* (1958), and later by others (e.g. Chance *et al.*, 1964; Klein & Halver, 1970; see also Mertz, 1972, who summarized studies on amino acid requirements by salmonids). Nose (1979) performed an extensive study on the quantitative amino acid requirements of the common carp. Ogino (1980) made a different approach in the study of the quantitative amino acid requirements of common carp and rainbow trout. He fed the fish a diet containing a protein of a high biological value (whole egg protein), and determined the amount of each essential amino acid deposited in the tissues. This was done by analysing fish tissues for these amino acids at the beginning and end of the experiment. The difference showed the amount deposited, from which the ratio between the essential amino acids could be calculated. However, since this method does not take into account the requirement of amino acids for maintenance, a certain element of error must exist. This error may be small when fish are growing rapidly and most amino acids are utilized for the building up of new tissues, but may be considerable when the fish are not growing. Ogino's experiment was carried out with growing fish and his results, therefore, do not materially differ from those obtained by the first method mentioned above.

The essential amino acid requirements of various fish species, as found by the above authors and others, are grouped together and presented in Table 37. Ketola (1982) points out that although the amino acid contents of fish eggs appear to differ from the reported dietary requirements, in some instances better fish growth rates have been obtained by supplementing amino acid deficient diets to the level of amino acids of fish eggs rather than the requirements based on individual studies. This may show that the knowledge on the amino acid requirements is still incomplete. Some of the experiments mentioned above showed that, as in higher vertebrates, interactions exist among some of the essential acids, and between them and non-essential amino acids. Kaushik & Luquet (1980), working with rainbow trout, and Jackson & Capper (1982), working with *Oreochromis mossambicus* have shown that excess free methionine has a depressive effect on growth. Jackson & Capper (1982) relate this to the inhibitory effect of methionine on the absorption through the intestinal wall of certain neutral amino acids (p. 43). Inhibition of leucine transport by methionine has been demonstrated in rainbow trout by Ingham & Arme (1977). As observed in higher vertebrates, this may be related to the high affinity of the methionine lipophilic side chain for the transport mechanism (Matthews, 1977). Common carp have shown a tendency to reduced growth as the ratio between leucine and isoleucine increased (Nose, 1979). Tyrosine, which is non-essential, showed a sparing effect on the essential phenylalanine, and cystine (non-essential) had a sparing effect on methionine. In both of these last cases the non-essential amino acids cannot completely replace the essential acids, but because of their capacities for partial replacement of the essential amino acids, they are often included with them. Robinson *et al.* (1980a) found that in channel catfish (*Ictalurus punctatus*) about 50% of the requirement for phenylalanine can be supplied by tyrosine.

As could be seen in Table 37 the arginine requirement of fish is similar to that of chicken, but considerably higher than that of mammals. Mertz (1968) pointed out that this may be because fish, like birds, have no, or a poorly developed, ornithine–urea cycle, which in mammals supplies most of the required arginine.

Where the information on amino acid requirement by fish is used for evaluating various dietary proteins efficiencies, it should be remembered that it is not only the amino acid composition of the protein that counts, but also their availability to the fish. The various acids in a protein may be digested to a different degree. Wilson *et al.* (1981) and Dabrowski & Dabrowska (1981) determined the apparent and true digestibilities of amino acids in different feeds and feed ingredients by channel catfish and common carp. Although there was a reasonable agreement between protein digestibility and the average amino acid digestibility, individual amino acid digestibilities varied within and among the various feedstuffs. Thus, while the apparent and true digestibilities of arginine from soybean meal by *Ictalurus punctatus* were found to be 95.4 and 96.8%, respectively, that of proline was only 77.1 and 77.9%, respectively (Wilson *et al.*, 1981). Moreover, even for similar digestibility coefficients, different *rates* of absorption of amino acids may be of importance (p. 44). It was demonstrated (Melnick *et al.*, 1946) that the *BV* of raw soybean protein was 53% (probably in human subjects), while that for heat processed soybean meal was 71%. Digestibility coefficients of the proteins were approximately the same (81 and 84% respectively). Calculations based upon methionine intake and excretion indicated that in both cases approximately 50% of the dietary methionine was absorbed. However, in the case of the raw soybean meal, absorption was much slower and occurred so late in the intestinal transit that this amino acid, as well as other amino acids, was not efficiently utilized. It is obvious that two proteins equal in amino acid composition may thus be of different *BV* for the fish. Fischer & Lipka (1983) have demonstrated differences in assimilation of amino acids by common carp and in weight gain when fed two diets containing proteins of different origin, but equal in percentage of their amino acids.

Since protein must supply the fish with certain amounts of essential amino acids, the lower the protein content in the diet the higher must be the concentration of these amino acids in the protein so as to supply them adequately. Similarly, the lower the digestibility of the protein, the higher the concentration of the essential amino acids must be. The true value of amino acid requirements as a percentage of the dietary protein will, therefore, be obtained only when the latter is fed at the minimal amount required for maintenance and maximal growth. Since fish require a much higher protein level than higher vertebrates, and since it is assumed that part of this protein is utilized for energy, it can be expected that the total concentration of essential amino acids in the protein in fish diets would be lower than for mammals or birds. From Table 37 it is apparent that this may be so, but not at the same proportions as the difference in protein requirement. While the amount of total essential amino acids required by the fish is 25–35.8% of the protein (11.3–14.8% of the food), that required by young pig is 30.4% of the protein (only 4% of the food), and that by chicken is 45.4% of the protein (8.2% of the food). This indicates that the higher requirement is not just for the protein, which is utilized in part for energy, but for the essential amino acids *per se*, which are probably utilized for replacing tissues and growth.

Information on the requirement of amino acids has been used in higher vertebrates to calculate the overall requirement for protein. Since in fish dietary protein must serve also as a source of energy, it is doubtful whether this method could be applied for fish. One can, however, attempt to calculate how much protein is required for maintenance and growth only, excluding requirements for energy. This can be done by taking into account the utilization rates of the protein for maintenance and growth as expressed by the *NPU*. The average endogenous nitrogen excretion of fish over 1 g was taken as 10.7 mg/100 g body weight/day (see Table 30). Protein requirement for maintenance is thus:

$$0.0107 \times 10^{-2} \times 6.25 \times \bar{W} = 0.00067\bar{W}\ \text{g/day}$$

The total protein requirement (*PR*) is then:

$$PR\ (\text{g/day}) = \frac{0.00067\bar{W} + \dot{W} \times \%\ \text{protein in tissue}}{NPU} \qquad (60)$$

Growth rate $\dot{W}$ can be calculated by equation (35).

For calculating in this way, as an example, the dietary protein requirement for maintenance and growth of common carp of 100 g, it is assumed that the protein content of its tissue is 16%, the *NPU* of the diet (containing fishmeal, soybean meal and cereals) is 45 (compare with Table 35), and growth rate is calculated by equation (37). The resulting protein requirement is then:

$$PR = \frac{0.00067 \times 100 + 0.176 \times 100^{0.66} \times 16}{45} = 1.3\ \text{g/day} \qquad (61)$$

Taking into account a feeding level of 4% of body weight/day, which is more or less that accepted for common carp of this size, the protein content of the diet is about 32%. This is higher than the level for most farm animals, but lower than the values in Table 32. It may be assumed that the difference results from the use of protein for energy. By tracing the hourly pattern of ammonia and urea excretion of fingerling sockeye salmon following a single daily meal, and comparing it with endogenous excretion of starved fish, Brett & Zalla (1975) showed that the deaminated fraction of the nitrogen was approximately 15 mg N/kg/day, or 27% of the nitrogenous intake. This leaves the retained portion at a similar level to that calculated above for the common carp (about 31%).

Dietary non-protein nitrogen

The greater part of the protein synthesized by fish in tissues is made of non-essential amino acids. These can be synthesized from carbon skeletons (α-keto acids), provided for by carbohydrates, and ammonia (see Chapter 3). The ammonia for this synthesis does not have to be of protein origin. Other sources, such as a single non-essential amino acid, ammonium citrate, or urea, can also supply the necessary ammonia. Studies on rats have shown that ammonium citrate, when added to the 10 essential amino acids, can be utilized as well as non-essential amino acids, but urea and glycine are apparently less efficient (Rose *et al.*, 1949; Lardy & Feldott, 1950; Mertz *et al.*, 1952). DeLong *et al.* (1959) performed a similar experiment on chinook salmon and found that urea was not utilized at all, diammonium citrate even had a negative effect, but as in higher vertebrates, arginine and to some extent glycine were utilized for protein production.

In contrast to the trout, grey mullet seems to be able to utilize urea as a source of nitrogen for protein synthesis. Vallet (1970, quoted from Cowey & Sargent, 1972) fed mullet (*Mugil* sp.) a diet containing 25% protein. He found that about half of this protein could be replaced by urea (on an isonitrogenous basis) without affecting growth rate. Also, when urea was added to a diet containing the entire 25% protein, at an amount equivalent to an additional 25% protein, a growth rate of 10% over that of fish fed the diet with protein only was obtained. Also Leray (1971) found that mullet can utilize non-protein nitrogen (NPN) and that urea could replace half of the dietary protein, without any reduction in growth rate. Similar results were obtained by Albertini-Berhaut & Vallet (1971). According to them, assimilation of urea seems to be due to activity of the bacterial flora in the digestion tract of the mullet. This is feasible, although it was not proven experimentally. Kakimoto & Mowlah (1980) isolated bacteria from the intestines of grey mullet. The predominating isolates were *Vibrio* and *Enterobacter*. All isolates utilized glycine as a sole carbon source, but no information was given on their ability to utilize NPN.

The results obtained with mullet led to the assumption that NPN can be also utilized by common carp and tilapia, which possess a rich intestinal bacterial flora. However, results did not confirm this assumption. While in an early study Dabrowski & Wojno (1978) found a positive effect of diets containing urea (2.5%) and ammonium citrate (9.3%) on growth rate and *NPU*, compared with common carp fed a control diet, without NPN, later experiments by Dabrowski (1979) as well as those by Hepher *et al.* (1979) could not find any such effect. No NPN utilization could either be found by Viola & Zohar (1984) with hybrid tilapia. The inclusion of urea in the diet of these fish reduced growth rate proportionally to the level of its inclusion.

The role of the intestinal bacterial flora in utilizing NPN or protein lacking essential amino acids was stressed by most of the above authors. Where NPN is utilized by fish, the bacteria probably absorb it for the synthesis of their own body protein, which is subsequently utilized by the fish.

Protein source

As can be seem from the preceding discussion, the biological value of a protein does not depend on the amino acid composition alone.

Also the presence in the diet of other nutritional factors such as energy, vitamins, minerals etc. which take part in the synthesis of tissue proteins, may affect the biological value. Proteins from different sources may thus be utilized by fish to a different extent and affect growth rate differently. Dupree & Sneed (1966) fed channel catfish diets containing equal concentrations of proteins of casein, wheat gluten or soybean. The growth rate obtained from the diet containing casein was considerably higher than those containing wheat gluten and soybean proteins. Cowey *et al.* (1974) carried out a similar study with plaice (*Pleuronectes platessa*), using five proteins. Even though the protein intake was high, significant differences in growth rate were evident between fish given the different proteins.

These studies and many others have highlighted the high dietary value of casein and fish meal proteins in promoting growth of fish. The effect of these proteins on growth may vary in different fish species. Tanaka *et al.* (1977b) improved the performance of a casein diet for common carp by supplementing it with free L-alanine, L-valine, L-tryptophan, L-methionine and L-proline. Rumsey & Ketola (1975) found casein to be inferior to an isolated fish protein in supporting growth of salmon fry. Similarly, Eckhardt *et al.* (1981) found defatted herring meal to be better than casein and egg albumin for common carp. Nevertheless, on the whole, these proteins are of a higher value in promoting fish growth than other protein sources, especially plant proteins. Most fish feeds, therefore, depend heavily on fish meal as a protein source. This dependence on fish meal may cause economic constraints in some geographical areas, especially in developing countries, where fish meal is less abundant, of lower quality and expensive. Also in the more developed countries, difficulties are sometimes encountered due to fluctuations in fish meal supply and its high price when supply is short.

Silkworm pupal protein was also found to have a high dietary value for fish, and was used extensively for carp diets in some regions of Japan. However, here too difficulties appeared with the development of synthetic silk, and the parallel decrease in the production of natural silk and the supply of silkworm pupae (Koyama *et al.*, 1961). In this case a replacement protein was also needed.

Meals of different oil-rich seeds, such as soybean, groundnuts, sunflower, rapeseed, copra, etc., after the extraction of the oil, are good and relatively inexpensive sources of protein. These meals are readily available in many developing countries. However, attempts to include them in fish diets, to replace fish meal or silkworm pupae, had only a limited success. The performances of such diets were satisfactory only when they contained a low proportion of these meals, but above a level of 25–50% of the total protein, their inclusion in the diet usually resulted in a reduced growth (Cowey *et al.*, 1971; Higgs *et al.*, 1979, 1982; Jackson *et al.*, 1982). This effect on growth was sometimes associated with the presence in the meal of some toxic compounds, such as gossypol and certain toxic fatty acids in cottonseed meal (Tiemeir & Deyoe, 1980; Ofojekwu & Ejike, 1984). Dorsa *et al.* (1982) found that growth of channel catfish was inhibited when fish were fed diets with more than 17.4% cottonseed meal. They attributed this to the gossypol contained in the meal. The addition of 0.09% of free gossypol to the diet had a similar effect. Rapeseed contains glucosinolates and erucic acid which act as antithyroid compounds. However, toxic compounds do not seem to be the sole cause for the low performance of oil meals. Robinson *et al.* (1984) fed tilapia (*Oreochromis aureus*) glandless cottonseed meal, which is low in gossypol, to substitute for soybean or peanut meals, and found it to be still inferior to those meals. They concluded that the inferior performance of fish fed the cottonseed meal did not appear to be related only to the dietary gossypol. Canola meal is derived from strains of rapeseed specially bred to contain low levels of toxic compounds. Higgs *et al.* (1979, 1983) found that canola meal may comprise no more than 25% of the dietary protein of juvenile chinook salmon diet without adversely influencing growth, provided the glucosinolate content is below 2.65 μmol/g dry diet, since feeding with this amount could be handled through the compensatory increase in thyroid activity (Higgs *et al.*, 1979). Although no toxic compounds were identified in many other oil meals, diets containing these oil meals were still inferior to those containing fish meal or silkworm pupae.

One of the most abundant of the above mentioned oil meals is that of soybean. Except for relatively low methionine and cystine contents, which can be made up by supplementary synthetic amino acids, the protein of soybean meal is of high biological value for most farm animals. However, the replacement of soybean meal for fish meal in fish diets also resulted in the same reduced fish growth in most fish species (Fowler, (1980) for chinook and coho salmon; Nose (1971), Reichle & Wunder (1974), Steffens & Albrecht (1976), Koops *et al.* (1976), Atack & Matty (1979), Reichle (1980), for rainbow trout; Krishnandhi & Shell (1967), Andrews & Page (1974), for channel catfish; Koyama *et al.* (1961), Hepher *et al.* (1971), Albrecht (1972), Viola (1975), Atack *et al.* (1979), for common carp; David & Stickney (1978), for *Sarotherodon aureus*). Koops *et al.* (1976) and Spinelli *et al.* (1979) found that when all fish meal was replaced by soybean meal, in a diet of rainbow trout, considerable fish mortality occurred. A certain controversy exists with respect to partial replacement of soybean meal for fish meal. While many researchers found that even a partial replacement reduced fish growth rate (e.g. Andrews & Page, 1974; Spinelli *et al.*, 1979), others showed that soybean meal can replace some of the fish meal without reducing growth rate substantially (Reinitz, 1980, for rainbow trout; Krishnandhi & Shell, 1967, for channel catfish; Kim *et al.*, 1984, for common carp). These last experiments, however, are usually associated with high protein levels, so that the remaining part of the fish meal may be adequate to sustain food growth, or that the inferior quality of the soybean meal is compensated by the higher protein level, as was found by Hepher *et al.* (1971) for common carp. In any case, at best, only part of the fish meal could be replaced (Cho *et al.*, 1974).

Three possible explanations could be given for the inferior value of soybean meal as compared to fish meal in fish diets: (a) lower digestibility of the soybean protein; (b) the presence in soybean meal of toxic substances affecting fish growth; (c) lack in the soybean of a nutrient present in fish meal.

(a) *Protein digestibility* In addition to the fact that the nutrients in soybean, as in any other plant, are enclosed in hard-walled cells, which make them sometimes less digestible by fish (see Chapter 2), soybean may also contain a number of compounds which reduce protein digestibility. These include trypsin inhibitors and carbohydrates which may be associated with the proteins. Also, the tertiary structure of some of the soybean proteins seems to resist digestion. Fukushima (1968) points out that the structure of the soybean globulin molecules, which are folded tightly with an inner hydrophobic part, makes this protein resistant to digestion. The protein cannot be completely digested until this inner portion is broken up. In most of the experiments reported above, however, a roasted soybean meal has been used. Roasting of soybean meal after oil extraction is today a common industrial practice. The meal is usually roasted at a temperature of about 120 °C for about 20 min. This is done to inactivate trypsin inhibitors or any other heat-sensitive growth inhibitors which may be present, and to convert some proteins to a more digestible form. Roasting temperature and time may be of great importance in this respect. Under-roasting may leave some of the trypsin inhibitors intact, while over-roasting may harm the nutritive value of the meal. Grabner & Hofer (1985), who measured the digestibility of proteins *in vitro*, found that heating soymeal by hydrothermic treatment to 140 °C, under pressure, inactivated trypsin inhibitors within 10–30 min. Smith (1977) stated that roasting soybean meal at a temperature of 190–240 °C for 12 min improves growth in fish fed this meal. Also Ketola (1982), who studied the effect of heat treatment of commercial soybean meal on the growth of Atlantic salmon (*Salmo salar*) fry, stated that additional heating improved the growth of fry by 22–30%. On the other hand, Kellor (1974) warns against over-roasting, which may cause destruction of proteins and thus lower the nutritional value of the meal for mammals. The major harm is caused to lysine which may undergo, when heated, a condensation with carbonyl carbohydrates to form a much less digestible compound. Fowler (1980) found that feeding chinook salmon a diet containing full fat soybean which was heated to a temperature of 218 °C for one minute by a 'jet sploder' process (in which dry heat is applied under pressure to explode the bean) resulted in a lower growth rate compared with soybean heated to only 178 °C. Viola (1981) and Viola *et al.* (1983) studied the effect of soybean meal, roasted to various degrees, on the growth rate of common carp. The meal in his experiments was roasted at 105 °C for various periods and was then included in the diet at a level of 30%. Trypsin inhibitors were completely inactivated with 60 min of cooking of the soybean meal. Lower growth rates were noticed in fish fed soybean meal roasted for less than 30 min, but no differences in growth were noticed among carp fed the soybean meals roasted between 30 and 120 min. It seems, therefore, that when soybean meal is roasted well, trypsin inhibitors and poorly digestible protein do not play an important role. Also, from the data on the digestibility of soybean meal protein given in Appendix I, it can be seen that though it has a somewhat lower digestibility coefficient than that of fish meal protein, the difference is not large. Poston (1965) found that the level of amino acids in the blood of trout fed soybean meal was equal to that of trout fed fish meal. It can thus be concluded that the difference in digestibility of the protein between soybean meal and fish meal is not enough to explain the difference in fish growth.

(b) *Toxic substances* Some studies on mammals have shown that soybean may contain growth inhibiting substances such as phytahaemagglutinins, phytic acid and its derivatives, saponins and phenolic compounds. Rackis (1974) showed that phytahaemagglutinin inhibited growth of rats fed soybean flour by 25%. Rainbow trout fed high soybean meal diets showed blood zinc, iron and copper abnormalities, as opposed to those fed diets that did not contain soybean (Spinelli *et al.*, 1979). Zinc was reduced by 33–40%, while copper increased by 30–50%. Spinelli *et al.* (1979) removed phytates from the soybean by mixing it with tuna entrails, which are rich in phytase, or with enzyme phytase itself, and heating the mixture to 50 °C for 16 h, or by mixing the soybean flour with 8% of its weight of wheat bran, which is also rich in phytase. Rainbow trout fed dephytinized soybean meal showed improved growth and feed conversion to those fed whole soybean flour. However, fish fed these dephytinized soybean diets had comparatively reduced zinc and elevated copper levels in the blood, indicating that substances other than phytates were affecting mineral availability. In a later study, Spinelli *et al.* (1983) found that phytates produce complexes with proteins rendering them less digestible. When 0.5% phytic acid was added to rainbow trout diet containing casein as the major protein (this amount may normally be encountered in diets containing soybean meal) growth was reduced by 8–10% over a 150 day period. This concentration of phytic acid did not reduce the levels of zinc or iron in the blood, liver and kidney, but reduced protein digestibility, as determined *in vitro* by pepsin, by 6.6%. The authors attributed the reduced growth of rainbow trout fed the phytic acid to this lowering of protein digestibility. However, Spinelli *et al.* (1979) worked with raw, not roasted, soybean flour. It seems that roasting also destroys most of these harmful compounds. The fact that the difference in growth rate between fish fed soybean meal diets and those fed fish meal diets can be compensated for and abridged by higher contents of soybean meal (Hepher *et al.*, 1971; Lovell *et al.*, 1975) indicates that the probability of the presence of toxic compounds in roasted soybean is very low.

(c) *Deficiency in nutrients* More attention has been given to the possible deficiency in the soybean meal of essential nutrients which are present in fish meal. Andrews & Page (1974) identified the sulphur containing amino acids methionine-cystine and lysine as the first limiting amino acids for catfish in soybean meal as compared to fish meal. However, supplementation of the diets containing soybean meal by these amino acids did not improve their performance. Similar results have been obtained for common carp (Hepher *et al.*, 1971; Albrecht, 1972; Viola, 1975). By contrast, other studies have shown marked improvement in growth rates of fish through supplementation with amino acids. Rumsey & Ketola (1975) supplemented rainbow trout diets containing soybean meal by methionine, leucine, lysine, valine and threonine to raise their level to that of trout egg protein. Trout groups fed the supplemented diet grew signficantly better than those fed unsupplemented soybean diet. Comparison between the supplemented soybean diet and the fish meal diet showed that the first is almost as good as, but not equal to, fish meal diets (Dabrowska & Wojno, 1977; Ketola, 1982). Improvement in performance of soybean diets by supplementation with amino acids was also noticed by Viola (1975), Wu & Jan (1977), Viola *et al.* (1982, 1983) and Murai *et al.* (1982, 1986). Andrews & Page (1974) postulated the presence of an unidentified growth factor in fish meal which is absent from soybean meal. They supplemented soybean meal diets with the fish oil, water soluble and mineral fractions of fish meal, but these had no effect on growth when all diets were isocalorific. This indicated, according to these authors, that such a factor must be associated with the lipid-free fish meal fraction.

Phosphorus may be a critical mineral in soybean meal because of the high requirement by fish (see p. 239). and low content of available phosphorus in soybean meal. Although soybean meal contains about 0.6% phosphorus, only about one third of it is available to fish, since the rest is associated with phytin. Soybean meal usually contains adequate amounts of trace minerals, but sometimes the level of availability of these elements especially zinc, may be too low for normal nutrition of fish.

Viola (1981) studied the problem of replacing soybean meal for fish meal in common carp diets. He found an interaction between the deficiency of amino acids, especially lysine, and the deficiency in energy in the soybean meal. The metabolizable energy value of soybean meal for fish seems to be lower than that given in tables for poultry, which are often used for formulating fish diets. This may be due to the low digestibility of some of the carbohydrates in the soybean meal, which include high proportions of pectins, galactans, cellulose and lignin (Table 38). Supplementing of soybean meal diets for common carp with energy (oil) improved their performance as compared to fish meal diets. However, only by supplementing the diet both with oil (8–10%) and amino acids (methionine, 0.5% and lysine, 0.5%) could soybean fully replace fish meal (Viola *et al.*, 1982).

Numerous studies were performed with many other prospective substitute proteins. Tiews *et al.* (1979) tested 43 proteinaceous feedstuffs for rainbow trout. Diets which included 23 of these performed as well as, or only slightly less well (up to 5% difference) than, fish meal diets. Various protein sources have been tested by others, such as leaf protein (Cowey *et al.*, 1971; Jan *et al.*, 1977), poultry by-product meal and hydrolysed feather meal (Gropp *et al.*, 1979), whey (Meske *et al.*, 1977), yeast (Appelbaum, 1977; Omae *et al.*, 1979; Atack *et al.*, 1979; Atack & Matty,

Table 38. *The composition of carbohydrates in soybean meal*

Proximate analysis	(%)	Carbohydrate analysis	(%)
Dry matter	90.0	Cellulose	7.7
Crude fibre	5.3	Pentosans	14.6
N-free extract	31.4	Lignin	5.0
		Other carbohydrates	9.4

Source: From Nehring & Norge (1942), as quoted from Becker & Nehring (1965).

1979; Gropp *et al.*, 1979), bacteria (Beck *et al.*, 1979, Spinelli *et al.*, 1979; Atack *et al.*, 1979; Atack & Matty, 1979; Kaushik & Luquet, 1980), blood meal (Reece *et al.*, 1975), corn gluten meal (Gropp *et al.*, 1979), rumen contents (Reece *et al.*, 1975), fly larvae (Spinelli *et al.*, 1979) and small crustacea (Lakshmanan *et al.*, 1967). In most cases results were quite satisfactory, but generally these feedstuffs are either in very short supply or their price is prohibitively high.

A proteinaceous feedstuff which attracted a considerable interest was yeast grown on hydrocarbons – petroleum yeast. A full replacement of petroleum yeast for fish meal usually resulted in a reduced fish growth and an increased feed conversion ratio (Iida *et al.*, 1970), though the difference was sometimes only slight (Beck *et al.*, 1977; Atack & Matty, 1979), and partial replacement usually compared well with fish meal diets (Andruetti *et al.*, 1973). Different supplements have been tried to improve the performance of petroleum yeast. Shimma & Nakada (1974) added oil to a petroleum yeast diet, but it did not perform as well as a commercial fish meal diet which was also supplemented with oil. Spinelli *et al.*, (1979) added methionine, but even then petroleum yeast could replace only about 25% of the fish meal in the diet.

Another feedstuff which recently aroused some interest as a possible substitute for fish meal was algae meal (Meske & Pruss, 1977; Matty & Smith, 1978; Hepher *et al.*, 1979; Atack *et al.*, 1979; Appler & Jauncey, 1983). Systems have been developed where microalgae can be cultured in domestic waste-water as part of the wastewater treatment process, which can reduce the algae production cost considerably. The dry meal which is obtained from the algae contains 50–60% protein. Fish feeding experiments have shown that the algae protein is inferior to that of fish meal, but superior to soybean meal and cereal proteins (Hepher *et al.*, 1979). The inclusion of algae meal at a rate of 25% in common carp diet (25% protein) replaced not only all of the fish meal, but also soybean meal and part of the cereals. The differences between the proteins were thus balanced and the resulting diet was as efficient as the commercial diet containing 15% fish meal. Reed *et al.* (1974) found this to be true for small channel catfish (< 10 cm) but not for the larger catfish (20–25 cm).

While most of the attention in the previous discussion was given to the protein utilization from protein-rich feedstuffs, attention should also be given to feedstuffs containing lesser amounts of protein. At low standing crops of fish in ponds, very often carbohydrate-rich and protein-poor feeds are used. These are often cereal grains containing 8–15% protein. Such feeds were used extensively in the culture of common carp in Central Europe. Since cereals are in these circumstances the only or the major supplementary feed, their protein constitutes an important part of the overall protein supply of the fish. However, due to the low *BV* of these proteins, their utilization is usually low. Lieder (1964) therefore suggested feeding the fish with a mixture of various cereals, so that amino acid contents would be more balanced and thus improve the *BV* and the utilization of the protein.

7

Other essential nutrients

Essential fatty acids

Fatty acids form part of a number of essential compounds in the animal body (e.g. membrane phospholipids). Most of the fatty acids can be synthesized by the animals *de novo* from acetate as a precursor. Fatty acids produced in this way are saturated, but they usually undergo a partial desaturation to form monoenoic acids (of one unsaturated bond) of varying chain length. The requirement for essential fatty acids (EFA) which cannot be synthesized *de novo* by the animal was demonstrated for rats by Burr & Burr (1929) and later for many other animals, and is probably general to all higher vertebrates. Signs of deficiency were avoided by supplementing the oil deficient diet with linoleic acid (18:2ω6)[1], an unsaturated lipid found in many vegetable oils such as those of corn, peanut and sunflower seeds. Supplementation with linolenic acid (18:3ω3) partially corrected the condition, but was not as efficient as linoleic acid. The dietary acids of the ω6 group have been found to be precursors of prostaglandins and also provide a substrate for the pre-oxidation of microsomal fats in the liver. Fatty acids of the ω3 group (e.g. linolenic acid) are non-essential for higher vertebrates, although they can replace to some extent the essential ω6 acids.

---

[1] Fatty acids can be designated by their common name, chemical nomenclature, or using a numerical convention. For the chemical nomenclature the Greek enumeration is used to designate the number of carbons in the chain and suffix -noic, -enoic, -dinoic, etc. for the degree of unsaturation. The numerical designation gives the number of carbons in the chain, the number of unsaturated bonds, and the 'omega' number designating after which carbon, counting from the methyl group end, the first unsaturated bond occurs. Thus: stearic acid = octadecanoic acid = 18:0 = $CH_3(CH_2)_{16}COOH$; oleic acid = octadecacenoic acid = 18:1ω9 = $CH_3(CH_2)_7CH{=}CH(CH_2)_7COOH$; linoleic acid = octadecadienoic acid = 18:2ω6 = $CH_3(CH_2)_4CH{=}CHCH_2CH{=}CH(CH_2)_7COOH$; linolenic acid = octadecatrienoic acid = 18:3ω3 = $CH_3CH_2CH{=}CHCH_2CH{=}CHCH_2CH{=}CH(CH_2)_7COOH$.

Animals can further desaturate and elongate the chains of the dietary unsaturated fatty acids to form polyunsaturated fatty acids (PUFA). Different animals and even different species have a different capacity to elongate fatty acids. Owen *et al.* (1975) fed turbot (*Scophthalmus maximus*) and rainbow trout labelled fatty acids. Six days later he examined the distribution of radioactivity in the tissue fatty acids. When rainbow trout was fed labelled linolenic acid, [1-^{14}C]18:3ω3, 70% of the radioactivity was present in 22:6ω3 fatty acid. In contrast, turbot fed [1-^{14}C]18:1ω9, 18:2ω6, or 18:3ω3 converted only small amounts of labelled fatty acids (3–15%) into longer chain fatty acids. Also, in the same animal there seems to be a sort of 'competition' or 'suppression' between certain groups of fatty acids. The higher the concentration of ω3 fatty acids in the diet, the lower the chain elongation of the fatty acids of the ω6 and ω9 groups, since these three groups compete for the same enzyme system. The substrate–enzyme affinity decreases from linolenic through linoleic to oleic acid. Thus, in turn, when ω6 fatty acids are present at a high concentration in the diet, the elongation of the ω9 fatty acid group is suppressed. Phospholipid analyses in animals showing EFA deficiency signs revealed a high concentration of polyenoic acids in the oleic (ω9) and palmitoleic (ω7) groups. In contrast, animals which did not show EFA deficiency signs had a proportion of the linoleic (ω6) and linolenic (ω3) groups. Thus, Castell *et al.* (1972c) and Watanabe *et al.* (1974a) found that when rainbow trout were fed a diet containing neutral fatty acids, an elevated level of eicosatrienoic acid (20:3ω9) was found in the body lipids, while it was depressed by dietary linoleate and linolenate. Methyl linoleate elevated the level of 20:4ω6 and 22:5ω6 acids in polar lipids of the liver, and methyl linolenate elevated the level of 22:6ω3. Similar results were found by Farkas *et al.* (1980) for common carp. This phenomenon indicates that animals cannot synthesize the fatty acids of the ω6 and ω3 groups and are dependent on their supply in the diet.

The importance of EFA has also been proven for various fish species. Nicolaides & Woodall (1962) determined their essentiality for chinook salmon. Deficiency signs were marked depigmentation and reduced growth. These symptoms, a greatly increased mitochondrial swelling, the degeneration of the caudal fin and higher mortality, have also been observed in rainbow trout (Higashi *et al.*, 1964, 1966; Castell *et al.*, 1972b; Watanabe *et al.*, 1974b). Takauchi *et al.* (1980a) showed that growth of eels and their feed utilization are both reduced due to the lack of EFA in the diet. No external signs due to EFA deficiency have been noticed in common carp, but growth was retarded (Watanabe *et al.*, 1975a).

Studies on intermediate metabolism of lipids in fish have also indicated the dependence of dietary supply in EFA. Klenk & Kremer (1960) incubated liver slices of various vertebrates, including fish (common carp and rainbow trout) with acetate-1-^{14}C. Analysis of the polyenoic acids showed that the label always predominated in the carboxyl ends of the acids, while the methyl ends were practically inactive. From this the authors concluded that in all the animals they investigated the synthesis of C_{20} and C_{22} polyenoic acids takes place, as in the rat, principally from exogenous precursors of linoleic and linolenic types of fatty acids. Several other studies with labelled acetate or fatty acids were also carried out in order to trace the synthesis of fatty acids and their conversion to highly unsaturated fatty acids (HUFA). Mead *et al.* (1960) used acetate-1-^{14}C, which they injected into *Sarotherodon mossambicus*, and Kanazawa *et al.* (1980c) injected it into *Tilapia zillii* and puffer fish, *Fugu rubripes*. Only saturated or monoenoic acids were synthesized *de novo* from the acetate. However, when linolenic acid-1-^{14}C was injected into *T. zillii* it was converted to eicosapentaenoic acid (20:5ω3) and docosahexaenoic acid (22:6ω3). The metabolism of linoleic-1-^{14}C to ω6 series of HUFA was only slight. Puffer fish converted linoleic acid to eicosadienoic acid (20:2ω6) and arachidonic acid (20:4ω6). From these studies it can be concluded that the high content of linoleic and arachidonic acids in fish is evidently derived from the diet.

In this respect it is interesting to compare the composition of fatty acids in various fish species with those of terrestrial animals. Castell (1979) found that the fatty acids of fish are much less saturated than those of terrestrial animals and contain less ω6 and more ω3 fatty acids. A comparison between freshwater and marine fishes showed that the former have a higher proportion of C_{16} and C_{18} fatty acids while the latter have more C_{20} and C_{22}. The ratio between total linoleic to total linolenic types of acid (ω6:ω3) is higher in freshwater than in marine fishes (Ackman, 1967; Castell, 1979). This is demonstrated in Table 39 by data taken from Castell (1979). These differences are in part due to the different foods which the fish get in their respective environments. Several studies have shown that the fatty acid composition of the lipid in the diet affects considerably the fatty aid composition of the body lipids (e.g. Worthington & Lovell, 1973; Dupree *et al.*, 1979; Farkas *et al.* 1980). According to these studies, a diet rich in ω6 fatty acids affects to some extent the ratio in body lipids, though the proportion of the ω6 acids is usually higher in the body lipids than in the diet. It seems, however, that the differences in fatty acid composition between fishes are due primarily to environmental factors, and to the different roles fatty acids have in physiological processes in the various environments. Farkas *et al.* (1980) have injected sodium-1-^{14}C-acetate to common carp exposed to two different temperatures (5 and 25 °C). They followed the incorporation of the labelled carbon into the fatty acids. The major product of fatty acid biosynthesis in warm injected fish was palmitic acid. Oleic acid also picked up a considerable proportion of radioactivity. Radioactivity appearing in long chain PUFA was low and did not seem to be a function of the amount and type of unsaturated acids in the diet. On the other hand, fish exposed to cold reorganized the pattern of fatty acid biosynthesis. There was a drastic reduction of labelling of saturated fatty acids and a higher proportion of the label was directed into the PUFA fraction. Essential fatty acid deficient carp were unable to increase the rate of production of long chain PUFA upon exposure to cold. When, however, fish acclimated to cold were fed sufficient linolenic acid, a higher proportion of PUFA, especially docosahexaenoic acid, was found in the phospholipids of the fish. Castell (1979) studied the fatty acid composition and the ratio between ω6 and ω3 in triglicetides and phospholipids of fish migrating

Table 39. *Average contents of the various fatty acids, as per cent of total fatty acids, in freshwater and marine fishes*

Fatty acids	Freshwater fishes	Marine fishes
14:0	4.1	4.7
16:0	14.6	19.0
16:1	14.2	8.3
18:0	2.9	2.9
18:1	22.8	19.7
18:2ω6	3.5	1.2
18:3ω3	3.4	0.8
18:4ω3	1.7	2.0
20:1	1.8	6.7
20:4ω6	2.5	1.5
20:4ω3	1.0	0.5
20:5ω3	5.9	8.1
22:1	0.9	7.7
22:5ω6	0.7	0.9
22:5ω3	2.3	1.4
22:6ω3	8.7	11.3
Total saturated	23.3	25.7
Total monoenoic	41.6	42.7
Total ω6	6.0	3.6
Total ω3	23.4	23.3
ω6:ω3	0.34	0.15

Source: Calculated from data of Castell (1979).

from the sea into freshwater and vice versa. The average ratios of ω6:ω3 during the freshwater stage of these fishes was 0.305 and 0.29 for the triglycerides and phospholipids respectively. During the marine stage the ratio ω6:ω3 decreased to 0.12 and 0.035 respectively. The larger difference was in the phospholipids, which have an important role in cell membranes.

The higher content of ω3 fatty acids, especially 20:5ω3 and 22:6ω3, in the membranes seems to be important for life in the marine environment or in cold water. Meister *et al.* (1974), in their study of the effect of salinity on the composition of fatty acids in the European eel (*Anguilla anguilla*) also found major differences in the relative concentrations of the fatty acids in the gill phospholipids between freshwater and seawater eels. The latter had a higher percentage of 22:6ω3 acid in phosphatidylcholine and phosphatidylethanolamine, which the authors related to variations in the membranes' enzyme activity. Indeed, several studies have shown that phospholipids, especially phosphatidylserine and phosphatidylglycerol, activate the membrane-bound enzyme (Na^+/K^+)ATPase, which is an important component of the marine fish osmoregulatory system (Kimelberg & Papahadjopoulos, 1972). These phospholipids show a significant discrimination for K^+ over Na^+permeability, and perhaps are also active as cation carriers themselves. To activate (Na^+/K^+)-ATPase and affect membrane permeability the phospholipids must be fluid. Ashe & Steim (1971) have indicated that a high proportion of the phospholipids in some natural membranes are in a fluid state. When membrane phospholipid undergoes a phase transition from a fluid to a gel state the permeability of the membrane drops considerably (Kimelberg & Papahadjopoulos, 1972). The temperature of such phase transition corresponds to the melting point of the fatty acid chain. Thus it seems that the incorporation of PUFA, especially of the ω3 type, which have a lower melting point than the ω6 and ω9 fatty acids, may be related to this phenomenon and makes the membrane more permeable and therefore more functional in seawater.

The above discussion may also explain the importance of fatty acids of the ω3 type for coldwater fish, where the incorporation of low melting point fatty acids in membrane phospholipids maintains the permeability of the membrane even at low water temperature. Lovern (1938) showed that European eels (*A. anguilla*) held at a low temperature contained a higher proportion of PUFA than eels kept at a high temperature. Hoar & Cottle (1952) found in goldfish (*Carassius auratus*) a change in the iodine value of the lipids (which indicates the degree of fatty acid saturation) of 0.5 units for a change of 1 °C in temperature. Similar experiments showed a decrease in melting point of fish lipids with the decrease in temperature (Lewis, 1962; Knipprath & Mead, 1965).

It is interesting to note that a similar effect of temperature on the composition of fatty acids has been observed in planktonic organisms which can serve as food for fish. Holton *et al.* (1964) found variations in the composition of fatty acids of blue-green algae with the change in temperature. The higher the temperature the higher the proportion of the saturated acids. Farkas & Herodek (1960, 1964) showed that the fatty acids of planktonic crustacea in Lake Balaton became less saturated with a decrease in temperature. Thus, in nature, the higher requirement for PUFA by fish seems to be met by a higher supply of these acids in fish food.

The essentiality of the ω3 type of fatty acids, such as linolenic acid, for marine fish and coldwater fish had also been proven experimentally. Early experiments were not specific, but indicated that fish oils (rich in ω3 fatty acids), when included in the diets of some fish, are superior to vegetable oils (rich in ω6 fatty acids) in their effect on fish growth. At first, this was related to the beneficial effects of vitamins A and D, but Phillips *et al.* (1951) pointed out that the effect of fish oil is greater than would be anticipated from the presence of the vitamins in the oil. Later experiments, however, elucidated the effect of the fatty acids of the ω3 type. Supplementing diets of rainbow trout with linolenic acid and linolenates positively affected fish growth (Higashi *et al.*, 1964, 1966; Lee *et al.*, 1967; Watanabe *et al.*, 1974b) and prevented physiological changes such as increased mitochondrial swelling rate, increased liver respiration rate, and lower haemoglobin content (Castell *et al.*, 1972b). This was even more pronounced in rainbow trout cultured in seawater (Lall & Bishop, 1979). The level of linolenic acid required for supplementing the diet of rainbow trout was found to be 1% of the diet (Castell *et al.*, 1972a; Watanabe *et al.*, 1974b; Yu & Sinnhuber, 1975). Supplementing rainbow trout diets with linoleic acid or linoleates only, improved growth rate and feed conversion, as compared to the EFA-deficient diet, but did not prevent the appearance of some of the deficiency signs, and resulted in lower growth rates than when linolenic acid was added (Castell *et al.*, 1972a). Similar results were obtained for the Japanese eel (Takeuchi *et al.*, 1980a) and red sea bream (Yone & Fujii, 1975a). Channel catfish also require a certain amount of linolenic acid (Stickney & Andrews, 1971).

Elevated dietary lipid levels increase the requirement of rainbow trout, and probably also that of the other species mentioned above, for linolenic acid. Takeuchi & Watanabe (1977b) found that feeding rainbow trout a diet containing 4% laurate (12:0) and 1% linolenic acid resulted in a good growth, but the same diet containing 9% and 14% laurate and 1% linolenic acid resulted in a reduced growth. With such elevated lipid levels, more than 2% linolenic acid was required for maximum growth, or about 20% of the dietary lipid. However, in a later study Takeuchi & Watanabe (1978) found that half of this amount (0.5% of the diet or 10% of the lipid) is sufficient. This can be added either as a triglyceride or in a methyl ester form. Once the requirement for EFA is fulfilled, hydrogenated oils can be used as an energy source.

Since, as was shown above, linoleic and linolenic acids are easily converted to C_{20} and C_{22} PUFA by desaturation and elongation of the fatty acid chains (see also Watanabe *et al.*, 1974a; Yone & Fujii, 1975b), Yu & Sinnhuber (1972) state that supplementing linolenic acid to rainbow trout diet is as efficient as 22:6ω3 acid. Similarly, no differences were found between the effect of linolenic acid and that of highly unsaturated acids when used to supplement the diet of eels (Takeuchi *et al.*, 1980a). In contrast, however, Takeuchi & Watanabe (1977a) found that supplementation of rainbow trout diets by 20:5ω3 and 22:6ω3 acids, and especially by the mixture of the two, resulted in a better growth rate than when linolenic acid was used. Similar results have been found by Yone & Fujii (1975a) for red sea bream. This may explain the beneficial effect of some fish liver oils, such as pollack and salmon oils, when added to the diet. These oils contain only small amounts of linolenic acid but large amounts of long chain PUFA (Lee *et al.*, 1967; Watanabe & Takeuchi, 1976).

Warmwater fishes, such as common carp and tilapia, seem to require less EFA of the ω3 type. Although initial experiments by Watanabe *et al.* (1975a) failed to show any advantage of linoleic or linolenic acid supplemented to diets containing non-EFA fatty acids when measured by common carp growth, later experiments with a number of warmwater fishes indicated the essentiality of both linoleic and linolenic acid. Best weight gains of common carp were obtained with 1% linoleic and 1% linolenic acid (Watanabe *et al.*, 1975b; Kanazawa *et al.*, 1980c; Watanabe, 1982). Takeuchi *et al.* (1980a) found that the Japanese eel also seems to require both ω6 and ω3 types of fatty acids. A mixture of corn oil (rich in ω6 fatty acids) and fish oil (rich in ω3 fatty acids) or a mixture of the two types of fatty acids at a level of 0.5% each, gave better results than each of these oils alone. Supplementation of the eel diet with fish oil alone resulted in the poorest growth rate. With channel catfish, while some experiments have shown that fatty acids of the linolenic type have no unique importance (Stickney & Andrews, 1972), subsequent experiments revealed a need for these fatty acids (Lovell, 1979).

Some fishes, however, seem to resemble mammals in requiring only linoleic acid. Kanazawa *et al.* (1980a) and Teshima *et al.* (1982) found that the addition of both linoleic and linolenic types of fatty acids (18:2ω6;

20:4ω6; 18:3ω3, and 20:5ω3) to a diet containing lauric acid (12:0) as a basal lipid, improved weight gain of *Tilapia zillii* and *Oreochromis niloticus*. However, the growth promoting effects of the ω6 acids were superior to those of the ω3 acids. They concluded that these tilapias require the linoleic type of fatty acids rather than the linolenic type.

Vitamins

While natural food is usually rich in vitamins, this may not be the case with supplementary feed. Vitamin deficiency appears, therefore, mainly in intensive culture systems, stocked at high densities of fish, where supplementary feed is the major, if not the only, source of food. According to Halver (1979) one of the first cases of vitamin deficiency in fish was recorded in 1941, when trout fed on common carp raw meat become paralysed. Wolf (1942) attributed this sign to thiaminase present in the raw common carp meat, which decomposes thiamine and causes a deficiency in this vitamin. Indeed, Schneberger (1941) cured the disease by injecting the trout with thiamine.

The study of the vitamin requirement of fish received a considerable boost by the development of a vitamin test diet, at first by Wolf (1951), and then with further improvement by Halver (1957a). This diet contains vitamin free casein and gelatin (50%), white dextrin (28%), oil (9%), mineral mixture (4%) and a mixture of 13 vitamins in an α-cellulose carrier (9%). This test diet has supported the growth of various salmonids for at least 24 weeks without any apparent deficiency signs. By omitting one or more vitamins from the mixture the syndromes caused by their deficiency could be studied. Halver (1972, 1979) and others (e.g. Wolf, 1951; Hashimoto *et al.*, 1970; Ashley, 1972; Murai & Andrews, 1975; Ketola, 1976; Lovell, 1979) have, by use of these diets, described many of the vitamin deficiency diseases in fish, which are well defined in their papers and do not require repetition. However, it should be mentioned that of all of the more common signs they describe the most damaging from the point of view of the fish farmer are growth retardation and increased mortality. Other common signs of deficiency for most vitamins are loss of appetite, discoloration, lack of coordination, nervousness, haemorrhages, lesions, fatty livers and increased susceptibilty to bacterial infections.

Vitamins can be classified according to their solubility into two groups: (a) water soluble vitamins, which include all the vitamins of B group (thiamine, riboflavin, pyrodixine, folic acid, pantothenic acid, nicotinic acid, and vitamin B_{12}), choline, inositol, ascorbic acid (vitamin C) and para-aminobenzoic acid (PABA); (b) fat soluble vitamins, which include

Table 40. *Vitamin requirement for growth of some fish species common in aquaculture, when fed at about their protein requirement. Superscripts designate sources given as a footnote, and bold script are the averages of all cited sources.*

Vitamin (mg/kg dry diet)	Salmon[a]	Trout[b]	Common carp	Channel catfish
Thiamine	10–15[15] 10[28] **11**	10–12[15] 10[28] 1–10[22] **9**	2–3[15] NR[c8] **1**	20[14] 1–3[15] 1[26] **8**
Riboflavin	20–25[15] 20[28] **21**	20–30[15] 20[28] 5–15[22] 6[35] 3[17] **13**	7–10[15] 6–8[31] 7[35] 4[6] **7**	20[14] 9[25] **15**
Pyridoxine	15–20[15] 10[28] **14**	10–15[15] 10[28] 1–10[22] **11**	5–10[15] 5[7] 4[30] **6**	20[15] 3[3] **12**
Pantothenic acid	40–50[15] 40[28] **43**	40–50[15] 40[28] 10–20[22] **33**	30–40[15] 25–35[31] **28**	250[24] (fry) 50[14] 25–50[15] 10[27] (fingerlings) **87**
Niacin	150–200[15] 150[28] **163**	120–150[15] 150[28] 1–5[22] **96**	30–50[15] 29[7] **35**	100[14] 14[2] **57**
Folic acid	6–10[15] 5[28] **7**	6–10[15] 5[28] 1–5[22] **5**	NR[9] **NR**	5[14] NR[13] **3**
Vitamin B_{12}	0.015–0.02[15] 0.02[28] **0.019**	0.02[28] **0.02**	NR[16] **NR**	0.02[14] **0.02**
Inositol	400[28] 300–400[15] **375**	250–500[22] 400[28] 200–300[15] **342**	440[5] 200–300[15] **345**	100[14] NR[11] NR[13] **33**
Choline	3000[28] 600–800[15] **1850**	3000[28] 1000[18] 50–100[22] **1358**	1500–3000[33] 500–600[15] **1400**	550[14] **550**
Biotin	1–1.5[15] 1[28] **1.1**	1–1.5[15] 1[28] 0.05–0.25[22] **0.8**	1–1.5[15] 1[32] **1.1**	0.1[14] NR[13] **0.1**
Ascorbic acid	100–150[15] 100[28] **113**	250–500[22] 100–150[15] 100[28] **200**	30–50[15] NR[34] **20**	30–100[14] 60[19] 50[1] NR[13] **44**

Table 40 (*cont.*)

Vitamin (mg/kg dry diet)	Salmon[a]	Trout[b]	Common carp	Channel catfish
Vitamin A (IU)	2500[28]	15000[21] 2500[28] 2000–2500[15]	4000–20 000[10] 1000–2000[15]	5500[14]
	2500	**6583**[d]	**6750**	**5500**
Vitamin D (IU)	2400[28]	2400[28]	NR[29]	1000–4000[4] 1000[14] 500[20]
	2400	**2400**	**NR**	**1333**
Vitamin E	40–50[15] 30[28]	50[12] 30[28]	200–300[36] 80–100[15]	100[23] 50[14]
	38	**40**	**170**	**75**
Vitamin K	10[28]	10[28]	NR[29]	10[14]
	10	**10**	**NR**	**10**

Notes: [a] Chinook and coho salmon.
[b] Rainbow, brook and brown trout.
[c] Not required.
[d] Maximum tolerance – 900 000 IU/kg (Hilton, 1983).

Sources: (1) Andrews & Murai (1975); (2) Andrews & Murai (1978); (3) Andrews & Murai (1979); (4) Andrews *et al.* (1980); (5) Aoe & Masuda (1967); (6) Aoe *et al.* (1967a); (7) Aoe *et al.* (1967b); (8) Aoe *et al.* (1967c); (9) Aoe *et al.* (1967d); (10) Aoe *et al.* (1968); (11) Burtle (1981, quoted from NRC, 1983); (12) Cowey *et al.* (1983); (13) Dupree (1966); (14) Dupree (1977); (15) Halver (1979, which sums up his previous works); (16) Hashimoto (1953); (17) Hughes *et al.* (1981); (18) Ketola (1976); (19) Lim & Lovell (1978); (20) Lovell & Li (1978); (21) Loyanich (1974); (22) McLaren *et al.* (1947); (23) Murai & Andrews (1974); (24) Murai & Andrews (1975); (25) Murai & Andrews (1978a); (26) Murai & Andrews (1978b); (27) Murai & Andrews (1979); (28) NRC (1981); (29) NRC (1983); (30) Ogino (1965); (31) Ogino (1967); (32) Ogino *et al.* (1970a); (33) Ogino *et al.* (1970b); (34) Sato *et al.* (1978); (35) Takeuchi *et al.* (1980b); (36) Watanabe *et al.* (1977).

vitamins A, D, E and K. Vitamins of the B group are required in small amounts, but have an important role in fish metabolism. The other water soluble vitamins, except PABA, are usually required in larger amounts. Since no deficiency signs were observed when PABA was omitted from the diet, this vitamin was considered as non-essential and was excluded from the test diet (Halver, 1979, 1980b; Aoe & Masuda, 1967). The fat soluble vitamins A, E and K were found to be essential for fish (e.g. Kitamura *et al.*, 1967) and deficiency signs appear when they are absent from the diet. Early studies found no signs of deficiency for vitamin D. Recently, however, the essentiality of this vitamin was established for rainbow trout (Barnett *et al.*, 1979; George *et al.*, 1979), catfish, *Clarias batrachus* (Swarup & Srivastav, 1982), eels, *Anguilla anguilla* (Chartier *et al.*, 1979) and tilapia, *Sarotherodon mossambicus* (Wendelaar Bonga *et al.*, 1983), and this is probably true for other, if not all, fish species. In all, about 15 vitamins have been found to be essential for most fish (Table 40).

Many studies were carried out to determine the levels of vitamins required in fish diets. In most cases this was done by feeding parallel groups of fish with diets containing increasing concentrations of the vitamin under study. The required level was defined as the minimal concentration which resulted in highest growth, highest concentrations of the tested vitamins in the liver, and no abnormalities with respect to blood constitutents and histology. Sometimes, other biochemical criteria were used in evaluating the requirement of vitamins, usually as an adjunct to growth measurements. Thus, Cowey (1976) made use of the fact that thiamine pyrophosphate serves as a coenzyme to erythrocyte transketolase. He measured the activity of this enzyme and the degree of its stimulation by thiamine pyrophosphate in blood samples from turbot (*Scophthalmus maximum*) given diets of different thiamine content. This served as an indicator of thiamine status and its requirement. Similarly, Hughes *et al.* (1981) determined riboflavin requirement of rainbow trout fingerlings by the activity of erythrocyte glutathione reductase (EGR), an enzyme which is dependent on the riboflavin derivative flavin adenine dinucleotide (FAD). This blood enzyme is very sensitive to depletion of riboflavin stores of the body and measurement of its ability to reduce oxidized gluthathione in the presence of reduced nicotinamide adenine dinucleotide phosphate (NADPH) gives, according to these authors, an accurate diagnosis of the riboflavin status of the fish. Other biochemical methods for diagnosing vitamin deficiencies have been used in higher vertebrates, but it is not known if they were ever used for fish. Thus, for instance, serum glutamic-oxalacetic transaminase activity and leucine transaminase activity in kidney were found to be highly sensitive indicators of pyridoxine level in rats and chicks respectively (Thiele & Brin, 1968; Shiflett & Haskell, 1969).

Vitamin requirements of several fish species, as determined by the methods discussed above are presented in Table 40. There are sometimes considerable differences among the researchers mentioned in this table, on the recommended vitamin level. Some of these differences are probably due to the different conditions in which the studies were performed. A considerable variation, however, is caused by specific factors such as fish species, absorption and metabolism of the vitamin in question, feedstuffs and availability of the vitamins, and the storing conditions. It is important to understand the effect of these factors on vitamin requirement, since feeding vitamin excessively may harm the fish as much as feeding them too little (Chudova, 1961; Halver, 1980a). Improving the knowledge on vitamin utilization and metabolism will permit a better estimation of the optimum requirement levels at given conditions.

Species specificity of vitamin requirement

The response of various fish species to different vitamins may be different, and therefore their requirements for the vitamins can also vary. Halver (1979) points out that coho salmon are more sensitive to deficiency of folic acid than chinook salmon or rainbow trout. On the other hand, rainbow trout are more sensitive than coho salmon to deficiency of biotin. Differences in vitamin requirements among fish species may arise from the fact that some species are able to synthesize certain vitamins. Thus, Ikeda & Sato (1964) have demonstrated that common carp was able to convert D-glucose-1-^{14}C and D-glucuronolactone-6-^{14}C, which were administered intraperitoneally, to L-ascorbic acid-^{14}C. Following this, Sato *et al.* (1978) found that ascorbic acid was not essential in the diet of common carp. Contrary to this, in salmonids, channel catfish, yellowtail and eel, severe deficiency signs developed when ascorbic acid was missing from the diet (Kitamura *et al.*, 1965; Halver *et al.*, 1969; Sakaguchi *et al.*, 1969; Arai *et al.*, 1972a; Lovell, 1973; Wilson & Poe, 1973; Andrews & Murai, 1975). Wilson (1973) even found that both channel catfish and blue catfish (*Ictalurus furcatus*) lack the enzyme involved in ascorbic acid synthesis.

Niacin can be synthesized by some animals, including fish, from tryptophan. The conversion efficiency of tryptophan to niacin varies widely among species. According to Halver (1972) this conversion may account for the slow development of niacin deficiency signs in fish. Halver also found that niacin requirements of salmon appeared to be approximately twice those of trout. He postulated that these differences may reflect differences in metabolism or imbalance of nutrients in the test diet, but they could also be explained as being due to a better conversion, by trout, of tryptophan to niacin. However, by measuring the activity of the enzymes involved in the metabolism of tryptophan in the livers of fish, Poston & Dilorenzo (1973) and Poston & Combs (1980) came to the conclusion that in general L-tryptophan is an efficient precursor of niacin for salmonids, although they have not excluded the possibility of intergeneric differences in this respect among the five salmonid species they have examined.

Vitamins can also be produced, in certain species, through the activity of intestinal bacteria. Deficiency in vitamin B_{12} causes anaemia and growth inhibition in salmon and trout, but it seems not to be required by the common carp (Hashimoto, 1953) and *Sarotherodon niloticus* (Lovell & Limsuwan, 1982), since these fishes can produce the vitamin in their own body by the activities of intestinal bacteria (Kashiwada & Teshima, 1966). The ability of intestinal bacteria to produce vitamin B_{12} was also found by Lebedeva *et al.* (1972) in some marine fish from the Mediterranean. Trepsiene *et al.* (1947b) found the intestinal bacteria of certain warmwater fishes such as common carp, white amur (*Ctenopharyngodon idella*) and tench (*Tinca tinca*) to be also capable of synthesizing other vitamins of the B group such as thiamine, nicotinic acid and biotin. The ability to synthesize these vitamins decreases their dependence on dietary sources, although this does not always free the fish from some requirement of the vitamins from exogenic sources. Thus, Lieder (1965b) showed that common carp fed a thiamine-deficient diet developed a deficiency disease after 12 weeks. The signs were convulsions, muscle atrophy and exudation, medium exophthalmus and accelerated breathing. These resulted in increased mortality. An intraperitoneal injection of 0.5 g of vitamin B_1 remedied all these signs. Lieder remarks, however, that low levels of thiamine, usually those naturally present in most cereals, are sufficient to satisfy the requirements of common carp for thiamine. This may partly explain results by Aoe *et al.* (1967c). They raised common carp fry for 16 weeks on a test diet wtih thiamine excluded, but no deficiency signs were observed. The synthesis of thiamine by intestinal bacteria probably supplied most, if not all, of the required thiamine.

Inositol seems also to be synthesized in carp intestines (Aoe & Masuda, 1967), but not enough to free the fish from exogenous sources of this vitamin. Similarly, Lovell (1979) found an appreciable intestinal synthesis of biotin in channel catfish. Additional biotin was required only on rare occasions as with high content of lipid in the diet.

It should be remembered, however, that intestinal bacteria can also consume vitamins and thus increase the overall requirements. Tomiyama & Ohla (1967) observed that biotin deficiency in goldfish was completely prevented by the addition of 1% succinyl sulphathiazole to a biotin-deficient feed. This, they assumed, was because the micrococci in the intestines of the goldfish utilized considerable amounts of biotin; the administration of the cocci-static succinyl sulphathiazole inhibited the growth of these micrococci and hence they did not utilize the biotin in the food.

With respect to vitamin production by intestinal bacteria warmwater fishes seem to differ markedly from salmon and trout. Coldwater fishes have a much lower ability to produce vitamins through their intestinal bacterial population, perhaps because their intestine is much shorter, and thus the influence of microorganisms is less important.

Variations in vitamin requirement by different species may also result from differences in the ability to absorb, transport and metabolize the vitamins present in the food. For instance, studies in mammals have shown that the absorption of thiamine and pyridoxine (B_6) depends to a high degree on the presence of hydrochloric acid in its stomach. When the formation of hydrochloric acid is low or absent, the absorption of these vitamins decreases drastically. This, if true also for fish, may affect the absorption, and consequently the requirements, for these vitamins by stomachless fishes lacking hydrochloric acid. This, however, has not yet been studied in fish.

Individual variation in vitamin requirement

Vitamin requirement can be different under different physiological conditions. Halver (1979) found that young coho salmon showed no difference in growth rate when their diets contained either medium or high levels of ascorbic acid, but when experimental skin and muscle wounds were made in these fish, the healing of the wounds was slower when their diet contained less than 400 mg ascorbic acid/kg dry feed, and the fastest healing occurred when the fish received at least 1 g/kg feed of ascorbic acid. Halver (1979) also pointed out that the ascorbic acid requirement by salmon and trout decreases when the fish are stressed. The physiological conditions of the fish may also affect the requirement for other vitamins. Yanase (1963) found a definite relationship between the content of vitamin B_{12} and the maturity of rainbow trout. With the advance of maturation vitamin B_{12} level in various organs (except the ovary) increased markedly. This suggests that vitamin B_{12} plays a role in the maturation process, and may be required during maturation at higher amounts.

Requirement for vitamins can vary with age. Thus, Aoe & Masuda (1967) found that younger common carp require higher amounts of inositol than do older fish. Water temperature and the size of the fish may also affect vitamin requirements, as does the season. Trepsiene *et al.* (1974a) studied the levels of biotin, pantothenic acid, thiamine, pyridoxine and nicotinic acid in the muscle, liver and intestine of pond fishes. They found higher vitamin levels in these organs during summer, when natural food was plentiful, and a sharp decrease in autumn.

In contrast to terrestrial animals, fish can absorb vitamins directly from their solution in water, at least during their initial development stages, the larval and post-larval stages. Das (1967) added a mixture of vitamins of the B complex or yeast (56 mg/l) to water in tanks holding Indian carp fry. These tanks shared a higher survival rate of fry, which was double that of the control tanks, where vitamins were not added. The fry in the treated tanks grew during the first 19 days after hatching by 4.7 times, while those in the controls grew only by 3.5 times. It seems, however, that when the fish grow larger this source of vitamins is not sufficient and they become dependent on the supply of vitamins in their food.

Vitamins in feedstuffs

Fish feeds contain various amounts of vitamins. Fresh animal viscera, especially liver, spleen and kidney, contain large amounts of vitamins and can serve as an important source of them. This supply of vitamins seems to be the main reason why such tissues were important, and for a time irreplaceable, ingredients of salmonid diets before the vitamin requirements for these fish were determined. The dependence on fresh tissues for food hindered salmonid culture for some time, since these tissues were not available in sufficient amounts, it was difficult to process them into the diet, and it was hard to preserve diets containing them for long periods. Only after the vitamin requirements of the salmonids were determined, and with the supplementation of these vitamins to the diets, could dry ingredients be used, and salmonid culture could become a fully viable industry.

Feedstuffs of vegetable origin are usually richer than those of animal origin in certain water soluble vitamins such as some of the B group vitamins (thiamin, riboflavin, pyridoxin and niacin). On the other hand, feedstuffs of animal origin are richer than vegetables in fat soluble vitamins such as vitamins A, D and K, as well as in pantothenic acid and vitamin B_{12}. Diets of most carnivorous fish contain a high proportion of ingredients from animal sources and only a limited level of vegetable material, since these fish do not digest well the carbohydrates in the latter. This may cause a deficiency in water soluble vitamins, especially those of the B group, which should then be supplemented to the diet. On the other hand, diets of omnivorous or herbivorous fish usually contain a higher proportion of cereal meals and a lower proportion of fish meal or ingredients from animal sources. If such fish are depending on supplementary feed and do not receive the vitamins they require from natural food, a deficiency in fat soluble vitamins, such as vitamin A, may develop.

The composition of the diet may affect the vitamin requirement of the fish. Woodall *et al.* (1964) and Watanabe *et al.* (1981a, b) showed that the requirement for vitamin E (α-tocopherol) by chinook salmon (*Oncorhynchus tshawytscha*) and common carp, respectively, depends on the level of dietary lipid and on the degree of unsaturation of the fatty acids. Elevated levels of dietary lipid and a higher degree of unsaturation increased the requirement for vitamin E. In higher vertebrates an enhanced supply of carbohydrates is known to increase the requirement for thiamine, and a higher supply of protein to increase the requirement for pyridoxine. Only a few studies have been carried out to clarify these relationships in fish. Aoe *et al.* (1969) found that no thiamine was required when the diet was rich in protein and low in carbohydrate, but when the carbohydrate level in the diet was elevated, a requirement for thiamine was created.

Vitamins can undergo hydrolysis and inactivation by various compounds in the feed. A good example is the thiaminase found is some raw fish meats, which has already been mentioned above. Ishihara *et al.* (1974) studied the effects of such thiaminases on yellowtail (*Seriola quinqueradiata*) fed on raw anchovy as a sole feedstuff. The contents of thiamine, cholesterol and triglyceride in various tissues of the yellowtail decreased significantly, while the level of pyruvic acid in the blood increased. These symptoms were corrected by the addition of thiamine to the diet. Thiaminases were also found in some legumes and mustard seeds (Goldsmith, 1964). The inclusion of these feedstuffs in fish diets may cause thiamine deficiency, unless the activity of the thiaminase is stopped by roasting, cooking or pasteurization. Also vitamin E, mainly α-tocopherol, can be activated by antagonists and antimetabolites present in the feed. Thus, for instance, torula yeast, sometimes used in fish diets, contains substances with anti-vitamin E activity. It is suggested that saponins in legumes (including soybean meal already discussed in Chapter 6) inhibit the absorption of tocopherols (Zintzen, 1972). Linseed meal also has an inhibitory effect on the utilization of pyridoxine. The antagonistic relationships have not yet been studied in fish. Halver (1972) mentioned a number of other metabolites and antagonistic compounds found in feedstuffs, which may inactivate vitamins. For instance, avidin, a heat labile glycoprotein found in fresh egg albumin, binds biotin to form a stable non-digestible complex and thus inactivates it. Such compounds, however, are rare and not found in feedstuffs usually used in fish diets.

Vitamins may be present in feedstuffs in fair amounts, but can be sometimes bound to different organic compounds to form a range of macromolecules which are less available to animals. Studies on chicks (Frigg, 1976) have shown that the availability of biotin (as per cent of total biotin content) varies considerably in different cereals. A very low availability has been found for wheat, barley, oats and milo (0–30%), whereas the biotin in maize and soybean meal was completely available. Considerable variation was also found in the availability of niacin. According to Luce *et al.* (1967) the niacin in milo and corn is almost unavailable to pigs. Manoukas *et al.* (1968) found that for laying hens the availability of niacin in yellow corn was 30%, in wheat 36%, but in soybean meal 100%. There is as yet, however, no information on the availability of vitamins from the various feedstuffs to fish.

Vitamin requirement and availability can be affected by their interactions among vitamins themselves or with certain amino acids and minerals present in the diet. In higher vertebrates interactions between vitamin E and thiamine, riboflavin, ascorbic acid, vitamin B_{12} and inositol are believed to exist, because these vitamins have beneficiial effects on certain vitamin E deficiency signs in animals (Zintzen, 1972). Also, in mammals, it is known that vitamin E has a 'vitamin A sparing effect' resulting from the fact that vitamin E as an antioxidant, protects vitamin A, which is sensitive to oxygen, from oxidative destruction (Weber & Wiss, 1966). In contrast to this synergistic effect of vitamin E toward vitamin A, the latter can also act as a vitamin antagonist, since massive doses of vitamin A increase the requirement for vitamin E, as found for rats by Brubacher *et al.* (1965). In fish, choline hydrochloride may act with vitamin E (α-tocopherol) and vitamin K and inactivate them (Halver, 1972). The former should therefore be added in aqueous solution, while the latter should be added later in fat solution to prevent their contact and interaction.

An interaction between vitamins and amino acids can be demonstrated by the relationship between niacin and tryptophan, which has already been discussed above. As was pointed out, tryptophan can be converted into niacin (Nishizuka & Hayaishi, 1963; Goldsmith, 1964). 60 mg of tryptophane are equivalent to 1 mg niacin. Halver (1972) found that niacin deficiency signs in chinook salmon could be partially relieved by feeding a diet with an excess of tryptophan. Tryptophan also seems to interact with ascorbic acid. West *et al.* (1966) state that animals fed adiet low in tryptophan have an increased requirement for ascorbic acid.

An example of an interaction between vitamins and minerals can be seen in the relationship between vitamin E and selenium. Vitamin E is a fat soluble intracellular antioxidant involved with stabilizing unsaturated fatty acids. It inhibits the formation of lipoperoxides by saturating the peroxide radicals with hydrogen protons. It is known that in mammals selenium can act, to a certain extent, as a substitute for vitamin E, probably forming selenium–protein compounds that can also be very effective antioxidants, such as glutathione peroxidases.

Most of the relationships discussed above, whether among the vitamins in the diet themselves, or between the vitamins and the amino acids or the minerals, have been studied very little in fishes.

Some of the vitamins in the diet may be lost before even reaching the fish. One way, specific to fish feed, by which vitamins may be lost is leaching. Phillips *et al.* (1945) point out that a large proportion of the pantothenic acid in fish feed can be leached and lost before the feed is consumed by the fish. This loss may be especially high when the feed particles are small, as in starter feeds. Murai (1966, quoted from Murai & Andrews, 1975) found that trout feed of particle diameter 1 mm, which contained 500 mg pantothenic acid/kg feed, lost half of this content within 10 sec of immersion in water, and all the pantothenic acid within one minute. Other water soluble vitamins are also prone to leaching. Goldblatt *et al.* (1979) found that some pelleted diets lost through leaching as much as 50% of their riboflavin content, 65% of the vitamin C, and over 80% of the choline in 20 min of immersion in water. This problem may be more severe with slow feeders such as common carp or other warmwater fishes. Murai & Andrews (1975) found that the requirement for pantothenic acid by small channel catfish is 250 mg/kg dry feed, which is 10 times higher than that required by larger fish (see Table 40). They explain this difference as being due to leaching of pantothenic acid from the feed prior to its consumption by the fish.

Some of the vitamins are sensitive to environmental factors stimulating oxidation and decomposition such as heat, moisture, light and the presence of air or various oxidants. Such vitamins can easily decompose during the processing of the diet or during its storage. This is especially true if the feed is pelleted, since this is associated with high moisture and heat. Heat affects especially the levels of vitamins B_{12}, ascorbic acid, pantothenic acid and pyridoxine, while light affects mainly riboflavin and pyridoxine. Feeds should therefore be stored in a cool and dark place. However, even then, some of the vitamins decompose during storage due to the presence of air or various oxidants. The more sensitive to these are pyridoxine, ascorbic acid, vitamin B_{12}, vitamin E and vitamin K. The practical conclusion is that fish feed should not be stored for too long before use; and that when feed is processed in a way which may cause a certain loss of vitamins, an excess of vitamins should be added to account for the loss, so that the remaining concentration should be sufficient for normal growth.

Minerals

Like all other animals, fish require minerals as essential factors in their metabolism and growth. However, in contrast to terrestrial animals, which are entirely dependent on a dietary supply of minerals, fish can absorb part of the required minerals directly from the water through their gills or even through their entire body surface. Absorbing minerals from water is a vital process for osmoregulation in freshwater fish, but no doubt it is also important from a nutritional point of view. The rate of mineral absorption varies among fish species and with variations in certain environmental factors, such as the mineral concentration, water temperature, pH etc. This makes it difficult to determine the overall requirement of minerals by fish, and how much should be supplemented through the diet. It seems, however, that minerals absorbed from water do not meet the total requirement and a certain supplementation through the diet is required, whether in the natural food or supplementary feed.

Fish food contains, of course, certain amounts of minerals, which compose the 'ash' part of the food. These amounts, however, are not always sufficient to meet the requirements. Some of the minerals may be leached during the processing of the food. For instance, most of the water soluble minerals in fish meal are washed out during processing and only minerals insoluble in water, or even weak acids, remain. Feeds may contain sufficient amounts of certain minerals, but not of others. Thus, petroleum yeasts are rich in phosphorus but poor in calcium. The yeast must, therefore, be supplemented by this mineral (Arai *et al.*, 1975; Takeuchi & Ishii, 1975 – quoted from Nose & Arai, 1979). In other foods minerals can be found, but are only partly available to animals. Thus, zinc in soybean meal is less available than that in casein due to the phytic acid in soybean forming a complex (chelate) which binds the zinc in an unavailable form (Cunha, 1967). In these cases, where food cannot supply the mineral requirement, the diet must be supplemented by a mineral mixture.

Nose (1972 – quoted from Nose & Arai, 1979), Ogino & Kamizono (1975) and Takeuchi *et al.* (1981) cultured rainbow trout on a test diet developed by Halver (1957b) from which the mineral mixture was omitted. After 2–10 weeks, evidence of deficiency appeared in the fish, e.g. poor appetite, sluggish movement, growth retardation, hypochromic-microcitic anaemia, and a certain percentage of the fish started convulsing and died. Only after the mineral mixture was added again, at a level of at least 2% of the diet (in Ogino & Kamizono, 1957 study – 4%), were the deficiency signs prevented. Most of the surviving fish showed scoliosis, lordosis and deformation of the head bones, especially the snout which became shorter and rounder. Mineral supplementation seems to be required also by common carp (Tacon *et al.*, 1984) and Japanese eel (*Anguilla japonica*). Arai *et al.* (1974) found that the addition to the diet of 2% of a mineral mixture appreciably increased the growth rate of the eels, although requirements for much higher concentrations were earlier reported by these authors. At concentrations of mineral mixture lower than this level the eels lost their appetite, their growth was inhibited and they even lost weight, and mortality was high. In contrast, no deficiency signs could be identified in common carp fed on a mineral deficient diet even after 50 days (Ogino & Kamizono, 1975). Cho & Schell (1980) summarized the deficiency signs and requirements by fish for 16 minerals. Deficiency of only four of these (phosphorus, magnesium, iron and iodine) resulted in clear signs, while deficiency in the other minerals did not result in defined signs. The requirement for the 16 minerals, as summarized by Cho & Schell (1980) is given in Table 41. It should be remembered that, as mentioned above, most of the requirement in mineral is supplied by the food and supplementation, if necessary, is partial.

The discussion above related to mineral mixture without distinguishing between the individual mineral elements, although, no doubt, the requirement, absorption, and need for supplementation may vary for each of

Table 41. *Summary of information on mineral requirements of fish*

Mineral	Requirement (per kg dry diet)
Calcium	5 g
Phosphorus	7 g
Magnesium	500 mg
Sodium	1–3 g
Potassium	1–3 g
Sulphur	3–5 g
Chlorine	1–5 g
Iron	50–100 mg
Copper	1–4 g
Manganese	20–50 mg
Cobalt	5–10 mg
Zinc	30–100 mg
Iodine	100–300 mg
Molybdenum	trace
Chromium	trace
Fluorine	trace

Source: From Cho & Schell (1980).

these mineral elements. The elements required for the metabolic processes in fish can be classified into three groups:

(a) Constructional: Calcium, phosphorus, fluorine and magnesium are all important for the construction of the bones; sodium and chlorine are the main electrolytes of blood plasma and the extracellular fluid, while sulphur, potassium and phosphorus are the main electrolytes of the intracellular fluid. These elements are necessary, therefore, for the production of the above mentioned tissues.

(b) Respiratory: Iron and copper are important elements in haemoglobin and, therefore, also in the transfer of oxygen in blood.

(c) Metabolic: Many mineral elements, including some of those already mentioned above, take part in the metabolic processes. Usually they are required in much smaller amounts than for the previous two functions, and some only in trace quantities.

Calcium

The contents of calcium and phosphorus are the highest among the inorganic constituents of the animal body. About 99% of the calcium and 80% of the phosphorus are found in the bones. The remaining 1% of the calcium plays a vital role in metabolism. It constitutes an important part of the intracellular fluid (in mammals calcium amounts to about 200 meq/100 g tissue – White *et al.*, 1959). The presence of calcium is essential for the activity of some enzymes, notably those responsible for muscle contraction and for the transmission of nervous impulses. Calcium concentration is important also in extracellular fluid, especially in association with the response of muscle to external stimuli and with the mechanism of blood clotting.

Experiments have shown that many, if not all, fishes absorb calcium from water, mainly through their gills (e.g. Phillips *et al.*, 1953, 1954, 1955, for rainbow trout; Toyama *et al.*, quoted from Phillips *et al.*, 1957, for goldfish; Pavlovski, 1958, for carp). Phillips *et al.* (1954, 1955) studied the absorption of labelled calcium (^{45}Ca) by rainbow trout. They found that the amount of absorbed calcium is high, and more or less equal to that assimilated from the food. Absorption of calcium from water continues even when the dietary calcium meets nutritional requirements. Calcium absorbed from water accumulates mainly in the bones, which indicates that it is used for construction rather than for metabolism. Calcium uptake, both by absorption from water and assimilation from food, depends on the metabolic rate of the fish. The higher the metabolic rate, the higher the calcium uptake, thus, starved fish, when metabolism is low, take up less calcium than fed fish (Phillips *et al.*, 1955). Temperature affects metabolism, and therefore calcium uptake: the higher the temperature, the higher the calcium uptake. Osmoregulation, which depends on the mineral concentration in water, affects metabolism and calcium uptake. Most important in this respect is the concentration of calcium in water. Eddy (1975) studied the effects of inorganic ions in water on the electrical potential and the fluxes of sodium and chloride in the gills of goldfish. The potential was affected by a number of inorganic ion species, but that with the most significant influence on transepithelial potential was calcium. Phillips *et al.* (1955, 1956, 1957) found that at concentrations of calcium in water below 50 mg/l, the lower its concentration, the higher the metabolism and uptake of calcium. Thus, a sort of compensatory system exists here, making calcium uptake constant irrespective of its concentration in water.

The amount of phosphorus available to the fish has a considerable effect on the absorption of calcium from the water. The higher the phosphorus concentration in water or its content in the diet, the higher the absorption of calcium (Phillips *et al.*, 1953, 1956). A number of experiments (Sakamoto & Yone, 1978a, for red sea bream; Ogino *et al.*, 1979, for common carp; Watanabe *et al.*, 1980a, for chum salmon) have shown that when fish have sufficient amounts of calcium both in water and diet, but are fed a diet low in phosphorus, the content of calcium in the bones and other tissues is lower than that in fish fed sufficient amounts of phosphorus. There was almost no difference in the ratio Ca:P in the fish body irrespective of the amounts of phosphorus taken up.

The absorption of calcium from water by fish makes them less dependent on, or even independent of, dietary calcium. McCay *et al.* (1931) found that rainbow trout, fed on a diet poor in calcium but rich in phosphorus, grew normally for many months. The addition of calcium to their diet did not affect the amount of calcium or phosphorus in the fish bodies. Similar results were obtained by Lovell (1979) for channel catfish and by Ogino & Takeda (1976) for common carp. When the water contained 16–20 ppm calcium and the diet contained sufficient amounts of phosphorus, the growth of the carp was normal though their diet contained only 30 mg Ca/100 g feed. Also Watanabe *et al.* (1980a) obtained similar results for chum salmon. However, since the requirement for calcium and its absorption rate may differ in different fish species, some fishes do require supplementation of dietary calcium. Thus, feeding Japanese eel on diet poor in calcium caused deficiency symptoms such as loss of appetite and retardation in growth. Arai *et al.* (1975, quoted from Nose & Arai, 1979) determined the requirement of the Japanese eel for supplemented dietary calcium and found it to be 270 mg/100 g feed. Also, red sea bream requires calcium supplementation. Sakamoto & Yone (1973) found that reducing the content of supplementary calcium in the diet of red sea bream from 0.34% to 0.136% (phosphorus in both cases was 0.68%) resulted in a considerable reduction in growth rate, from 43 g/fish during 76 days to 27.1 g/fish during the same period, respectively.

While dietary calcium may be essential in the cases quoted above, excess dietary calcium may be harmful in certain circumstances. Cowey (1976) found that when magnesium is deficient in rainbow trout diet, high dietary calcium (2.7 g/100 g diet) may cause renal nephrocalcinosis and reduction of growth. This, however, did not occur when sufficient magnesium was present in the diet.

Phosphorus

The content of phosphorus in the fish body is more or less equal to that of calcium. In addition to its role in bone construction, phosphorus also comprises an important part of the intracellular fluid. It is found in phospholipids, which are important elements in the cell membranes, and it plays an essential role in the transformation of energy in the cells. Phosphorus, like calcium, is absorbed directly from water (Mullins, 1950; Phillips *et al.*, 1953), but the rate of its absorption is much lower than that of calcium. McCay *et al.* (1936) point out that brook trout (*Salvelinus fontinalis*) fry after hatching contain more phosphorus than calcium, but after the fry start feeding they rapidly absorb calcium from water until the concentrations of phosphorus and calcium equalize. Phillips *et al.* (1957) studied the absorption of labelled phosphorus ($^{32}PO_4$) by brook trout. They found that absorption was relatively small. In 96 h the fish absorbed only 1/30 000 of the total phosphorus content in their body, while during the same time they absorbed calcium amounting to a quarter of their total calcium content. These authors also studied the distribution of the absorbed phosphorus in the trout's body. They found a higher proportion of absorbed phosphorus in the intestine than in the gills or any other tissue. Also Brichon (1973) found, in European eel at its 'yellow' stage, held for 30 min in water containing labelled ^{32}P as sodium orthophosphate, that the labelled phosphate was accumulated especially in the mucous tissue of the intestine. This suggests that the site of absorption is mainly the intestine. However, such a conclusion would require that the fish drinks water, which is opposed by the common concept that freshwater fish do not drink (Black, 1957) or to experimental results showing a very small rate of drinking (Mullins, 1950). Phillips *et al.* (1957) quote Tomiyama *et al.*, who carried out a similar experiment with common carp, using ^{32}P, but blocking the oesophagus to prevent any possibility of drinking. In spite of this, most of the absorbed labelled phosphorus concentrated in the intestine. These authors only hinted at a possible explanation, according to which the phosphorus is bound by a metabolite, possibly ATP, whose concentration is highest in the intestine, although the main absorption site may be the gills. Since the target tissues are not the bones, but rather the viscera, it seems that, contrary to calcium, phosphorus absorbed from the water is used for metabolism and not for bone construction.

Phosphorus absorption rate is greatly affected by the concentration of phosphorus in water (Phillips *et al.*, 1957). The higher the phosphorus concentration in water, the higher its absorption by the fish. Since phosphorus concentration also affects calcium absorption (see section on calcium), the ratio Ca:P in fish is independent of the phosphorus concentration in water (Watanabe *et al.*, 1980a). In contrast, the concentration of calcium in water does not affect the absorption of phosphorus, nor the total amount of calcium absorbed (though it affects the rate of calcium absorption: see section on calcium). The ratio Ca:P is therefore also independent of the calcium concentration in water.

Most of the experiments of Phillips *et al.* (1957) on the absorption of phosphorus from water were carried out at a concentration of about 3 ppm phosphorus in water. In such conditions a brook trout of 7.5–10.1 g absorbed 0.23–0.28 μg P/g live weight/48 h. However, phosphorus concentrations in natural waters are usually much lower than those given above, and generally are in the range of 0.005–0.05 ppm. It can be assumed, therefore, that in natural conditions the absorption of phosphorus from water is relatively small and probably not enough to meet the nutritional requirements of the fish. There is, therefore, a need for a dietary phosphorus supply, especially for the construction of bones. Indeed, according to Phillips *et al.* (1957) the assimilation of phosphorus from food is 200 times higher than that absorbed from water. Experiments have shown that when fish are fed a phosphorus-deficient diet their growth is retarded and their mortality is high (Andrews *et al.*, 1973; Ogino & Takeda, 1976; Yone & Toshima, 1979; Watanabe *et al.*, 1980a). Other signs of phosphorus deficiency were: in common carp – lordosis deformation in the skull bones and decalcification; in trout – poor appetite, sluggish movements and low contents of both calcium and phosphorus in the whole body (Murakami, 1967; Watanabe *et al.*, 1980a); and in channel catfish – lower food utilization and low haematocrit level (Andrews *et al.*, 1973).

Sakamoto & Yone (1978a) and Ogino *et al.* (1979) found that when phosphorus is deficient in the diet, more lipid is accumulated in the body, especially in the intestine. The phosphorus content of the diet affects not only the absorption of calcium, as was already mentioned above, but also that of other minerals, such as cobalt and magnesium. As a consequence, when phosphorus in the diet is low, the total ash content of the fish is also low (Ogino *et al.*, 1979; Watanabe *et al.*, 1980a).

Phillips *et al.* (1957) studied the assimilation of dietary labelled phosphorus and its distribution among the tissues, such as those of the stomach, intestine and gills. Since the accumulation rate of labelled phosphorus in the intestine was parallel to that in the gills and other body tissues, but higher than that of the stomach, they concluded that the main site of phosphorus assimilation is the intestine. The distribution of assimilated phosphorus in the body is similar to that of calcium, and it seems that in contrast to the phosphorus absorbed from water, it is utilized mainly for the construction of bones and other body tissues.

Phosphorus assimilation rate is not affected by the phosphorus content in the diet. Therefore, the total amount of phosphorus assimilated is linearly related to the content of the phosphorus in the diet (Nakamura, 1982). The assimilation rate of phosphorus, from the minerals added to the diet, is quite high and is usually around 90–95% (Ogino *et al.*, 1979; Nakamura, 1982). However, lower rates, of only about 40%, were recorded for the assimilation of phosphorus contained in vegetable matter, such as peas or peanut oil meal, by common carp (Shcherbina *et al.*, 1970). As with absorption of phosphorus from water, assimilation from the diet is also affected by the content of calcium in the diet. The higher the calcium content, the lower the assimilation rate (Nakamura, 1982). Also, temperature affects dietary phosphorus assimilation: the higher the temperature the higher the phosphorus assimilation rate (and its metabolism in the fish body). Nevertheless, since at low temperatures the intestine evacuation time is longer, the total amount of phosphorus assimilated from unit weight of food is independent of temperature (Podoliak, 1961).

In view of the variation in phosphorus absorption and assimilation rates among different fish species, their requirements for dietary phosphorus may also vary. Some data on phosphorus requirement by various species of fish, collected from the literature, are presented in Table 42.

Phosphorus is found in most feedstuffs, but its availability to various fish species may differ. It should be remembered that most phosphorus compounds in nature are not soluble in water but only in acids. Some, as apatite (phosphate rock) or tri-calcium phosphate, which appears in the bone, will dissolve only in strong acids. In this respect a difference exists between fish with a stomach (and acid secretion) and fish without a stomach, which also lack acid. Takeda & Ogino (1975 – quoted from Nose & Arai, 1979) and Ogino *et al.* (1979) studied the rate of phosphorus assimilation from various feedstuffs and compounds by common carp and rainbow trout (Table 43). As can be seen from the table, common carp, which lacks a stomach, assimilates the water soluble mono-calcium phosphate just as well as the trout, but not tri-calcium phosphate which is acid soluble. Since a large proportion of the phosphorus in fish meal is bone phosphate (tri-calcium phosphate), common carp cannot assimilate well the phosphate from this source, while the trout can. The organic phosphorus of casein and petroleum yeast is assimilated with high efficiency and equally well by both common carp and rainbow trout, while phosphorus from rice bran, wastewater sludge and phytin is hardly assimilated by either. The same is true for other fish species. Thus, red sea bream (*Chrysophrys major*) assimilates water soluble phosphates better than water insoluble phosphates (Sakamoto & Yone, 1979), but chum salmon (*Oncorhynchus keta*) assimilates 71% of the phosphates from fish meal, and *Sarotherodon niloticus* can assimilate 65% of bone phosphate (Watanabe *et al.*, 1980b).

In this respect it is worth mentioning that phosphate may become insoluble in water when it reacts with calcium. Therefore, when a mixture of phosphates and calcium compounds is added to the diet, even when the former is water soluble, it may become insoluble before it is taken up by the fish, as was found experimentally by Phillips *et al.* (1957).

Table 42. *Dietary phosphorus requirement for maximum growth of different fish species*

Fish species	Phosphorus requirement (% of diet)	Source
Chum Salmon (*Oncorhynchus keta*)	0.5–0.6	Watanabe *et al.* (1980b)
Rainbow trout (*Salmo gairdneri*)	0.65–1.09	Ogino & Takeda (1976)
Common carp (*Cyprinus carpio*)	0.6–0.7	Ogino & Takeda (1976)
Channel catfish (*Ictalurus punctatus*)	0.8	Andrews *et al.* (1973)
	0.4	Lovell (1979)
Red sea bream (*Chrysophrys major*)	0.68[a]	Sakamoto & Yone (1973, 1979)
Black sea bream (*Mylio macrocephalus*)	0.11	Yone & Toshima (1979)
Japanese eel (*Anguilla japonica*)	0.29	Arai *et al.* (1975)[b]

Notes: [a] At the presence of −0.34% calcium.
[a] Quoted from Nose & Arai (1979).

From the discussion above it is clear that diets, though they may contain high amounts of phosphorus, may sometimes need supplementation with phosphates, since naturally occurring phosphorus is not available to certain fish species. Murakami (1967) found that the addition of 4% of Na_2HPO_4 to a commercial diet of common carp prevented deficiency signs such as skull bone deformation, and resulted in a better growth rate, better food utilization and improved survival of fish during the winter. It is interesting to note that phosphorus supplementation in this case also resulted in an increased protein content and a decrease in lipid content of the fish, which indicates the role of phosphorus in lipid metabolism and in sparing protein. Ogino *et al.* (1979) found that there was almost no difference in the growth of rainbow trout fed diets supplemented by various phosphate compounds, but common carp fed diets supplemented by mono-sodium phosphate, mono-potassium phosphate or mono-calcium phosphate grew twice as fast as carp fed on diets supplemented by di- or tri-calcium phosphate. Similar results were obtained by Sakamoto & Yone (1979) with red sea bream.

Sulphur

Like phosphorus, sulphur is an important element in the intracellular fluid. It constitutes an essential part of some amino acids (methionine, cystine and cysteine). The content of sulphur in the muscle almost equals that of phosphorus.

Sulphur, like calcium and phosphorus, can be absorbed directly from water (Phillips *et al.*, 1960, 1963), but in spite of the fact that its

Table 43. *Availability of phosphorus from some feedstuffs and compounds to rainbow trout and common carp*

Feedstuffs	Phosphorus content (%)	Availability (%) Common Carp	Rainbow Trout
Tri-calcium phosphate	0.65	3	51
Mono-calcium phosphate	0.79	80	61
Phytin	1.65	8	19
Casein	0.47	100	90
Fish meal[a]	0.99	26	60
Petroleum yeast[a]	0.46	99	91
Wheat seedlings[a]	0.58	57	58
Activated sludge[a]	0.84	12	49
Rice bran[a]	0.79	25	19

Note: [a] With no fat and after heat treatment.

Source: From Takeda & Ogino (1975, quoted from Nose & Arai, 1979).

concentration in water is usually much higher than that of phosphorus, the absorption of sulphur from water is appreciably lower than that of phosphorus. Moreover, only a small fraction of the sulphur absorbed from the water is retained in the organic compounds of the body. Most of the sulphur required by the fish must, therefore, be supplied through the diet. Assimilation rate of sulphur from food is about 100 times as high as the rate of its absorption from water. Phillips *et al.* (1962) studied the uptake of labelled sulphur (^{35}S) by brook trout. At a concentration of 4 mg/l labelled sulphur in water, the fish absorbed from the water 9.4 μg S/day. When, however, the diet contained 4 mg labelled sulphur the fish assimilated 480.8 μg, which were 1% of the total sulphur in the fish body. An examination of the labelled sulphur distibution showed that most of it accumulated in the blood, and only a small portion was contained in muscle as organic sulphur. It seems that the sulphur was supplied in excess of requirements, and the excess was excreted within a few days of uptake. The higher the sulphur content in the diet, the higher the amount retained by the body. However, the utilization rate (retained sulphur/ingested sulphur) was higher at lower dietary sulphur contents (Page *et al.*, 1978).

Iron

Iron is an essential element in the formation of haemoglobin in blood erythrocytes. Hevesy *et al.* (1964) found that upon injecting labelled iron (^{59}Fe) to tench at 18 °C, up to 70% was taken up by the erythrocytes. Iron deficiency was found to cause hypochromic-microcytic anaemia in some fishes such as the Japanese eel (Arai & Hashimoto, 1975, quoted from Nose & Arai, 1979), yellowtail (Ikeda *et al.*, 1973), red sea bream (Yone, 1976) and common carp (Sakamoto & Yone, 1978b). Ikeda *et al.* (1973) determined the requirement of iron by yellowtail as 57 mg/100 g feed. Arai & Hashimoto (1975 – quoted from Nose & Arai, 1979) studied the iron requirement of Japanese eel by counting the erythrocytes, determining the concentration of haemoglobin in them, and measuring the haematocrit level. Using these methods they found the minimum requirement for iron to be 17 mg/100 g feed.

Trace elements

As was pointed out above, some elements taking part in the metabolism or its regulation are required in minute quantities. These elements, most of which are metals, often have a trimodal effect: severe deficiency can result in deficiency signs, retarded growth and even death; a higher level of intake, after satisfying the essential requirement, may have, in the case of some elements, a stimulatory effect on specific physiological processes (see sections on cobalt and chromium); at a certain level of intake, however, these elements usually become toxic and may result again in retarded growth and mortality. The primary areas of concern in the present discussion will be the nutritional and stimulatory levels of intake rather than the toxic level. All trace elements are important for normal metabolism of the fish. However, for some elements, information relating to fish is quite scarce and these will be dealt with here only briefly.

Cobalt

Cobalt is an essential constituent of vitamin B_{12}. This vitamin is produced, in some fishes, by the intestinal bacterial flora. Cobalt must, however, be provided for this production. Cobalt also takes part in other metabolic processes, such as enhancing the activity of certain peptidases and by being a co-factor to several enzymes as aldolase, which catalyses the transformation of fructose 1,6-diphosphate to triphosphate, and phosphoglucomutase which catalyses the transformation of glucose 1-phosphate to glucose 6-phosphate. Of special interest in fish is its effect on the regulation of carbohydrate and amino acid metabolic pathways, which will be discussed below.

Cobalt can be absorbed by fish directly from solution in water in a manner similar to the absorption of calcium. Like calcium, cobalt absorption rate depends on the metabolic rate. Phillips *et al.* (1956) used the rate of labelled cobalt (^{60}Co) absorption as an indicator for the metabolic rate. However, the relationship between cobalt absorption and metabolism is not too clear, since according to Frolova (1957), metabolism is affected by the uptake of cobalt rather than the other way round as suggested by Phillips *et al.* (1956). Since calcium concentration in water affects metabolism (see section on calcium), it also affects the absorption of cobalt. Phillips *et al.* (1956) found that as calcium concentration in water was reduced from 50 mg/l to 5.0, 0.5 and 0.0 mg/l, the absorption of cobalt by brook trout (*Salvelinus fontinalis*) from water containing 1 mg/l ^{60}Co increased by two, three, and ten times respectively. Later experiments (Phillips *et al.*, 1957) showed this to be also true for rainbow trout. As with calcium, the absorption site seems to be the gills. Reed (1971) studied the absorption of cobalt by black bullhead, *Ictalurus melas*. During uptake the gills contained 19–30% of the whole body's ^{60}Co activity. Most of the cobalt taken up was eliminated within a few days, but some was retained for a longer period.

A number of experiments have shown that the addition of cobalt to the fish diet or to the water affected some physiological processes within the fish and resulted in appreciable increases in growth rate, feed utilization and survival. The first to study these effects in fish was Frolova (1957). In this and later studies (Frolova, 1961, 1970) she found that cobalt added to the diet or to aquarium water affected the composition of the blood of common carp. The concentrations of erythrocytes and haemoglobin increased, while that of leucocytes decreased, although their capacity to multiply in case of need was not reduced. Following the first studies by Frolova a number of experiments were carried out, mainly in the Soviet Union, on the effect of cobalt on fish growth (Sukhoverkhov & Krymova, 1961; Vinogradov & Erokhina, 1961, 1962; Sukhoverkhov *et al.*, 1963; Shabalina, 1964; Anon.; 1964, Tomnatik & Batyr, 1965; Anon., 1971). These studies have shown that addition, through the diet, of 0.02–0.08 mg cobalt/kg live weight/day increased yields of common carp in ponds by 16–30% and reduced feed conversion ratio by 20–25%, compared with controls in which the feed did not contain cobalt. Schaeperclaus (1964, 1965), on the other hand, found only a slight, or no effect, of cobalt on fish growth, which may be due to the specific conditions of his experiments. Joshev & Grozev (1971) obtained best results by adding 0.4 mg Co/kg live weight/day, as $Co(NO_3)_2{\cdot}6H_2O$, to the diet of common carp. This increased average yield in ponds by 30.8% and reduced feed conversion ratio by 0.7 as compared to control ponds where fish were fed a diet without added cobalt. Das (1967) added $Co(NO_3)_2$ to water in clay jars holding fry of Indian carp. Survival rate of these fry increased to 49.75% compared with 30.33% in the control jars without cobalt. Ghosh (1975) obtained similar results when he added cobalt to containers holding mullet (*Mugil parsia*), fry. The best results were obtained when 0.1 ppm cobalt was added. Growth rate of the fry increased during the first month by 129% over the control, and survival rate reached 95% compared with only 60% in the control. Sen & Chatterjee (1979) and Chatterjee (1979) found that cobalt affected the survival and growth of young fry of Indian carp (rohu, *Labeo rohita*; catla, *Catla catla*; and mrigal, *Cirrhinus mrigala*) after their hatching. He added cobaltous chloride to their diet at a rate of 0.01 mg/fry/day. Cobalt was also effective when added directly to pond water and increased fish yields (Kovalskii *et al.*, 1965; Anon., 1967). As in the experiments by Das (1967) quoted above, here too it seems that the fish absorbed the cobalt from the water.

No explanation was given in the papers cited above as to the mode of action of cobalt in fish. Frolova (1961) assumed that the cobalt affected cytopoietic activity and haemopoiesis in the bone marrow. Das (1959) assumed that the cobalt was necessary for the production of vitamin B_{12}. Another possible explanation is that cobalt takes part in the insulin–glucagon balance and thus affects glycolysis and amino acid metabolism. Several studies have shown that cobalt, if administered to vertebrates, accumulates in the Islets of Langerhans tissue (Falkmer & Knutson, 1963; Falkmer *et al.*, 1964; Khanna & Bhatt, 1974). The main sites of accumulation in fish were the parenchymal cells, especially β-cells of the islets and possible also α_1-cells. However, contrary to the case in mammals, cobalt is not accumulated at all in the α_2-cells (Falkmer *et al.*, 1964). Cobalt has two biochemical properties which may be of importance here. Like zinc it crystallizes insulin to form a more stable complex (Cunningham *et al.*, 1955; Graig, 1962; Butt, 1975). Cobalt is also known to produce complexes with substances containing sulphydryl (SH-) groups such as cysteine and glutathione (GSH). In order to be concentrated in the cells, the cobalt ions must pass the cell membranes. Cells with a low content of SH groups in their cell membranes, such as the islet cells, probably have a greater ability to concentrate cobalt than cells with high content. β-cells synthesizing or storing insulin might tend to absorb the cobalt ions. The absorbed cobalt could also combine with SH groups in the cell membrane or in the mitochondria, thus causing severe metabolic derangement and leakage of insulin to the blood with ensuing hypoglycaemia. Indeed, some studies have shown histological changes in the islet cells and cell lesions due to a necrotizing activity of the accumulated metal ions (Khanna & Bhatt, 1974). The effect of cobalt on the levels of blood glucose and amino acids seems to depend on a delicate balance between these effects which is dependent on the concentration of administered cobalt. Khanna & Bhatt (1974) found that administration of cobaltous chloride to catfish (*Clarias batrachus*) caused hypoglycaemia and decreased liver glycogen within five hours. This was accompanied by histological changes in the Islets of Langerhans. No doubt further study on the effects of cobalt on fish may prove valuable.

Chromium

The role of chromium as an essential trace element for normal growth and development of mammals and poultry has been well documented (Shamberger, 1980; Anderson, 1981; Steele & Rosebrough, 1981). Tacon & Beveridge (1982) studied the essentiality of chromium to fish, but could not reach a conclusive result, which may have been due, as the authors pointed out, to absorption of chromium from water and contamination of the control diet by chromium. Mertz & Schwarz (1955) observed that rats fed certain kinds of laboratory diet had impaired tolerance to glucose loads. They found this to be a result of chromium deficiency. Experimental animals raised on a chromium-deficient diet devleoped a fasting hyperglycaemia, impaired growth, elevated serum cholesterol and decreased fertility. The glucose tolerance and other symptons were reversed by feeding with brewer's yeast and other chromium-rich feedstuffs.

Biologically active chromium functions in a number of physiological reactions, especially those occurring in the peripheral tissue, e.g. glucose uptake, oxidation of glucose to carbon dioxide and its incorporation into fat, liver lipogenesis (Steele & Rosebrough, 1981), glycogen synthesis (Roginski & Mertz, 1969; Rosebrough & Steele, 1981), and amino acid transport and uptake (Mertz, 1969; Roginski & Mertz, 1969). Mertz *et al.* (1974) proposed, as a possible mechanism for chromium action, that it catalyses the sulphide interchange between insulin and sulphydryl membrane receptor sites. This enhances the binding of the insulin to specific receptors on the surface of the plasma membranes of the cells of target tissues and thereby acts synergistically to potentiate insulin-sensitive metabolic pathways. Polarographic studies with rat liver mitochondria supported this postulate.

According to Anderson (1981), inorganic chromium compounds display little or no *in vitro* insulin potentiating activity, but certain organic chromium complexes acquire significant insulin potentiating activity. However, experimental animals raised on chromium-deficient diets responded to supplementation with inorganic chromium, which seems to be converted physiologically to a biologically active form (Schwarz & Mertz, 1959). Intestinal microorganisms are apparently not necessary for this conversion, since animals responded to injected as well as to orally administered chromium.

The richest known source for biologically active organic chromium complexes is brewer's yeast. Toepfer *et al.* (1977) found a method to synthesize a biologically active form of chromium similar to the natural complex isolated from the yeast. This is a low molecular weight compound, with Cr^{3+} coordinated to two nicotinic acid molecules with the remaining coordinates protected by amino acids such as glycine, cystine and glutamic acid. Different amino acids have little effect upon the insulin potentiating effect of the synthetic chromium complex, but nicotinic acid does, and may be crucial for production of these complexes.

Tacon & Beveridge (1982) found that fish fed a diet containing 45.5% crude protein displayed the best growth when 1 mg supplementary Cr^{3+} was added per 1 kg of the diet (for comparison – 20 mg/kg feed were found to cause best growth of turkey poults by Steele & Rosebrough, 1981). This rather low level of chromium is usually present in most commercial fish feeds and it seems unlikely that fish would suffer from lack of dietary chromium. However, in view of the diabetic-like nature of fish, well planned studies are needed to determine the role and effects of chromium on the carbohydrate utilization from low protein diets.

Magnesium

Magnesium is an essential constituent of bone in fish and its metabolic activity is interrelated with calcium and phosphorus metabolism. Magnesium is also an essential element in some enzymatic processes in the metabolism of carbohydrates, especially those associated with phosphorus-rich metabolites and ATP (White *et al.*, 1959). While Sakamoto & Yone (1974 – quoted from Nose & Arai, 1979) did not find any magnesium deficiency signs in red sea bream, such signs were found in some freshwater fishes. Magnesium deficiency caused poor appetite, growth retardation, reduced ash content, lordosis, deformations, convulsions, cataracts and death in rainbow trout and common carp (Ogino *et al.*, 1978; Satoh *et al.*, 1983a). In the Japanese eel poor appetite and growth retardation were observed, but not the other signs. When the diet contained more than 25 mg/100 g feed these signs were prevented, but to achieve best growth higher levels had to be added. Cowey (1976) showed that intake of calcium and phosphorus may aggravate magnesium deficiency in rainbow trout. When diets deficient in magnesium, but high in calcium and phosphorus (27 g/100 g diet of each) were fed to trout, in addition to reduced growth rate renal nephrocalcinosis also occurred, i.e. deposition of calcium in the nephrons. Fish fed on control diets containing either sufficient magnesium (100 mg/100 g diet) or lower calcium and phosphorus contents (1.3 g/100 g diet or less) had histologically normal kidneys and renal calcium levels were not elevated.

Magnesium levels added to the diet to achieve best growth were; for Japanese eel, 40 mg/100 g feed (Arai *et al.*, 1975, quoted from Nose & Arai, 1979); for common carp, 40–50 mg/100 g feed; and for rainbow trout, 60–70 mg/100 g (Ogino & Chiou, 1976; Ogino *et al.*, 1978).

Iodine

Iodine is an important component of the thyroid hormones. The iodine content in the thyroid gland is 0.5–1.0%; it is associated mainly with thyroglobulin. The main metabolites formed by the hydrolysis of thyroglobulin are the iodothyronines and thyroxine. Woodall & LaRoch (1964) examined the dietary iodine requirement of chinook salmon by feeding them diets containing 0.1–10.1 mg iodide/kg diet. After six months no significant differences had occurred in growth, feed efficiency and body composition, except for lower iodine storage in the thyroid glands in the fish fed 0.1 mg iodide/kg diet compared with those fed 0.6 mg/kg and above. It seems that the requirement for iodine is quite low and that high concentrations of iodine may even be toxic. Ikeda *et al.* (1972) studied the effect of diet supplementation with lower concentrations of iodine (added as potassium iodide) than those mentioned above, on the growth and scale development of goldfish. Maximum growth and scale development were obtained at a concentration of 0.0025–0.25 mg I/g body weight/day. However, when the goldfish were fed a diet with a higher concentration than 2.5 mg I/g body weight/day, growth was inhibited.

Manganese

Manganese is associated with the metabolism of arginase in the liver and with some peptidases. It activates many enzymes such as phosphoglucomutase, cholinesterase, oxidative α-keto-decarboxylases and ATPase in the muscle. Manganese-deficient rainbow trout displayed cataracts, growth inhibition and malformation of the backbone and tail (Satoh *et al.*, 1983b). Common carp and rainbow trout fingerlings were found to have higher growth rates when fed 12–13 mg Mn/kg dry diet, compared with those fed a diet containing only 4 mg/kg (Ogino & Yang, 1980), but channel catfish require less, and 2.4 mg Mn/kg contained in the basal diet tested was apparently sufficient to meet the manganese requirement of fingerlings (Gatlin & Wilson, 1984). It seems, however, that age affects the requirement for manganese, since Knox *et al.* (1981) did not find any deficiency signs in rainbow trout over 15 g even when fed a low manganese diet (1.3 mg Mn/kg).

Zinc

Zinc is a component of some enzymes such as carbonic anhydrase, the dehydrogenases of alcohol and lactic acid, and some peptidases. Zinc has been shown to be an essential trace element for rainbow trout (Ogino & Yang, 1978, 1979a; Ketola, 1979a) and common carp (Ogino & Yang, 1979b; Satoh *et al.*, 1983a, b). 15–30 mg Zn/kg dry diet promoted satisfactory growth of rainbow trout and common carp, while 5 mg/kg or less produced lower growth, and 1 mg/kg resulted in poor growth, high mortality, high incidence of cataracts, high incidence of fin erosion and reduced protein and carbohydrate digestibility (Ogino & Yang, 1978, 1979a, b).

Copper

Dietary copper requirements have been investigated for channel catfish (Murai *et al.*, 1981b), common carp, (Satoh *et al.*, 1983a) and rainbow trout (Ogino & Yang, 1980; Satoh *et al.*, 1983b). Murai *et al.* (1981b) suggested that since fish can absorb copper from the surrounding water, the requirement for dietary copper is lower than that of most terrestrial animals. They determined the dietary copper requirement for fingerling channel catfish as 1.5 mg/kg feed, or maybe even less. Ogino & Yang (1980) found that common carp fed 3 mg Cu/kg feed showed a better growth than those fed 0.7 mg/kg, but no such difference was observed with rainbow trout. Increasing the copper concentration in the diet over these values seemed to intoxicate the fish. According to Murai *et al.* (1981b) channel catfish fed 9.5 mg Cu/kg feed grew signficantly slower than those fed diets containing 3.5 mg/kg, and further reductions in weight gains occurred when these fish were fed diets containing 17.5 and 33.5 mg Cu/kg feed.

Selenium

Selenium takes part in many metabolic processes, although the role of selenium in metabolism is still largely unexplained. It is known that selenium can act, to a certain extent, as a substitute for vitamin E. It is doubful whether it can function as an antioxidant, but selenium–protein compounds, such as glutathione peroxidase, can be very effective antioxidants. In the process of converting reduced glutathione to the oxidized form, damaging peroxidases are rendered into harmless alcohols. Thus glutathione peroxidase, in destroying hydrogen peroxidase, prevents the formation of hydroxyl free radicals which are the result of superoxide free radicals reacting with hydrogen peroxide (King *et al.*, 1975). Due to the fact that selenium and vitamin E appear to be, to some extent, mutually replaceable in metabolism, the typical selenium deficiency signs also show similarities to vitamin E deficiencies. This is usually expressed in muscular dystrophy.

It appears that selenium also has a very specific action upon growth and cell function, in which it cannot be replaced by any other active substance. Regardless of the level of vitamin E, selenium was found to be required to prevent impaired growth, pancreatic fibrosis and myopathies of the heart in poultry and rats (Scott *et al.*, 1967; McCoy & Wesig, 1969; Thompson & Scott, 1970). In mammals, when selenium is deficient, the secretory cells of the pancreas degenerate, resulting in a decline in the secretion of digestive enzymes (Zintzen, 1972), and in chicks it was found that selenium is involved in the metabolic conversion of methionine to cystine (Bunk & Combs, 1980). Selenium has also been isolated as a component of enzyme proteins such as β-galactosidase and cytochrome C. The synthesis and activities of other enzymes are also influenced by selenium (e.g. glutamic–oxalacetic transaminase). However, in toxic doses selenium is an enzyme inhibitor (e.g. blocking dehydrogenases). Dietary supplementation with selenium levels above those recommended as nutritional requirements enhances the primary immune response in mice. This was measured by the number of antibody-forming cells and levels of sheep-erythrocyte-agglutinating antibody. Stimulation of selenium was independent of vitamin E nutrition (Spallholz *et al.*, 1973).

Although Schwarz (1961) suggested a possible need for selenium by trout, by showing that simultaneous dietary supplementation of both vitamin E and selenium was necessary to significantly reduce the incidence of mortality shown in fish fed a basal diet for six weeks, very little work has been reported that conclusively documented such a requirement, or the interrelation between requirements for vitamin E and selenium. Post *et al.* (1976) demonstrated the dietary essentiality of selenium for Atlantic salmon fry and fingerlings. Deficiency of dietary selenium suppressed glutathione peroxidase activity, while supplementation of both vitamin E (500 IU DL-α-tocopherol acetate/kg dry diet) and selenium (0.1 mg/kg dry diet) prevented muscular dystrophy. Recent studies have shown that rainbow trout can readily take up selenium from water at ambient concentrations as low as 0.4 μg/l (Hilton *et al.*, 1980). The contribution of both dietary and water soluble selenium should, therefore, be considered when evaluating selenium requirement. This is probably the reason for the low requirement for dietary selenium by fish. Dietary selenium concentrations as low as 0.07 mg/kg dry diet prevented deficiency signs such as degeneration of the liver and muscle in rainbow trout fingerlings fed 400 IU vitamin E/kg dry diet (Hilton *et al.*, 1980).

In mammals sulphur inhibits the absorption of selenium, and probably displaces it completely in the passage through the cell membrane, but there is no information on the relationship between these two minerals in fish.

Part II

Food sources and their utilization

8

Natural food and its estimation

In spite of many gaps in the knowledge of nutrition of fish, as pointed out in the preceding chapters, sufficient information is now available to define, however crudely, the nutritional requirements of various fish species. In certain fish culture systems, such as flow or cage culture systems, where natural food has a minimal role, and most, if not all, food is provided by the farmer, this knowledge enables the formation of complete diets and the computation of feeding rates. Diets and rates will, no doubt, improve in time with the filling in of the gaps in our knowledge of fish nutrition. Formulation of complete diets removes one of the major factors limiting fish growth, and since the water current in these systems supplies the necessary oxygen and removes noxious metabolites, fish can be stocked at very high densities with a resulting high yield per unit area.

This is different in less intensive systems, where natural food plays some role in fish feeding. In extensive fish culture systems, where little or no supplementary feed is used, and fish depend largely on natural food, fish density must be adjusted to the amount of available natural food. If the density is too high, the amount of natural food per fish will not be enough to sustain the potential growth rate and the yield can be low. If on the other hand, the density is too low some of the food will not be utilized. Fish density becomes, therefore, a factor of prime importance in such systems. The relationship between the density and the individual growth rate is not linear but rather a curved regression. Since yield per unit area is a product of the average individual growth rate and the number of fish per unit area (density), the effect of density on yield is also not a simple one. As long as the rate of increase in fish density is higher than the rate of decrease in individual growth rate, yield increases. When, however, the decrease in growth rate exceeds the increase in fish density, yield decreases (Figure 31). The optimum density is therefore that in which the fish utilize the natural food to give the highest possible yield per unit area. This does not necessarily coincide with the density resulting in the highest individual growth rate. For generations fish farmers have determined this optimum density in their farms through a long process of trial and error. Their experience, however, is not easily adaptable either to different environments or to change of management procedures, such as the introduction of chemical fertilization, manuring, supplementary feeding, new species or polyculture. The determination of the optimum fish density in these cases requires a better knowledge of the amounts of available natural food in the pond, on the one hand, and the relationships between the amount of food, fish density, individual growth rate and yield, on the other.

This is also true for the semi-intensive culture systems, where feed is added to supplement the natural food. Fish densities in these systems are usually much higher than in the extensive ones, and the natural food ration per fish is, therefore, lower and requires supplementation. However, natural food still constitutes an important part of the overall food supply and fish density affects the amounts available per fish, and thus

Figure 31. Schematic presentation of the relationships between the stocking density, the short interval growth rate and the short interval yield per unit area, with (broken line) and without (solid line) supplementary feeding (from Hepher, 1978).

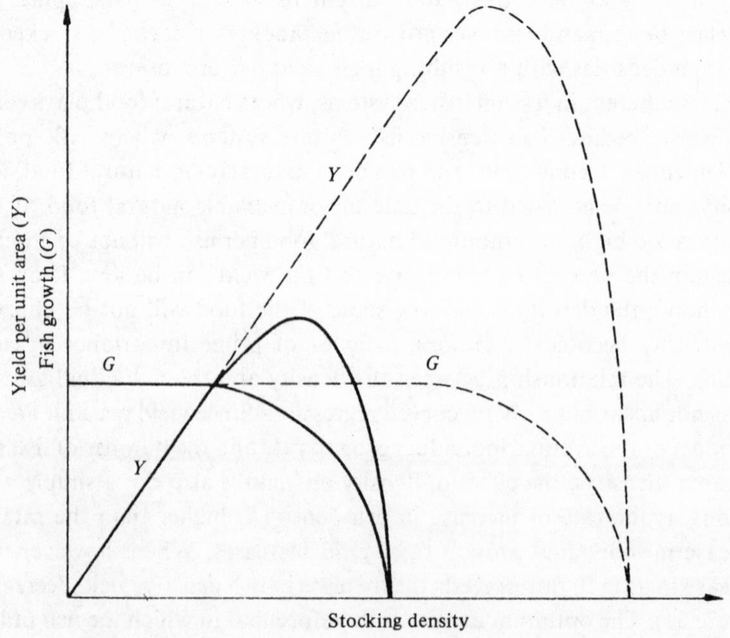

also the level of supplementary feeding. In order to determine the feeding level, the amount of natural food available in the pond must again be determined, as well as the relationships between this amount and fish density, growth rate and yield. These will be dealt with in greater detail in the present and following chapters.

The nature of natural food

The amount of natural food available to fish in ponds (or any other water body) can be estimated in three ways: (a) estimating the natural food stock in the pond; (b) estimating the amount of food consumed by the fish; and (c) estimating the food intake indirectly through a bioenergetic balance analysis, taking into account fish weight, fish growth rate and energy expenditure for maintenance.

The first method seems to be the most direct one. However, difficulties involved become immediately apparent. The first question to resolve is: what is the natural food of a particular fish species?

All organisms, plants and animals, in the pond (or any other biotope) form the 'biocenose' of the pond and can serve as food for various fishes. These organisms interact with each other, mainly through predator–prey relationships, but also through others, such as competition for food, space, etc. These relationships have been described in many papers in various ways such as 'food chains', trophic levels creating a 'food pyramid' (in which the biomass of the lower trophic levels, especially primary producers, is much larger than that of the upper trophic levels, the consumers), as having intricate interrelationships of a 'food web', or otherwise. Knowledge of these relationships, no matter how they are described, is important to understand better the processes involved in natural food production, the factors affecting it, and the possible ways of increasing production, such as by chemical fertilization or manuring. However, from the point of view of fish feeding on natural food, it is the resultant composition of the natural food and the amount available for the fish which is important rather than the processes by which they were produced.

From the point of view of fish food the biocenose of any aquatic biotope can be divided into several different groups by their nature, plant or animal, and by their size. In Figure 32 eight such groups are given. The detritus, which is the non-living organic particles (usually small in size) suspended in water or accumulated on the bottom, is also included since not only can these particles serve as food, but also they are populated with a large number of bacteria and protozoa which are of high nutritional value. The other groups included are (in order of their average sizes): from the plants – phytoplankton, periphyton and macrophytes, and from the animals – zooplankton, zoobenthos, small nekton (which includes water bugs, mosquito larvae, tadpoles, etc.), and fish. Each of these assemblages includes many types of organisms, not all of which appear in each particular biotope. Ecological conditions will determine which organism will appear and its abundance, as well as the ratios among the groups.

The biomass of each of the groups mentioned above can be determined and the 'biocenose profile' of the particular aquatic biotope thus obtained. However, from the nutritional point of view of the fish it is the production rate of these groups which counts, rather than the biomass at any given moment. When the production rates of each group of the biocenose is determined, the 'biocenose production profile' can be obtained. This is a long and tedious work and only very few studies gave the full biocenose profile, let alone biocenose production profile. Those available are for larger biotopes, usually lakes, where ecological conditions and thus also the biocenose profile are more stable, rather than for fishponds where conditions and biocenose profiles change rapidly. Figure 33 represents an almost complete biocenose production profile of Mikolajskie Lake in Poland, drawn from data taken from Kajak (1975). Another biocenose

Figure 32. The division of the biocenose to food organism groups according to their nature and size.

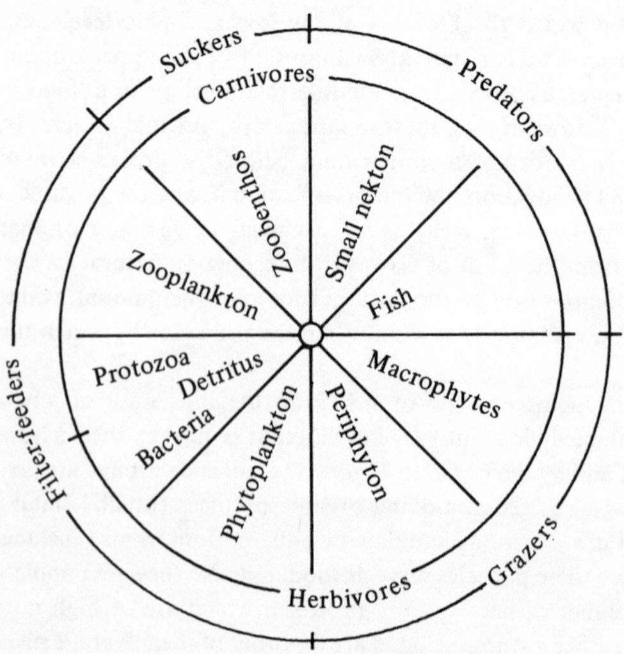

profile, of a fishpond in Taiwan (though in a different schematic presentation) is given by Tang (1970). The biocenose profile can be more detailed, giving sub-groups as orders, genera and even species, which of course will demand much more work. This may be especially tedious and difficult in fish ponds, where the changes in the biocenose profile is much more rapid.

Each of the fish species in the pond can feed on a certain portion of the biocenose. This portion, which includes *all* organisms the fish can eat, is the 'trophic basis'[1] of the fish. The trophic basis is determined by the feeding habits of the fish and the anatomy of its alimentary system, such as the food filtering system, the sizes of the mouth and stomach, etc. The trophic basis of a fish may change throughout the different developmental stages of its life. The most noticeable differences are between the larval stage, fry, and larger fish. Most fish feed in their young (late larval and fry stages) on zooplankton, even when the larger fish becomes herbivorous. Grass carp (*Ctenopharyngodon idella*), for instance, change from almost

Figure 33. Biocenose production profile of Mikolajskie Lake, Poland, drawn from data of Kajak (1975). Values in the sections are kcal/m²/yr

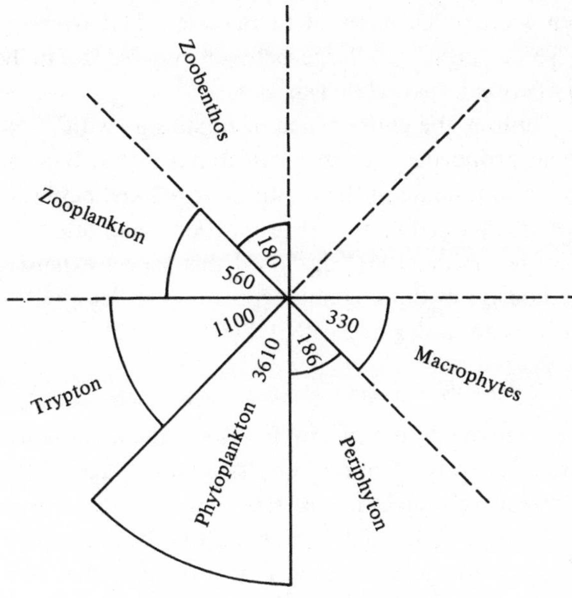

[1] The terms 'trophic basis' and 'trophic niche' which will be discussed below have been used extensively in the literature for variable, very often overlapping and even conflicting, meanings. An attempt is made here to define these two terms more clearly from the point of view of fish feeding.

exclusively carnivorous to herbivorous feeding at a length of 25–30 mm (De Silva & Weerakoon, 1981). The same size was also found to mark the change in diet in gilthead bream, *Sparus aurata*, sea bass, *Dicentrarchus labrax* and the mullets *Liza ramada, L. aurata* and *L. saliens* (Ferrari & Chieregato, 1981). Also the diet of Indian carp (*Labeo rohita*) changes from zooplankton in the fry stage to phytoplankton in the adult stage (Khan & Siddiqui, 1973). At a later stage, however, the trophic basis becomes more defined. Some of the more conventional divisions of fishes according to their feeding habits and their relationship to the biocenose are given in Figure 32. The trophic basis of the individual species does not always correspond to these divisions. It can be wide in 'euryphags' which can consume a wide variety of food organisms, such as the common carp which can feed on benthic organisms, zooplankton, detritus, and sometimes also on small nekton and fry (Spataru & Hepher, 1977). The trophic basis is narrow in the 'stenophags' with a limited assortment of food organisms, such as in the silver carp which feeds mainly on phytoplankton larger than 20–30 μm and zooplankton (Milstein *et al.*, 1985a, b). The trophic basis can be very narrow in the 'monophags' which feed on only one sort of food e.g. crustacean or mollusc feeders. Often there is an overlap between the trophic bases of two species. This overlap can be small between some species, or large between others. It can be a full overlap, usually between two related species.

A fish seldom utilizes the entire range of organisms within its trophic basis at the same proportions as they are found in the biocenose, i.e. without discrimination. Some of the organisms are liked better, or easier to get, and these are ingested in greater proportions than others less liked or more hard to get. Ivlev (1961) discussed this aspect extensively and developed a method for evaluating and expressing the selectivity of food items numerically as an 'index of electivity' (E).

$$E = (r_i - p_i) / (r_i + p_i) \tag{62}$$

Where r_i = the relative content of any food ingredient (organism) or a part of the ration (a group of organisms) as a percentage of the whole ration; p_i = the relative abundance of the same ingredient in the food complex of the biotope. Ivlev did not define the food complex of the biotope, but from the foregoing discussion it is obvious that this should include only the trophic basis of the fish, since there is no point in including organisms which *a priori* cannot be consumed by the particular fish species.

Ivlev (1961) presented the results of a series of experiments which show that selectivity is dependent not only on the inherent preference of the fish for its food, but also on the overall abundance of food, the quantitative relationships between the food items, the accessibility of these food items, and the degree of satiation of the fish. With the increase in the availability of food (increase in abundance or accessibility) or with the decrease in requirement (higher degree of satiation) the fish become more selective, while they discriminate less as they become hungry. Satiation, in turn, depends upon the density of the fish and their weight (the product of which gives their standing crop) in relation to the amounts of available natural food. The more dense the fish population at a given total amount of food, the less food is available per fish, and the larger the fish the more food is required to satiate it. It follows that selectivity depends mainly on the production of natural food and the variety of organisms produced, on the one hand, and fish standing crop on the other.

The variety of organisms actually consumed by the fish and the quantitative relations among them may be called the 'trophic niche'[1] of the fish species, to be differentiated from the total variety of organisms the fish *can* feed on, which is the 'trophic basis'. This conforms to the definition of the niche as proposed by Weatherly (1963): 'The niche is the nutritional role of the animal in its ecosystem, that is, its relations to all the food available to it'. The trophic niche depends, to a large extent, on the selectivity and, like selectivity, varies, therefore, with natural food productivity, the standing crop of fish and the plasticity of feeding behaviour. The flexibility of the trophic niche is well illustrated in the study of Werner & Hall (1976). They studied the feeding behaviour of three species of sunfishes: the bluegill (*Lepomis macrochirus*), pumpkinseed (*L. gibbosus*) and green sunfish (*L cyanellus*), stocked together in a common pond, or each species in a separate pond. When the species were stocked separately the resource utilization pattern was similar. However, when stocked together the diets differed. That of the green sunfish changed only a little, but that of the bluegill and pumpkinseed shifted. The first took mainly prey from the open water column, whereas the second fed largely on prey associated with the sediments. Significant shifts of the trophic niches were also related to the seasonal pattern and resource availability. De Silva *et al.* (1984) studied the feeding habits of *Oreochromis mossambicus* from nine man-made reservoirs in Sri Lanka and found them to vary considerably, changing from herbivory to total carnivory. In one reservoir, detritus constituted 88.4% of the gut contents, in another 90.6% consisted of plant material, and in a third cladocerans and copepods constituted 73.5% of the gut content.

[1] See footnote on p. 259

A graphic illustration of the change in the trophic niche with changes in standing crop of fish and/or natural food production is given in Figure 34. When the biocenose profile is considered, giving the production of each organism or groups of organisms within the trophic basis, the profile of the trophic niche will not be as regular as given in Figure 34, but rather follow, to a certain extent, that of the biocenose production profile for the relevant food organisms. Nevertheless, it can be seen from Figure 34 that two fish species may have overlapping trophic bases, but still different and separate trophic niches, thus not competing with each other as long as food is sufficient for each. However, when food production is low and standing crop of fish is high, the trophic niches may overlap, resulting in a partial competition for food. Tang (1970) evaluated the natural food available to each of a variety of fish species grown together in a polyculture system in a pond heavily manured with organic matter (12 600 kg dry matter/ha/yr). He characterized silver carp as planktophagic (feeding mainly on phytoplankton), grey mullet as detritophagic (although a considerable part of its food was planktonic), bighead carp as zooplanktophagic, common carp as benthophagic, and sea perch as nektophagic. As pointed out by Tang (1970), when the densities of the fish

Figure 34. A graphic illustration of the change in the trophic niche with increase in standing crop of fish and natural food production. Sections in the trophic bases represent single food organisms and the length of the trophic niches along these sections represent the amount of these organisms consumed. (*a*) Low standing crop, high food production; (*b*) high standing crop, high food production; (*c*) high standing crop, low food production; (*d*) overlap of trophic niches between two fish species resulting in a partial competition.

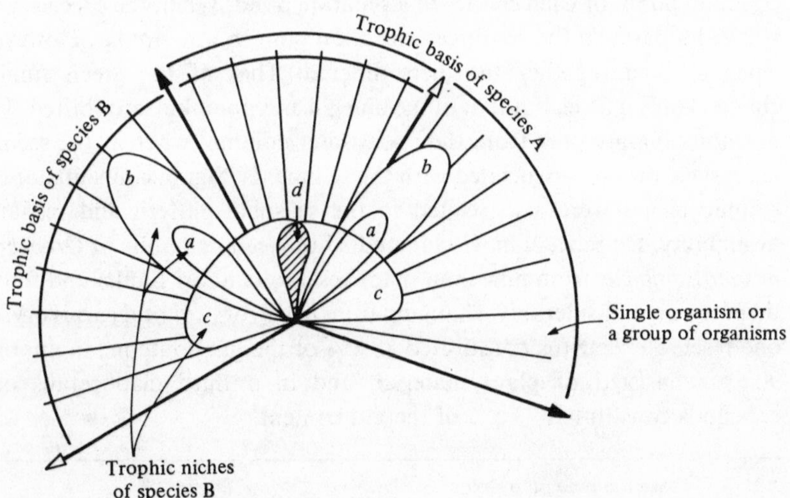

species were balanced, or as called by the Chinese, 'harmonious', competition among species was minimal. However, as Schroeder (1980) points out, when density increases the trophic niche of the fishes shifts, and they can compete for the same food source. Summerfelt *et al.* (1970) observed that in the absence of zoobenthos, the main component of the gut contents of common carp was organic detritus. Detritus was also found by Spataru *et al.* (1980) to be a prominent component of the gut contents of common carp stocked at high density in manured ponds. Odum (1970) showed that when algal concentrations were inadequate, grey mullet fed on detritus, and Terrell & Fox (1974) found that in the absence of macrophytes grass carp consumed large amounts of detrital matter. Thus, those species which have distinct and separate niches at relatively low densities, become competitors when their densities increase.

From the foregoing discussion, the first difficulty in determining the amount of available natural food directly from its concentration in the pond becomes quite apparent. What trophic niche should be considered? Should it include only the more favourable food organisms, as was done by many authors, considering a state when food is ample and sufficient to sustain maximum growth rates? Or, should a certain shortage in natural food be taken into account, which will broaden the trophic niche to include additional organisms, thus resulting in a better utilization of the trophic basis and in a higher yield per unit area? It should be also remembered that with the continuous increase in standing crop of fish, the trophic niche changes constantly. Weatherley (1963) states that fish in a polyculture system change their diet readily and that few species will be likely to occupy persistently the same niche. There is, however, a paucity of information on the relationships between the trophic niche and the productivity of natural food and the standing crop of fish.

Other factors may also affect the trophic niche. Supplementary feeding affects it in various ways. Spataru *et al.* (1980) showed that supplementary feeding has affected both feeding habits and the selection of food items by common carp. When fed supplementary feed, the fish selected a narrower range of natural food organisms from the higher, easier to reach, water layers. The nature of the pond and the protection it can give to some of the food organisms against fish predation can also affect the trophic niche. Spataru (personal communication) has found that after stones were used to protect pond embankments from erosion, the proportion of chironomid larvae in the intestines of common carp stocked into these ponds decreased and that of zooplankton increased. This was explained by the refuges the chironomid larvae found between the stones.

The trophic niche is usually determined by gut content analysis. This is tedious work in which the food organisms eaten by the fish are identified from their remaining fragments. However, one may question to what extent gut analysis really reflects the true trophic niche. Some food organisms, such as bacteria and protozoa, are so delicate that soon after being ingested they are completely digested without leaving any identifiable traces. Also, the time of sampling may affect the results of the gut analysis. Thus, Mathur (1970) found differences in food organisms consumed by channel catfish during the day. Between 0600 and 0900 h fishes formed the bulk of the diet, insects were the major item between 1600 and 2400 h, whereas zooplankton was the principal food at other periods. The above discussion clearly indicates the difficulty in determining the nature of the natural food consumed by fish in ponds.

Quantitative estimation of natural food

In spite of the difficulties, pointed out above, in determining the trophic niches of pond fishes, some attempts have been made to estimate the biomass[1] of a number of natural food organisms in the pond and to take this estimation as a measurement of the natural food available for the fish. Merla (1966a, b, 1968) found that if fish ponds are stocked with common carp at an equal density, and the same amount of supplementary feed is added to all ponds, there is a positive and significant correlation between the biomass of both bottom fauna and zooplankton, during the period from July to October, and the growth rate of the fish. However, here too a number of constraints should be considered.

Estimates of the biomass of food organisms in the pond do not necessarily reflect the amount of the available food. The biomass is the resultant of an equilibrium between the rate of production of the organisms, on the one hand, and the rate of its mortality, both natural and due to predation, on the other. Conditions may develop where both production of food organisms and their consumption by fish are high, while their biomass remains low; or, consumption by fish may be low, leading to a high biomass of food organisms in the pond. In both cases data on biomass may be irrelevant to the amount of available or consumed natural food organisms. Lyakhnovich (1967) measured the biomass of zooplankton and benthos in a pond stocked densely with fish

---

[1] Biomass is here meant in the sense given by Winberg's (1971) definition as the mass of organisms per unit of surface area or volume of water. It can be expressed in units of wet or dry weight, carbon or nitrogen, but from the point of view of nutrition the best expression would be in units of energy.

and compared it with that of another pond with no fish, but could not find any difference between the two. Another example can be cited from Schroeder (1974), who studied the effect of organic manures on the production of natural food organisms and on the growth rates of common carp in ponds. He compared the biomass of groups of organisms in manured and non-manured ponds, with and without fish. He found that common carp, when stocked at a density of 5000/ha, grew 25–100% faster in the manured ponds than in the non-manured ponds. This indicated that more food was available to fish in the manured ponds. However the effect of the manure on the biomass of food organisms was apparent mainly in ponds without fish (Table 44), while there was only a slight difference in the biomass in ponds stocked with fish, due to the consumption of these organisms by the fish.

Lellak (1957) tried to overcome this difficulty by estimating the biomass of bottom fauna in small areas (1.5 m × 1.5 m) of the pond which were isolated by a wire screen. He determined the amount of consumed natural food as the difference in biomass between those areas protected from fish predation and non-isolated areas open to fish predation. However, Hruska (1961) in re-examining this method criticized it on the grounds that there is a difference in production rate between the protected and non-protected areas, and that there is a migration of larvae from (and into) the protected areas. It is obvious that the method cannot apply to planktonic organisms since these can move freely between isolated and non-isolated areas. From the above discussion it is quite clear that information on natural food *production* rate or, even better, on food intake, is more relevant than that on biomass.

A number of methods have been developed for estimating production

Table 44. *Biomass of zooplankton and counts of chironomids and bacteria in manured and non-manured ponds, with and without fish*

	Without fish		With fish	
	Manured	Non-manured	Manured	Non-manured
Zooplankton (g dry matter/m^3)	3.3–424	<0.055	0.34–1.3	<0.055
Chironomids (100's/m^2)	79.0–215	1.0–7.0	1.0–4.0	1.0–2.0
Bacteria (1000's/ml)	17.0–27	0.7–4.3	1.6–6.7	

Source: From Schroeder (1974).

of aquatic organisms. Those concerned with the production of phytoplankton ('primary production') are used extensively in the oceans, lakes and sometimes in ponds (Strickland, 1960; Hepher, 1962b; Talling, 1982). One common method is estimating the uptake of labelled carbon by phytoplankton during photosynthesis. Another method is measuring the volume of oxygen produced during photosynthesis in transparent bottles and comparing it with the oxygen taken up by respiration in dark bottles. Still another method estimates the primary production by the change in pH in water, caused by the uptake of carbon dioxide during photosynthesis. In all these methods the measured parameters are related to production, but independent of the estimation of the biomass of phytoplankton. Though the range of error involved in these methods is still high, they can give a fair approximation of the phytoplankton production in the water. This, however, is not always the case with the estimation of production by the heterotrophic organisms ('secondary production').

According to Winberg (1971) the production of the population of a species consists of the sum of growth increments of the individuals forming that population, including the production of sexual products and other organic substances which became separated from the body during the period being considered. This is why any estimation of production requires quantitative data about growth, duration of development of different stages, and fecundity, as well as information on how these are influenced by environmental conditions. Collecting all these data is very tedious, and therefore production is usually estimated through estimating changes in biomass. Thus, according to Boysen-Jensen's method for estimating production of population without a continuous recruitment (quoted from Winberg, 1971), the production (P) is estimated by directly determining the biomass at the beginning of the year (B_1) and at its end (B_2), as well as the new recruitment of the young:

$$P = B_e + B_2 - B_1 \tag{63}$$

where B_e is the biomass of organisms eliminated during the period (consumed or died) which is taken to be the difference between the initial number (N_1) and the final number (N_2) of organisms multiplied by the arithmetical means of the initial and final individual weights:

$$B_e = (N_1 - N_2) \times \tfrac{1}{2} (B_1/N_1 + B_2/N_2) \tag{64}$$

Though this method seems to be quite crude, the basic principles of the computations used by Boysen-Jensen have received a wide recognition and have been used and further developed by a number of investigators. The main improvement of these investigators was to sample the biomass more frequently and take into account the different developmental stages of the organisms.

The methods for estimating production of populations of organisms with continuous reproduction are more complicated than the previous ones, but these are also based on measuring the biomass of the organisms in the water body studied. One of the most important approaches to measuring productivity of organisms is that of Elster (1954). He considered the change in the number of a species population as the resultant of two opposing processes – reproduction and elimination – from which he calculates the 'coefficient of renewal'. In Elster's method, or in those using the same concept, the estimation of the rate of elimination of the population, as well as other indices of duration of life stages and fecundity, are of great significance. However, these are usually determined from the number of the different stages of the organisms found in water during sampling, i.e. from the biomass. When the relationship between production (P) and biomass (B) has been established, the production of a population is often estimated using the coefficient P/B by multiplying it by the mean biomass during a given period of time.

The determination of the biomass of the different organisms in a water body is very laborious work. Moreover, Winberg (1971) writes: 'Estimates of production only become realistic when based upon good quantitative data of abundance and biomass of species populations inhabiting a body of water. However, in practice, such data are difficult to obtain because our methods for estimating the density of aquatic populations are very imperfect.' It seems, however, that the difficulties in estimating production by relating it to biomass measurements go far beyond the mere imperfection of quantitative estimation, especially in fishponds. Most methods for calculating production of an organism which is reproduced continuously are valid only for populations in a steady state, and whose mortality is similar for the different age classes or developmental stages. This assumption is not always true even for large bodies of water such as lakes, although the density of fish in them is small and therefore predation is small in comparison to the biomass of the organisms. Archibald (1975) found that predation by goldfish in a small saline lake was size selective, causing distribution of the principal prey organisms to be skewed toward the smaller-size classes. At a mild predation pressure of the fish, fecundity and abundance of the prey organisms increased, but with increasing predation pressure the fecundity and abundance declined. In fishponds, where the density of fish is much higher than in natural waters, predation of fish becomes so important in relation to biomass that it is doubtful whether the assumption on a non-selective predation is true at all.

The changing conditions in the pond usually cause a selective predation. Thus, Spataru *et al.* (1980) have shown that common carp offered supplementary feed selected more chironomids of the older larval stages and pupae, than did carp which had to depend solely on natural food. Polyculture can also cause selective predation. Grygierek (1973) found that when young phytophagous Chinese carp (*Ctenopharyngodon idella, Hypopthalmichthys molitrix* and *Aristichthys nobilis*) were stocked in a pond together with common carp, the larger species of crustacean were found in the pond at higher frequency than in ponds stocked with common carp alone. The main problem, however, is the fact that with the growth of the fish and with the increase in their standing crop in the pond the predation pressure increases; and with it, not only does elimination rate increase significantly, but the selection, both for different species of prey organisms and for different stages of these organisms, also changes considerably. This makes the estimation of secondary production in fishponds, based on biomass measurements, very unreliable.

Food intake

A direct way to determine the amount of natural food available to a fish in a pond is to estimate the quantity of food actually consumed at different fish densities, pond productivity and environmental conditions. Several methods have been used for such estimations. Some of these are based on laboratory experiments in which the food intake by a single fish or a group of fish is measured. Others are field methods usually based on the estimation of food contents in the stomach and intestine.

It is obvious that laboratory results cannot be applied directly to the pond. Not only does the food given in the laboratory not necessarily correspond with that found in the pond, but also the food in the laboratory is usually given in excess, while in the pond food may be short and subject to competition among fish. Several problems, however, can be solved best in controlled laboratory experiments, such as: What are the factors affecting food consumption? What is the maximum amount of food which can be consumed in a single feeding (satiation level), and per day? What is the turnover rate of the food in the stomach and the intestine, and its relationship to feeding rate? What is the relationship between food consumption and food requirement? Can fish overconsume food irrespective of their requirement? Can conditions develop in which food, though consumed at maximum rate, cannot meet nutritional requirement?

Satiation level is the quantity of food required to satiate a fish in a single feeding. Since individual fish may vary in their satiation levels, Ishiwata

(1970) prefers to satiate a group of fish to obtain the average satiation level. The rate of food intake during a single feeding decreases as the amount eaten approaches satiation level (Hotta & Nakashima, 1968; Ishiwata, 1970).

When food is abundant, long term feeding rate depends on the satiation level and food turnover in the digestive tract of the fish. Since the latter is related to gastric and intestinal evacuation rates, a correlation seems to exist between gastric and intestinal evacuation rate and the long term food intake (Brett, 1971; Elliot, 1975c). However, much still remains to be learned about this relationship. Factors affecting evacuation rate have already been discussed above (Chapter 3). These factors, as well as additional ones, also affect the satiation level, and through it food intake.

Fish species Moore (1941) studied the food intake in three species – a sunfish (*Lepomis cyanellus*), bluegill (*Helioperca incisor*), and a perch (*Perca flavescens*). Daily food intake was variable, but when considered over a long period, the consumption rate proved to be constant for a given species and size. Also, Hunt (1960) who worked with three other fish species – *Chaenobryttus gulosus, Micropterus salmoides* and *Lepidosteus platyrhincus* – found day-to-day fluctuations in the amount of food eaten, but each species could be characterized by the average quantity consumed. Beukema (1968) drew up a table of food intake for seven species (weighing 1.3–4.6 g), which ranged from 5 to 13% of body weight.

Physiological state and feeding habits The physiological state of the fish may affect food intake. Ishiwata (1970) noted that the degree of hunger affects the satiation level. As the time of food deprivation increases, the satiation level increases at first and then levels off to a constant value. Also, Moore (1941, working with *Apomotis cyanellus*), Pegel (1950), Krayukhin (1963), and Hotta & Nakashima (1968, working with jack mackerel) found that after a period of starvation fish tend to consume more food than before starvation period. However, there seems to be an adjustment of the food taken in subsequent feeding. Beukema (1968) found a consistent positive influence of the length of the previous food deprivation period on the quantity eaten during the first feeding hour, but there was no significant influence of the length of deprivation period (16–88 h) on the total amount eaten during the whole subsequent eight hour feeding period. Thus, the length of the deprivation period affected the pattern of food intake rather than the total daily intake.

The rate of food intake can decrease before and during spawning and nursing of fry. Sehgal (1966) found that the feeding rate of *Labeo calbasu* declined during the breeding season. Fasting occurs in striped bass (*Morone saxatilis*) for a brief period before and during spawning

(Trent & Hassler, 1966). Mouth breeding tilapia feed very little, if at all, during mouth incubation and brooding of fry.

Stress may reduce food intake. This is more often caused by diseases, which may result in complete cessation of feeding. However, environmental factors, such as water quality, oxygen concentration and salinity, may also cause stress and affect food intake. The effect of oxygen concentration, because of its special importance, will be treated separately below. The first apparent reaction of fish to high ammonia concentrations or that of toxic substances in water is reduced feeding rate or cessation of feeding. De Silva & Perera (1976) found that food intake by young grey mullet (*Mugil cephalus*) is salinity dependent when food is present in excess. Stress can have a behavioural cause. For instance, according to Ishiwata (1968, 1970) fish do not take food well immediately after their introduction into a new tank. This may be due to their need to be acclimated to the new environment or to the new food. Ishiwata (1968) found that such acclimation is faster with some diets, probably tastier to the fish, than others.

Feeding habits also affect the satiation level and the rate of food intake. Ishiwata (1970) found that satiation level of a school of fish declines when the number of fish in the school falls below a certain point. This seems to indicate a socio-behavioural effect which makes the fish increase their food intake. The association in a school also helps in location of the food. Pitcher *et al.* (1982) showed that schools of goldfish and minnows (*Phoxinus phoxinus*) found food faster than did individual fish. The larger the school (2–20), the more rapidly the fish located the food.

Fish weight Food intake depends on the weight of the fish. As the body weight increases satiation level increases, although the relative satiation ratio (satiation level/body weight) decreases (Ishiwata, 1970; Garber, 1983).

Temperature With the increase in temperature satiation level increases (Ishiwata, 1968). It is difficult to distinguish here between the direct effect of temperature on satiation from the indirect one. Higher temperature accelerates metabolism and food requirement, and thus causes increased hunger. Winberg (1956) has pointed out, when discussing the standardization of conditions for feeding experiments, that equal periods of starvation at different temperatures are not at all equivalent physiologically. Since temperature also affects gastric activity there is a strong overall effect of temperature on food intake. With the increase in temperature, up to an optimum level, food intake increases. Studies on some fishes have shown that Q_{10} for food intake, below the optimum temperature, was 2–3 (Hathaway (1927); Rozin & Mayer (1961), for goldfish; De Silva &

Balbontin (1974), for young herring). However, in other species a higher Q_{10} was observed. Bokova (1938) found that at a temperature range of 1–5 °C, the food intake of Caspian roach (*Rutilus rutilus caspius*) was 0.5% of its body weight per day, and this increased to 7.8% at the temperature range of 10–15 °C, and to 12.8% at 15–20 °C. Baldwin (1956) fed minnows to brook trout at various temperatures and found that the weekly consumption approximately doubled for each 4 °C rise in temperature up to 13 °C. The higher increases in feeding rate with temperature rise in the last cases may be explained by the stenothermic nature of the fish.

When temperature exceeds the optimum level, feeding decreases. In Baldwin's (1956) experiment quoted above, at 13 °C the brook trout consumed minnows at 7% of its body weight per day, but only 1% at 21 °C. Also, Brett *et al.* (1969), working with young sockeye salmon, found that at 23 °C, which is close to the upper lethal temperature limit of the species, the fish lost their appetite. A similar phenomenon, though at a different temperature range, has been observed by Caulton (1978) with *Tilapia rendalli.* Between the temperature of 18 and 30 °C, daily food ingestion increased curvilinearly with increasing temperature, but between 30 and 34 °C temperature change had little influence on the amount of food consumed. Feeding activity decreased rapidly when the temperature exceeded 34 °C and fish acclimated to 38 °C, which is near the upper lethal limit, rarely fed at all.

Oxygen concentration Lowering ambient oxygen concentration below 3–4 mg/l has been found to suppress appetite and food intake (Herrmann *et al.*, 1962; Chiba, 1965; Stewart *et al.*, 1967; Adelman & Smith, 1970; Andrews *et al.*, 1973). This naturally also affects the growth of the fish.

Composition of the diet Fish seem to feed, up to a certain level, to supply their nutrient requirements rather than to a volume or weight of the food. Rozin & Mayer (1961) found that goldfish increased their food intake significantly in response to dilution of their normal diet with kaolin. Similar results have been obtained for rainbow trout by Bromley & Adkins (1984). When the diet was diluted with indigestible α-cellulose, the trout compensated for up to 30% of the cellulose by increasing the total weight of food eaten, but failed to compensate for levels of 40–50% cellulose. The regulation of food intake by nutrient content in the diet has been found in higher vertebrates and humans. Hunt (1980) discusses two sets of duodenal receptor cells in humans. One is stimulated by the osmotic properties of the digestion products of carbohydrate and protein, and the other by the digestion products of fat. These osmoreceptors usually respond to properties associated with the energy content of the digestion products of food and slow gastric emptying into the duodenum, and thus, indirectly, food intake. No such receptors have yet been found in fish.

Taste also plays an important role in determining food intake. Ishiwata (1968) found that the satiation level of file fish (*Stephanolepis cirrhifer*) was much higher when fed oyster meat than when fed mackerel meat (Figure 35). When fed first with mackerel meat and then with oyster meat the satiation level increased to that of the second diet.

What is the relationship between the satiation level and the daily food intake? Or, in other words, what is the turnover of food through the digestive tract? This is dependent on two factors: (a) the number of meals per day; (b) the amount fed in each meal, or the satiation level when food is abundant. Elliott (1975a) fed brown trout (*Salmo trutta*) to satiation in each of four meals at seven different temperatures. The period from the start of one meal to the start of the following meal was not significantly affected by the weight of the fish, but was negatively correlated with water temperature. The number of meals per day ranged from one at about 4 °C to three meals at about 18 °C. Ishiwata (1979) fed fish (filefish, puffer, yellowtail and rainbow trout) a number of times per day. As the frequency

Figure 35. Satiation level for file fish. Solid line and black points – fish fed on oyster meat; broken line – fish fed first on mackerel meat (triangles) and then on oyster meat (black points) (from Ishiwata, 1968).

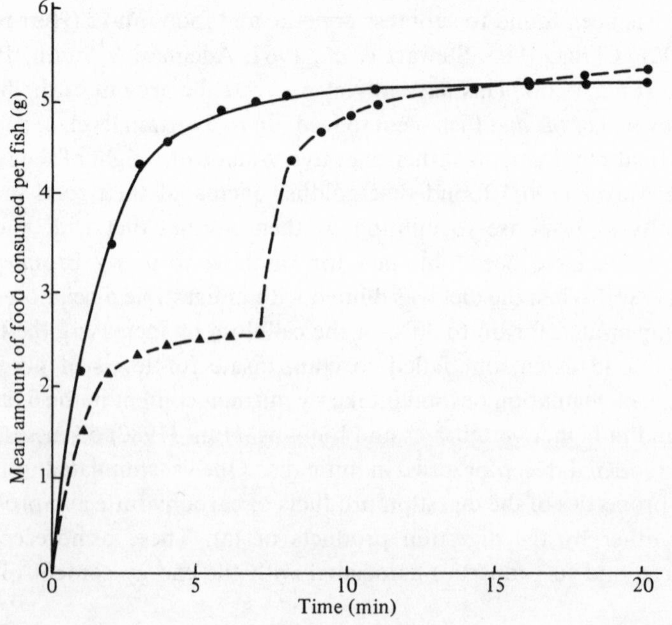

of feeding increased from one to two feedings per day, the daily ration increased, but it was not double that of the single feeding. Only a slight further increase has been noted when fish were fed three times a day. This was due to the decrease in satiation level which was highest with one meal per day. Similar results have been obtained by Grayton & Beamish (1977) working with rainbow trout. Feeding frequency in their study varied from one meal every second day to six meals per day. Maximum daily food intake occurred with just two feedings to satiation per day. According to Brett (1971), who worked with sockeye salmon (*Oncorhynchus nerka*), the satiation level increased in a sigmoid pattern with starvation time, approaching a plateau after 25–30 h. Ishiwata (1969) expressed the relationship between the frequency of feeding and the mean daily food intake by the following formula:

$$R_d = S' T (1 - \alpha)^{T-1} \quad (65)$$

where R_d = mean daily food intake per fish (g); S' = mean daily food intake per fish fed once a day (satiation level); T = frequency of feeding per day; α = rate of decrease in satiation level. A similar equation has been developed by Kono & Nose (1971) who studied the feeding behaviour of a number of fish species:

$$R_d = C (1 - e^{-mT}) \quad (66)$$

where C is the maximum daily rate of feeding and m is a constant which represents the rate of approach of the daily rate of feeding to its maximum, which is determined by the ratio of the satiation level of one meal per day to the maximum daily ration. It can be seen from the above equations that with increasing frequency of feeding daily food intake increases. In Ishiwata's (1969) study the average daily food intakes by filefish at two and three meals per day were 1.47 and 1.57 times higher than the mean satiation level at one meal a day. In Kono & Nose (1971) study the number of meals to reach maximum ration differed with fish species. It was two meals for goby (*Chasmichthys gulosus*), three for rockfish (*Sebastes inermis*), four for jack mackerel (*Trachurus japonicus*), and 12 for goldfish (*Carassius auratus*). Similar relationships were found by Brett (1971) for sockeye salmon (*O. nerka*). The mean daily intake when fed every 22 h was 4.83% of dry body weight, but when fed every 11 h it was 6.52% of dry body weight, i.e. 1.35 times higher.

The foregoing discussion illustrates the difficulty in determining both satiation level and turnover rate of food in fish in nature, where feeding is regulated by the fish itself, rather than by man as in the laboratory. Feeding rate in these conditions is determined mainly by gut analyses and daily intake is calculated by assuming a certain rate of turnover. Bajkov

(1935) proposed the estimation of the total food intake by a fish by using the formula:

$$R_d = A\frac{24}{n} \tag{67}$$

where R_d is the daily food ration during the time of the estimation; A is the average amount of food present in the stomach at the time of analysis; and n is the gastric evacuation time for the species at ambient temperature. This formula, sometimes with slight modifications, is still used in many studies (e.g. Bialokoz & Krzywosz, 1981; Ross & Jauncey, 1981). However, Bajkov's equation is based on a number of assumptions which are open to severe criticism. The following are some of these arguments.

Since at the time of analysis fish are caught at various stages of stomach (or intestinal) evacuation, the value A – the average amount of food in the stomach does not represent the average satiation level. If we consider a normal distribution of stomach fullness between satiation and complete evacuation, and we assume, as Bajkov's formula implies, that fish do not take up a new meal until they have completely evacuated the previous meal, the level of stomach fullness found in natural populations should be about 50% satiation. Indeed, Bajkov (1935) himself, estimated the daily food consumption of whitefish (*Coregonus clupeaformis*) in Manitoban lakes by setting gill nets at sunset and lifting them at sunrise. Since the fish is a night feeder and during the summer has a rapid rate of digestion, he assumed that the average stomach content showed a 'half-feed' condition.

The second assumption on which Bajkov's formula rests is that fish feed throughout the day at intervals corresponding to their gastric evacuation rates, and that the satiation level is equal for each feeding. Although Bokova (1938) showed that for Caspian roach (*Rutilus rutilus caspius*) the time lapse between two meals equals the evacuation time of the fore-gut, and also Pegel (1950) observed that food intake of some fish species coincided with a certain degree of emptiness of the fore-gut, many other observations showed a different pattern. Western (1971) studied the feeding habits of the marine fishes *Cottus gobio* and *Enophrys bubalis*, and found that if food is available these fishes do not wait until the stomach is emptied before taking additional food. Similar feeding activities in other carnivorous fishes have been reported by Moore (1941), Windell (1966) and Hotta & Nakashima (1968).

As for equal feeding rates for each meal, this assumption does not accord with the large variation in daily food intake among fish of the same species in the same habitat, nor does it fit the concept of a daily feeding rhythm in fish. Brett (1971) found in sockeye salmon (*O. nerka*), starved for 11 h before meals, that the sizes of two successive meals, consumed voluntarily, were inversely correlated. A relatively large meal was usually followed by a relatively small meal and vice versa. Similar findings were reported by Darnell & Meierotto (1962) for young black bullhead (*Ictalurus melas*), where two distinct feeding periods were recognized. A major peak occurred just before dawn and a minor peak shortly after dark. Little food of any kind was taken during the middle of the day. Smagula & Adelman (1982) have shown a substantial day-to-day variation in food consumption by groups of laboratory held young largemouth bass (*Micropterus salmoides*), though they were provided with an unrestricted ration of live fathead minnows (*Pimephales promelas*). Such long term cycles of feeding rate were also found by Yoshida & Sakurai (1984) for walleye pollock (*Theragra chalcogramma*) with a daily food intake peak every five days. It is thus obvious that the estimation of the daily food intake is not as simple as assumed by Bajkov, and it is doubtful if his formula can result in a reliable estimation.

Several attempts have been made to get around the difficulties of estimating food consumption in natural waters. Ishiwata (1979) estimated food consumption by jack mackerel (*Trachurus japonicus*) in natural waters by comparing its growth rate with that of fish held in aquaria and receiving graded amounts of live food. Equal growth rates were taken as equal food consumption. The weak point of this method is that although fish in the aquaria were held in a similar temperature as in nature, in other respects these two environments are completely different, and feed conversion does not necessarily have to be equal for both. Another method has been used by Kolehmainen (1974). He estimated the daily food intake of blue sunfish (*Lepomis macrochirus*) from a small impoundment by measuring the intake of ^{137}Cs which was added at low levels to the impoundment through radioactive effluents flowing into the impoundment. The calculations of ^{137}Cs intake were based on fish growth, the seasonal fluctuation of ^{137}Cs concentration in the fish, and the temperature dependent elimination rate. Feeding rates were obtained by dividing the daily intake of ^{137}Cs by the concentration of ^{137}Cs in the diet. It is obvious that such conditions are very specific to the particular impoundment receiving the radioactive effluent and cannot be applied generally.

The foregoing discussion focused on the difficulties of determining food intake from gut analyses and estimating gut turnover. No wonder that Conover (1978), who reviewed these problems, concluded: 'It is relatively easy to obtain samples of gut contents from fishes and their quantitative and qualitative description constitute a vast amount of literature, but conversion of this information to an estimated feeding rate is rarely possible. To turn a static measurement of weight of gut content to a reliable measure of ingestion requires knowledge of rate of filling or emptying. The problem is variously complicated by feeding rhythms, temperature, season, previous nutritional experience, etc.' These difficulties still perplex the estimation of food uptake.

Indirect estimation of food consumption

In view of the difficulties pointed out in the foregoing discussion on the direct estimation of food consumption, no wonder that many have sought indirect methods for the estimation of food consumption by fish. Two such methods have been developed, one using nitrogen (= protein) and the other energy as the basis for calculation. The principle in both methods is the same. The requirements for maintenance and growth are combined and the efficiency of nitrogen or energy utilization for these purposes, usually found through laboratory experiments, is taken into account to find the digestible nitrogen or energy consumed. Once the nitrogen or energy content of the food and its digestibility are known, the total food consumption can be estimated.

One of the first to use nitrogen as a basis for estimating food ration was Ivlev (1939b), who estimated the daily food consumption of carp in both nitrogen and energy balances. A number of other studies in which the nitrogen balance method has been used were quoted by Winberg (1956). However, a more complete version of this method has been worked out for bluegill sunfish (*Lepomis macrochirus*) by Gerking (1962, 1972). The weakest component in the nitrogen balance is the estimation of the nitrogen used for maintenance and for catabolism (endogenous nitrogen, see p. 175).

The most common indirect way to estimate food consumption is by using the energy balance, as proposed by Winberg (1964) (Using Winberg's designations):

$$A = P + T \tag{68}$$

where A is the assimilated part of the ration (ME according to the designation adopted here); P = growth increment of the individual fish; T = metabolic loss, expressed in energy units. Winberg (1971) writes: 'In principle T represents the loss in body weight of the organism when it has no food.' It is obvious that this definition of T is not sufficient since it does not account for losses of energy due to heat increment or SDA. Mann (1965) corrected this fault by doubling the energy expenditure for metabolism, as in fact was suggested by Winberg (1965) himself. In this way Mann (1965) calculated food consumption by some fishes in the River

Thames. Some later studies (e.g. Caulton, 1978; see also a review by Conover, 1978), however, corrected this somewhat by modifying the previous equation. According to Majkowski & Waiwood (1981) and Majkowski & Hearn (1984) it took the form:

$$\Delta R = (\Delta W + \Delta E + \Delta T)/\beta \quad (69)$$

where ΔW is the increment in body weight during Δt; ΔE denotes losses or reproductive products during Δt; ΔT is the sum of all metabolic losses during Δt; and β is the relative fraction of food intake assimilated but not excreted. Since in most cases the culture of pond fishes is done prior to their sexual maturity, ΔE can be omitted and ΔW is equivalent to the calorific value of the observed somatic growth. One of the difficulties in solving the above equation is the determination of the various components of ΔT. While the standard metabolism (or at least the routine metabolism in laboratory conditions) can be estimated to a fair degree of accuracy, the estimation of energy expenditure for *SDA* and activity in natural conditions are much more difficult. Attempts to account for these expenditures were made by some authors (e.g. Kerr, 1982) but not always successfully (see p. 138). In nature there is also the difficulty of estimating the instantaneous growth rate ΔW. In spite of these difficulties, acceptable estimation of food rations have been done in a number of cases such as for cod (*Gadus morhua*) and haddock (*Melanogrammus aeglefinus*) by Kerr (1982). The situation is much easier in ponds. Since most of the energy expenditure other than routine metabolism is associated with food consumption, especially the increased food consumption related to growth, it seems more profitable to divide the losses which are related to maintenance from those due to growth. While the former fits well Winberg's rule of doubling the routine metabolism (see p. 145), energy loss associated with growth can be estimated by the use of the efficiency value E_{pg} (see p. 169). Both these corrections correspond to that by the value β in equation (69). It seems, therefore, that equation (48) can be used successfully for estimating the metabolizable energy consumed by fish in ponds, once the constants for calculating the calorific value of the gained body weight and the E_{pg} for the species are known. The main parameters one then needs for calculating food consumption in units of energy are the average weight of the fish between two sample weighings ($\bar{W}$) and the rate of growth during this time ($\frac{dw}{dw}$).

No less important than food consumption by the individual fish is the overall amount of natural food available for the entire population of fish in the pond. This information is essential for calculating the optimum density of fish to be stocked or the amount of supplementary feed to be added. Attempts have been made at calculating protein or energy budgets for populations by extension of the technique for calculating these budgets for a single fish from the information on its protein turnover or energetic metabolism and growth rate. For such calculations information on number, size and growth rate of fish in the relevant water body is required. Gerking (1954) estimated the food turnover of the bluegill (*Lepomis macrochirus*) population of Lake Gordig, Indiana, USA by determining the protein turnover of a single fish and combining this information with that from measurements on growth, mortality and population of the bluegills in the lake. Mann (1965) used the energy budget method. After collecting the relevant information on the different fish populations in the River Thames, he calculated their total food requirement. A similar way to calculate food consumption of a population of goldfish (*Carassius auratus*) in a small pond was used by Miura (1973). Here too information on the metabolism was obtained first in the laboratory for a single fish and was applied to a known population in the small concrete pond. However, for calculating the total available natural food resources in a fishpond more information is needed on the relationships between the amount of food consumed by the individual fish, the overall food resources, growth rate, and production per unit area. These will be discussed in the next chapter.

9

Food, fish growth and fish yield relationships

The biomass of fish food organisms in a pond, the proportion of each species, and their chemical composition vary constantly due to changes in environmental conditions and the effect of the selective grazing and predation by fish. Nevertheless, at any moment, the biomass of natural food in a pond is finite and can, therefore, support a finite standing crop of fish. When the standing crop of fish is low, the amount of available natural food exceeds the population requirement. Each fish then gets an adequate amount of food for maintenance and for maximum growth under the existing environmental conditions (temperature, water quality, oxygen, etc.). The average instantaneous growth rate of the individual fish ($\frac{dw}{dt}$) will then depend on its average weight. The larger the fish the more it can grow.

Yield per unit area is a product of the individual growth rate and the number of fish per unit area (density). Since the individual growth rate is physiologically limited, the only way to increase yield per unit area is through increasing the density. As long as the amount of natural food exceeds requirements for maintenance and maximum growth, an increase in fish density (and thus also in standing crop) should not affect the individual growth rate of the fish. Weatherley (1963), discussing competition in nature and the effect of population density on growth pointed out that there is some critical level of population density below which the effect of competition for food is not present, and fish are assimilating food at their maximum rate for that particular set of environmental conditions. The instantaneous yield increases in these conditions linearly with the increase in density. However, with the increase in standing crop the food requirement of the population also increases, until at a certain density/standing crop food resources will be overtaxed and will not suffice for both maintenance and growth. Since maintenance is vital, less food will be diverted for growth, and individual growth rate will decrease. Hepher (1978) defined this standing crop as 'critical standing crop' (*CSC*). When standing crop reaches a level at which natural food is sufficient only for maintenance and no food is left for growth, growth ceases entirely. This is the 'carrying capacity' of the pond for the particular species.

The rate of decrease in individual growth rate as the standing crop increases over the *CSC* is at first smaller than the increase in fish density. Yield therefore continues to increase, although not in proportion to the increase in density. At a certain standing crop, food demand for maintenance becomes so high that the decrease in individual fish growth rate becomes faster than the increase in density, and yield falls to reach zero at carrying capacity.

If at standing crops above *CSC* the fish are fed supplementary feed of adequate nutritional quality, maximum growth rate will be maintained up to a point where some limiting factor in the feed will inhibit growth. Yield per unit area will thus continue to increase linearly with increasing density until the new *CSC* is reached (Figure 31).

The larger the individual weight of the stocked fish, the higher the absolute food requirement for maintenance and growth. Therefore, while the instantaneous growth rate of large fish below the *CSC* is higher than that of small fish, a given amount of food will suffice for a smaller number of fish and with increasing body weight *CSC* and carrying capacity will be reached at lower densities. The relationships between *CSC*, carrying capacity and body weight are, however, quite obscure. One may expect that since the *relative* requirement of food for maintenance and growth decreases with increasing body weight, the available food will suffice for a larger standing crop (kg/ha) of large fish than of small fish. However, farmers' experience does not support this. It has been noticed in practical fish farming that growth of fish ceases at a carrying capacity characteristic to the pond and method of its management irrespective of the average weight of the fish. Thus, for instance, if the carrying capacity of a pond is 150 kg/ha, at a density of 1500/ha fish will cease growing when they reach an average weight of 100 g, but 15 000 fish/ha will cease growing when they reach 10 g. In many cases it has even been observed that the smaller fish reach a higher carrying capacity than do large fish. This discrepancy may perhaps be explained by the higher efficiency of grazing and predation as the density increases (see p. 284). Using the above example, 15 000 fish each of 10 g have a greater capacity to graze or seek and catch prey than 1500 fish/ha of 100 g. From the above discussion it is clear that the main factor determining the *CSC* and the carrying capacity is the productivity of the pond, the treatment it gets (fertilization and manuring) and supplementary feeding.

In most cases a fish pond is stocked at a pre-determined density with small fish which are expected to grow during the culture period. As they grow their standing crop increases and eventually reaches the *CSC*, at which point growth rate drops. If not fed, the fish may reach carrying capacity and stop growing altogether, but if they are fed they continue to grow at their maximum rate. This is well illustrated in Figure 23 in which results of experiments with common carp, carried out at the Fish and Aquaculture Research Station, Dor, Israel, are presented. When the fish were small they grew at maximum rate, whichever treatment the pond received. However, when the fish in the untreated ponds (stocked at 1200/ha) reached about 54 g, they attained *CSC* and their growth rate dropped, while those in the other ponds continued to grow at the maximum rate. When the fish in the untreated ponds reached about 108 g their growth ceased at the carrying capacity. In the fertilized ponds (stocking rate 1200/ha), where more food was available growth continued unhampered up to an average weight of about 117 g, and in those ponds where fish were fed, *CSC* was attained at higher standing crop. The *CSC*s and carrying capacities for the treatments presented in Figure 23 could thus be calculated. This has shown the following:

Treatment	*CSC* (kg/ha)	Carrying capacity (kg/ha)
No fertilization, no feeding	65	130
Fertilization but no feeding	140	480
Fertilized and fed sorghum	550	2500 (estimated)
Fertilized and fed protein-rich pellets	2400	—

Figure 36 illustrates the relationship between standing crop and instantaneous yield per unit area for common carp fed sorghum and stocked at two densities. Growth rate for any standing crop was calculated from the average growth rates in Figure 23. It can be seen that there is an increase in yield with increase in standing crop up to a peak attained when the standing crop is between *CSC* and carrying capacity. Since the relative growth rate (i.e. growth rate per unit body weight) gets smaller as fish become larger, the yield from a given standing crop is higher when it consists of small fish rather than large ones. The overall yield of a unit area during the culture period is the area enclosed under the curve, and this is larger when the peak in yield is higher due to a higher *CSC*, as when a protein-rich diet is used.

It should be noted that supplementary feeding has no effect below the *CSC* since fish receive all their nutritional needs from the natural food. The amount of food to be supplemented above the *CSC* depends on the available natural food on the one hand and the standing crop of fish on the other. The higher the standing crop, the less natural food can satisfy the fish nutritional requirement, and more supplementary food is needed to bridge the gap. It is obvious that for calculating the amount of supplementary feed required one must first estimate the amount of available natural food.

Available natural food

In previous chapters the energy requirements for maintenance and growth have been discussed, and appropriate equations were given for calculating the amount of food, in terms of energy, required to satisfy these requirements. These equations can be used to calculate the amount of food consumed by the fish. Once the weight of the fish and its growth rate are known, the amount of energy consumed can be back-calculated.

Figure 36. The relationship between standing crop and yield per unit area for fish receiving sorghum in ponds stocked at two fish densities: (*a*) 2000/ha; (*b*) 4000/ha. Solid line gives calculated values based on actual average growth rates (Figure 22). Broken line gives extrapolated possible growth when food is not limiting according to equation (37) (from Hepher, 1978).

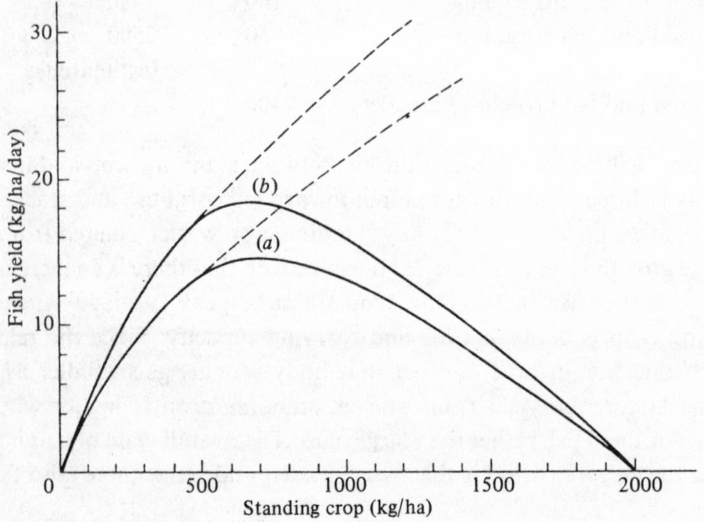

It should be taken into account that when the fish consume natural food, fat accumulation is minimal and the energetic value of gained tissues does not increase with increasing weight as was assumed in equation (43). Here the value of 5 kcal/g dry weight, or 1 kcal/g wet weight, as was assumed by Winberg (1956) and others, seems to be more appropriate. Equation (43) can, therefore, be modified for the purpose of calculating food consumption in this case, as follows:

$$F_n \text{ (kcal } ME\text{/day)} = 0.07\ W^{0.08} + \frac{G}{0.45} \qquad (70)$$

Multiplying the value obtained by the number of fish per unit area (density) gives the consumption of the entire fish population in this area.

It is obvious that below *CSC* the available natural food exceeds population consumption, but at or above *CSC* the available natural food is utilized to its maximum, and then natural food consumption reflects the availability of natural food in the pond. Equation (70) can be used, therefore, to estimate the available natural food.

From the above discussion it is clear that a number of factors will affect the amount of available natural food. One of the major determinants is, of course, the productivity of the pond, whether natural or induced by chemical fertilization and organic manuring. The effect of productivity can be demonstrated from data presented in Figure 23 for ponds with no supplementary feeding. When *CSC* has been reached in these ponds the average individual weights in the non-fertilized and fertilized treatments were 54 and 117 g respectively. Growth rates were 2.4 and 3.9 g/day, respectively. Both treatments had a density of 1200 common carp/ha. Calculating population consumption from these data through equation (70) gives 8400 kcal/ha/day (*ME*) in non-fertilized ponds and 14 700 kcal/ha/day (*ME*).

From another experiment carried out at Dor, in which the ponds were heavily manured with chicken manure or with liquid cowshed manure[1] the *CSC* for common carp (density 2000/ha) in a polyculture with tilapia (2500/ha), silver carp (1000/ha) and grass carp (500/ha) was determined. Average individual weight and growth rate at *CSC* were 205 g and 5.9 g/day respectively. Calculated amount of natural food available for the common carp population was 36 000 kcal/ha/day (*ME*).

The effect of increasing the standing crop above the *CSC* on the availability of natural food was studied by calculating the latter for the different points on the curves above the *CSC* in Figure 23. The results are

---

[1] Raw data from this experiment were provided by Dr G. Wohlfarth to whom thanks are due.

presented in Figure 37. It can be seen that at first, with the increase in standing crop and the associated increase in food requirement, grazing/predation becomes more efficient and the total population consumption (= available natural food) increases. However, at a certain standing crop overgrazing/predation impairs the production of food organisms and the replenishment of natural food resources, and the amount of available food decreases.

The fact that increase in demand for food can increase the efficiency of grazing is not new (see, for instance, Paloheimo & Dickie, 1965). The increase in standing crop and demand for food can also be due to increase in density. For a given change in standing crop grazing efficiency may increase more with increasing density than with increasing individual weight, since more fish can cover a larger area in search for food. Information on the effect of fish density on the amount of natural food consumed in a pond is scarce. Schaeperclaus (1966b) showed that increasing the density of two-year-old common carp in ponds without

Figure 37. The effect of increasing standing crop on the available natural food in fertilized and non-fertilized ponds. Raw data for calculation were taken from Figure 22.

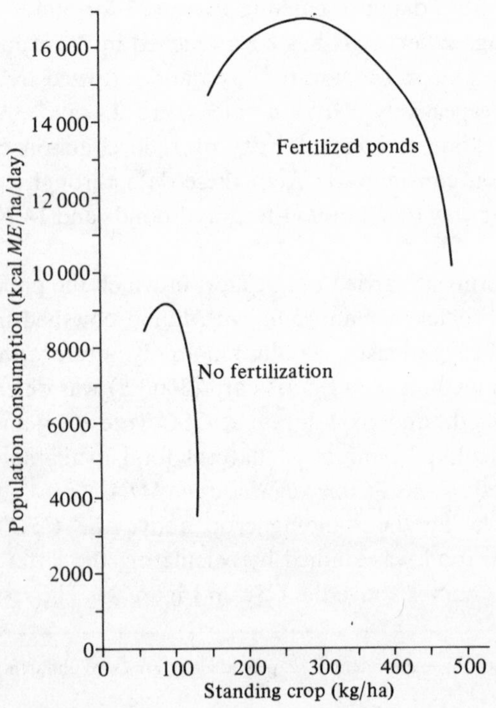

supplementary feeding from 240–250/ha to 900–1200/ha increased yields by 48–64%. Ivlev (1961) studied the relationship between fish and the amount of food available to it in laboratory experiments, in which the amount of food per fish varied. He suggested the following equation to describe the relationship between the amount of food consumed (r) and the concentration of food in the water body (p):

$$\frac{\mathrm{d}r}{\mathrm{d}p} = \xi\,(R - r) \tag{71}$$

where ξ is the coefficient of proportionality and R is the satiation ration. The integration of the above equation gives:

$$r = R\,(1 - \mathrm{e}^{\xi p}) \tag{72}$$

or if decimal logarithms are used instead of the natural ones:

$$r = R\,(1 - 10^{-kp}) \tag{73}$$

The effect of food density on the amount of food consumed by different species was described by Ivlev graphically and this is presented in Figure 38. If one assumes a fixed amount of food but a variable number of fish to correspond to the same relationship, one can calculate the amount of food consumed by the population at varying densities. This is given in Figure 38. It can be seen that with the increase in fish density the amount of food consumed from a given biomass of food per unit area also increases because the efficiency of grazing/predation increases. However, Ivlev has considered the biomass of food at a given moment rather than its production rate and availability with time. At high fish densities overgrazing affects production rate of the food organisms and thus decreases food supply and replenishment of stocks. Gurzeda (1965) noted that as the increase in standing crop of fish in a pond becomes higher, the increase in grazing/predation efficiency becomes less apparent until, at high standing crops, the utilization of natural food becomes uniform or even decreases, due to the consumption of young stages of the food organisms which reduces their reproduction rate.

The association of species in a polyculture can affect production of natural food and, through interspecific competition, its availability to the fish. Hepher *et al.* (unpublished data) found that silver carp at high densities (1300 and 2600/ha) in a polyculture with bottom feeding fish (common carp, 1000/ha; tilapia hybrids, 1500/ha) grew better than silver carp stocked at the same density in monoculture, although the concentration of algae was higher in the latter. Milstein *et al.* (1985a, b) explains this by overgrazing on the larger algae and zooplankton in the monoculture ponds, which gave rise to the development of a large number of very small algae which pass the silver carp's filtering apparatus. In the polyculture system, however, the bottom feeding fish, by their burrowing action, caused an 'upwelling' of nutrients. This stimulated the production of the large algae to the advantage of the silver carp, which therefore grew better in the presence of the bottom feeding fish.

On the other hand, the same experiments showed that silver carp at high densities depressed the growth of common carp and tilapia in the polyculture ponds, compared with the growth of these fish stocked at the same density in ponds without silver carp. This was probably due to the reduction in the amount of suspended organic matter in water, which can serve, directly or indirectly, as food for the bottom feeding fish after it settles to the bottom.

Figure 38. (*a*) The relationship between food concentration and average food consumption by an individual fish according to Ivlev (1961). (*b*) The relationship between fish density and population food consumption from a given food concentration using the same ratios as in (a).

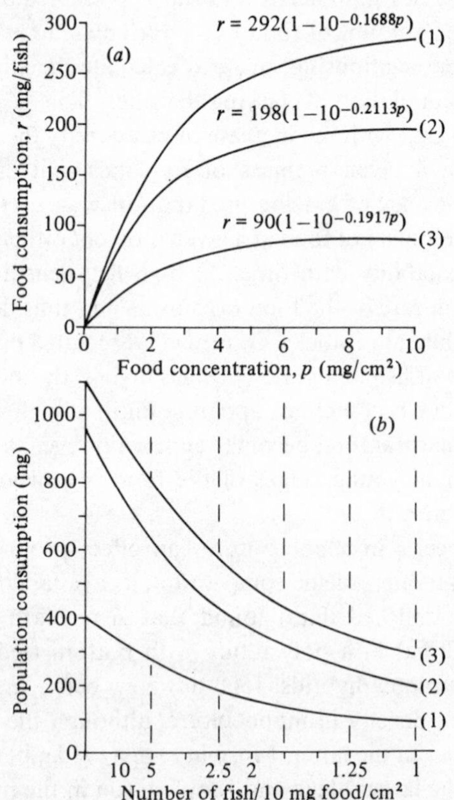

One of the major factors affecting production of natural food is the supplementary feed. It should be remembered that about 20–25% of the feed remains undigested and is emitted into the water as fish faeces. These and feed residues serve as manure, the effect of which has already been discussed above. Zur (1981) has shown that primary production increased with the increase in density of tilapia fed on supplementary feed. It was 3.8 g C/m^2/day at a density of 20 000/ha and 8.15 g C/m^2/day at a density of 100 000/ha. The difference seems to be due to the manuring effect of feed residues and wastes. Prikryl & Janecek (1982) found increased biomasses of zooplankton and zoobenthos, in particular chironomid larvae, with increasing feed ration, although these increases in biomass were probably masked, to some extent, by the increasing predation pressure of the fish (see p. 264). The higher the amount of feed per unit area, the more residues and wastes remain in the water and the more natural food is produced through their manuring effect.

Any attempt to estimate the amount of natural food in a pond receiving supplementary feed by subtracting the calorific value of the feed from consumption calculated from the body weight and gain, as was done above for fertilized ponds, should take into account a number of pitfalls. Since energy is the first limiting factor of natural food, while protein is in excess, the first stage of supplementary feeding involves carbohydrate feeds such as cereals (see p. 290). A new *CSC* is created when protein, and later other nutrients, are lacking for normal growth. Excessive energy does not result, therefore, in growth increment, both below the *CSC*, since the maximal growth is anyhow attained, and above *CSC*, since nutrients other than energy may limit growth. No indication for excessive energy feeding is evident except for high feed conversion rates, and their evaluation is usually done on an economic basis rather than a nutritional one. When excessive energy is added by over-feeding, the subtraction of the added energy from the overall consumption will result in an underestimate of the natural food contribution. Nevertheless, an attempt is made here to evaluate the contribution of natural food in the treatment receiving sorghum given in Figure 23, assuming that over-feeding, if it occurred, was small, and the estimation will therefore give the minimum value for natural food contribution in such conditions.

A note of warning should also be given as to the calculation of the amount of supplementary feed. It is customary to adjust feeding rate for a certain period according to the weight of fish determined at the beginning of the period. Since growth during the period may vary, the actual feeding rate, relative to the average fish weight during the period, may also vary. It is always lower than that calculated for the fish at the beginning of the period.

In the ponds fed sorghum (Figure 23) *CSC* has been reached when fish attained an average weight of 250 g. Growth rate at this stage was 6.7 g/day. In calculating the calorific equivalent of the gain one should take into account that feeding sorghum causes some fat accumulation (see p. 165). From the analyses made during some of the experiments, the data of which are presented in Figure 23 (see Hepher *et al.*, 1971), the content of fat in the gain was calculated as 12.8% and the content of protein was 16%. The calorific value of the gain was thus 2.11 kcal/g. The energy consumed by the fish in this pond at the stage of *CSC* was thus: $0.07 \times 250^{0.8} + (6.7 \times 2.11/0.45) = 37.2$ kcal *ME*/day, or for a population of 2200 fish/ha $37.2 \times 2200 = 81\,840$ kcal *ME*/ha/day. Back-calculating the amount of sorghum fed to the fish during this period showed 8.02 g/day. The metabolic energy value of sorghum was considered as 2.54 kcal/g. The total 2200 fish/ha have thus received with the supplementary feed $2.54 \times 8.02 \times 2200 = 44\,800$ kcal/ha/day, and the calculated contribution of natural food was thus $81\,840 - 44\,800 = 37\,040$ kcal/ha/day. This is roughly equivalent to the energy produced when manure is added to the pond with no supplementary feeding. The amount of natural food produced due to the manuring effect of the supplementary feed increases with increasing feeding rate, but its estimation above *CSC* becomes difficult.

Some early studies on the effects of supplementary feed on fish yield in ponds (e.g. Walter, 1934) attempted to determine its net effect, apart from that of natural food. For that they have subtracted the yield received in ponds without supplementary feeding from that received in the same or similar ponds when supplementary feed was added, assuming that the amount of natural food in both is equal. From the discussion above it is clear that due to the manuring effect of the feed residues and fish wastes this assumption is not valid. It can be argued, of course, that the effect of supplementary feed on natural food, although indirect, cannot be separated from that of the feed itself, and therefore it is a part of the overall nutritional effect of supplementary feeding. In this case, however, it should be accepted that the nutritional value of the feed *in ponds* is greater than that attributed to its nutrients directly absorbed by the fish. It is difficult to assess this extra value of supplementary feed.

Table 45 summarizes the contribution of metabolizable energy by natural food in the conditions discussed above.

Another major group of factors affecting the abundance of natural food in ponds consists of the environmental factors. In temperate and subtropical climates these are usually manifested as seasonality in the amount of natural food per unit water volume or area. Spring is the season when natural food is more abundant, while in summer and autumn there is a sharp decline in available food to reach a low in winter (Merla, 1966a). Since in these climates the ponds are very often stocked in spring with fingerlings which grow during the summer and are harvested in the autumn, the standing crop of fish increases while the biomass of natural food decreases. This may result in a surplus of natural food in spring. Natural food is sometimes so abundant in spring that fish do not come to supplementary feed at all. In contrast, however, in autumn, when the standing crop of fish is high, natural food may be in very short supply in spite of the manuring effect of the supplementary feed.

Composition of natural food

The composition of the organisms which serve as food for fish may vary not only among species, but also within species at different developmental stages, seasons and feeding conditions (Prus, 1970; Wissing & Hasler, 1971; Schindler *et al.*, 1971; Kosiorek, 1979). Nevertheless, when the composition of these organisms is reviewed, some general conclusions may be drawn.

Average values of nutrients found in proximate analyses of many samples of freshwater organisms were collected from the literature and are presented in Appendix II. It is apparent that except for macro-vegetation, most organisms are rich in protein. Protein content found in the analyses quoted in Appendix II was in the range of 16–70% of the dry matter, and even higher if the non-energetic components of the food, such as ash and chitin, are excluded. This is true not only of the aquatic fauna but also of

Table 45. *Contribution of natural food (kcal metabolizable energy/ha) for common carp, as calculated for ponds at the Fish and Aquaculture Research Station, Dor, Israel, receiving different treatments*

Treatment	Density/ha	kcal/ha/day
No fertilization, no feeding	1200	8400
Chemical fertilization, no feeding	1200	13 900
Manuring with liquid cowshed manure or chicken manure, no feeding	2000[a]	36 000
Feeding with sorghum	2200	37 000

Note: [a] Together with tilapia (2500/ha), silver carp (1000/ha) and grass carp (500/ha) in a polyculture.

algae, the protein in which may reach well over 40%. Macro-vegetation, however, contains less protein. The overall average protein content calculated for the organisms in Appendix II (excluding macro-vegetation) was about 49% of the dry matter. Albrecht & Breitsprecher (1969), who reviewed a considerable volume of literature on the composition of fish food organisms in ponds, gave the following average composition (percentage of dry matter in parenthesis): water 85.8%, protein 7.4% (52.1), carbohydrate 3.8% (27.3), lipid 1.1% (7.7) and ash 1.1% (7.7).

The calorific value of natural food resembles that of the fish. According to Appendix II it fluctuates in the range of 1.6–5.7 kcal/g dry matter, the average being 3.9 kcal/g. Here also, fluctuations are due mainly to the content of ash and are proportional to the content of organic matter (Vijverberg & Frank, 1965; Salonen *et al.*, 1976). Prus (1970) has calculated the average calorific value of ash-free dry weight for a collection of 64 species of aquatic animals and found it to be 5.5 kcal/g. The ratio between protein content in natural food and metabolizable energy (taken here as 75% of total energy, see Chapter 1) is thus about 130 mg/kcal.

The protein requirement of fishes and the proper protein:energy ratio for maximum growth have already been discussed in Chapter 6. According to Table 30 most fishes require 35–45% dietary protein, with a protein:energy ratio of about 96 mg/kcal. This ratio, which is higher than that of poultry and mammals, is still lower than that of the natural food. This indicates an excess of protein in the natural food.

It is obvious that when *CSC* for natural food is reached, and this food cannot support maximum growth any more, it is energy which limits the growth rather than protein. The deficit in energy can be satisfied by carbohydrate feeds, in species where it can be digested and utilized. Common carp and tilapia can be fed at this stage with cereals. Thus, if the proper amount of energy-rich supplementary feed is added, growth continues at the maximum rate. With the increase in standing crop of fish and the demand for protein, protein becomes depleted and inhibits further growth. The *CSC* observed in ponds in which fish are fed energy-rich diets is due to this limitation of protein. Growth is resumed at its maximum rate when sufficient protein is added to the diet. It should be remembered, however, that due to the manuring effect of the supplementary feed (see p. 288) new natural protein is produced which affects the *CSC*. This effect cannot, of course, be separated from the overall effect of the supplementary feed.

Natural food in ponds seems also to be rich in vitamins and minerals. Only above a certain *CSC* does a deficit in vitamins and minerals develop, which necessitates the addition of these nutrients to the supplementary feed. Lovell (1979) found that when the standing crop of channel catfish in ponds was 3750 kg/ha or less, there was no need for supplementing the feed with ascorbic acid. However, at a standing crop of 6270 kg/ha the absence of supplementary ascorbic acid in the feed resulted in reduced fish growth rate and the appearance of fish with deformed backs. In an another experiment Lovell (1979) found that stocking tilapia with 9800 channel catfish/ha, both competing for natural food, resulted in the catfish showing signs of ascorbic acid deficiency if the vitamin was not added to the catfish feed. Hepher (1979) found that common carp could be cultured at maximum growth rates on a pelleted diet containing 25% protein but no vitamins up to a *CSC* of about 2.4 t/ha. Only above this *CSC* vitamins affected the growth rate of the fish. Similarly, supplementing the diet with phosphorus was effective in increasing the growth rate of common carp at a standing crop exceeding 1.8 t/ha (Hepher & Sandbank, 1984).

10

Supplementary feed and its utilization

Kinds of supplementary feed

Supplementary feed can be classified according to several criteria. Classification by energy and protein contents is quite common since energy-rich feedstuffs, such as cereals, are usually poor in protein, while those rich in protein, such as tissue meals and oil meals, are often poor in energy. Feeds can also be classified into 'natural', such as grains, grasses etc., and 'processed', such as meals or yeasts. However, from the standpoint of nutrition, the most signficant classification is that of simple feedstuffs and compounded ones, because of the greater availability of the former, but the greater ability to control feed composition with the latter.

Simple feedstuffs are usually homogeneous, although their nutrient content may vary widely according to their origin. Thus, fishmeal of different species and production processes may contain 40–74% protein, and protein and energy contents of oil meals may vary according to the oil extracting process. The most important advantage of simple feedstuffs is their availability even in areas where industry is scarce. Because of the abundance of some simple feedstuffs in certain areas, such as rice-bran and broken rice in South East Asia, or sorghum in tropical regions, it is of relatively low cost in comparison to other feedstuffs. Many of the simple feedstuffs are agricultural by-products such as rice-bran, broken rice or relatively inexpensive grains such as sorghum. Even grass is used as fish feed in certain areas (mainly China) and for certain fish (mainly grass carp, *Ctenopharyngodon idella*).

Many of the simple feedstuffs are also rich in carbohydrate. These are usually cereals or of cereal origin, such as grains, middlings and brans. However, other carbohydrate-rich feeds have also been used for fish, such as potatoes (Wunder, 1937; Demoll, 1940a, b) or lupins.

Carbohydrate-rich feedstuffs can be used only as long as they supplement and balance protein from natural food. When this is not sufficient, protein-rich feeds should be used.

Protein-rich simple feedstuffs may be of two groups according to their origin: (a) of vegetable source, such as oil meals; (b) of animal source, such as trash fish, meat and poultry by-products (discarded entrails, shanks, etc.) or tissue meals (fish, meat or shrimp meals).

Simple feedstuffs serve as ingredients for compounded feed. Mixing these feedstuffs at various proportions results in feeds of different composition, which may be more suitable for the fish than the simple feedstuffs as they are. It is obvious that just mixing the ingredients is not enough since the mixture will immediately disintegrate in water and separate again, so that the fish will not receive the intended diet composition. Swingle (1958) compared the efficiency of such dry mixture with that of the same diet when processed into hard pellets. A pond stocked with 7400/ha channel catfish fingerlings, which were fed on pellets, produced during 252 days 2648 kg/ha, with a food conversion ratio (*FCR*) of 1.6. Another pond, with identical conditions but where fish were given dry mixture produced only 1247 kg/ha, with a *FCR* of 3.3.

Two ways are commonly used to bind the mixture together, producing dough or pellets. Erokhina (1959) conducted studies to compare the efficiency of these two forms of compounded feed for fish. She found that when served as dough, much of the feed becomes inaccessible to the fish, either due to disintegration and dispersion or due to leaching. Although pellets also undergo disintegration and leaching, these processses are much slower than with the dough, and therefore better affect fish growth. She found that common carp fed the same diet at the same rate either as dough or pellets, grew better (reaching 600–650 g) on pellets than on dough (reaching 520–540 g). Feed conversion ratio was also better with the pellets (2.2) than with the dough (3.0). In practice, today, pellets are used much more widely than dough.

In order to reduce losses from disintegration and leaching, as much as possible, care should be taken in the preparation of the pellets. Two kinds of pellets are currently used for fish: the compressed (sinking) pellet, and the extruded (usually floating) pellet. The first is prepared by wetting the mixture with water (cold pelleting) or with steam (hot pelleting) and then compressing the wetted mixture through a die. The temperature of the pellets increases while passing through the die, due to friction. The moisture and heat gelatinize part of the starch of the mixture, especially at the surface of the pellet where the heat is highest during the friction, which contributes to the stability of the pellet in water. Higher water stability of the pellets can be achieved by: (a) using finely ground ingredients so that the small cavities on the surface of the pellets are filled up preventing the penetration of water into the pellet; (b) increasing the temperature during pelleting, thus increasing the proportion of the gelatinized starch; (c) avoiding high concentrations of fat, water repellent or highly hygroscopic ingredients in the diet mixture; (d) using binders (Hastings, 1971; Hastings *et al.*, 1971). Different binding agents have been used for fish feed pellets, such as Guam gum, agar, gelatine, etc. with varying extent of success (Joyner, 1965; Slinger *et al.*,1974). Some, such as alginates, require a complex production technology, which increases the cost of the feed.

Extruded pellets are pre-cooked at higher moisture and temperature than the compressed pellets, and then they are compressed through the extruder under high pressure. As a result, most of the starch is gelatinized and the pellet comes out of the extruder partly puffed. The expanded pellets are water stable up to 24 h. Part of the resistance in water is due to the mesh of air pockets created by the expansion and to stabilization by case hardening which occurs during high-temperature drying. The air pockets usually give the pellet the floating properties, although when puffing is regulated the pellets can be made to sink. The fact that the pellets float has some management advantages since they allow the feeder to observe the fish feeding. He can thus determine whether and when fish take the feed and thus tell the effect of low dissolved oxygen and water temperature (or high temperature) on feeding (Anon., 1972). Extruded pellets, however, cost considerably more than sinking pellets, although this is not reflected in higher returns from fish growth. Tiemeier & Deyoe (Anon., 1972) found that production of channel catfish fingerlings fed the same feed in either sinking or floating pellets, was similar. Percy & Avault (1974) even found an increase in growth rate, a better feed conversion ratio and higher production per unit area with channel catfish fed sinking pellets compared with floating pellets. This may be due to the fact that the feed expands so much that the fish may fill its stomach without consuming enough nutrient for maximum growth rate. Lovell (1977) found in feeding tests where channel catfish were offered either extruded pellets *ad libitum*, that they consumed 17% more of the compressed pellets and consequently the fish consuming this feed gained appreciably more weight.

It should be remembered that increasing the temperature during processing, of either compressed or expanded pellets, can result in a higher loss of labile nutrients such as certain vitamins. This must be taken into consideration when formulating the diet.

Feeding levels

Since the level of supplementary feed required by the fish depends on the available natural food, and no reliable methods for its rapid estimation have yet been developed, most tables for feeding pond fishes are based on some nutritional assumptions and much empirical experience. Some of these tables are given in Appendix IV.

As has been mentioned above, with the increase in the weight of the fish, the relative calorific nutritional requirement, i.e. requirement per unit fish weight, decreases. However, the amount of supplementary feed given to the fish depends not only on the food requirement but also on the availability of the natural food and the change in it with the change in fish weight. If the relative amount of natural food per fish decreases during the growing period at a higher rate than the decrease in the nutritional requirement of the fish, more supplementary feed must be given to balance the difference. This means a more moderate decrease in the level of supplementary feeding than if it were the complete only feed. The level of supplementary feeding may even remain constant relative to the weight of the fish, i.e. set a certain percentage of body weight per day, irrespective of the fish weight. This, indeed, is often the practice in many fish farms. If, however, the rate of decrease in the relative amount of natural food is lower than that of the nutritional requirement, less supplementary feed is required and feeding level can decrease at a faster rate.

When the standing crop of fish in a pond increases over the critical standing crop (*CSC*), a nutritional deficit develops, which has to be made up by supplementary feed. With the increase in standing crop and nutritional deficit the relative amount of supplementary feed, as the percentage of fish body weight per day, increases. After a while, the relative amount of natural food reaches an equilibrium in relation to the requirement and the amount of supplementary feed levels off. Empirical experience seems to indicate that although the absolute amount of natural food increases with the increase in standing crop of fish, due to the greater efficiency of grazing and the manuring effect of the added fish wastes (p. 287), as well as due to the effect of seasonal factors and the increased feed consumption by the fish (p. 289), the relative amount of natural food to requirement decreases. Therefore, after reaching a peak, supplementary feeding level decreases at rather a moderate rate, and in some cases, when the amount of natural food decreases at a higher rate, feeding level (in per cent body weight per day) remains constant during most of the culture period. Nevertheless, also in this case, when fish become larger, at a certain stage supplementary feeding level is usually decreased.

Temperature must, of course, be taken into consideration when feeding charts are prepared. At lower temperatures the requirements for both maintenance and growth are lower, and so, therefore, is food requirement. However, the increase in temperature may affect these two functions differently. Due to adaptation maintenance may be less affected by the rise in temperature than growth, at least at a certain temperature range (p. 123). However, not much is known about the effect of temperature on feeding level, and most feeding charts are based, in this respect, on empirical experience.

In one of the earliest studies on the culture of channel catfish, Swingle (1958) gave the rate of feeding as decreasing from between 5 to 3% of the body weight per day. These limits were set through observing the growth of the fish and the feed conversion ratio. Lovell (1977) recommended lower feeding rates for fingerling to market size channel catfish (Table 46). In setting these rates he also took into account the season and the water temperature in the southeastern USA, and based them on the feeding of compressed pellets containing 36% protein and 2880 kcal/kg feed. Small fry require a higher percentage of feed. According to Lovell (1977) fry at the swim-up stage require 6–10% of the body weight per day and when they reach 3.6–14 cm long, 3–4% of their body weight per day. Foltz

Table 46. Feeding schedule for channel catfish in ponds stocked as 12.7 cm fingerling and harvested as 0.5 kg fish in southeastern United States

Date	Water temperature (°C)	Fish weight (g)	Feeding level (% of body wt)
April 15	20.0	20	2.0
April 30	22.2	30	2.5
May 15	25.5	50	2.8
May 30	26.7	70	3.0
June 15	28.3	100	3.0
June 30	28.9	130	3.0
July 15	29.4	160	2.8
July 30	29.4	190	2.5
August 15	30.0	270	2.2
August 30	30.0	340	1.8
September 15	28.3	400	1.6
September 30	26.1	460	1.4
October 15	22.8	500	1.1

Source: From Lovell (1977).

(1982) has calculated the regression between feeding level (per cent body weight per day) and body weight for channel catfish and gave the formula:

$$R = 11.957 - 4.1789 \log_{10} \text{length (mm)} \quad (74)$$

the actual feeding table derived from this equation is given in Appendix IV.

Marek (1975) composed feeding charts for common carp and tilapia. In his charts he took into account a certain estimation of the natural food in the pond and subtracted it from the calculated requirements for maintenance and expected growth (according to experience gained in Israeli fish ponds) in order to determine the amount of supplementary feed to be given. To do so, he had to find a common denominator for the nutritional requirement, natural food and supplementary feed. Marek used protein, assuming the productive protein value (*PPV*, see p. 180) to be 4, i.e. four units weight of dietary protein required to produce one unit weight of fish body protein. Fish yield obtained in ponds in which common carp were fed supplementary feed sufficient for maintenance only, was taken as an estimate of natural food. This was converted to supplementary feed equivalent and subtracted from the amount required by the fish. Since with the increase in density of fish the natural food is divided among more fish, leaving less per fish and therefore higher requirement of supplementary feed, the density becomes an important determinant in the feeding chart composed by Marek (Appendix IV). For all its crudeness, experience by fish farmers has proved Marek's chart to be quite efficient, especially with common carp weighing 150–400 g. With smaller fish Marek's chart may be somewhat excessive and with larger fish it may be somewhat short in supplementary feed.

Since feeding charts are based on the weight (or size) of the fish, and this changes every day as the fish grow, rations should also be changed on a daily basis. In most cases, however, rations are fixed for a longer time period, very often in relation to periodical sample weighing of the fish. Feeding charts should take into account the growth during the periods between two successive sample weighings.

Feed composition

From Table 32 it is clear that most pond fishes require 35–45% protein in their diet. Natural food, however, contains 50–60% protein (dry matter basis). When the fish live on natural food alone, this excessive amount of protein is utilized for energy. This may be advantageous from a biological point of view since protein has a higher calorific value than supplementary carbohydrate, but it is wasteful from an economic point of view since protein is much more expensive than carbohydrate. When *CSC* for natural food is reached, it is energy that is lacking rather than protein or vitamins. Supplementary feed must fill this energy deficit first. At this stage a high energy–low protein diet can be used. If the fish have no difficulty in digesting and utilizing starch, as is the case with common carp, a starchy diet is usually the most economical. Cereal grains or compounded feed containing a high proportion of starch are suitable in these cases. Where difficulties in digesting the starch exist, lipid can be a proper source of energy. From the discussion above it may be concluded that it is a good economical practice to increase the density of the fish in the pond so that the *CSC* for natural food is reached early in the culture cycle.

With the increase in standing crop, a point is reached where natural protein is no longer sufficient to sustain maximum growth. This is the *CSC* for the energy-rich diet, and growth rate starts to fall off. Supplementary protein must then be added for maximum growth rate. The higher the standing crop above the new *CSC*, the larger the deficit in natural protein, and the higher the protein content in the supplementary diet should be. Lovell (1975) brings results of an experiment in which earthen ponds were stocked with channel catfish at densities of 4940 and 9880/ha, and fed supplementary diet containing either 32 or 45% protein. Yields obtained were:

Stocking rate/ha	Protein (%)	Yield (kg/ha)	*FCR*
4940	32	2275	1.45
4940	45	2466	1.34
9880	32	2742	1.65
9880	45	3824	1.16

The difference in weight gain and *FCR* between the two diets was appreciably greater at the higher density when natural food did not provide sufficient protein for growth.

The increase in requirement for supplementary protein with increasing standing crop is gradual. Hepher *et al.* (1971) and Hepher (1975) have shown that with a monoculture of common carp, up to a standing crop of 800 kg/ha (*CSC* for the cereal diet at the conditions of the experiment) there was no difference in growth rate when fish were fed either pelleted sorghum (9% protein), a pelleted diet containing 22.5% protein, or a pelleted diet containing 27.5% protein. Above this standing crop the growth rate of fish fed sorghum fell behind that of the other two groups, while there was no difference between the two until the standing crop reached 1400 kg/ha. Above this standing crop the diet containing 27.5% protein gave better results than the diet containing 22.5% protein.

The same arguments are also true for the inclusion of vitamins and minerals to the supplementary feed. As long as these are supplied at adequate amounts by natural food, there is no need for their presence in the supplementary feed. Experiments in monoculture of common carp in Israel (Hepher, 1975) have shown the *CSC* for these to be about 2.4 t/ha. Below this standing crop feeding fish diets supplemented with a vitamin mixture did not result in a higher growth rate compared with a non-supplemented diet. Their performance was, however, superior at higher standing crops.

It is obvious that one cannot speak of a fixed composition of supplementary diet for pond fishes, since it differs not only with fish species but also with increasing standing crop of the same species. Three important points should be remembered in this respect:

(a) Since the composition of supplementary feed, at any standing crop, depends on the amount of available natural food, productivity of the pond plays an important role here. The higher the productivity, the less protein or vitamins is required at a given standing crop. It is often worthwhile to treat the pond with relatively inexpensive fertilizers and manure to increase productivity, than to supplement the feed with costly proteins and vitamins.

(b) 'Protein' and 'vitamins' were treated here as homogeneous substances. However, one should remember that although natural food protein is of a high biological value, i.e. contains amino acids much at the same proportion as required by the fish, still a certain amino acid of the protein may be limiting growth and not necessarily 'protein' as such. Supplementing the diet with a protein rich in this particular amino acid may be more efficient than with another protein. This may be even more pronounced with vitamins. It is not known, however, which amino acid or vitamin becomes exhausted first.

(c) Since the composition of the diet is changing continuously with increasing standing crop, it would require an infinite number of diets, if maximum feed utilization is sought. This is, of course, impossible. A partial solution can be found in one of the following ways:

(1) A complete diet, or a diet rich in protein, is 'diluted' with a diet rich in carbohydrates. This method is practised in some of the Israeli fish farms. 'Dilution' of the feed is done stepwise and the

changeover from one step to another is usually done at the following standing crops:

Up to 700 kg/ha	Sorghum only
700–1200 kg/ha	75% sorghum:25% pellets (25% protein)
1200–1500 kg/ha	50% sorghum:50% pellets (25% protein)
1500–1800 kg/ha	25% sorghum:75% pellets (25% protein)
Over 1800 kg/ha	Pellets only (25% protein)

While the principal of 'dilution' is one of the bases of Marek's feeding chart, it does not adhere to the above steps and even at very high standing crops some sorghum is given to keep the energy in the diet at high level.

(2) A limited number of diets of increasing nutritional value are prepared. Each of these covers the nutritional requirements up to a given critical standing crop, when it is exchanged by another diet of a higher nutritional value.

Frequency of feeding

Supplementary feed can be served to the fish in single or multiple meals, or given continuously. When the ration is served in a single meal per day, the amount of feed may exceed the satiation dose, and part of the feed may remain uneaten for some time, even hours. During this time the feed can disintegrate and be dispersed in water, and nutrients can be leached out. Thus the efficiency of the feed may often be reduced. When whole cereal grains are given, as often is the case with common carp in Europe, the loss is smaller. A certain soaking time is even beneficial since the grains become more soft and easier to ingest by the fish.

Many studies have shown that multiple feeding results in a more efficient utilization of the feed than a single feeding. The number of feedings per day and the time of feeding vary with species, size of fish and environmental conditions. When feed is given *ad libitum* a larger total daily amount of feed is usually consumed by the fish with multiple feeding than with a single one. The effect on feed conversion is usually small, so that the yield is also higher with multiple feedings. Andrews & Page (1975) and Lovell (1977) reported on feeding experiments with channel catfish. Pond fed catfish consumed more feed and produced more gain when fed twice daily than when fed once a day. Feeding more than two times per day did not improve food consumption or growth rate. However, when channel catfish were smaller than 1.5 g, they did benefit from more frequent feeding. According to Murai & Andrews (1976) feeding fingerlings of this size eight times per day resulted in a better growth rate than feeding them four times a day. Also Palmer *et al.* (1951), working with the salmon *Oncorhynchus nerka* found that fingerlings of less than 4.5 g fed eight times per day gained significantly more weight than those fed five times per day and overwhelmingly more than fish fed three times or once daily. However, with larger fingerlings the differences in growth rate with different feeding frequencies decreased. Channel catfish of over 1.5 g grew as well when fed four times daily as those fed eight times daily (Murai & Andrews, 1976), and feeding blueback salmon larger than 4.5 g five times daily was almost as good as eight times a day, although still better than three times or once a day (Palmer *et al.*, 1951). These results suggest that the beneficial effect of high feeding frequency on growth of small fish is related to their higher feed intake per unit weight. This also may be the reason for the different effect of feeding frequency at different temperatures. Elliott (1975d) fed brown trout (*Salmo trutta*) to satiation in each of four meals at seven different temperatures between 3.8 and 18.1 °C. The higher the temperature, the more beneficial was the multiple feeding. Highest food ingestion and growth rate was at one meal at about 4 °C to three meals at about 18 °C.

Feeding can be spread throughout the day in many meals when an automatic feeder is used, dispensing feed at fixed times and intervals. It can be given continuously when a demand feeder is used. The effect of such feeding may vary with species and conditions. Andrews & Page (1975) state that channel catfish fed 24 times/day with an automatic feeder had significantly poorer weight gain and food conversion than fish fed four to eight times per day. With common carp, however, continuous feeding by a demand feeder did not greatly affect the food conversion ratio and increased the yield of fish by about 14% (Hepher, unpublished data).

Feed utilization and feed conversion ratio

It is appropriate here to remind the reader of the two interrelated ways of expressing the efficiency of conversion of feed to body weight gain. The one, usually expressed as a percentage, is the 'food utilization ratio' or 'food conversion efficiency', which is the ratio between the gain in weight and the ingested feed ($\frac{G \times 100}{R}$). The reciprocal of this expression is 'food conversion ratio' (*FCR*) which designates the amount of feed required to obtain a weight unit of body gain ($\frac{R}{G}$). Some of the difficulties in *FCR* using for evaluating diets with different protein contents have been already discussed above (p. 179). Attempts have been made to improve the evaluation by using calories for both food and body weight gain. Another modification has been suggested by Ivlev (1939b) and later others (e.g. Davis & Warren, 1968; Klekowski, 1970). They distinguished three levels of food conversion efficiency according to the basis of reference. The first (which Ivlev called 'energy coefficient of growth of the first order', K_1) referred to the gross calorific value of the food; the second ('the coefficient of the second order', K_2) referred to the metabolizable energy of the food; and the third (the 'coefficient of the third order', K_3) referred to the net energy after deducting the calorific expenditure for maintenance. K_3 is thus equivalent to E_{pg} which was already discussed above (p. 169). These coefficients, however, are not widely used. *FCR*, for all its faults, is used much more often by biologists and fish farmers alike to evaluate the efficiency of supplementary feed in ponds, since it gives a simple yardstick for estimating how much feed is required to produce a unit weight of fish, and whether the use of this amount of feed is economically worthwhile.

The problems involved in using *FCR* in ponds are much greater than in a system with a complete feed as the sole source of food. In most cases *FCR* of supplementary feed is calculated ignoring the natural food consumed by the fish. The resulting *FCR* is therefore an apparent rather than the true value. Early studies on pond fish feeding in Central Europe paid special attention to this point and tried to distinguish between the two (Walter, 1934; Schaeperclaus, 1961). The 'true' *FCR* was calculated by subtracting the yield produced in the same, or similar, ponds when no supplementary feed was added. The ratio between the amount of feed added to the pond (assuming that it was all consumed by the fish) and the 'net' yield was considered as the true *FCR*. Swingle (1958) pointed out the difficulties of calculating the true *FCR* in this way.

Since true *FCR* is the ratio between total consumed food and growth, any factor affecting growth and its proportion to maintenance requirement also affects *FCR*. Food ration and composition are amongst the most important factors affecting growth and *FCR*. When food ration is below or at maintenance level, *FCR* is infinite since no growth occurs. With increasing food level more dietary energy is channelled to growth. Niimi & Beamish (1974), who worked with largemouth bass (*Micropterus salmoides*), state that utilization of energy for growth increased from zero at maintenance ration to 40% of the ingested energy at satiation. Thus *FCR* decreases with increasing feeding level until it reaches its lowest value (or highest utilization efficiency). When feeding level is increased further, *FCR* increases (utilization efficiency decreases) gradually, until food becomes excessive and *FCR* increases sharply (Figure 39). From an economic point of view it is important to consider here the marginal *FCR*. This is the ratio between the extra amount of feed added over that of the preceding feeding level, and the extra weight gain of fish, gained due to this additional feed, over that gained by the preceding feeding level. The marginal *FCR* increases at a much steeper slope than *FCR* calculated for the total amount of feed and the total weight gain. The economic limit of *FCR* is when the cost of the marginal feed is equal to the income from the marginal gain.

Environmental factors affect growth, and therefore also *FCR*. Andrews & Stickney (1972) have shown that *FCR* for channel catfish decreased with the increase in water temperature over the range of 18–30 °C. This may be due to the fact that while growth rate increases with the increase in temperature, maintenance may become adapted to the increase in temperature and the energy expenditure associated with maintenance remains relatively low. Energy is thus channelled more to growth than to maintenance. The relationship between *FCR* and temperature in Andrews & Stickney's (1972) experiment was slightly curvilinear, which may indicate that adaptation of maintenance to increase in temperature is effective only at a limited range of temperatures.

In a fishpond both true and apparent *FCR* of the supplementary feed are affected by the amount of available natural food, i.e. the productivity of the pond, on the one hand and the standing crop of fish on the other. It is obvious that below *CSC* when supplementary feed is not required, growth of fish is independent of supplementary feed. If such feed is added, *FCR* will be directly related to the amount of feed, but true *FCR* is infinite since no growth can be attributed to the supplementary feed. With increasing standing crop above *CSC*, assuming that suitable amount and composition of supplementary feed is added to make up the deficit in natural food, true *FCR* of supplementary feed decreases. It reaches its

Figure 39. The effect of feeding level and feed composition on the *FCR*. *MR* = maintenance ration.

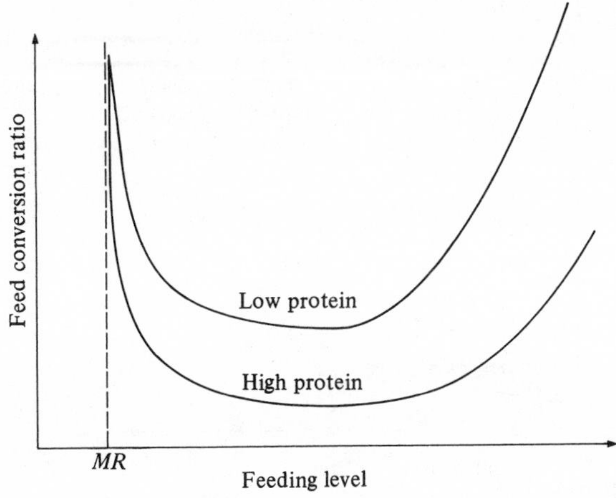

lowest value (highest utilization efficiency) much in the same way as with the increase in the feeding level when no natural food is present (Figure 40, compare with Figure 39). At a fixed fish density, standing crop increases with the increase in fish weight. *FCR* also increases gradually with increasing standing crop (and fish weight) until the *CSC* of the supplementary feed is reached.

The gradual increase in *FCR* with increasing weight has been observed by many researchers (e.g. Parker & Larkin, 1959; Davis & Warren, 1968; Brett & Groves, 1979). This may also be associated with the different rate of increase in energy requirement for maintenance and growth with the increase in fish weight. While maintenance metabolism increases at the power of 0.8 of the weight, growth rate, and therefore also the energy required for it, increases only by the power of 0.66 of the body weight. This means that the ratio between these two functions changes in favour of maintenance, which leads to an increase in *FCR*. Some researchers tried to formulate the relationship between fish weight and *FCR* in the form of a regression (Yoshida, 1970a, b). The regression coefficient seems to be dependent on temperature. It is higher at higher temperature, but at low temperature, when growth is in any case small, the effect of body weight is small. De Silva & Balbontin (1974) fed young herring (*Clupea harengus*) to satiation at two temperatures (14.5 and 6.5 °C) and determined the resulting feed conversion efficiency. This showed a drastic decrease with increasing weight at 14.5 °C, but at 6.5 °C no such trend was observed.

With the increase in standing crop (and fish weight) the critical standing crop for the specific supplementary feed is reached. The feed cannot

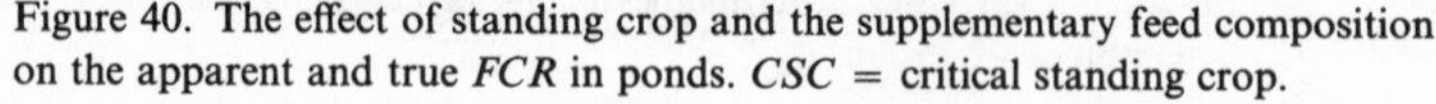
Figure 40. The effect of standing crop and the supplementary feed composition on the apparent and true *FCR* in ponds. *CSC* = critical standing crop.

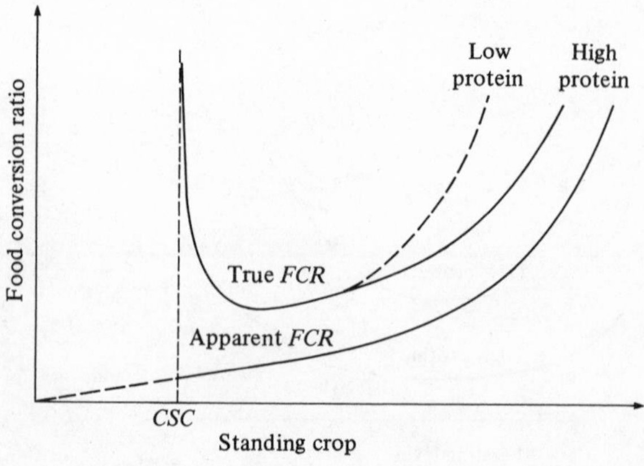

support maximum growth rate any more, and growth falls off this rate and at higher standing crops actually decreases. *FCR* then increases sharply until, when growth ceases entirely at carrying capacity, *FCR* reaches infinity (Figure 40). This is clearly illustrated by Hepher (1978) with a population of common carp in ponds fed on either sorghum or pellets containing 25% protein (Figure 41). At a standing crop of about 500 kg/ha sorghum became inadequate to meet nutritional requirements and the *FCR* increased sharply, while the pellets were still sufficient. The sharp increase in *FCR* is usually critical from an economic point of view. The fish farmer has three ways to avoid it: (a) Improve the diet and add the missing nutrients. This means an increase in cost for feed, which is not always paid by return from the extra gain in fish production. (b) Decrease density. This is associated with a decrease in yield (p. 256). (c) Increase, if possible, the productivity of the pond by fertilization or manuring.

The use of food conversion ratio for comparing the performance of diets of different composition is even more problematic than its use for evaluating the performance of the same diet at different conditions. The performance of a protein-rich diet and a carbohydrate-rich diet may be similar below the *CSC* of the latter diet, but different above it. Moreover, there is no common denominator for comparing diets of different

Figure 41. Supplementary feed conversion ratio at different standing crops of common carp in ponds (stocking density 2000/ha). Solid circles – fish fed on sorghum; open circles – fish fed on protein rich pellets. Each point is an average of three replications (from Hepher, 1978).

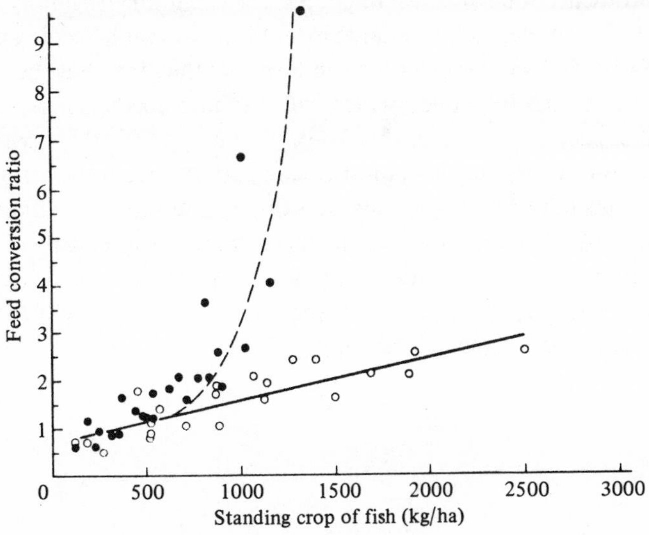

composition, since they can differ in various nutrients (energy, protein, vitamins, minerals) which have sometimes very little in common, and which become important at different levels of standing crop. Lehmann (quoted from Schaeperclaus, 1968b) suggested such a common denominator based on 'total nutrients' ('Gesamtnaerstoff' – *GN*), as was attempted in other animal nutrition studies. *GN* equals to the sum of the digestible protein, the digestible nitrogen free extracts, and the digestible fat $\times$ 2.3. The '*FCR*' suggested by Lehmann was *GN* $\times$ 100/weight gain. However, in addition to the fact that *GN* does not express the true energetic ratios in fish, the suggested '*FCR*' does not remedy the problems pointed out above, while it makes the *FCR* meaningless to the farmer in terms of feed consumed.

One way to solve the last problem is carrying it to its extreme. Since the main purpose of the fish farmer is to evaluate the economic efficiency of the feed, *FCR* can be expressed in terms of cost, where the cost of each feed is multiplied by the *FCR*. This expresses the cost of feed required to gain a weight unit of fish. 'Monetary *FCR*' or 'weighed *FCR*' can be also used to compare the efficiency of feeds with that of fertilizers and manures which increase natural food in the pond and thus save supplementary feed. Monetary *FCR* seems, therefore, a practical way to express food inputs compared with other costs of the aquacultural management system.

Conclusion

An attempt has been made in this book to lead the reader from the theoretical aspects of nutritional anatomy and physiology to the practical problems of feeding fish in ponds. The economic feasibility of an aquaculture system depends to a large extent on reasonable feeding costs. Feeding charts that will best suit the conditions in the farm must be based on sound nutritional principles, which will take into account most of the theoretical aspects reviewed in this book. Some of the problems raised have been answered by pertinent information gathered from the literature. Other problems require further study. It is hoped that indicating these last problems will draw more attention to them, which will lead to a more intensive study. Solving the still outstanding problems will, no doubt, contribute toward the construction of better and more efficient feeding charts and more profitable fish farming.

Appendix I

Average apparent digestibility coefficients (ADC, of ingested nutrients) and digestible energy (kcal/kg) of various feeds and feed ingredients

Feed	Protein	Carbohyd	Fat	Fibre	Dry matter	Energy	Source
	(*n*)[a] *ADC*	(*n*) *ADC*	(*n*) *ADC*	(*n*) *ADC*	(*n*) *ADC*	(*n*) *ADC*	
(a) Salmonids							
Albumin – egg	(1) 98.0						75
Alfalfa meal (dehydrated)	(2) 54.8		(1) 70.7			(2) 31.4	19, 92
Algae – *Spirulina maxima* (dry)	(1) 83.1						3
Bacteria – methanophilic	(1) 93.5						3
Blood meal	(1) 39.8					(1) 51.4	19
– spray dried	(1) 65.2					(1) 65.8	92
Casein	(20) 92.4						2, 3, 53, 75, 90, 94
Cellulose				(1) 13.7			75
Corn gluten meal	(2) 92.6	(1) 62.0	(1) 91.1			(2) 55.1	19
Cotton seed meal	(2) 79.2					(1) 56.4	43, 92
– without hulls	(1) 77.7					(1) 63.3	92
– flour	(1) 83.5					(1) 67.0	92
Dextrin		(3) 84.9					19, 89, 90
Egg– whole	(1) 95.0						75
Egg– yolk	(1) 93.0						75
Fatty acids:							
14:0			(1) 96.5				29
16:0			(1) 91.5				29
16:1			(1) 92.0				29
18:1			(1) 90.5				29
20:1			(1) 96.0				29
22:1			(1) 90.5				29
20:5			(1) 95.0				29
22:6			(1) 97.0				29
Feather meal	(1) 62.3		(1) 68.0			(1) 73.7	19
Fish – fresh	(10) 94.4						2, 53, 62, 75
Fish meal (average)	(211) 84.8		(9) 90.4		(8) 88.2	(6) 91.8	3, 19, 29, 43, 53, 71, 85, 90, 92, 96, 110
– anchovy	(9) 62.1		(1) 88.6			(1) 91.3	29, 72, 92
– brown (Peru)						(1) 87.6	29, 96
– capelin	(2) 89.4		(1) 86.3				29, 96
– flatfish	(55) 87.6						20, 71
– herring	(15) 88.7		(7) 91.3		(8) 88.2	(3) 95.9	3, 19, 20, 29, 72, 90, 92, 110
– pilchard	(13) 77.8						72
– pollack	(1) 80.0						72
– saury pike	(75) 85.6						72
– whitefish	(38) 86.2					(1) 84.0	43, 52, 53, 69, 92
Fish solubles (dried)	(1) 69.1					(1) 78.8	92
Fly larvae						(1) 85.0	5
Glucose		(8) 96.9					19, 78, 89, 90
Lactose		(6) 90.1					78, 89
Linseed meal	(1) 76.8					(1) 71.1	92
Maltose		(1) 93.0					78
Meat – fresh (average)	(15) 87.4		(4) 94.2				53, 66, 85, 102
– beef heart	(4) 96.1		(4) 94.2				66
– beef liver	(5) 74.9						53, 66, 102
– beef spleen	(3) 94.0						53
– pork spleen	(3) 89.9						85, 102
Meat meal (average)	(11) 72.1		(1) 83.6			(3) 72.6	19, 43, 92
– poulty by-product	(2) 71.7		(1) 83.6			(2) 73.1	19, 92
– whale (fin)	(2) 64.9						43
– whale (sperm)	(6) 75.0						43

Feed	Protein (n)[a] ADC	Carbohyd (n) ADC	Fat (n) ADC	Fibre (n) ADC	Dry matter (n) ADC	Energy (n) ADC	Source
Okara[b]	(1) 44.3						43
Pond fauna (average)							
– chironomids	(1) 91.0					(1) 85.5	14, 62
– *Gammarus* sp.						(4) 81.0	28
– mysids (dried)	(19) 85.6						53, 69
– *Tubifex* sp.						(3) 78.2	13
Rapeseed oil meal	(1) 63.8		(1) 61.4			(1) 21.4	19
Silkworm pupa meal	(25) 86.9				(1) 71.1		41, 42, 53
– defatted	(7) 83.7						42, 52
Soybean meal (full fat)	(8) 65.0					(8) 66.9	92
Soybean oil meal	(11) 86.3	(1) 54.0			(2) 77.6	(3) 76.8	19, 44, 53, 92
Soybean protein (pure)	(1) 43.6						3
Starch– corn – raw	(14) 35.2						7, 19, 78, 90, 102
– corn – autoclaved	(6) 57.4						7, 19, 78, 90, 102
– potato – alpha	(1) 69.2						89
– pre-cooked	(4) 48.2						57
Sucrose	(7) 95.5						78, 89, 101
Wheat germ	(1) 76.8					(1) 60.0	92
Wheat middlings	(2) 80.3	(4) 31.5	(1) 89.1			(2) 39.9	19, 92, 102
Whey – low lactose (dried)	(2) 87.8		(1) 91.8			(2) 84.5	19, 92
Yeast – Brewers	(3) 76.1					(2) 67.8	3, 19, 92
– petroleum	(1) 91.6						3
Compounded feed[c]	(87) 86.5	(30) 52.0	(37) 89.9		(79) 64.5	(37) 75.3	7, 17, 19, 20, 29, 43, 48, 53, 55, 62, 70, 79, 80, 94, 96, 99, 104, 105, 106, 109, 110
(b) Common carp							
Alfalfa meal	(2) 95.0	(2) 81.0	(1) 95.0	(2) 50.0			10, 11
Algae (average)	(34) 65.8						4, 40, 81
– *Ankistrodesmus* (dried)	(1) 85.6						40
– *Euglena* spp.	(18) 63.5						40, 81
– *Oocystis* spp.	(6) 57.1						40, 81
– *Scenedesmus* spp.	(7) 72.4						40, 81
– *Spirulina maxima*	(1) 87.1						4
Alginic acid		(1) 53.0					65
Amino acids (average)	(132) 82.9						22
– arginine	(12) 89.6						22
– cystine	(12) 73.6						22
– histidine	(12) 82.7						22
– isoleucine	(12) 86.5						22
– leucine	(12) 87.2						22
– lysine	(12) 90.9						22
– methionine	(12) 88.2						22
– phenylalanine	(12) 87.2						22
– tyrosine	(12) 79.2						22
– tyrptophan	(12) 62.0						22
– valine	(12) 84.1						22
Bacteria – methanophilic	(2) 95.5						4, 73
Barley feed[d]	(1) 66.0	(1) 51.0	(1) 86.0	(1) 51.0			11
Barley – grain	(2) 79.7	(2) 74.0		(2) 31.3	(1) 57.7		11, 88
Blood meal – whale	(3) 84.7						96
Casein	(18) 98.3						4, 73
Cassava		(3) 95.5					103
Castor bean oil meal	(1) 91.0	(1) 81.0					11
Corn – grain	(1) 88.0	(1) 84.3	(1) 79.0	(1) 38.0			10, 11
Corn gluten meal	(3) 91.2						74
Cotton seed oil meal	(12) 77.0	(12) 39.4	(2) 87.0	(2) 67.0	(10) 46.0		11, 84, 85, 86, 87
Duckweed (*Lemna* sp.)	(1) 80.0						105
Egg – yolk	(13) 95.3						74
Fish meal (average)	(44) 90.3						
– brown (Peru)	(12) 79.9						40, 50, 81
– herring	(2) 80.3						4, 74
– whitefish	(27) 95.0						51, 73
Gelatine	(3) 96.8						73
Lupin							
– bitter	(2) 97.5	(2) 80.5	(2) 86.5	(2) 78.0			10, 11
– sweet	(3) 96.0	(4) 80.7	(2) 91.0	(2) 81.5			10, 11, 60
Pea	(2) 71.8	(2) 60.0		(2) 40.8	(1) 44.3		11, 88
Peanut oil meal	(3) 92.1	(3) 73.9	(2) 90.5	(3) 57.2	(1) 62.4		10, 11, 88
Phalaris – seed	(1) 88.0	(1) 67.0		(1) 42.0			11
Pond fauna (average)					(8) 85.8		44
– chironomids					(5) 89.0		44
– cladocera					(1) 72.9		44
– copepoda					(2) 84.3		44
Rice meal	(2) 89.5	(5) 90.0	(2) 91.5	(2) 91.5			10, 11, 103
Rye – grain	(17) 72.8	(1) 67.5			(20) 80.9		60, 68
Safflower oil meal	(1) 90.0	(1) 42.0		(1) 25.0			11
Silkworm pupa meal	(1) 63.9						50
Soybean oil meal (defatted)	(2) 78.1	(2) 83.5	(2) 86.0	(2) 52.5	(1) 44.1	(1) 53.6	10, 11, 73, 81, 84
Soybean protein (pure)	(3) 84.7						4, 50, 74

Feed	Protein (n)[a] ADC	Carbohyd (n) ADC	Fat (n) ADC	Fibre (n) ADC	Dry matter (n) ADC	Energy (n) ADC	Source
Starch							
– potato – alpha		(9) 85.0					18, 65
– potato – beta		(15) 54					18
Sunflower oil meal	(11) 76.6	(10) 26.7	(2) 86.4	(2) 46.5	(8) 46.1		84, 85, 86, 87
Tobacco – seed	(1) 92.0	(1) 71.0		(1) 71.0			11
Wheat bran	(1) 92.0	(2) 76.5	(2) 78.5	(2) 68.0			10, 11
Wheat germ	(4) 93.6						50, 74
Wheat – grain	(1) 85.8			(1) 51.3			85
Xylan		(1) 66.0					65
Yeast – petroleum	(13) 90.9						4, 74
Compounded feed[c]	(36) 87.2	(18) 73.7	(14) 82.8	(15) 60.6	(13) 88.2	(12) 91.6	22, 33, 46, 51, 65, 74, 87, 88, 98, 111
(c) Channel catfish							
Alfalfa meal (dehydrated)	(3) 12.5		(1) 50.5		(1) 15.7		21, 58
Casein	(8) 92.6						58, 90, 94
Cellulose				(6) 9.0			90, 94
Corn – grain							
– raw		(1) 66.0	(1) 76.0			(1) 26.1	58
– cooked		(1) 78.0	(1) 96.0			(1) 58.5	58
Cotton seed meal		(1) 17.0	(1) 81.0			(1) 56.2	58
Dextrin		(1) 73.0					58
Distillers solubles (dry)	(1) 67.0						58
Feather meal	(2) 68.5		(1) 83.0			(1) 66.6	58
Fish meal (average)	(5) 85.2					(1) 84.5	58
– anchovy	(2) 87.5		(1) 97.0				58
– menhaden	(2) 80.5						58
Glucose		(2) 90.0					58
Meat – fresh	(1) 82.0						21
Meat meal	(1) 75.0					(1) 80.5	58
Oil							
– corn			(6) 97.3				90
– fish			(1) 97.0				58
Rice bran	(1) 71.0						58
Soybean oil meal	(4) 79.2		(1) 81.0			(1) 56.4	58
Starch – corn – raw		(10) 82.6					90, 94
Wheat bran	(1) 82.0					(1) 56.2	58
Wheat – grain	(1) 83.8	(1) 59.0	(1) 96.0			(1) 60.4	58
Wheat shorts	(2) 72.8						58
Compounded feed[c]	(14) 91.3	(6) 80.7	(6) 97.3				58, 76, 93
(d) Other fishes							
European eel, *Anguilla anguilla*							
Bacteria – methanophilic	(1) 89.0				(1) 88.0		83
Casein	(1) 99.0				(1) 98.0		83
Fish meal	(1) 94.0				(1) 87.0		83
Gelatine	(1) 94.0						83
Soybean oil meal	(1) 94.0				(1) 68.0		83
Soybean protein (pure)	(1) 96.0				(1) 76.0		83
Compounded feed[c]					(2) 87.4		82
Japanese eel, *Anguilla japonica*							
Fish – fresh	(3) 92.0						73
Starch							
– potato – alpha		(1) 88.0					73
– potato – beta		(1) 52.0					73
Blennius pholis							
Squid					(1) 96.0	(1) 96.0	107
Goldfish, *Carassius auratus*							
Algae – *Chlorella* spp.	(14) 52.1						69
Cambaroides sp.	(5) 62.8						36
Fish meal (average)	(6) 67.4						35
– sardine	(1) 90.4						35
Meat – fresh (average)	(4) 95.0						34
– horse	(4) 95.0						34
Silkworm pupae meal	(13) 92.1						35, 36, 69
– defatted	(6) 88.0						35, 36
Taeuita sp.	(9) 67.6						36
Tecticepts sp.	(1) 71.0						36
Compounded feed[c]	(5) 91.4						36
Milkfish, *Chanos chanos*							
Algae (average)					(2) 42.5		100
– Diatoms					(1) 50.0		100
Red Sea bream, *Chrysophris major*							
Fishmeal (average)	(7) 78.3						31, 32
– whitefish	(7) 78.3						31, 32
Glucose		(3) 63.7					32
Compounded feed[c]	(3) 87.5	(3) 71.9					33

Feed	Protein (n)[a] ADC	Carbohyd (n) ADC	Fat (n) ADC	Fibre (n) ADC	Dry matter (n) ADC	Energy (n) ADC	Source
Cichlasoma bimacularum							
Tubifex						(5) 82.1	108
African catfish, *Clarias gariepinus*							
Compounded feed[c]	(11) 79.7		(6) 86.8		(5) 50.2	(6) 57.9	39, 59
Sculpin, *Cottus perplexus*							
Chironomids						(1) 81.9	14
Grass carp, *Ctenopharyngodon idella*							
Copra cake	(1) 65.4		(1) 100		(1) 9.1	(1) 18.7	56
Duckweed (*Lemna* sp.)					(2) 59.0	(1) 61.0	105
Fish meal	(1) 90.8	(1) 37.4	(1) 100		(1) 68.1	(1) 83.4	56
Grass meals							
– Carpet grass meal	(1) 73.4		(1) 40.1		(1) 21.0	(1) 22.1	56
– Napier grass meal	(1) 75.5		(1) 93.8		(1) 16.7	(1) 11.3	56
Lettuce						(1) 86.8	30
Maize	(1) 50.6	(1) 87.9	(1) 19.5		(1) 64.8	(1) 21.5	56
Rice bran	(1) 71.1		(1) 73.4		(1) 4.8	(1) 21.5	56
Soy bean meal	(1) 96.2	(1) 63.4	(1) 98.8		(1) 81.8	(1) 82.7	56
Cynoglossus sp.							
Polychaetes					(1) 81.0	(1) 95.0	26
Red hind, *Epinephelus guttatus*							
Fish – fresh						(3) 94.6	64
Cod, *Gadus morhua*							
Fish – fresh						(1) 98.7	27
Etroplus suratensis (Cichlidae)							
Hydrilla verticellata	(12) 64.3		(12) 67.2		(12) 41.3		23
Haplochromis nigripinnis							
Algae (average)					(10) 55.8		67
– *Microcystis* spp.	(7) 71.0						67
Histrio histrio							
Decapods						(3) 76.2	91
Roho, *Labeo rohita*							
Compounded feed[c]	(2) 93.3	(2) 65.2	(2) 90.0	(2) 15.9			46
Green sunfish, *Lepomis cyanellus*							
Mealworm (*Tenebrio molitor*)	(1) 95.7						37
Bluegill sunfish, Lepomis macrochirus							
Mealworm (*Tenebrio molitor*)	(2) 96.7						38
Longear sunfish, *Lepomis*							
Mealworm (*Tenebrio molitor*)	(1) 97.4						37
Megalops cyprinoides							
Fish – fresh	(2) 94.6				(1) 92.6		77
Shrimp – *Metapenaeus monoceros*	(2) 97.2					(2) 92.8	77
Largemouth bass, *Micropterus salmoides*							
Fish – fresh	(2) 96.5				(3) 80.7	(5) 90.6	6, 9
Mudfish, *Ophicephalus striatus*							
Shrimp – *Metapenaeus monoceros*	(2) 97.1					(2) 90.7	77
Blue Tilapia, *Oreochromis aureus*							
Compounded feed[c]					(20) 70.8		17
Oreochromis mossambicus							
Algae							
– *Spirulina maxima* (dry)					(1) 78.5		63
Fish meal							
– whitefish	(1) 90.0						45
Meat – fresh							
– goat liver					(1) 94.5		63
Pond fauna and algae	(3) 43.9	(3) 53.0	(3) 45.7		(3) 37.5		25
Periphytic detritus	(1) 76.9	(1) 63.4			(1) 62.2		12
Tadpoles					(1) 93.5		63
Compounded feed[c]					(28) 72.3		17
Nile tilapia, *Oreochromis niloticus*							
Algae (average)	(1) 72.3				(31) 58.8		1, 67
– *Anabaena* spp.					(4) 75.0		67
– *Chlorella* spp.					(4) 75.0		67
– *Cladophora glomerata*	(1) 72.3						1

– *Microcystis* spp.					(6) 70.0		67
– *Mitzschia* spp.					(2) 79.0		67
Fish meal – herring	(1) 87.4						1
Compounded feed[c]	(12) 82.5				(26) 66.4	(12) 77.5	24
Perch, *Perca fluviatilis*							
Gammarus pulex					(2) 86.1		95
Oligochaetes	(1) 96.3						8
Plaice, *Pleuronectes platessa*							
Polychaetes	(1) 91.8						27
Roach, *Rutilus rutilus*							
Algae – periphyton	(2) 65.7				(2) 39.9		47
Pond fauna – copepods	(4) 83.1				(4) 76.7		47
Yellowtail, *Seriola quinqueradiata*							
Fish – fresh	(1) 91.0						54
Fish meal (average)	(2) 66.0						54
– whitefish	(1) 58.0						54
Starch – potato – alpha		(4) 84.5					54
Rabbit fish, *Siganus spinus*							
Algae					(2) 26.0		15
Sole, *Solea solea*							
Oligochaetes	(1) 96.0						8
Polychaetes	(1) 85.0						27
Pike-perch (walleye), *Stizostedion vitreum*							
Fish – fresh					(2) 97.4	(18) 96.2	8, 49
– Perch						(14) 96.0	49
– Emerald shiner						(4) 97.0	49
Amphipods – *Gammarus lacustris*						(8) 81.0	49
Crayfish – *Orconectes virilis*						(4) 82.0	49
Tilapia melanopleura							
Macrophytes							
– *Elodea canadensis*	(2) 50.0			(2) 50.0			61
– *Spirodella polyrrhiza*	(2) 48.5			(2) 60.0			61
Tilapia zillii							
Macrophytes – *Najas quadalupensis*	(1) 75.1		(1) 75.9			(1) 45.4	16

Notes: [a] Number of analyses.
[b] A residue of soybean from which soluble proteins are extracted.
[c] Various diet compositions, most of which contain quite a high proportion of fish meal.
[d] A by-product of barley treated industrially for human consumption, contains a relatively large amount of hulls.

Sources: 1. Appler & Jauncey; 1983; 2. Asgard & Austreng, 1985; 3. Atack & Matty, 1979; 4. Atack *et al.*, 1979; 5. Averett, 1969 (quoted from Brett & Groves, 1979); 6. Beamish, 1972; 7. Bergot & Breque, 1983; 8. Birkett, 1969; 9. Blackburn, 1968 (quoted from Brett & Groves, 1979); 10. Bondi & Spandorf, 1953; 11. Bondi *et al.*, 1957; 12. Bowen, 1981; 13. Brocksen & Bugge, 1974; 14. Brocksen *et al.*, 1968; 15. Bryan, 1975; 16. Buddington, 1979; 17. Buddington, 1980; 18. Chiou & Ogino, 1975; 19. Cho & Slinger, 1979; 20. Cho *et al.*, 1974; 21. Cruz, 1967; 22. Dabrowski & Wojno, 1978; 23. De Silva & Perera, 1983; 24. De Silva & Perera, 1984; 25. De Silva *et al.*, 1984; 26. Edwards *et al.*, 1971; 27. Edwards *et al.*, 1972; 28. Elliot, 1976; 29. Ellis & Smith, 1984; 30. Fischer, 1970; 31. Furuichi & Yone, 1971a; 32. Furuichi & Yone, 1971b; 33. Furuichi & Yone, 1982b; 34. Furukawa & Ogasawara, 1952; 35. Furukawa & Ogasawara, 1953; 36. Furukawa *et al.*, 1953; 37. Gerking, 1952; 38. Gerking, 1955a; 39. Henken *et al.*, 1985; 40. Hepher *et al.*, 1979; 41. Hirao *et al.*, 1960; 42. Inaba *et al.*, 1962; 43. Inaba *et al.*, 1963; 44. Jablonski (quoted from Karsinkin, 1935); 45. Jauncey, 1982; 46. Jayaram & Shetty, 1980; 47. Karsinkin, 1935; 48. Kaushik, 1980; 49. Kelso, 1972; 50. Kim, 1974; 51. Kirchgessner *et al.*, 1986; 52. Kitamikado *et al.*, 1964a; 53. Kitamikado *et al.*, 1964b; 54. Kitamikado *et al.*, 1965; 55. Lall & Bishop, 1979; 56. Law, 1986; 57. Lindsay *et al.*, 1984; 58. Lovell, 1977; 59. Machiels & Henken, 1985; 60. Mann, 1948; 61. Mann, 1967; 62.

Mann, 1969; 63. Mathavan *et al.*, 1976; 64. Menzel 1960; 65. Migita *et al.*, 1937; 66. Morgulis, 1918; 67. Moriarty & Moriarty, 1973; 68. Nehring, 1963; 69. Nose, 1960; 70. Nose, 1967. 71. Nose & Mamiya, 1963. 72. Nose & Tayama, 1966. 73. Ogami, 1964. 74. Ogino & Chen, 1973; 75. Ogino & Nanri, 1980; 76. Page & Andrews, 1973; 77. Pandian, 1967; 78. Phillips *et al.*, 1948; 79. Refstie & Austreng, 1981; 80. Rychly & Spannhof, 1979; 81. Sandbank, 1979; 82. Schmitz *et al.*, 1983; 83. Schmitz *et al.*, 1984; 84. Shcherbina, 1964; 85. Shcherbina, 1975; 86. Shcherbina & Kazlauskene, 1971; 87. Shcherbina & Sorvachev, 1967; 88. Shcherbina *et al.*, 1970; 89. Singh & Nose, 1967; 90. Smith, 1971; 91. Smith, 1973; 92. Smith, 1976; 93. Smith & Lovell, 1971; 94. Smith & Lovell, 1973; 95. Solomon & Brafield, 1972; 96. Storebakken, 1985; 97. Syazuki *et al.*, 1953; 98. Syazuki *et al.*, 1956; 99. Tacon & Rodrigues, 1984; 100. Tang & Hwang, 1966; 101. Tunison *et al.*, 1939; 102. Tunison *et al.*, 1942; 103. Ufodike & Matty, 1983; 104. Ulla & Gjedrem, 1985; 105. Van Dyke & Sutton, 1977; 106. Vens-Cappell, 1984; 107. Wallace, 1973; 108. Warren & Davis, 1967; 109. Watanabe *et al.*, 1979; 110. Windell *et al.*, 1978; 111. Zeitler *et al.*, 1983.

Appendix II

Essential amino acid composition of proteins of fishes, fish eggs and some feeds ingredients (in % of total amino acids), and calculated chemical scores and EAAIs of the feeds as related to fish egg protein.

Fish species	Amino acids Arg	His	Iso	Leu	Lys	Met	Cys	Phe	Thy	Thr	Try	Val	Source
(a) fish muscle proteins													
Anarichas lupus	5.8	2.1	6.1	7.8	8.9	2.9	1.1	4.0	3.2	5.0	1.0	5.7	2
Brosme brosme	6.3	2.0	6.3	9.1	9.4	3.2	1.0	4.1	3.4	4.6	1.0	6.2	2
Clupea harengus	7.3	3.7	5.1	9.6	10.3	2.0	1.4	4.7	4.1	5.3	1.4	6.0	4
Cyprinus carpio	7.1	3.1	6.2	7.7	6.2	3.2	1.7	4.5	—	4.9	1.3	5.6	7
	6.0	2.7	—	—	8.9	3.1	1.9	4.1	—	5.3	1.3	5.0	10
	6.0	2.2	5.1	9.2	11.6	3.3	—	5.1	3.8	5.9	1.1	6.8	9
Engraulis japonicus	6.5	2.6	5.4	9.5	10.3	3.4	1.0	4.2	3.8	5.2	1.3	5.5	1
Gadus morhua	6.8	2.3	4.3	8.8	10.3	3.5	1.2	5.0	4.0	5.1	—	5.2	11
	6.7	3.5	4.9	9.3	10.3	2.1	1.4	4.7	4.0	5.3	1.3	5.8	4
	5.8	1.8	5.5	8.1	8.5	3.1	0.9	3.7	3.3	4.5	0.9	5.6	2
Hippoglossus hippoglossus	6.0	2.2	5.9	8.1	7.9	2.8	1.1	3.8	3.3	4.6	1.0	6.0	2
Koreius bicoloratus	6.2	2.5	5.1	9.0	10.3	3.1	—	4.4	4.2	5.6	1.2	7.0	5
Solea nasuta	6.8	3.6	5.5	8.7	10.7	3.4	1.4	4.9	4.1	5.7	1.6	5.9	12
Melanogrammus aeglefinus	7.0	3.7	4.7	9.3	11.0	3.2	1.4	4.7	4.3	5.8	1.4	5.7	2
Molva molva	5.9	1.9	6.1	9.0	9.5	3.1	1.1	4.1	3.3	4.6	0.9	6.1	2
Murenoesox cinereus	6.7	2.8	7.3	8.3	11.3	3.5	—	4.6	3.8	5.9	1.2	5.8	9
Mustelus manazo	7.0	2.9	7.7	8.3	10.8	3.0	—	4.4	3.6	6.2	1.3	6.1	9
Oncorhynchus masou	6.0	2.3	3.8	7.3	8.5	3.0	1.3	4.5	3.5	4.5	0.8	4.7	14
Oncorhynchus nerka	5.8	2.9	5.2	9.1	10.6	3.4	—	4.8	4.1	6.0	1.3	7.4	9
Pacrosomus major	7.0	2.4	6.4	9.1	10.0	3.7	—	5.0	4.4	5.7	1.4	6.9	9
Pleuronectes platessa	6.3	2.1	5.9	8.3	8.7	2.8	1.1	3.9	3.3	4.6	1.0	6.0	2
Pollachius virens	6.0	2.0	6.2	8.1	9.1	3.0	1.0	3.8	3.3	4.4	1.0	6.1	2
	5.7	2.0	6.1	8.2	8.5	2.8	1.0	4.0	3.4	4.4	1.0	6.0	12
Salmo gairdneri	5.9	3.1	3.7	6.7	8.0	2.6	0.7	4.0	3.4	3.3	—	4.2	16
Salmo salar	5.9	3.3	5.1	7.8	8.6	3.3	0.6	4.5	3.1	3.9	—	5.6	8
(migrating)	12.6	4.7	3.8	6.3	12.3	1.9	1.0	2.5	2.0	3.8	1.0	4.8	5
(spent)	12.1	4.3	3.5	5.8	13.0	2.0	1.0	2.7	1.9	3.9	0.9	4.7	5
Sardina melanosticta	6.9	2.5	6.0	9.3	11.0	3.6	—	4.7	4.4	5.6	1.3	7.4	9
Sebastes marinus	6.0	2.0	6.3	8.6	9.2	3.0	1.1	4.0	3.2	4.8	1.0	5.9	2
Scomber scombrus	5.7	4.5	6.0	8.9	8.5	3.1	1.0	3.9	3.3	4.7	0.9	6.1	2
Theragra chalcoramma	7.0	2.4	7.2	8.3	11.0	3.6	—	4.3	3.0	5.8	1.2	5.7	14
Trachurus japonicus	6.8	2.3	6.1	9.1	10.7	3.6	—	4.9	4.6	5.6	1.3	7.1	9
Average	6.8	2.8	5.6	8.4	9.8	3.0	1.2	4.2	3.6	5.0	1.2	5.9	
(b) Chicken, whole egg													
	6.4	2.4	6.3	8.8	6.8	3.2	2.4	5.7	4.2	4.9	1.6	7.2	13
(c) Fish eggs													
Ictalurus punctatus	5.4	2.2	5.1	9.7	7.7	3.4	—	3.9	3.9	6.0	—	5.9	8
Menidia menidia	6.5	4.7	5.0	7.7	7.9	3.9	0.5	3.6	4.6	5.2	—	7.2	18
Oncorhynchus gorbuscha	7.2	2.9	6.9	9.4	8.9	3.0	—	4.9	—	5.1	1.1	8.1	19
Oncorhynchus kisutch	7.0	2.8	7.5	10.0	8.8	2.7	—	4.9	—	5.9	0.9	7.1	19
Oncorhynchus masou	6.2	2.7	5.6	9.5	8.8	3.1	1.8	6.0	5.3	4.9	0.8	6.6	14
Oncorhynchus nerka	7.2	2.7	7.5	10.2	8.5	2.8	—	4.8	—	5.8	0.9	7.3	19
Oncorhynchus tschawytcha	7.7	2.6	6.8	9.4	8.8	3.0	—	4.8	—	5.8	0.9	7.0	19

Salmo gairdneri	5.7	2.5	4.7	9.5	7.3	2.9	1.0	5.5	4.2	4.8	1.0	6.2	16
	5.7	2.5	4.7	9.5	7.3	2.9	1.0	5.5	4.2	4.8	1.0	6.2	20
	7.6	2.8	4.7	7.6	8.3	2.1	—	5.2	—	6.0	—	8.9	17
Salmo salar	6.4	2.7	6.4	10.3	8.8	2.7	1.8	5.3	1.1	5.9	1.1	7.9	6
(immature eggs)	12.8	4.5	4.3	6.9	10.5	1.6	1.3	2.8	2.3	4.3	1.0	5.9	5
Stizostedion vitreum	5.7	2.9	6.4	8.3	7.8	3.1	—	4.8	4.1	5.2	1.2	7.1	8
Average	7.0	3.0	5.8	9.1	8.4	2.9	1.2	4.8	3.7	5.4	1.0	7.0	

(d) Natural food organisms[a]

	Amino acids													
Organism	Arg	His	Iso	Leu	Lys	Met	Cys	Phe	Thy	Thr	Try	Val	*CS*[c]	*EAAI*[c]
Algae:														
Cyanophyta	5.4	[0.7]	6.5	9.8	6.2	0.7	0.2	4.8	3.9	6.5	—	5.9	23.3	69.2
Diatoms	6.9	[1.6]	4.6	7.1	5.0	2.3	—	4.7	3.7	4.3	1.3	5.3	53.3	78.1
Aquatic vegetation	5.0	1.9	5.1	9.3	5.4	[1.2	0.3]	6.7	4.0	5.1	-	5.9	36.6	77.0
Rotifers:														
Branchionus plicatilis	4.6	1.5	3.5	6.0	5.9	[0.9	0.8]	3.8	3.0	—	1.3	6.2	41.5	68.9
Oligochaetes	8.7	2.6	4.8	11.1	9.6	2.5	1.5	4.3	3.9	4.5	1.4	[4.6]	65.7	98.9
Crustaceans:	7.1	2.2	4.2	[5.2]	6.7	1.8	1.2	4.6	4.8	4.3	1.1	4.9	57.1	81.0
Anostraca:														
Artemia salina	5.0	1.3	2.6	6.1	6.1	[0.9	0.4]	3.2	3.7	—	1.0	3.2	31.7	58.4
Cladocera:	5.5	[1.9]	4.6	7.2	7.4	2.0	0.7	4.2	3.7	4.2	1.3	5.2	63.3	81.3
Daphnia sp.	5.7	[1.9]	5.6	7.4	7.5	2.5	—	4.2	3.6	—	1.3	6.2	63.3	85.1
Moina sp.	5.1	1.6	[2.5]	6.0	5.8	1.0	0.6	3.6	3.3	—	1.2	3.2	43.1	61.8
Copepoda:	6.2	2.8	[4.9]	7.7	7.7	2.3	1.3	4.1	6.2	4.9		5.8	84.5	91.1
Tigriopus japonicus	5.2	1.6	[2.5]	5.0	5.7	1.1	0.7	3.5	4.0	—	1.1	3.2	43.1	61.4
Gammaridae	6.7	2.6	4.5	[5.6]	6.4	1.3	3.5	5.2	—	4.0	1.1	4.1	61.5	83.3
Insects:	5.1	2.9	4.3	[4.9]	5.4	1.3	3.2	4.1	—	3.5	1.2	4.7	53.8	78.8
Ephemeridae (Mayflies):														
(*Hexagenia bilineata*)	6.9	2.7	5.4	[5.9]	6.2	1.4	3.6	4.6	—	4.1	1.3	5.4	64.8	90.0
Odonata (Dragonflies)	7.5	4.9	3.8	6.8	6.6	[1.1	0.7]	3.8	7.1	4.7		6.5	43.9	87.7
Trichoptera (Caddisflies)	[2.3]	6.3	4.5	11.0	—	1.8	5.7	5.5	—	3.5	1.2	2.9	32.9	90.4
Chironomidae	9.0	2.7	5.3	13.0	8.8	[2.4]	—	5.1	3.8	—	1.4	6.0	58.3	100.3
Acortia clausi	4.3	1.9	3.5	5.5	5.4	[1.5	0.8]	3.7	3.6	—	1.1	4.5	56.1	68.0
Average	5.8	2.0	4.0	6.6	6.2	1.5	1.7	4.2	3.6	3.9	1.2	4.7		

Sources: 3, 15, 21.

(e) Feedstuffs[a]

Casein	[4.2]	3.1	6.8	10.5	8.5	3.3	0.4	5.7	5.8	4.7	1.3	8.0	60.0	103.0
Corn – grain	4.4	2.7	3.6	12.6	[2.6]	1.7	2.3	5.0	3.9	3.7	0.8	4.6	30.9	74.9
Fish meal (average)	6.2	[2.4]	4.6	7.3	7.6	3.1	1.0	4.0	3.2	4.1	1.1	5.4	74.7	85.5
Anchovy	5.8	2.5	4.8	7.6	7.7	3.8	0.9	4.3	3.4	4.2	1.2	[5.4]	77.1	89.4
Herring	6.4	2.3	4.4	7.2	7.5	2.9	1.0	3.8	3.1	[4.0]	1.1	6.0	74.1	85.1
Menhaden	6.1	[2.4]	4.7	7.3	7.7	2.9	0.9	4.0	3.2	4.1	1.1	5.3	74.7	84.8
White FM	6.5	[2.2]	4.4	7.0	7.3	2.7	1.2	3.7	2.9	4.1	1.1	4.9	73.3	82.6
Poultry by-product meal	6.4	1.7	4.0	6.8	4.9	1.8	1.6	[3.1	1.6]	3.3	0.8	4.9	55.2	69.0
Rice bran	5.6	1.8	3.6	5.5	[3.8]	1.8	0.8	3.5	5.4	3.3	0.8	5.4	45.2	67.7
Rye – grain	4.4	2.1	3.8	5.8	[3.5]	1.4	1.5	4.6	2.2	3.0	0.9	4.6	41.7	65.4
Shrimp meal	6.3	2.4	4.2	6.7	[5.5]	2.1	1.5	4.0	0.7	3.6	0.9	4.6	65.5	73.3
Sorghum – grain	3.5	2.1	4.0	12.9	[2.3]	1.2	1.8	5.0	3.7	3.2	1.0	4.7	27.4	69.9
Soybean meal	6.8	2.4	4.6	7.3	6.0	1.2	1.7	4.7	3.0	3.7	1.4	[4.5]	64.3	81.5
Wheat – grain	5.1	2.4	4.0	6.9	[2.9]	1.7	2.5	4.9	3.4	2.9	1.3	4.7	34.5	73.5
Yeast														
– Brewers	5.0	2.5	5.1	7.4	7.1	[1.7	1.1]	4.2	3.4	4.8	1.2	5.4	68.3	84.3
– Torula	5.4	2.7	5.8	7.2	7.6	[1.6	1.2]	5.8	4.1	5.4	1.1	6.0	68.3	90.6

Notes: [a] First limiting amino acid relative to fish egg protein is given in brackets.

[b] Calculated from tables in source 13. First limiting amino acid relative to fish egg protein is given in brackets.

[c] Chemcial Score and Essential Amino Acid Index relate to fish egg protein.

Source: 1. Arakaki & Syama, 1966 (quoted from Maslennikova, 1974); 2. Braekkan & Boge, 1962 (quoted from Maslennikova, 1974); 3. Block, 1959 (quoted from Lieder, 1965*a*); 4. Connel & Howgate, (quoted from Maslennikova, 1974); 5. Cowey *et al.*, 1962; 6. Cowey *et al.*, 1972; 7. Dupont, 1958 (quoted from Lieder, 1965a); 8. Ketola, 1982; 9. Konossu *et al.*, 1956 (quoted from Maslennikova, 1974); 10. Lindner *et al.*, 1960 (quoted from Lieder, 1965a), 11. Maslennikova, 1968 (quoted from Maslennikova, 1974); 12. Maslennikova, 1974; 13. NRC, 1983; 14. Ogata *et al.*, 1983; 15. Ogino, 1963; 16. Rumsey & Ketola, 1975; 17. Satia *et al.*, 1974; 18. Schauer *et al.*, 1979; 19. Seagran *et al.*, 1954; 20. Suyama & Ogino, 1958; 21. Watanabe *et al.*, 1978b.

Appendix III

Proximate analyses of organisms serving as food for pond fishes.

		Dry matter composition[a]					
Organism group	Dry matter (n)[b] %	Protein (n) %	Carbohydrate (n) %	Lipid (n) %	Ash (n) %	Energy (n) kcal/kg	Source
Bacteria					(2) 5.4	(2) 4710	5
Algae							
Cyanophyta (blue greens)		(1) 31.3			(5) 46.7	(8) 2213	5
Chlorophyta (greens)	(13) 16.8	(8) 17.6		(7) 3.7	(21) 26.9	(29) 3773	2, 5, 11, 14, 21
Phaeophyta (brown algae)	(15) 14.1				(24) 32.3	(25) 3056	5
Bacillariacaea (diatoms)		(4) 30.7		(3) 9.9	(5) 38.3	(4) 3654	5, 10, 15, 19
Rhodophyta (red algae)	(21) 21.7				(39) 32.1	(39) 3170	5
Aquatic macro-vegetation	(51) 15.8	(64) 14.6		(29) 4.5	(62) 13.9	(107) 3906	2, 3, 5, 14
Protozoa						(1) 5938	5
Rotifers	(15) 11.2	(11) 64.3		(13) 20.3	(5) 6.2	(4) 4866	13, 17
Oligochaetes	(4) 7.3	(2) 49.3		(1) 19.0	(1) 5.8	(9) 5569	5, 10, 14
Leeches	(1) 24.0	(2) 61.0			(2) 5.1	(3) 5432	5, 7, 12
Crustacea							
Anostraca (*Artemia*)	(6) 11.0	(6) 61.6		(6) 19.5	(6) 10.1	(2) 5835	5, 17
Cladocera	(18) 9.8	(30) 56.5	(20) 28.2	(28) 19.3	(57) 7.7	(75) 4800	5, 6, 8, 9, 10, 11, 12, 13, 16, 17, 20
Copepoda	(10) 10.3	(35) 52.3	(36) 9.2	(29) 26.4	(29) 7.1	(39) 5445	5, 6, 8, 12, 13, 14, 16, 18, 21
Ostracoda	(2) 35.0	(1) 41.5				(2) 5683	5, 14
Malacostraca (higher Crustacea)	(32) 24.6	(16) 49.9	(7) 18.4	(10) 20.3	(50) 19.6	(112) 5537	4, 5, 6, 7, 8, 9, 12, 13, 14, 18, 20, 21
Insects	(1) 23.2	(1) 55.94	(1) 20.1	(1) 18.6	(6) 4.9	(6) 5075	9, 20
Plecoptera (stone flies)						(1) 4900	1
Ephemeridae (mayflies)	(3) 17.6	(5) 50.2			(8) 3.7	(19) 5646	1, 5, 7, 12, 14, 15, 20
Odonata (dragon flies)	(6) 21.1	(4) 51.9			(12) 5.8	(14) 4985	5, 7, 12, 14
Hemiptera (water bugs)	(2) 26.0	(2) 68.8				(2) 5150	7
Trichoptera (caddisflies)	(2) 14.8	(3) 34.7			(6) 11.8	(11) 5019	1, 5, 12, 14
Diptera	(1) 16.0	(1) 55.3			(1) 6.9	(3) 5177	12
Chironomids (larvae)	(5) 19.1	(4) 59.0	(1) 22.5	(2) 4.9	(17) 5.8	(31) 5034	1, 5, 7, 9, 10, 12, 14, 21
Molluscs	(15) 32.2	(19) 39.5	(15) 7.5	(15) 7.8	(26) 32.9	(84) 3889	1, 4, 5, 14
Aquatic detritus	(59) 91.5				(68) 12.4	(67) 4701	5

Notes: [a] Since the values are averages of figures collected from different sources they do not necessarily add up to 100%.
[b] The numbers in brackets are those of data collected from the literature. Each may be an average of a number of analyses.

Sources: 1. Alimov & Shadrin, 1977; 2. Boyd, 1968; 3. Boyd & Goodyear, 1971; 4. Conover, 1978; 5. Cummins & Wuycheck, 1971; 6. Dabrowski & Rusiecki, 1981; 7. Driver *et al.*, 1974; 8. Farkas, 1958; 9. Lieder, 1965a; 10. Ogino, 1963; 11. Richman, 1958; 12. Salonen *et al.*, 1976; 13. Schindler *et al.*, 1971; 14. Stiaramaiah, 1967; 15. Trama, 1957; 16. Vijverberg & Frank, 1965; 17. Watanabe *et al.*, 1978a; 18. Watanabe *et al.*, 1978b; 19. Wikfors, 1986; 20. Wissing & Hasler, 1968; 21. Wissing & Hasler, 1971.

Appendix IV

Feeding charts for various fish species cultured in ponds or intensive systems.

(a) Recommended amounts of food to feed rainbow trout (percentage of body weight per day) for different sizes and water temperatures at intensive culture systems

Water Temp. (°C)	Weight of fish in grams: 0.18 or less	0.18–1.49	1.50–5.25	5.26–11.9	12.0–23.1	23.3–38.4	38.5–62.4	62.5–90.8	90.9–124	125–166	167 or more
5.5	3.5	2.8	2.4	1.8	1.4	1.2	0.9	0.8	0.7	0.6	0.5
6.0	3.6	3.0	2.5	1.9	1.4	1.2	1.0	0.9	0.8	0.7	0.6
6.5	3.8	3.1	2.5	2.0	1.5	1.3	1.0	0.9	0.8	0.8	0.6
7.0	4.0	3.3	2.7	2.1	1.6	1.3	1.1	1.0	0.9	0.8	0.7
7.5	4.1	3.4	2.8	2.2	1.7	1.4	1.2	1.0	0.9	0.8	0.7
8.0	4.3	3.6	3.0	2.3	1.7	1.4	1.2	1.0	0.9	0.8	0.7
9.0	4.5	3.8	3.0	2.4	1.8	1.5	1.3	1.1	1.0	0.9	0.8
9.5	4.7	3.9	3.2	2.5	1.9	1.5	1.3	1.1	1.0	0.9	0.8
10.0	5.2	4.3	3.4	2.7	2.0	1.7	1.4	1.2	1.1	1.0	0.9
10.5	5.4	4.5	3.5	2.8	2.1	1.7	1.5	1.3	1.1	1.0	0.9
11.0	5.4	4.5	3.6	2.8	2.1	1.7	1.5	1.3	1.1	1.0	0.9
11.5	5.6	4.7	3.8	2.9	2.2	1.8	1.5	1.3	1.1	1.1	1.0
12.0	5.8	4.9	3.9	3.0	2.3	1.9	1.6	1.4	1.3	1.1	1.0
13.0	6.1	5.1	4.2	3.2	2.4	2.0	1.6	1.4	1.3	1.1	1.0
13.5	6.3	5.3	4.3	3.3	2.5	2.0	1.7	1.5	1.3	1.2	1.0
14.0	6.7	5.5	4.5	3.5	2.6	2.1	1.8	1.5	1.4	1.2	1.1
14.5	7.0	5.8	4.8	3.6	2.7	2.2	1.9	1.6	1.4	1.3	1.2
15.0	7.3	6.0	5.0	3.7	2.8	2.3	1.9	1.7	1.5	1.3	1.2
15.5	7.5	6.3	5.1	3.9	3.0	2.4	2.0	1.7	1.5	1.4	1.3

Source: From Post (1975).

(b) Feeding chart for common carp cultured in ponds during summer (water temperature over 20 °C). Amounts are given as grams per fish

Fish weight (g)	Density of fish per hectare: 2000–4000		4000–6000		6000–8000		8000–12 000		12 000–15 000		15 000–20 000		20 000–50 000	
	g	% prot[a]	g	% prot[a]	g	% prot[a]	g	% prot[a]	g	% prot[a]	g	% prot[a]	g	% prot[a]
20–50	1	12	2	12	2	12	3	12	3	12	3	18	3	18
50–100	2	12	3	12	4	12	4	18	4	18	4	18	4	18
100–200	6	12	6	18	9	18	9	18	9	18	9	25	9	25
200–300	10	12	11	18	11	18	12	25	12	25	12	25	12	25
300–400	11	18	13	25	14	25	15	25	15	25	15	25	15	25
400–500	14	18	15	25	16	25	17	25	17	30	17	30	17	30
500–600	15	25	17	25	18	25	19	30	19	30	19	30	19	30
600–700	15	25	18	25	19	30	19	30	19	30	19	30	19	30
700–800	16	25	18	25	20	30	19	30	19	30	19	30	19	30
800–900	17	25	18	25	19	30	20	30	20	30	20	30	20	30
900–1000	17	25	19	25	20	30	21	30	21	30	21	30	21	30
1000–1100	18	25	20	30	21	30	22	30	22	30	22	30	22	30
1100–1200	18	25	20	30	21	30	22	30	22	30	22	30	22	30

Note: [a] Based on 4 pelleted diets of increasing protein content: 12, 18, 25 and 30%.

Source: Modified from Marek (1975).

(c) Feeding chart for tilapia cultured in semi-intensive, intensive and polyculture systems (in grams per fish per day)

Fish weight (g)	Polyculture[b]	Semi-intensive[a] (less than 20 000/ha)	Intensive[b] (more than 20 000/ha)
1–3	0.2	0.3	
3–5	0.3	0.4	
5–10	0.5	0.6	
10–20	0.8	1.0	
20–30	1.2	1.5	
30–50	1.4	2.0	
50–70	1.6	2.5	
70–100	1.8	3.0	1.8
100–150	2.3	3.5	2.7
150–200	2.8	4.0	3.7
200–250	3.5	5.0	4.5
250–300			5.5
300–350	4.3	6.0	6.3
350–400			6.7
400–500	5.0	7.0	7.2
500–600	6.0	8.0	8.3
600–700	7.0	9.0	

Notes: [a] From Marek (1974).
[b] From Zohar (1986).

(d) Feeding chart for channel catfish as percentage of body weight per day

Fish weight (g)	Temperature (°C)															
	15	16	17	18	19	20	21	22	23	24	25	26	27	28	29	30+
4	2.0	2.2	2.4	2.5	2.7	2.9	3.1	3.2	3.4	3.5	3.7	3.8	4.0	4.1	4.3	4.4
6	1.9	2.1	2.2	2.4	2.6	2.8	2.9	3.1	3.2	3.4	3.5	3.7	3.8	3.9	4.1	4.2
8	1.8	2.0	2.1	2.3	2.5	2.6	2.8	2.9	3.1	3.2	3.4	3.5	3.6	3.8	3.9	4.0
10	1.7	1.9	2.1	2.2	2.4	2.5	2.7	2.8	3.0	3.1	3.2	3.4	3.5	3.6	3.7	3.8
13	1.7	1.8	2.0	2.1	2.3	2.4	2.6	2.7	2.8	3.0	3.1	3.2	3.3	3.5	3.6	3.7
16	1.6	1.7	1.9	2.0	2.2	2.3	2.5	2.6	2.7	2.8	3.0	3.1	3.2	3.3	3.4	3.6
20	1.5	1.7	1.8	2.0	2.1	2.2	2.4	2.5	2.6	2.7	2.9	3.0	3.1	3.2	3.3	3.4
24	1.5	1.6	1.8	1.9	2.0	2.2	2.3	2.4	2.5	2.6	2.8	2.9	3.0	3.1	3.2	3.3
29	1.4	1.6	1.7	1.8	2.0	2.1	2.2	2.3	2.4	2.5	2.7	2.8	2.9	3.0	3.1	3.2
35	1.4	1.5	1.6	1.8	1.9	2.0	2.1	2.2	2.3	2.5	2.6	2.7	2.8	2.9	3.0	3.1
41	1.3	1.5	1.6	1.7	1.8	1.9	2.0	2.2	2.3	2.4	2.5	2.6	2.7	2.8	2.9	3.0
48	1.3	1.4	1.5	1.6	1.7	1.9	2.0	2.1	2.2	2.3	2.4	2.5	2.6	2.7	2.8	2.9
56	1.2	1.4	1.5	1.6	1.7	1.8	1.9	2.0	2.1	2.2	2.3	2.4	2.5	2.6	2.7	2.8
64	1.2	1.3	1.4	1.5	1.6	1.7	1.8	1.9	2.0	2.1	2.2	2.3	2.4	2.5	2.6	2.7
73	1.2	1.3	1.4	1.5	1.6	1.7	1.8	1.9	2.0	2.1	2.2	2.2	2.3	2.4	2.5	2.6
83	1.1	1.2	1.3	1.4	1.5	1.6	1.7	1.8	1.9	2.0	2.1	2.2	2.3	2.3	2.4	2.5
94	1.1	1.2	1.3	1.4	1.5	1.6	1.7	1.8	1.9	1.9	2.0	2.1	2.2	2.3	2.3	2.4
106	1.0	1.2	1.2	1.3	1.4	1.5	1.6	1.7	1.8	1.9	2.0	2.0	2.1	2.2	2.3	2.3
119	1.0	1.1	1.2	1.3	1.4	1.5	1.6	1.7	1.7	1.8	1.9	2.0	2.0	2.1	2.2	2.3
133	1.0	1.1	1.2	1.3	1.4	1.4	1.5	1.6	1.7	1.8	1.8	1.9	2.0	2.1	2.1	2.2
148	1.0	1.0	1.1	1.2	1.3	1.4	1.5	1.6	1.6	1.7	1.8	1.9	1.9	2.0	2.1	2.1
163	0.9	1.0	1.1	1.2	1.3	1.3	1.4	1.5	1.6	1.7	1.7	1.8	1.9	1.9	2.0	2.1
180	0.9	1.0	1.1	1.1	1.2	1.3	1.4	1.5	1.5	1.6	1.7	1.7	1.8	1.9	1.9	2.0
198	0.9	1.0	1.0	1.1	1.2	1.3	1.3	1.4	1.5	1.6	1.6	1.7	1.7	1.8	1.9	1.9
218	0.8	0.9	1.0	1.1	1.2	1.2	1.3	1.4	1.4	1.5	1.6	1.6	1.7	1.8	1.8	1.9
238	0.8	0.9	1.0	1.0	1.1	1.2	1.3	1.3	1.4	1.5	1.5	1.6	1.6	1.7	1.8	1.8
260	0.8	0.9	0.9	1.0	1.1	1.2	1.2	1.3	1.3	1.4	1.5	1.5	1.6	1.6	1.7	1.8
283	0.8	0.8	0.9	1.0	1.0	1.1	1.2	1.2	1.3	1.4	1.4	1.5	1.5	1.6	1.7	1.7
307	0.7	0.8	0.9	1.0	1.0	1.1	1.1	1.2	1.3	1.3	1.4	1.4	1.5	1.5	1.6	1.6
333	0.7	0.8	0.8	0.9	1.0	1.0	1.1	1.2	1.2	1.3	1.3	1.4	1.4	1.5	1.5	1.6
360	0.7	0.8	0.8	0.9	0.9	1.0	1.1	1.1	1.2	1.2	1.3	1.3	1.4	1.4	1.5	1.5
388	0.7	0.7	0.8	0.9	0.9	1.0	1.0	1.1	1.1	1.2	1.3	1.3	1.4	1.4	1.5	1.5
418	0.6	0.7	0.8	0.8	0.9	0.9	1.0	1.1	1.1	1.2	1.2	1.3	1.3	1.4	1.4	1.5
449	0.6	0.7	0.7	0.8	0.9	0.9	1.0	1.0	1.1	1.1	1.2	1.2	1.3	1.3	1.4	1.4
482	0.6	0.7	0.7	0.8	0.8	0.9	0.9	1.0	1.0	1.1	1.1	1.2	1.2	1.3	1.3	1.4
517	0.6	0.7	0.7	0.8	0.8	0.9	0.9	1.0	1.0	1.1	1.1	1.1	1.2	1.2	1.3	1.3
553	0.6	0.6	0.7	0.7	0.8	0.8	0.9	0.9	1.0	1.0	1.1	1.1	1.1	1.2	1.2	1.3

Source: From Foltz (1982).

References

Abel, H., Pieper, A. & Pfeffer, E. (1978). Vergleich der energetischen Verwertung von Protein und Kohlenhydraten bei Regenbogenforellen. *Z. Tierphysiol., Tierernaehrg., Futtermittelk.*, **40**, 127–8.

Abel, H., Pieper, A. & Pfeffer, E. (1979). Untersuchungen an wachsenden Regenbogenforellen (*Salmo gairdneri* R.) ueber die intermediaere Anpassung an Protein oder Kohlenhydrate als Energietraeger im Futter. *Z. Tierphysiol., Tierernaehrg., Futtermittelk.*, **41**, 325–34.

Ablett, R. F., Sinnhuber, R. O., Holmes, R. M. & Selivonchick, D. P. (1981). The effect of prolonged administration of bovine insulin in rainbow trout (*Salmo gairdneri R.*). *Gen. Comp. Endocrinol.*, **43**, 211–17.

Ablett, R. F., Taylor, M. J. & Selivonchick, D. P. (1983). The effect of high carbohydrate diets on [^{125}I]iodoinsulin binding in skeletal muscle plasma membranes and isolated hepatocytes of rainbow trout (*Salmo gairdneri*). *Br. J. Nutr.*, **50**, 129–39.

Ackman, R. G. (1967). Characteristics of the fatty acid composition and biochemistry of some fresh-water fish oils and lipids. *Comp. Biochem. Physiol.*, **22**, 907–22.

Adelman, I. R. & Smith, L. L. (1970). Effect of oxygen on growth and food conversion efficiency of northern pike. *Prog. Fish Cult.*, **32**, 93–106.

Adron, J. W., Blair, A., Cowey, C. B. & Shanks, A. M. (1976). Effects of dietary energy level and dietary energy source on growth, food conversion and body composition of turbot (*Scophthalmus maximus* L.). *Aquaculture*, **7**, 125–32.

Ahmad, M. M. & Matty, A. J. (1975). Effect of insulin on the incorporation of ^{14}C-leucine into goldfish muscle protein. *Biologia (Lahore)*, **21**, 119–24.

Aiso, K., Simidu, U. & Hasuo, K. (1968). Microflora in the digestive tract of inshore fish in Japan. *J. Gen. Microbiol.*, **52**, 361–4.

Alabaster, J. S. & Lloyd, R. (1980). *Water quality criteria for freshwater fish.* London: Butterworths, 297 pp.

Albertini-Berhaut, J. & Vallet, F. (1971). Utilization alimentaire de l'urée chez les muges. *Tethys*, **3**, 677–80.

Albrecht, M-L. (1972) Untersuchungen ueber die Auswirkungen eines Zusates von Lysin zu Mischfuttermittelen fuer Karpfen. *Z. Binnenfisch. DDR*, **19**(6), 181–4.

Albrecht, M.-L. & Breitsprecher, B. (1969). Untersuchungen ueber die chemische Zusammensetzung von Fischenaehrtieren und Fischfuttermitteln. *Z. Fisch., N.F.*, **17**, 143–63.

Aleev, Yu.G. (1963). Function and gross morphology in fish. *Izdatel'stvo Akad. Nauk SSSR, Moskva*, 222 pp. English transl: Israel Program for Sci. Transl., Jerusalem, 1969, 268 pp.

Al-Hussaini, A. H. (1947). The feeding habits and the morphology of the alimentary tract of some teleosts living in the neighbourhood of the Marine Biological Station, Ghardaqa,

Red Sea. *Publ. Mar. Biol. Stn, Fouad I Univ.*, **5**. 1–61.

Al-Hussaini, A. H. (1949). On the functional morphology of the alimentary tract of some fish in relation to their feeding habits: cytology and physiology. *Quart. J. Micros. Sci.*, **90**, 323–54.

Al Hussaini, A. H. & Kholy, A. A. (1953). On the functional morphology of the alimentary tract of some omnivorous teleost fish. *Proc. Egypt. Acad. Sci.*, **4**, 17–39.

Alimov, A. F. & Shadrin, N. V. (1977). The calorific value of some representatives of freshwater benthos. *Hydrobiol. J.*, **13**, 68–73.

Allen, K. R. (1969). Application of the Bertalanffy growth equation to problems of fisheries management: a review. *J. Fish. Res. Bd Can.*, **26**, 2267–81.

Alliot, E., Febvre, A. & Metailler, R. (1974). Les proteases digestive chez un teleosteen carnivore *Dicentrarchus labrax*. *Ann. Biol. Anim. Biochim. Biophys.*, **14**, 229–37.

Alliot, E., Pastoureaud, A. & Nedelec, J. (1979). Etude de l'apport calorique et du rapport calorico-azote dans l'alimentation du bar, (*Dicentrarchus labrax*). In J. E. Halver and K. Tiews (Eds), *Finfish nutrition and fishfeed technology, vol. 2*, pp. 229–35. Berlin: Heenemann Verlagsgesel.

Ameer Hamsa, K. M. S. & Kutty, M. N. (1972). Oxygen consumption in relation to spontaneous activity and ambient oxygen in five marine teleosts. *Indian J. Fish.*, **19**, 76–85.

Amlacher, E. (1960). Das Verhalten der inneren Organe und der Muskulatur drei soemmriger Karpfen aus dem Teich des Dresdner Zwingers bei extremer Kohlenhydratfutterung, 2e teil. *Z. Fisch., N.F.*, **9**, 749–61.

Ananichev, A. V. (1959). Digestive enzymes of fish and seasonal changes in their activity. *Biokhimiya*, **24**, 1033–40. (English transl: *Biochemistry*, **24**. 952–8, 1959.)

Anderson, J., Jackson, A. J., Matty, A. J. & Capper, B. S. (1984). Effects of dietary carbohydrates and fibre on the tilapia *Oreochromis niloticus* (Linn.). *Aquaculture*, **37**, 303–14.

Anderson, R. A. (1981). Nutritional role of chromium. *The Science of the Total Environment.*, **17**, 13–29.

Anderson, R. J., Kienholz, E. W. & Flickinger, S. A. (1981). Protein requirements of smallmouth bass and largemouth bass. *J. Nutr.*, **111**, 1085–97.

Ando, K. (1968). Biochemical studies on the lipids of cultured fishes. *J. Tokyo Univ. Fish.*, **54**(2), 61–98.

Andrews, J. W. (1972). Stocking density and water requirements for high-density culture of channel catfish in tanks or raceways. *Feedstuffs*, **44**(6), 40–1.

Andrews, J. W. & Murai, T. (1975). Studies on the vitamin C requirement of channel catfish (*Ictalurus punctatus*). *J. Nutr.*, **105**, 557–61.

Andrews, J. W. & Murai, T. (1978). Dietary niacin requirements of channel catfish. *J. Nutr.*, **108**, 1508–11.

Andrews, J. W. & Murai, T. (1979). Pyridoxine requirements of channel catfish. *J. Nutr.*, **109**, 533–7.

Andrews, J. W., Murai, T. & Gibbons, G. (1973). The influence of dissolved oxygen on the growth of catfish. *Trans. Am. Fish. Soc.*, **102**, 835–8.

Andrews, J. W., Murai, T. & Page, J. W. (1980). Effects of dietary cholecalciferol and ergocalciferol on catfish. *Aquaculture*, **19**, 49–54.

Andrews, J. W. & Page, J. W. (1974). Growth factors in fish meal components of catfish diets. *J. Nutr.*, **104**, 1091–6.

Andrews, J. W. & Page, J. W. (1975). The effect of feeding frequency on culture of catfish. *Trans. Am. Fish. Soc.*, **104**, 317–21.

Andrews, J. W. & Stickney, R. R. (1972). Interactions of feeding rates and environmental temperature on growth, food conversion, and body composition of channel catfish. *Trans. Am. Fish. Soc.*, **101**, 94–9.

Andruetti, S., Vigliani, E. & Ghittino, P. (1973). Possible use in trout pellets formulation of proteins from yeast grown on hydrocarbons ('BP proteins'). *Riv. Ital. Piscic. Ittiopatol.*,

8(4), 97–100.

Anonymous. (1964). The effect of cobalt on the growth and hematologic indices of carp. (In Polish.) *Gospod. Rybna*, **16**(2), 8–9.

Anonymous. (1967). A little cobalt makes fish grow faster. *New Scientist*, **36**(570), 344.

Anonymous. (1971). Cobalt in carp feed. *FAO Aquacult. Bull.*, **3**(4), 4.

Anonymous. (1972). Floating or sinking feed...what's best for you? *Fish Farming Industries*, **3**(3), 18–20.

Anonymous. (1973). Protein requirements of fish. *Nutr. Rev.*, **31**(5), 164–5.

Antalfi, A. & Toelg, I. (1971). *Graskarpfen*. Donau: Guenzburg, 205 pp.

Aoe, H. & Masuda, I. (1967). Water-soluble vitamin requirements of carp. II. Requirement for *p*-aminobenzoic acid and inositol. *Bull. Jap. Soc. Sci. Fish.*, **33**, 674–80.

Aoe, H., Masuda, I., Saito, T. & Komo, A. (1967a). Water-soluble vitamin requirements of carp. I. Requirement for vitamin B_2. *Bull. Jap. Soc. Sci. Fish.*, **33**, 355–60.

Aoe, H., Masuda, I. & Takeda, T. (1967b). Water-soluble vitamin requirements of carp. III. Requirement for niacin. *Bull. Jap. Soc. Sci. Fish.*, **33**, 681–5.

Aoe, H., Masuda, I., Saito, T. & Komo, A. (1967c). Water-soluble vitamin requirements of carp. IV. Requirement for thiamine. *Bull. Jap. Soc. Sci. Fish.*, **33**, 970–4.

Aoe, H., Masuda, I., Saito, T. & Takeda, T. (1967d). Water-soluble vitamin requirements of carp. V. Requirement for folic acid. *Bull. Jap. Soc. Sci. Fish.*, **33**, 1068–71.

Aoe, H., Masuda, I., Mimura, T., Saito, T. & Komo, A. (1968). Requirement of young carp for vitamin A. *Bull. Jap. Soc. Sci. Fish.*, **34**, 959–64.

Aoe, H., Masuda, I. & Takeda, T. (1969). Water-soluble vitamin requirements of carp VI. Requirement for thiamine and effects of antithiamines. *Bull. Jap. Soc. Sci. Fish.*, **35**, 456–65.

Aoe, H., Masuda, I., Abe, I., Saito, T., Toyoda, T. & Kitamura, S. (1970). Nutrition of protein in young carp. I. Nutritive value of free amino acids. *Bull. Jap. Soc. Sci. Fish.*, **36**, 407–13.

Aoe, H., Ikeda, K. & Saito, T. (1974). Nutrition of protein in young carp. II. Nutritive value of protein hydrolyzates. *Bull. Jap. Soc. Sci. Fish.*, **40**, 375–9.

Appelbaum, S. (1977). Geeigneter Ersatz fuer Lebendnahrung von Karpfenbrut? *Arch. Fischereiwiss.*, **28**, 31–43.

Appler, H. N. & Jauncey, K. 1983. The utilization of filamentous green algae [*Cladophora glomerata* (L) Kutzin] as a protein source in pelleted feeds for *Sarotherodon* (Tilapia) *niloticus* fingerlings. *Aquaculture*, **30**, 21–30.

Arai, S., Mueller, R., Shimma, Y. & Nose, T. (1975). Effect of calcium supplement to yeast grown on hydrocarbons as a feedstuff for rainbow trout. *Bull. Freshwater Fish. Res. Lab., Tokyo*, **25**, 33–40.

Arai, S., Nose, T. & Hashimoto, Y. (1971). A purified test diet for the eel, *Anguilla japonica*. *Bull. Freshwater Fish. Res. Lab., Tokyo*, **21**, 161–78.

Arai, S., Nose, T. & Hashimoto, Y. (1972a). Qualitative requirements of young eels *Anguilla japonica* for water-soluble vitamins and their deficiency symptoms. *Bull Freshwater Fish. Res. Lab., Tokyo*, **22**, 69–83.

Arai, S., Nose, T. & Hashimoto, Y. (1972b). Amino acids essential for the growth of eels, *Anguilla angullia*, and *A. japonica*. *Bull. Jap. Soc. Sci. Fish.*, **38**. 753–9.

Arai, S., Nose, T. & Kawatsu, H. (1974). Effect of minerals supplemented to the fish meal diet on growth of eel, *Anguilla japonica*. *Bull. Freshwater Fish. Res. Lab., Tokyo*, **24**, 95–100.

Archibald, C. P. (1975). Experimental observation on the effects of predation by goldfish (*Carassius auratus*) on the zooplankton of small saline lake. *J. Fish. Res. Bd Can.*, **32**, 1589–94.

Asgard, T. & Austreng, E. (1985). Casein silage as feed for salmonids. *Aquaculture*, **48**, 233–52.

Ash, R. (1980). Hydrolytic capacity of trout (*Salmo gairdneri*) intestinal mucosa with respect to three specific dipeptides. *Comp. Biochem. Physiol.*, **658**, 173–6.

Ashe, G. B. & Steim, J. M. 1971. Membrane transitions in gram-positive bacteria. *Biochim. Biophys. Acta*, **233**, 810–14.

Ashley, L. M. (1972). Nutritional pathology. In J. E. Halver (Ed.), *Fish nutrition*, pp. 439–537. New York: Academic Press.

Atack, T. H., Jauncey, K. & Matty, A. J. (1979). The utilization of some single-celled proteins by fingerling mirror carp (*Cyprinus carpio* L.) *Aquaculture*, **18**, 337–48.

Atack, T. & Matty, A. J. (1979). The evaluation of some single-cell proteins in the diet of rainbow trout: II. The determination of net protein utilization, biological value and true digestibility. In J. E. Halver & K. Tiews (Eds), *Finfish nutrition and fishfeed technology, vol. 1*, pp. 261–73. Berlin: Heenemann Verlagsgesel.

Atema, J. (1971). Structures and functions of the sense of taste in catfish (*Ictalurus natalis*). *Brain, Behaviour and Evolution*, **4**, 273–94.

Austic, R. E. & Calvert, C. C. (1981). Nutritional interrelationships of electrolytes and amino acids. *Fed. Proc., Fed. Am. Soc. Exp. Biol.*, **40**, 63–7.

Austreng, E. (1978). Digestibility determination in fish using chromic oxide marking and analysis of contents from different segments of the gastrointestinal tract. *Aquaculture*, **13**, 265–72.

Austreng, E. & Refstie, T. (1979). Effect of varying dietary protein level in different families of rainbow trout. *Aquaculture*, **18**, 145–56.

Babkin, B. P. & Bowie, D. J. (1928). The digestive system and its function in *Fundulus heteroclitus*. *Biol. Bull.*, **54**, 254–77.

Bachand, L. & Leray, C. (1975). Erythrocyte metabolism in the yellow perch (*Perca flavescens* Mitchill). I. Glycolytic enzymes. *Comp. Biochem. Physiol.*, **50B**, 567–70.

Backiel, T. (1977). An equation for temperature dependent metabolism or for the 'normal curve' of Krogh. *Pol. Arch. Hydrobiol*, **24**, 305–9.

Bagenal, T. B. (1967). A short review of fish fecundity. In S. D. Gerking (Ed.), *The biological basis of fish production*, pp. 89–111. Oxford: Blackwell Sci. Publ.

Bagenal, T. B. (1973). Fish fecundity and its relations with stock recruitment. *Rapp. Proc. Verbaux. Cons. internat. l'Explor. Mer.*, **164**, 186–98.

Bainbridge, R. (1958). The speed of swimming fish as related to size and to frequency and amplitude of the tailbeat. *J. Exp. Biol.*, **35**, 109–33.

Bajkov, A. D. (1935). How to estimate the daily food consumption of fish under natural conditions. *Trans. Am. Fish Soc.*, **65**, 288–9.

Bakhtin, E. K. & Filiushina, E. E. (1974). Some ultrastructural peculiarities of secondary cells and of the olfactory epithelium secretion in some fish species, related to a possible mechanism of olfactory perception. (In Russian.) *Tsitologiya*, **16**, 1089–94.

Bakos, J. (1979). Crossbreeding Hungarian races of common carp to develop more productive hybrids. In T. V. R. Pillay & W. A. Dill (Eds), *Advances in Aquaculture*, pp. 633–5. Farnham, England: Fishing News Books.

Baldwin, N. S. (1956). Food consumption and growth of brook trout at different temperatures. *Trans. Am. Fish. Soc.*, **86**, 323–8.

Barnett, B. J., Cho, C. Y. & Slinger, S. J. (1979). The requirement for vitamin D_3, and relative biopotency of dietary vitamins D_2 and D_3 in rainbow trout. *J. Nutr.*, **109**, xxiii (Abstr.).

Barrett, I. & Robertson-Connor, A. (1964). Muscle glycogen and blood lactate in yellowfin tuna, *Thunnus albacares*, and skipjack, *Katsuwonus pelamis*, following capture and tagging. *Bull. Inter-Amer. Trop. Tuna Comm.*, **9**, 219–68.

Barrington, E. J. W. (1957). The alimentary canal and digestion. In M. E. Brown, (Ed.), *The physiology of fishes*, pp. 109–161. New York: Academic Press.

Barrington, E. J. W. (1962). Digestive enzymes. In O. Lowenstein (Ed.), *Advances in comparative physiology and biochemistry, vol. 1*, pp. 1–65. New York: Academic Press.

Basu, S. P. (1959). Active respiration of fish in relation to ambient concentration of oxygen and carbon dioxide. *J. Fish Res. Bd Can.*, **16**, 175–212.

Beamish, F. W. H. (1964a). Influence of starvation on standard and routine oxygen consumption. *Trans. Am. Fish. Soc.*, **93**, 103–7.

Beamish, F. W. H. (1964b). Respiration of fishes with special emphasis on standard oxygen consumption. II. Influence of weight and temperature on respiration of several species. *Can. J. Zool.*, **42**, 177–88.

Beamish, F. W. H. (1964c). Seasonal changes in the standard rate of oxygen consumption of fishes. *Can. J. Zool.*, **42**, 189–94.

Beamish, F. W. H. (1964d). Respiration of fishes with special emphasis on standard oxygen consumption. III. Influence of oxygen. *Can. J. Zool.*, **42**, 355–66.

Beamish, F. W. H. (1966). Swimming endurance of some Northwest Atlantic fishes. *J. Fish. Res. Bd Can.*, **23**, 341–7.

Beamish, F.W.H. (1968). Glycogen and lactic acid concentrations in Atlantic cod (*Gadus morhua*) in relation to exercise. *J. Fish. Res. Bd Can.*, **25**, 837–51.

Beamish, F. W. H. (1970). Oxygen consumption of largemouth bass, *Micropterus salmoides*, in relation to swimming speed and temperature. *Can. J. Zool.*, **48**, 1221–8.

Beamish, F. W. H. (1972). Ration size and digestion in largemouth bass, *Micropterus salmoides* Lacepede. *Can. J. Zool.*, **50**, 153–64.

Beamish, F. W. H. (1974). Apparent specific dynamic action of largemouth bass, *Micropterus salmoides*. *J. Fish. Res. Bd Can.*, **31**, 1763–9.

Beamish, F. W. H. (1978). Swimming capacity. In W. S. Hoar & D. J. Randall (Eds), *Fish Physiology, vol. 7* pp. 101–187. New York: Academic Press.

Beamish, F. W. H. & Dickie, L. M. (1967). Metabolism and biological production in fish. In S. D. Gerking (Ed.), *The Biological basis of freshwater fish production*, pp. 215–42. Oxford: Blackwell Sci. Publ.

Beamish, F. W. H. & Mookherjii, P. S. (1964). Respiration of fishes with special emphasis on standard oxygen consumption. I. Influence of weight and temperature on respiration of gold fish, *Carassius auratus* L. *Can. J. Zool.*, **42**, 161–75.

Beauvalet, H. (1933). Physiologie de l'hepato-pancreas chez quelques teleosteens. *C.R. Seances Soc. Biol. Paris*, **113**, 242–4.

Beck, H., Gropp, J., Koops, H. & Tiews, K. (1979). Single cell proteins in trout diets in J. E. Halver & K. Tiews (Eds), *Finfish nutrition and fishfeed technology, vol. 2*, pp. 269–80 Berlin: Heenemann Verlagsgesel.

Beck, H., Koops, H., Tiews, K. & Gropp, J. (1977). Weitere Moeglichkeiten des Fischmehl-Ersatzes in Futter fuer Regenbogenforellen: Ersatz von Fischmehl durch Alkanhefe und Krillmehl. *Arch. Fischereiwiss*, **28**, 1–17.

Becker, M. & Nehring, K. (1965). *Handbuch der Futtermittel, vol. II*. Hamburg: Paul Parey, 475 pp.

Bell, G.H., Emslie-Smith, D. & Peterson, C. (1980). *Textbook of physiology and biochemistry* (10th edn). Edinburgh: Livingston, 563 pp.

Bergot, F. (1979a). Effects of dietary carbohydrates and of their mode of distribution on glycaemia in rainbow trout (*Salmo gairdneri* Richardson). *Comp. Biochem. Physiol.*, **64A**, 543–7.

Bergot, F. (1979b). Carbohydrate in rainbow trout diets: effects of the level and source of carbohydrate and number of meals on growth and body composition. *Aquaculture*, **18**, 157–67.

Bergot, F. (1981). Digestibility of a purified cellulose by the rainbow trout (*Salmo gairdneri*) and the common carp (*Cyprinus carpio*). *Reprod. Nutr. Dev.*, **21**, 83–93.

Bergot, F. & Breque, J. (1983). Digestibility of starch by rainbow trout: effects of the physical state of starch and of the intake level. *Aquaculture*, **34** 203–12.

Bergot, P. & Flechon, J.-E. (1970). Forme et voie d'absorption intestinale des acides gras à chaine longue chez la truite arc-en-ciel (*Salmo gairdneri* Rich): I. Lipides en particules. *Ann. Biol. Anim. Biochim. Biophys.*, **10**, 459–72.

Berman, S. A. (1967). Data on contact digestion in carp of two age groups. *Trud. Karelskogo*

Otdeleniya Gos. NIORKH, **5**, 455–7. Transl. Ser., Fish. Res. Bd Can., (2874), 1974.

Bertalanffy, L.von. (1938). A quantitative theory of organic growth (inquiries on growth laws II). *Human Biol.*, **10**, 181–213.

Bertalanffy, L.von. (1964). Basic concepts in quantitative biology of metabolism. *Helgol. Wiss. Meeresunters.* **9**, 5–37.

Beukema, J. J. (1968). Predation by the three-spined stickleback (*Gasterosteus aculeatus* L.). The influence of hunger and experience. *Behaviour*, **31**, 1–126.

Bever, K., Chenoweth, M. & Dunn, A. (1977). Glucose turnover in Kelp bass (*Paralabrax* sp.): In vivo studies with [6–^{3}H, 6–^{14}C] glucose. *Am. J. Physiol.*, **232**, R66–R72.

Bialokoz, W. & Krzywosz, T. (1981). Feeding intensity of silver carp (*Hypophthalmichthys molitrix* Val.) from the Paproteckie Lake in the annual cycle. *Ekol. Pol.*, **29**, 53–61.

Bier, M. (1955). Lipases. In S. P. Colowick & N. O. Kaplan (Eds), *Methods in Enzymology, vol. 1*, pp. 627–42. New York: Academic Press.

Bilinski, E. (1963). Utilization of lipids by fish. I. Fatty acid oxidation by tissue slices from dark and white muscle of rainbow trout (*Salmo gairdneri*). *Can. J. Biochem. Physiol.*, **41**, 107–12.

Bilinski, E. (1969). Lipid catabolism in fish muscle. In O. W. Neuhaus & J. E. Halver (Eds), *Fish in research*, pp. 135–51. New York: Academic Press.

Bilinski, E. (1974). Biochemical aspects of fish swimming. In D. C. Malins & J. R. Sargent (Eds), *Biochemical and biophysical perspectives in marine biology, vol. 1*, pp. 239–88. London: Academic Press.

Birkett, L. (1969). The nitrogen balance in plaice, sole and perch. *J. Exp. Biol.*, **50**, 375–86.

Black, E. C. (1955). Blood levels of haemoglobin and lactic acid in some freshwater fishes following exercise. *J. Fish. Res. Bd Can.*, **12**, 917–28.

Black, E. C. (1958). Hyperactivity as a lethal factor in fish. *J. Fish. Res. Bd Can.*, **15**, 573–86.

Black, E. C., Robertson, A. C. & R. R. Parker, (1961). Some aspects of carbohydrate metabolism in fish. In A. W. Martin (Ed.), *Comparative physiology of carbohydrate metabolism in heterothermic animals*, pp. 89–124. Seattle: Univ. of Washington Press.

Black, E. C., Robertson Conner, A., Lam, K.-C. & Chiu, W.-G. (1962). Changes in glycogen, pyruvate and lactate in rainbow trout (*Salmo gairdneri*) during and following muscular activity. *J. Fish. Res. Bd Can.*, **19**, 409–36.

Black, E. C., Manning, G. T. & Hayashi, K. (1966). Changes in the levels of haemoglobin, oxygen, carbon dioxide, pyruvate and lactate in venous blood of rainbow trout during and following severe muscular activity. *J. Fish. Res. Bd Can.*, **23**, 783–95.

Black, V. S. (1957). Excretion and osmoregulation. In M. E. Brown (Ed.), *The physiology of fishes, vol. 1, Metabolism*, pp. 163–205. New York: Academic Press.

Blazka, P. (1958). The anaerobic metabolism of fish. *Physiol. Zool.*, **31**, 117–28.

Blazka, P. & Kopecky, M. (1961). Basic mechanisms of adaptations to anoxia and hypoxia. In *Proc. 5th Natl. Congr., Czechoslovak Physiol. Soc., III. The physiology of adaptive phenomena*, pp. 103–6.

Boddeke, R., Slijper, E. J. & Van der Stett, A. (1959). Histological characteristics of the body musculature of fishes in connection with their mode of life. *Proc. K. Ned. Akad, Wet. Amsterdam, Ser. C*, **62**, 576–88.

Boge, G. & Pérès, G. (1983). Acquisition récentes concernant l'absorption intestinale de protides chez les poissons. *Ichthyophysiol. Acta.* **7**, 114–8.

Boge, G., Rigal, A. & Pérès, G. (1979). A study of intestinal absorption *in vivo* and *in vitro* of different concentrations of glycine by the rainbow trout (*Salmo gairdneri* Richardson). *Comp. Biochem. Physiol.*, **62A**, 831–6.

Boge, G., Rigal, A. & Pérès, G. (1981). Rates of *in vivo* intestinal absorption of glycine and glycylglycine by rainbow trout (*Salmo gairdneri* R.). *Comp. Biochem. Physiol.*, **69A**, 455–9.

Bokdawala, F. D. & George, J.C. (1967a). A histochemical study of the red and white muscle of the carp, *Cirrhina mrigala*. *J. Anim. Morphol. Physiol.*, **14**, 60–8.

Bokdawala, F. D. & George, J. C. (1967b). A quantitative study of fat, glycogen, lipase and succinic dehydrogenase in fish muscle. *J. Anim. Morphol. Physiol.*, **14**, 223–30.

Bokova, E. (1938). Daily food consumption and digestive rate in *Rutilus rutilus*. (In Russian.) *Rybnoye Khoz.*, **6**.

Bondi, A. & Spandorf, A. (1953). The activity of digestion enzymes of the carp. (In Hebrew.) *Bamidgeh*, **5** (7–8), 116–30.

Bondi, A. & Spandorf, A. (1954). The action of the digestive enzymes of the carp. *Br. J. Nutr.*, **8**, 240–6.

Bondi, A., Spandorf, A. & Calmi, R. (1957). The nutritive value of various feeds for carp. *Bamidgeh*, **9**(1), 13–18.

Bone, Q. (1966). On the function of the two types of myotomal muscle fibre in elasmobranch fish. *J. Mar. Biol. Assoc. UK*, **46**, 321–49.

Bottesch, A. (1958). Variatti cantitative sicalitative ale grasimii crapului de cultura in timpul unui an. (in Romanian with French summary.) *Bul. Inst. Cerc. Pisc.*, **17**(3), 45–54.

Bouck, G. R. (1979). Histochemistry of leucine aminonaphthyl-amidase (LAN) in rainbow trout (*Salmo gairdneri*). *Trans. Am. Fish. Soc.*, **108**, 57–62.

Bowen, S. H. (1976). Mechanism for digestion of detrital bacteria by the cichlid fish *Sarotherodon mossambicus* (Peters). *Nature* (*Lond.*) **260**, 137–8.

Bowen, S. H. (1978). Chromic acid in assimilation studies – a caution. *Trans. Am. Fish. Soc.*, **107**, 755–6.

Bowen, S. H. (1981). Digestion and assimilation of periphytic detrital aggregate by *Tilapia mossambica*. *Trans. Am. Fish. Soc.*, **110**, 239–45.

Boyd, C. E. (1968). Fresh-water plants: a potential source of protein. *Econ. Bot.*, **22**, 359–68.

Boyd, C. E. & Goodyear, C. P. (1971). Nutritive quality of food in ecological systems. *Arch. Hydrobiol.*, **69**, 256–70.

Braekkan, O. R. (1956). Function of the red muscle in fish. *Nature (Lond.)*, **178**, 747–8.

Brafield, A. E. & Solomon, D. J. (1972). Oxy-calorific coefficients for animals respiring nitrogenous substrates. *Comp. Biochem. Physiol.*, **43A**, 837–41.

Brett, J. R. (1956). Some principles in the thermal requirements of fishes. *Quart. Rev. Biol.*, **31**, 75–87.

Brett, J. R. (1962). Some considerations in the study of respiratory metabolism in fish, particularly salmon. *J. Fish. Res. Bd Can.*, **19**, 1025–38.

Brett, J. R. (1964). The respiratory metabolism and swimming performance of young sockeye salmon. *J. Fish. Res. Bd Can.*, **21**, 1183–26.

Brett, J. R. (1965). The relation of size to rate of oxygen consumption and sustained swimming speed of sockeye salmon (*Oncorhynchus nerka*). *J. Fish. Res. Bd Can.*, **22**, 1491–501.

Brett, J. R. (1970). Fish – the energy cost of living. In W. J. McNeil (Ed.), *Marine aquaculture*,pp. 37–52. Oregon State Univ. Press.

Brett, J. R. (1971). Satiation time, appetite and maximum food intake of sockeye salmon (*Oncorhynchus nerka*). *J. Fish. Res. Bd Can.*, **28**, 409–15.

Brett, J. R. (1973). Energy expenditure of sockeye salmon, *Oncorhynchus nerka*, during sustained performance. *J. Fish. Res. Bd Can.*, **30**, 1799–809.

Brett, J. R. (1979). Environmental factors and growth. In W. S. Hoar, D. J. Randall & J. R. Brett (Eds), *Fish physiology, vol. 8*, pp. 599–675. New York: Academic Press.

Brett, J. R. & Glass, N. R. (1973). Metabolic rates and critical swimming speeds of sockeye salmon (*Oncorhynchus nerka*) in relation to size and temperature. *J. Fish. Res. Bd. Can.*, **30**, 379–87.

Brett, J. R. & Groves, T. D. D. (1979). Physiological energetics. In W. S. Hoar, D. J. Randall & J. R. Brett (Eds), *Fish physiology, vol. 8*. pp. 279–352. New York: Academic Press.

Brett, J. R. & Higgs, D. A. (1970). Effect of temperature on the rate of gastric digestion in fingerling sockeye salmon *Oncorhynchus nerka*. *J. Fish. Res. Bd Can.*, **27**, 1767–79.

Brett, J. R., Shelbourn, J. E. & Shoop, C. T. (1969). Growth rate and body composition of fingerling sockeye salmon, *Oncorhynchus nerka*, in relation to temperature and ration size.

J. Fish. Res. Bd Can., **26**, 2363–94.

Brett, J. R. & Sutherland, D. B. (1965). Respiratory metabolism of pumpkinseed (*Lepomis gibbosus*) in relation to swimming speed. *J. Fish. Res. Bd Can.*. **22**, 405–9.

Brett, J. R. & Zala, C. A. (1975). Daily pattern of nitrogen excretion and oxygen consumption of sockeye salmon (*Oncorhynchus nerka*) under controlled conditions. *J. Fish. Res. Bd Can.*, **32**, 2479–86.

Brichon, G. (1973). Phosphorus uptake by the intestine of the eel (*Anguilla anguilla* L.). I. Demonstration and details on in vitro phosphate transport in the fresh water eel. (In French.) *C.R. Seances Soc. Biol. Fil.*, **167**(8–9), 1142–5.

Brocksen, R. W. & Bugge, J. P. (1974). Preliminary investigations on the influence of temperature on food assimilation by rainbow trout *Salmo gairdneri* Richardson. *J. Fish Biol.*, **6**, 93–7.

Brocksen, R. W., Davis, G. E. & Warren, C. E. (1968). Competition, food consumption and production of sculpins and trout in laboratory stream communities. *J. Wildl. Manage.*, **32**, 51–75.

Brody, S. (1945). *Bioenergetics and growth, with special reference to the efficiency complex in domestic animals.* New York: Reinholds Publ. Corp., 1023 pp.

Bromley, P. J. & Adkins, T. C. (1984). The influence of cellulose filler on feeding, growth and utilization of protein and energy in rainbow trout, *Salmo gairdneri* Richardson, *J. Fish. Biol.*, **24**, 235–44.

Brouwer, E. (1964). Report of the sub-committee on constants and factors. In K. L. Blaxter (Ed.), *European Association for Animal Production, Publication 11 – Energy metabolism.* Proceedings of the 3rd symposium held at Troon, Scotland, 1964, pp. 441–5. New York: Academic Press.

Brown, D., Fleming, N. & Balls, M. (1975). Hormonal control of glucose production by *Amphiuma means* liver in organ culture. *Gen. Comp. Endocrinol.*, **27**, 380–8.

Brown, J. A. G., Jones, A. & Matty, A. J. (1984). Oxygen metabolism of farmed turbot (*Scophthalmus maximus*). I. The influence of fish size and water temperature on metabolic rate. *Aquaculture*, **36**, 273–81.

Brown, M. E. (1946a). The growth of brown trout (*Salmo trutta* Linn.). I. Factors influencing the growth of trout fry. *J. Exp. Biol.*, **22**, 118–29.

Brown, M. E. (1946b). The growth of brown trout (*Salmo trutta* Linn.) II. The growth of two-year-old trout at a constant temperature of 11.5 °C. *J. Exp. Biol.*, **22**, 130–44.

Brown, M. E. (1946c). The growth of brown trout (*Salmo trutta* L.). The effect of temperature on the growth of two-year-old trout. *J. Exp. Biol.*, **22**, 145–55.

Brown, M. E. (1957). Experimental studies on growth. In M. E. Brown (Ed.), *The physiology of fishes, vol. 1, Metabolism*, pp. 361–400. New York: Academic Press.

Brown, W. D. (1960). Glucose metabolism in carp. *J. Cell. Comp. Physiol.*, **55**, 81-5.

Brown, W. D. & Tappel, A. L. (1959). Fatty acid oxidation by carp liver mitochondria. *Arch. Biochem. Biophys.*, **85**, 149–58.

Brubacher, G., Schaerer, K., Studer, A. & Wiss, O. (1965). The mutual influence of vitamin E, vitamin A and carotenoids. (In German.) *Z. Ernaehrungswiss.*, **5**, 190–202.

Brungs, W. A. (1971). Chronic effects of low dissolved oxygen concentrations on the fathead minnow (*Pimephales promelas*). *J. Fish. Res. Bd Can.*, **28**, 1119–23.

Bryan, P. G. (1975). Food habits, functional digestive morphology, and assimilation efficiency of the rabbitfish *Siganus spinus* (Pisces, Siganidae) on Guam. *Pacific Sci.*, **29**, 269–71.

Buclon, M. (1974). Bioelectric potentials and the transfer of amino acids across the digestive epithelium of the tench (*Tinca tinca* L.). *J. Physiol. (Paris)*, **68**, 157–80.

Buddington, R.K. (1979). Digestion of an aquatic macrophyte by *Tilapia zilli* (Gervais). *J. Fish. Biol.*, **15**, 449–56.

Buddington, R. K. (1980). Hydrolysis-resistant organic matter as a reference for measurement of fish digestive efficiency. *Trans. Am. Fish. Soc.*, **109**, 653–6.

Buhler, D. R. & Benville, P. (1969). Effect of feeding and of DDT on the activity of hepatic glucose-6-phosphate dehydrogenase in two salmonids. *J. Fish. Res. Bd Can.*, **26**, 3209–16.

Buhler, D. R. & Halver, J. E. (1961). Nutrition of salmonoid Fishes. IX. Carbohydrate requirements of chinook salmon. *J. Nutr.*, **74**, 307–318.

Bullock, T. H. (1955). Compensation for temperature in metabolism and activity of poikiloterms. *Biol. Rev.*, **30**, 311–42.

Bunk, M. & Combs, G. F. (1980). The role of appetite, glutathione peroxidase activity and sulfur-containing amino acids in the etiology of nutritional pancreatic atrophy in selenium deficient chicks. Seminar, Cornell University, 1980.

Burr, G. O. & Burr, M. M. (1929). A new deficiency disease produced by the rigid exclusion of fat from the diet. *J. Biol. Chem.*, **82**, 345–67.

Burtle, G. J. (1981). Essentiality of dietary inositol for channel catfish. Ph.D. Dissertation, Auburn Univ., Alabama, USA.

Buterbaugh, G. L. & Willoughby, M. (1967). A feeding guide for brook, brown and rainbow trout. *Prog. Fish Cult.*, **29**, 210–15.

Butler, D. G. (1968). Hormonal control of gluconeogenesis in the North American eel (*Anguilla rostrata*). *Gen. Comp. Endocrinol.* **10**, 85–91.

Butt. W. R. (1975). *Hormone chemistry, vol. 1, Protein polypeptide and peptide hormones* (2nd edn). Chichester: Ellis Horwood Ltd, 272 pp.

Buttkus, H. (1963). Red and white muscle of fish in relation to rigor mortis. *J. Fish. Res. Bd Can.*, **20**, 45–58.

Cahill, G. F., Aoki, T. T. & Marliss, E. B. (1972). Insulin and muscle protein. In D. F. Steiner & N. Freinkel (Eds), *Handbook of physiology, section 7: Endocrinology, vol. 1, Endocrine pancreas*, pp. 563–77. Washington D.C.: Am. Physiol. Society.

Caillouet, C. W. (1964). Blood lactic acid concentration of unexercised and exercised mature carp in winter and summer. *Iowa State J. Sci.*, **38**, 309–22.

Caillouet, C. W. 1968. Lactic acidosis in channel catfish. *J. Fish. Res. Bd Can.*, **25**, 15–23.

Caldwell, R. S. & Vernberg, F. J. (1970). The influence of acclimation temperature on the lipid composition of fish gill mitochondria. *Comp. Biochem. Physiol.*, **34**, 179–91.

Castell, J. D. (1979). Review of lipid requirements of finfish. In J.E. Halver & K. Tiews (Eds), *Finfish nutrition and fishfeed technology, vol. 1*, pp. 59–84. Berlin: Heenemann Verlagsgesel.

Castell, J. D., Sinnhuber, R. O., Wales, J. H. & Lee, D. J. (1972a). Essential fatty acids in the diet of rainbow trout (*Salmo gairdneri*): Growth, feed conversion and some gross deficiency symptoms. *J. Nutr.*, **102**. 77–85.

Castell, J. D., Sinnhuber, R. O., Lee, D. J. & Wales, J. H. (1972b). Essential fatty acids in the diet of rainbow trout (*Salmo gairdneri*): Physiological symptoms of EFA deficiency. *J. Nutr.*, **102**, 87–92.

Castell, J. D., Lee, D. J. & Sinnhuber, R.O. (1972c). Essential fatty acids in the diet of rainbow trout (*Salmo gairdneri*): lipid metabolism and fatty acid composition. *J. Nutr.*, **102**, 93–100.

Castilla, C. & Murat, J.-C. (1975). Influence of insulin on protein metabolism in carp liver. (In French with English summary.) *C.R. Seances Soc. Biol. Fil.*, **169**(6), 1605–8.

Caulton, M. S. (1977). The effect of temperature on routine metabolism in *Tilapia rendalli* Boulenger. *J. Fish Biol.*, **11**, 549–53.

Caulton, M. S. (1978). The importance of habitat temperatures for growth in the tropical cichlid *Tilapia rendalli* Boulenger. *J. Fish. Biol.*, **13**, 99–112.

Chance, R. E., Mertz, E. T. & Halver, J. E., (1964). Nutrition of salmonoid fishes XII. Isoleucine, leucine, valine and phenylalanine requirements of chinook salmon and interrelations between isoleucine and leucine for growth. *J. Nutr.*, **83**, 177–85.

Chang, V. M. & Idler, D. R. (1960). Biochemical studies on sockeye salmon during spawning migration. XII. Liver glycogen. *Can. J. Biochem. Physiol.*, **38**, 553–8.

Chartier, M. M., Milet, C., Martelly, E., Lopez, E. & Warrot, E. (1979). Stimulation par la vitamine D_3 et la 1,25-dihydroxyvitamine D_3 de l'absorption intestinale du calcium chez l'anguilla (*Anguilla anguilla* L.). *J. Physiol.*, **75**, 275–82.

Chatterjee, D. K. (1979). Increased production of major carp fry by addition of growth promoting substances. In J. E. Halver & K. Tiews (Eds), *Finfish nutrition and fishfeed technology, vol. 1*, pp. 189–95. Berlin: Heenemann Verlagsgesel.

Chavin, N. & Singley, J. A. (1972). Adrenocorticoids of the goldfish *Carassius auratus* L. *Comp. Biochem. Physiol.*, **42B**, 547–62.

Chavin, W. & Young J. E. (1970). Factors in the determination of normal serum glucose levels of goldfish, *Carassius auratus* L. *Comp. Biochem. Physiol.*, **33**, 629–53.

Chekunova, V. I. (1983). Ecological groups of cold-water fishes and their energy metabolism. *J. Ichthyol.*, **23**(5), 111–21.

Chepik, K. (1964). Activity of carp digestive enzymes at different seasons of the year. (In Russian.) *Izv. Akad. Nauk Latv. SSR*, (5). 73–9. Ref. Zh. Biol., 7164: 1965.

Chiba, K. (1965). A study of the influence of oxygen concentration on the growth of juvenile common carp. *Bull. Freshwater Fish. Res. Lab., Tokyo*, **15**, 35–47.

Chiou, J. Y. & Ogino, C. (1975). Digestibility of starch in carp. *Bull. Jap. Soc. Sci. Fish.*, **41**, 465–6.

Chiu, Y. N. & Benitez, L. V. (1981). Studies on the carbohydrates in the digestive tract of the milkfish *Chanos chanos*. *Mar. Biol.*, **61**, 247–54.

Cho, C. Y., Bayley, H. S. & Slinger, S. J. (1974). Partial replacement of herring meal with soybean meal and other changes in the diet for rainbow trout (*Salmo gairdneri*). *J. Fish. Res. Bd Can.*, **31**, 1523–8.

Cho, C. Y., Slinger, S. J. & Bayley, H. S. (1976a). Influence of level and type of dietary protein, and the level of feeding on feed utilization by rainbow trout. *J. Nutr.*, **106**, 1547–56.

Cho, C. Y., Bayley, H. S. & Slinger, S. J. (1976b). Energy metabolism in growing rainbow trout: partition of dietary energy in high protein and high fat diets. In M. Vermorel (Ed.), *Energy metabolism of farm animals (7th Symp. Vichy, EAAP no. 19)*, pp. 299–302. Clermont-Ferrand: G. de Bussac.

Cho, C. Y. & Slinger, S. J. (1979). Significance of digestibility measurement in formulation of needs for rainbow trout. In J. E. Halver & K. Tiews (Eds), *Finfish nutrition and fishfeed technology, vol. 2*, pp. 239–47. Berlin: Heenemann Verlagsgesel.

Chow, K.W. & Schell, W. R. (1980). The minerals. In *Aquaculture Development and Coordination Programme, Fish Feed technology*, pp. 104–8. UNDP/FAO.

Christensen, H. N. (1977). Amino acid transport system in animal cells: interrelations and energization. *J. Supramol. Struct.*, **6**, 205–13.

Chudova, Z. I. (1961). The role of Vitamins A and B_1 in the rearing of salmon fry. (In Russian.) *Tr. Nauchn.-issled. Inst. Rybn. Khoz. Latviiski SSR*, **3**, 421–9.

Cismeros Moreno, J. A. (1981). Preliminary report on the raw protein requirements of red tilapia fingerlings. (In Spanish with English summary.) *Rev. Latinoam. Acuicult.*, **7**, 18–21.

Cockson, A. & Bourn, D. (1972). Enzymes in the digestive tract of two species of euryhaline fish. *Comp. Biochem. Physiol.*, **41A**, 715–18.

Cockson, A. & Bourn, D. (1973). Protease and amlyase in the digestive tract of *Barbus paludinosus*. *Hydrobiol.*, **43**, 357–63.

Cohen, P. P. & Brown, G. W. (1960). Ammonia metabolism and urea biosynthesis. In M. Florkin & H. S. Mason (Eds), *Comparative biochemistry*, Vol. 2, pp. 161–244. New York: Academic Press.

Cohen, T. (1981). Pancreatic proteolytic enzymes from carp *Cyprinus carpio*. (In Hebrew with English summary.) Ph.D. Thesis, Hebrew University of Jerusalem; 80 + III pp. (mimeographed).

Collins, R. A. & Delmendo, M. N. (1979). Comparative economics of aquaculture in cages, raceways and enclosures. In T. V. R. Pillay & W. A. Dill (Eds), *Advances in aquaculture*, pp. 472–7. Farnham, England: Fishing News Books.

Conover, R. J. (1978). Transformation of organic matter. In O. Kinne (Ed.), *Marine ecology*

IV. Dynamics, pp. 221–489. New York: John Wiley & Sons.

Cooper, E.L. (1953). Periodicity of growth and change of condition of brook trout (*Salvelinus fontinalis*) in three Michigan trout streams. *Copeia*, **2**, 107–14.

Cooper, E. L. (1961). Growth of wild and hatchery strains of brook trout. *Trans. Am. Fish. Soc.*, **90**, 424–38.

Corey, P. D., Leith, D. A. & English, M. J. (1983). A growth model for coho salmon including effects of varying ration allotments and temperature. *Aquaculture*, **30**, 125–43.

Cowey, C. B. (1975). Aspects of protein utilization by fish. *Proc. Nutr. Soc.*, **34**, 57–63.

Cowey, C. B. (1976). Use of synthetic diets and biochemical criteria in the assessment of nutrient requirements of fish. *J. Fish Res. Bd Can.*, **33**, 1040–5.

Cowey, C. B., Pope, J. A., Adron, J. W. & Blair, A. (1971). Studies on nutrition of marine fish. Growth of plaice (*Pleuronectes platessa*) on diets containing proteins derived from plants and other sources. *Mar. Biol.*, **10**. 145–53.

Cowey, C. B., Pope, J. A., Adron, J. W. & Blair, A. (1972). Studies on the nutrition of marine flatfish. The protein requirement of plaice (*Pleuronectes platessa*). *Brit. J. Nutr.*, **28**, 447–56.

Cowey, C. B. & Sargent, J. R. (1972). Fish nutrition. *Adv. Mar. Biol.*, **10**. 383–493.

Cowey, C. B., Adron, J., Blair, A. & Shanks, A. M. (1974). Studies on the nutrition of marine flatfish. Utilization of various dietary proteins by plaice (*Pleuronectes platessa*). *Br. J. Nutr.*, **31**, 297–306.

Cowey, C. B., Adron, J. W., Brown, D. A. & Shanks, A. M. (1975). Studies on the nutrition of marine flatfish. The metabolism of glucose by plaice (*Pleuronectes platessa*) and the effect of dietary energy source on protein utilization in plaice. *Br. J. Nutr.*, **33**, 219–31.

Cowey, C. B., De La Higuera, M. & Adron, J.W. (1977a). The effect of dietary composition and insulin on gluconeogenesis in rainbow trout (*Salmo gairdneri*). *Br. J. Nutr.*, **38**, 385–96.

Cowey, C. B., Knox, D., Walton, M. J. & Adron, J. W. (1977b). The regulation of gluconeogenesis by diet and insulin in rainbow trout (*Salmo gairdneri*). *Br. J. Nutr.*, **38**, 463–70.

Cowey, C. B., Adron, J. W. & Youngson, A. (1983). The vitamin E requirement of rainbow trout (*Salmo gairdneri*) given diets containing polyunsaturated fatty acids derived from fish oil. *Aquaculture*, **30**, 85–93.

Craig, J. F. (1977). The body composition of adult perch, *Perca fluviatilis* in Windermere, with reference to seasonal changes and reproduction. *J. Anim. Ecol.*, **46**, 617–32.

Creach, Y. (1966). Thiols proteines et acides amines libres des tissus chez la carpe (*Cyprinus carpio* L.) au cours du jeune prolonge. *Arch. Sci. Physiol.*, **20**, 115–21.

Cruz, E. M. (1967). Determination of nutrient digestibility in various classes of natural feed materials for channel catfish. *CLSU Sci. J.*, **12**, 87–96.

Cummins, K. W. & Wuycheck, J.C. (1971). Caloric equivalents for investigations in ecological energetics. *Mitt. Int. Ver. Limnol.*, **18**, 1–158.

Cunha, T. J. (1967). *Present status on swine feeding and nutrition*. Hoffmann-La-Roche, Basle; 20 pp.

Cunningham, L.W., Fischer, R.L. & Vestling, C.S. (1955). The binding of zinc and cobalt by insulin. *J. Am. Chem. Soc.*, **77**, 5703–7.

Dabroswka, H. & Wojno, T. (1977). Studies on the utilization by rainbow trout (*Salmo gairdneri* Rich.) of feed mixtures containing soya bean meal and an addition of amino acids. *Aquaculture*, **10**, 297–310.

Dabrowska, H., Grudniewski, C. & Dabrowski, K. (1979). Artificial diets for common carp: effect of the addition of enzyme extracts. *Prog. Fish Cult.*, **41**, 196–200.

Dabrowski, K. (1977). Protein requirements of grass carp fry (*Ctenopharyngodon idella* Val). *Aquaculture*, **12**, 63–74.

Dabrowski, K. (1979). Energy conversion in carp given feed with different protein contents. *Rocz. Nauk Roln. H*, **99**(3), 79–89.

Dabrowski, K. (1983). Comparative aspects of protein digestion and amino acid absorption in fish and other animals. *Comp. Biochem. Physiol.*, **74A**, 417–26.

Dabrowski, K. & Dabrowska, H. (1981). Digestion of protein by rainbow trout (*Salmo gairdneri* Rich.) and absorption of amino acids within the alimentary tract. *Comp. Biochem. Physiol.*, **69A**, 99–111.

Dabrowski, K. & Kaushik, S. J. (1984). Rearing coregonid (*Coregonus schinizi palea* Cuv. et Val.) larvae using dry and live food II. Oxygen consumption and nitrogen excretion. *Aquaculture*, **41**, 333–44.

Dabrowski, K. & Rusiecki, M. (1981). Content of total and free amino acids in zooplanktonic food of fish larvae. *Aquaculture*, **30**, 31–42.

Dabrowski, K. & Wojno, T. (1978). Use of non-protein nitrogen components for feeding carp (*Cyprinus carpio* L.). II. Digestibility of nutrients, feed protein utilization, absorption of amino acids. (In Polish with English summary.) *Zesz. Nauk ART Olszt.*, (7), 101–20.

Dahlberg, M. L., Shumway, D. L. & Doudoroff, P. (1968). Influence of dissolved oxygen and carbon dioxide on the swimming performance of largemouth bass and coho salmon. *J. Fish. Res. Bd Can.*, **25**, 49–70.

Daniels, W. H. & Robinson E. H. (1986). Protein and energy requirements of Juvenile red drum (*Sciaenops ocellatus*). *Aquaculture*, **53**, 243–52.

Danulat, E. & Kausch, H. (1984). Chitinase activity in digestive tract of the cod, *Gadus morhua* (L.). *J. Fish Biol.*, **24**, 125–33.

Darnell, R. M. & Meierotto, R. R. (1962). Determination of feeding chronology in fishes. *Trans. Am. Fish. Soc.*, **91**, 313–20.

Das, B. C. (1959). Comparative effects of vitamin B_{12}, cobalt nitrate and ruminant stomach extract on the survival rate of Indian carp during the first 3 weeks of life. *Indian J. Fish.*, **6**, 211–21.

Das, B. C. 1967. Effects of micro-nutrients on the survival and growth of Indian carp fry. *FAO Fish. Rep.*, **44**(3), 241–56.

Dave, G. 1976. The art of starvation. The lipid metabolism of the eel. *Zool. Revy*, **37**, 37–40.

Dave G., Johansson-Sjoebeck, M.-L., Larsson, A., Lewander, K. & Lidman, U. (1975). Metabolic and hematological effects of starvation in European eel, *Anguilla anguilla* L. I. Carbohydrate, lipid, protein and inorganic ion metabolism. *Comp. Biochem. Physiol.*, **52A**, 423–30.

Davis, G. E. & Warren, C. E. (1965). Trophic relations of a sculpin in laboratory stream communities. *J. Wildl. Manage.*, **29**, 846–71.

Davis, G. E. & Warren, C. E. (1968). Estimation of food consumption rates. In W. E. Ricker (Ed.), *Methods for assessment of fish production in fresh waters. IBP Handbook no. 3* pp. 204–25. Oxford: Blackwell Sci. Publ.

Davis, T. A. & Stickney R. R. (1978). Growth response of *Tilapia aurea* to dietary protein quality and quantity. *Trans. Am. Fish. Soc.*, **107**, 479–83.

Davis, P. M. C. (1967). The energy relations of *Carassius auratus*, L. III. Growth and the overall balance of energy. *Comp. Biochem. Physiol.*, **23**, 59–63.

Davison, H. (1970). *A textbook of general physiology, vol. I* (4th edn). London: J. & A. Churchill, xiv + 1029 pp.

Dawes, B. (1930). Growth and maintenance in plaice. (*P. platessa*, L.) Part II. *J. Mar. Biol. Assoc. UK*, **17**, 877–947.

Dawson, A. S. & Grimm, A. S. (1980). Quantitative seasonal changes in the protein, lipid and energy content of carcass, ovaries, and liver of adult female plaice, *Pleuronectes platessa* L. *J. Fish Biol.*, **16**, 493–504.

De Jagers, S. & Dekkers, W. J. (1975). Relations between gill structure and activity in fish. *Neth. J. Zool.*, **25**, 276–308.

DeLong, D. C., Halver, J. E. & Mertz, E. T. (1957). Protein requirement of chinook salmon at two water temperatures. *Fed. Proc., Fed. Am. Soc. Exp. Biol.*, **16**, 384(1644).

DeLong, D. C., Halver, J. E. & Mertz, E. T. (1958), Nutrition of salmonoid fishes. VI.

Protein requirements of chinook salmon at two water temperatures. *J. Nutr.*, **65**, 589–600.

DeLong, D. C., Halver, J. E. & Mertz, E. T. (1959). Nutrition of salmonoid fishes. VII. nitrogen supplements for chinook salmon diets., *J. Nutr.*, **68**, 663–9.

Demael-Suard, A., Garin, D., Brichon, G., Mure, M. & Peres, G. (1974). Neoglycogenesis from glycine-^{14}C in the tench (*Tinca vulgaris*) during asphyxiation. *Comp. Biochem. Physiol.*, **47A**, 1023–33.

Demoll, R. (1940a). Kartoffelpuelpe als Karpfenfutter. *Allg. Fisch.-Ztg.*, **23**, 2pp.

Demoll, R. (1940b). Vorlaufige Mitteilung ueber die Kartoffel-Fuetterung-Versuche des Jahres 1939 in Vielenbach, *Fisch.Ztg.*, **43**(8), 1 pp.

Denton, J.E., Yousef, M. K., Yousef, I. M. & Kuksis, J. E. (1974). Bile acid composition of rainbow trout, *Salmo gairdneri*. *Lipids*, **9**, 945–51.

De Silva, S. S. & Balbontin, F. (1974). Laboratory studies on food intake, growth and food conversion of young herring *Clupea harengus* L. *J. Fish Biol.*, **6**, 645–58.

De Silva, S. S. & Perera, M. K. (1983). Digestibility of an aquatic macrophyte by the cichlid *Etroplus suratensis* (Bloch) with observations on the relative merits of three indigenous components as markers and daily changes in protein digestibility. *J. Fish Biol.*, **23**, 675–84.

De Silva, S. S. & Perera, M. K. (1984). Digestibility in *Sarotherodon niloticus* fry: effect of dietary level and salinity with further observations of variability in dietary digestibility. *Aquaculture*, **38**, 293–306.

De Silva, S. S., Perera, M. K. & Maitipe, P. (1984). The composition, nutritional status and digestibility of the diets of *Sarotherodon mossambicus* from nine man-made lakes in Sri Lanka., *Environ. Biol. of Fishes*, **11**, 205–19.

De Silva, S. S. & Perera, P. A. B. (1976). Studies on the young grey mullet, *Mugil cephalus* L. I. Effects of salinity on food intake, growth and food conversion. *Aquaculture*, **7**, 327–38.

De Silva, S. S. & Weerakoon, D. E. M. (1981). Growth, food intake and evacuation rates of grass carp, *Ctenopharyngodon idella* fry. *Aquaculture*, **25**, 67–76.

Dorsa, W. J., Robinette, H. R., Robinson, E. H. & Poe, W. E. (1982). Effects of dietary cottonseed meal and gossypol on growth of young channel catfish. *Trans. Am. Fish. Soc.*, **111**, 651–5.

Driedzic, W. R. & Hochachka, P. W. (1975). The unanswered question of high anaerobic capabilities of carp white muscle. *Can. J. Zool.*, **53**, 706–12.

Driedzic, W. R. & Hochachka, P. W. (1978). Metabolism in fish during exercise in W. S. Hoar & D. J. Randall (Eds), *Fish physiology, vol. 7*, pp. 503–44. New York: Academic Press.

Driedzic, W. R. & Kiceniuk, J. W. (1976). Blood lactate levels in free-swimming rainbow trout (*Salmo gairdneri*) before and after strenuous exercise resulting in fatigue. *J. Fish. Res. Bd Can.*, **33**, 173–6.

Driver, E. A., Sugdew, L. G. & Kovach, R. J. (1974). Calorific, chemical and physical values of potential duck foods. *Freshwater Biol.*, **4**, 281–92.

Dupree, H. K. (1966). Vitamins essential for growth of channel catfish *Ictalurus punctatus*. *US Bureau of Sport Fish, Wildl., Tech. Pap.*, **7**, 12pp.

Dupree, H. K. (1977). Vitamin requirements. In *Nutrition and feeding of channel catfish*. Southern Cooperative Series; pp. 26–29.

Dupree, H. K. & Halver, J. E. (1970). Amino acid essential for the growth of channel catfish, *Ictalusus punctatus*. *Trans. Am. Fish. Soc.*, **99**, 90–2.

Dupree, H. K., & Sneed, K. E. (1966). Response of channel catfish fingerlings to different levels of major nutrients in purified diets. *US Bureau of Sport Fish Wildl., Tech. Pap.*, **9**, 21 pp.

Dupree, H. K., Gauglitz, E. J., Hall, A. S. & Houle, C. R. (1979). Effects of dietary lipids on the growth and acceptability (Flavour) of channel catfish (*Ictalurus punctatus*). In J.H. Halver & K. Tiews (Eds), *Finfish nutrition and fishfeed technology vol. II*, pp. 87–98. Berlin: Heenemann Verlagsgesel.

Eckhardt, O., Becker, K. & Guenther, K.-D. (1981). Zum Protein-und Energiebedarf wachsender Spiegelkarpfen (*Cyprinus carpio* L.) I. Qualitativer Vergleich verschiedener Protein und Kohlenhydrattraeger fuer Versuchsdiaeten. *Z. Tierphysiol. Tierernaehr. Futtermittelk.*, **46**, 124–31.

Eckhardt, O., Becker, K., Gunther, K.-D. & Meske, Ch. (1982). Zum Protein-und Energiebedarf wachsender Spiegelkarpfen (*Cyprinus carpio* L.) II. Einfluss von Fettzulagen bei unterschiedlichen Protein und Kohlenhydratgehalten der Ration auf Wachstum und Korperzusammensetzung. *Z. Tierphysiol. Tierernaehr. Futtermittelk.*, **47**, 186–96.

Eclancher, B. (1975). Respiratory control in teleost fish: respiratory responses to abrupt changes of ambient oxygen pressure. *C.R. Hebd. Seances Acad. Sci., Paris, Ser. D.*, **280**, 307–10.

Eddy, F. B. (1975). The effect of calcium on gill potentials and on sodium and chloride fluxes in the goldfish *Carassius auratus*. *J. Comp. Physiol.*, **96**, 131–42.

Edwards, D. J. (1971). Effects of temperature on the rate of passage of food through the alimentary tract of the plaice. *J. Fish Biol.*, **3**, 433–7.

Edwards, D. J., Austreng, E., Risa, S. & Gjedrem, T. (1977). Carbohydrate in rainbow trout diets. I. Growth of fish of different families fed diets containing different proportions of carbohydrate. *Aquaculture*, **11**, 31–8.

Edwards, R. R. C., Blaxter, J. H. S., Gopalan, U. K., Mathew, C. V. & Finlayson, D. M. (1971). Feeding, metabolism and growth of tropical flatfish. *J. Exp. Mar. Biol. Ecol.*, **6**, 279–300.

Edwards, R. R. C., Finlayson, D. M. & Steele, J. H. (1969). The ecology of 0-group plaice and common dabs in Loch Ewe. II. Experimental studies on metabolism. *J. Exp. Mar. Biol. Ecol.*, **3**, 1–17.

Edwards, R. R. C., Finlayson, D. M. & Steele, J. H. (1972). An experimental study of the oxygen consumption, growth and metabolism of the cod (*Gadus morhua* L.). *J. Exp. Mar. Biol. Ecol.*, **8**, 299–309.

Ehrlich, K. F. (1974). Chemical changes during growth and starvation of herring larvae. In J. H. S. Blaxter (Ed.), *The early life history of fish*, pp. 301–23. Springer-Verlag.

Ekberg, D. R. (1958). Respiration in tissues of goldfish adapted to high and low temperatures. *Biol. Bull.*, **114**, 308–16.

Elliott, J. M. (1972). Rate of gastric evacuation in brown trout, *Salmo trutta*. *Freshwater Biol.*, **2**, 1–8.

Elliott, J. M. (1975a). The growth rate of brown trout (*Salmo trutta* L.) fed on maximum rations. *J. Anim. Ecol.* **44**, 805–21.

Elliott, J. M. (1975b). The growth rate of brown trout (*Salmo trutta* L.) fed on reduced rations. *J. Anim. Ecol.*, **44**, 823–42.

Elliott, J. M. (1975c). Weight of food and time required to satiate brown trout, *Salmo trutta* L. *Freshwater Biol.*, **5**, 51–64.

Elliott, J. M. (1975d). Number of meals in a day, maximum weight of food consumed in a day and maximum rate of feeding for brown trout, *Salmo trutta* L. *Freshwater Biol.*, **5**, 287–303.

Elliott, J. M. (1976). Energy losses in the waste products of brown trout (*Salmo trutta* L.). *J. Anim. Ecol.*, **45**, 561–80.

Elliott, J. M. & Davison, W. (1975). Energy equivalents of Oxygen consumption in animal energetics. *Oecologia (Berl.)*, **19**, 195–201.

Ellis, A. E., Roberts, R. J. & Tytler, P. (1978). The anatomy and physiology of teleosts. In R. J. Roberts (Ed.), *Fish pathology*, pp. 13–54. London: Bailliere Tindall.

Ellis, R. W. & Smith, R. R. (1984). Determining fat digestibility in trout using a metabolic chamber. *Prog. Fish Cult.*, **46**, 116–19.

Elster, H. I. (1954). Ueber der Populationsdynamik von *Eudiaptomus gracilis* Sars und *Heterocope borealis* Fischer in Bodense-Obersee. *Arch. Hydrobiol., Suppl.* **20**, 546–614.

Erokhina, L. (1959), Experiments on the use of pelleted feed. (In Russian.) *Rybovod.*

Rybolov., **2**, 8–10.

Epple, A. (1969). The endocrine pancreas. In W. S. Hoar & D. J. Randall (Eds), *Fish physiology, vol II, The endocrine system*, pp. 275–319. New York: Academic Press.

Escoubet, P., Boge, G. & Rigal, A. (1974). Comparison de l'absorption intestinal du glycocolle et du glucose chez la truite (*Salmo gairdneri* R.) à differentes temperatures. *Ann. Inst. Michel Pacha*, 1973(6). 37–44.

Evans. R. M., Purdie, F. C. & Hickman, C. P. (1962). The effect of temperature and photoperiod on the respiratory metabolism of rainbow trout (*Salmo gairdneri Can. J. Zool.*, **40**, 107–18.

Fabian, G. Y.., Molnar, G. Y. & Tolg, I. (1963). Comparative data and enzyme kinetic calculations on changes caused by temperature in the digestion of some predatory fishes. *Acta Biol. Acad. Sci. Hung.*, **14**, 123–9.

Falge, R., Schpanof, L. & Jurss, K. (1978). Amylase, esterase and protease activity in the intestine content of rainbow trout *Salmo gairdneri* Rich., after feeding with feed containing different amounts of starch and protein. *Vopr. Ikhtiol.*, **18**, 314–19; *J. Ichthyol.*, **18**, 283–7.

Falkmer, S. (1961). Experimental diabetes research in fish. On the morphology and physiology of the endocrine pancreatic tissue of the marine teleost *Cottus scorpius* with special reference to the role of the glutathione in the mechanism of alloxan diabetes using a modified nitroprusside method. *Acta Endocrinol.*, **37**, Suppl. 59, 1–22.

Falkmer, S. (1966). Quelques aspects comparatifs de cellules pancreatiques et du glucagon. *Ann Endocrinol.*, **27**, 321.

Falkmer, S. & Knutson, F. (1963). Is cobalt concentrated in pancreatic islets tissue? *Acta Endocrinol.*, **42**, 309–20.

Falkmer, S., Knutson, F. & Voight, G. E. (1964). On the cobalt concentrating ability of pancreatic islet tissue. A histological, histochemical and autoradiographical investigation. *Diabetes*, **13**, 400–7.

Fänge, R. & Grove, D. (1978). Digestion. In W. S. Hoar, D. J. Randall, & J. R. Brett (Eds), *Fish physiology, vol. viii*, pp. 161–260. New York: Academic Press.

Farkas, T. (1958). Comparative studies on the chemical composition of lower and higher crustaceans. (In Hungarian with German summary.) *Ann. Inst. Biol. Tihany, Hung.*, **25**, 179–86.

Farkas, T. (1970). The dynamic of fatty acids in the aquatic food chain, Phytoplankton, Zooplankton, fish. *Ann. Inst. Biol., Tihany, Hung.*, **37**, 165–76.

Farkas, T., Csengeri, I., Majoros, F. & Olah, J. (1980). Metabolism of fatty acids in fish, 3: Combined effect of environmental temperature and diet on formation and deposition of fatty acids in the carp, *Cyprinus carpio* Linnaeus 1758. *Aquaculture*, **20** 29–40.

Farkas, T. & Herodek, S. (1960). Seasonal changes in the fat content of the crustacea plankton in Lake Balaton. *Ann. Inst. Biol., Tihany, Hung.*, **27**, 3–7.

Farkas, T. & Herodek, S. (1964). The effect of environmental temperature on the fatty acid composition of crustacea plankton. *J. Lipid Res.*, **5**, 369–73.

Farkas, T., Herodek, S., Csaki, L. & Toth, G. (1961). Incorporation of acetate-1-^{14}C into the liver fatty acids in the fish *Amiurus nebulosus*. *Acta Biol. Acad. Sci. Hung*, **12**, 83–6.

Farmer, G. J. & Beamish, F. W. H. (1969). Oxygen consumption of *Tilapia nilotica* in relation to swimming speed and salinity. *J. Fish. Res. Bd Can.*, **26**, 2807–71.

Fauconneua, B., Choubert, G., Blanc, D., Breque, J. & P. Luquet, (1983). Influence of environmental temperature on flow rate of food stuffs through the gastrointestinal tract of rainbow trout. *Aquaculture*, **34**, 27–39.

Feldmeth, C. R. & Jenkins, T. M. (1973). An estimate of energy expenditure by rainbow trout (*Salmo gairdneri*) in small mountain stream. *J. Fish. Res. Bd Can.*, **30**, 1755–9.

Ferrari, I. & Chieregato, A. R. (1981). Feeding habits of juvenile stages of *Sparus auratus* L., *Dicentrarchus labrax* L. and Mugilidae in a brackish embayment of the Po River delta. *Aquaculture*, **25**, 243–57.

Finger, T.E. (1983). Two gustatory systems – implications for pollution research. In K. Doevig (Ed.), *Chemoreception in studies of marine pollution, reports from a workshop at Oslo, July 13 and 14, 1980*. Rep. Forskning-spragram Havforurens (Norway), no. 1, 33–5.

Fischer, Z. (1970). The elements of energy balance in grass carp (*Ctenopharyngodon idella* val.), Part I. *Pol. Arch. Hydrobiol.*, **17**, 421–34.

Fischer, Z. (1978). Selected problems of fish bioenergetics. *FAO/EIFAC Symposium of finfish nutrition and feed technology, Hamburg*. EIFAC/78/Symp., R/7, 29.

Fischer, Z. & Lipka, J. (1983). The role of amino acids in the nitrogen metabolism of artificially fed carp. *Pol. Arch. Hydrobiol.*, **30**, 363–80.

Fish, G. R. (1951). Digestion in *Tilapia esculenta. Nature (Lond.)*, **167**, 900.

Fish, G. R. (1960). The comparative activity of some digestive enzymes in the alimentary canal of tilapia and perch. *Hydrobiol.*, **15**, 161–78.

Foltz, J. W. (1982). A feeding guide for single cropped channel catfish (*Ictalurus punctatus*). *J. World Maricult. Soc.*, **13**, 274–81.

Fontaine, M. & Marchelidon, J. (1971). Amino acid contents of the brain and the muscle of young salmon (*Salmo salar* L.) at parr and smolt stages. *Comp. Biochem. Physiol.*, **40A**, 127–34.

Fontaine, Y-A. (1975). Hormones in fishes. In D. C. Malins & J. R. Sargent (Eds), *Biochemical and biophysical perspectives in marine biology, vol. 2*, pp. 139–212. London: Academic Press.

Forster, R. P. & Goldstein, L. (1969). Formation of excretory products. In W. S. Hoar & D. J. Randall (Eds), *Fish physiology, Vol. I. Excretion, ionic regulation and metabolism*, pp. 313–50. New York: Academic Press.

Fowler, L. G. (1980). Substitution of soybean and cottonseed products for fish meal in diets fed to chinook and coho salmon. *Prog. Fish Cult.*, **42**, 87–91.

Fraser D. I., Dyer, W. J., Weinstein, H. M., Dingle, J. R. & Hines, J. A. (1966). Glycolytic metabolites and their distribution at death in the white and red muscle of cod following various degrees of antemortem muscular activity. *Can. J. Biochem. Physiol.*, **44**, 1015–33.

Freeman, R. I., Haskell, D. C., Longacre, D. L. & Stiles, E. W. (1967). Calculation of amounts to feed in trout hatcheries. *Prog. Fish Cult.*, **29**, 194–209.

Frigg, M. (1976). Bio-availability of biotin from cereals. *Poultry Sci.*, **55**, 2310–18.

Frolova, L. K. (1957). The effect of inorganic cobalt on the growth and metabolism of young carp. (In Russian.) *Inf. Sb. Vsesoj. Nauch-issled. Inst. Morsk. Ryb. Khoz. Okeanogr.* (VNIRO), **1**.

Frolova, L. K. (1961). Effect of cobalt on some haematological characteristics of carp. (in Russian). *Tr. Vses. Nauch-issled. Inst. Morsk. Ryb. Khoz. Okeanogr. (VNIRO)*, **44**, 48–59: (English Transl: Referat Zhur. Biol., 8 I 72, 1983.)

Frolova, L. K. (1970). Effect of cobaltous chloride on certain physiological indices of young carp (*Cyprinus carpio*) in conditions of commercial breeding (in Russian). *Tr. Vses, Nauch-issled. Inst. Morsk. Ryb. Khoz. Okeanogr. (VNIRO)*, **69**, 158–62. Transl. Ser., Fish. Res. Bd Can. (1631) 1971, 13pp.

Fromm, P. O. (1963). Studies on renal and extra-renal excretion in freshwater teleost *Salmo gairdneri. Comp. Biochem. Physiol.*, **10**, 121–8.

Fry, F. E. J. (1947). Effects of environment on animal activity. Univ. Toronto Studies, Biol. Ser., 55. *Publ. Ontario Fish. Res. Lab.*, **62**, 1–68.

Fry, F. E. J. (1957). The aquatic respiration of fish. In M. E. Brown (Ed.), *The physiology of fishes, vol. 1*, pp. 1–63. New York: Academic Press.

Fry, F. E. J. (1958). Temperature compensation. *Ann. Rev. Physiol.*, **20**. 207–24.

Fry, F. E. J. (1971). The effect of environmental factors on the physiology of fish. In W. S. Hoar & D. J. Randall (Eds), *Fish physiology, vol. 6. Environmental relations and behaviour*, pp. 1–98. New York: Academic Press.

Fry, F. E. J. & Hart, J. L. (1948a). Cruising speed of goldfish in relation to water temperature. *J. Fish. Res. Bd Can.*, **7**, 169–75.

Fry, F. E. J. & Hart, J. L. (1948b). The relation of temperature to oxygen consumption in the goldfish. *Biol. Bull.*, **94**, 66–7.

Fryer, G. & Iles, T. D. (1972). *The cichlid fishes of the great lakes of Africa, their biology and evolution*. Edinburgh: Oliver & Boyd, 641 pp.

Fryer, J. N. (1975). Stress and adrenocorticosteroid dynamics in the goldfish, *Carassius auratus*. *Can. J. Zool.*, **53**. 1012–20.

Fukushima, D. (1968). Internal structure of 7s and 11s globulin molecules in soybean proteins. *Cereal Chem.*, **45**(3), 203–24.

Furuichi, M. & Yone, Y. (1971a). Studies on nutrition of red sea bream – IV. Nutritive value of dietary carbohydrate. *Rep. Fish. Res. Lab., Kyushu Univ.*, **1**, 75–81.

Furuichi, M. & Yone, Y. (1971b). Studies on nutrition of red sea beam – VI. A study of carbohydrate utilization by glucose and insulin-glucose tolerance tests. *Rep. Fish. Res. Lab., Kyushu Univ.*, **1**, 101–6.

Furuichi, M. & Yone, Y. (1981). Change of blood sugar and plasma insulin levels of fishes in glucose tolerance test. *Bull. Jap. Soc. Sci. Fish.*, **47**, 761–4.

Furuichi, M. & Yone, Y. (1982a). Changes in activities of hepatic enzymes related to carbohydrate metabolism of fishes in glucose and insulin-glucose tolerance tests. *Bull. Jap. Soc. Sci. Fish.*, **48**, 463–6.

Furuichi, M. & Yone, Y. (1982b). Availability of carbohydrate in nutrition of carp and red seam bream. *Bull. Jap. Soc. Sci. Fish.*, **48**, 945–8.

Furuichi, M., Shitanda, K. & Yone, Y. (1971). Studies on nutrition of red sea bream – V. Appropriate supply of dietary carbohydrate. *Rep. Fish. Res. Lab., Kyushu Univ.*, **1**, 91–100.

Furukawa, S. & Ogasawara, Y. (1952). Studies on nutrition of fish. II. Digestibility of protein diet with cellulose. *Bull. Jap. Soc. Sci. Fish.*, **17**, 45–8.

Furukawa, A. & Ogasawara, Y. (1953). Studies on nutrition of fish. III. Digestibility of protein in dried, and fat free dried silkworm pupa and fish meal. *Bull. Naikai Regional Fish. Res. Lab.*, **5**, 25–9.

Furukawa, A., Ogasawara, Y. & Hoshiguchi, H. (1953). Studies on nutrition of fish. IV. Digestibility of protein in artificial and several natural diets. *Bull. Naikai Regional Fish. Res. Lab.*, **5**, 31–6.

Garber, K. J. (1983). Effect of fish size, meal size and dietary moisture on gastric evacuation of pelleted diets by yellow perch, *Perca flavescens*. *Aquaculture*, **34**, 41–9.

Garling, D. L. & Wilson, R. P. (1976). Optimum dietary protein to energy ratio for channel catfish fingerlings, *Ictalurus punctatus*. *J. Nutr.*, **106**, 1368–75.

Garling, D. L. & Wilson, R. P. (1977). Effects of dietary carbohydrate-to-lipid ratios on growth and body consumption of fingerling channel catfish. *Prog. Fish Cult.*, **39**, 43–7.

Gatlin, D. M. & Wilson, R. P. (1984). Studies on the manganese requirement of fingerling channel catfish. *Aquaculture*, **41**, 85–92.

Gaulitz, E. J., Hall, A. S. & Houle, C. R. (1979). Effects of dietary lipids on the growth and acceptability (flavour) of channel catfish (*Ictalurus punctatus*). In J. E. Halver & K. Tiews (Eds), *Finfish nutrition and fish feed technology, vol. II*, pp. 87–98. Berlin: Heenemann Verlagsgesel.

George, J. C. (1962). A histophysiological study of the red and white muscles of the mackerel. *Am. Midl. Nat.*, **68**, 487–94.

George, J. C. & Bokdawala, F. D. (1964). Cellular organization and fat utilization in fish muscle. *J. Anim. Morphol. Physiol.*, **11**, 124–32.

George, J. C., Barnett, B. J., Cho, C. Y. & Slinger, S. J. (1979). Impairment of white skeletal muscle in vitamin D_3 deficient rainbow trout. *J. Nutr.*, **109**(6), xxiii (Abstr.).

Gerking, S. D. (1952). The protein metabolism of sunfishes of different ages. *Physiol. Zool.*, **25**, 358–72.

Gerking, S. D. (1954). The food turnover of a bluegill population. *Ecology*, **35** 490–8.

Gerking, S. D. (1955a). Influence of rate of feeding on body composition and protein metabolism of bluegill sunfish. *Physiol. Zool.*, **28**, 267–82.

Gerking, S. D. (1955b). Endogenous nitrogen excretion of bluegill sunfish. *Physiol. Zool.*, **28**, 283–9.

Gerking, S. D. (1962). Production and food utilization in a population of bluegill sunfish. *Ecol. Monogr.*, **32**, 31–78.

Gerking, S. D. (1971). Influence of rate of feeding and body weight on protein metabolism of bluegill sunfish. *Physiol. Zool.*, **44**, 9–19.

Gerking, S. D. (1972). Revised food consumption estimate of a bluegill sunfish population in Wyland Lake, Indiana, U.S.A. *J. Fish. Biol.*, **4**, 301–8.

Gershanovich, A. D. (1983). Effects of temperature on metabolism, growth and food requirement of young beluga, *Huso huso*, and sheep sturgeon, *Acipenser nudiventris* (Acipenseridae). *J. Ichthyol.*, **23**, 55–60.

Ghosh, K. K. (1974). On the choice of the growth curve for Indian major carps – von Bertalanffy or Gompertz? *Jr. Ind. Soc. Agr. Statistics*, **26**(2), 57–70.

Ghosh, S. R. (1975). Preliminary observation on the effect of cobalt on survival and growth of *Mugil parsia*. *Bamidgeh*, **27**(4), 110–11.

Glass, N. R. (1969). Discussion of calculation of power function with special reference to respiratory metabolism in fish. *J. Fish. Res. Bd Can.*, **26** 2643–50.

Goel, K. A. (1975). Histochemical study of the activity of acid and alkaline phosphatases and lipase in the gastrointestinal tract of *Cirrhinus mrigala*. *Acta Histochem.*, **54**, 48–55.

Goldblatt, M. J., Conklin, D. E. & Brown, W. D. (1979). Nutrient leaching from pelleted rations. In J. E. Halver & K. Tiews (Eds), *Finfish nutrition and fishfeed technology, vol. II*, pp. 117–124. Berlin: Heenemann Verlagsgesel.

Goldsmith, G. A. (1964). The B vitamins: Thamine, riboflavin, niacin. In G. H. Beaton & E. W. McHenry (Eds), *Nutrition, a comprehensive treatise, vol. II*, pp. 102–206. New York: Academic Press.

Gomazkov, O. A. & Krayukhin, B. V. (1961). Role of the vagus in regulating digestion in fish. *Fiziol. Zhur. SSSR*, **47**(10), 54–8.

Gorbman, A. (1969). Thyroid function and its control in fishes. In W. S. Hoar & D. J. Randall (Eds), *Fish physiology*, pp. 241–74. New York: Academic Press.

Gordon, M. S. (1968). Oxygen consumption of red and white muscles from tuna fishes. *Science*, **159**, 87–90.

Grabner, M. (1985). An in vitro method for measuring protein digestibility of fish feed components. *Aquaculture*, **48**, 97–110.

Grabner, M. & Hofer, R. (1985). The digestibility of the proteins of broad bean (*Vicia faba*) and soya bean (*Glycine max*) under in vitro conditions simulating the alimentary tracts of rainbow trout (*Salmo gairdneri*) and carp (*Cyprinus carpio*). *Aquaculture*, **48**, 111–22.

Grafflin, A. L. & Gould, R. G. (1936). Renal function in marine teleosts. II. The nitrogenous constituents of urine of sculpin and flounder with particular reference to trimethylamine oxide. *Biol. Bull.*, **70**, 16–27.

Graig, F. A. (1962). Role of zinc and other trace metals in diabetes mellitus. *NY J. Med.*, **62**, 75–81.

Grant, P. T. & Reid, K. M. B. (1968). Biosynthesis of an insulin precursor by islet tissue of cod (*Gadus collarias*). *Biochem. J.*, **110**, 281–8.

Grant, P. T., Coombs, T. L. & Frank, B. H. (1972). Differences in the nature of the interaction of insulin and proinsulin with zinc. *Biochem. J.*, **126**, 433–40.

Grayton, B. D. & Beamish, F. W. H. (1977). Effects of feeding frequency on food intake, growth and body composition of rainbow trout (*Salmo gairdneri*). *Aquaculture*, **11**, 159–72.

Greer-Walker, M. & Pull, G. A. (1975). A survey of red and white muscle in marine fish. *J. Fish. Biol.*, **7**, 295–300.

Grollman, A. (1929). The urine of the goosefish (*Lophius piscatorius*) its nitrogenous constituents with special reference to the presence of trimethylamine oxide. *J. Biol. Chem.*,

81, 267–78.

Gropp, J. (1979). Standard fish husbandry designs of experiments and their evaluation. In J. E. Halver & K. Tiews (Eds), *Finfish nutrition and fish feed technology, vol. II*, pp. 445–52. Berlin: Heenemann Verlagsgesel.

Gropp. J., Koops, H., Tiews, K. & Beck, H. (1979). Replacement of fish meal in trout feeds by other feedstuffs. In T. V. R. Pillay & W. A. Dill (Eds), *Advances in aquaculture*, pp. 596–601. Farnham, England: Fishing News Books.

Gross, W. L., Roelofs, E. W. & Fromm, P. O. (1965). Influence of photoperiod on growth of green sunfish *Lepomis cyanellus*. *J. Fish. Res. Bd Can.*, **22** 1379–86.

Grygierek, E. (1973). The influence of phytophagous fish on pond zooplankton. *Aquaculture*, **2**, 197–208.

Guerin-Ancey, O. (1976a). Etude experimentale de l'excretion azotée du bar (*Dicentrarchus labrax*) en cours de croissance. I. Effects de la temperature et du poids du corps sur l'excretion d'ammoniac et d' urée. (In French with English summary.) *Aquaculture*, **9**, 71–80.

Guerin-Ancey, O. (1976b). Etude experimentale de l'excretion azotée du bar (*Dicentrarchus labrax*) en cours de croissance. II. Effects du juene sur l'excretion d'ammoniac et d' úrée. (In French with English summary.) *Aquaculture*, **9**, 187–94.

Gurzeda, A. (1965). Density of carp populations and their artificial feeding and the utilization of food animals. *Ekol. Pol. A*, **13**(7), 73–99.

Halver, J. E. (1957a). Nutrition of salmonoid fishes. III. Water-soluble vitamin requirements of chinook salmon. *J. Nutr.*, **62**, 225–43.

Halver, J. E. (1957b). Nutrition of salmonoid fishes. IV. An amino acid test diet for chinook salmon. *J. Nutr.*, **62**, 245–54.

Halver, J. E. (1965). Tryptophan requirements of chinook, sockeye and silver salmon. *Fed. Proc., Fed. Am. Soc. Exp. Biol.*, **24**, 169 (Abstr. 229).

Halver, J. E. (1972). The vitamins. In J. E. Halver (Ed.), *Fish nutrition*, pp. 29–103. New York: Academic Press.

Halver, J. E. (1979). Vitamin requirements of finfish. In J. E. Halver & K. Tiews (Eds), *Finfish nutrition and fish feed technology, vol. I*, pp. 45–58. Berlin: Heenemann Verlagsgesel.

Halver, J. E. (1980a). Vitamin, fat and mineral requirements. In W. Santos, N. Lopes, J. J. Barbosa, D. Chaves & J. V. Valente (Eds), *Nutrition and food science, present knowledge and utilization, vol. 2*, pp. 177–80. New York: Plenum Press.

Halver, J. E. (1980b). Vitamin requirements of finfish. In W. Santos, N. Lopes, J. J. Barbosa, D. Chaves & J. C. Valente (Eds), *Nutrition and food science, present knowledge and utilization, vol. 2*, pp. 191–8. New York: Plenum Press.

Halver, J. E., Ashley, L. M. & Smith, R. R. (1969). Ascorbic acid requirements of coho salmon and rainbow trout. *Trans. Am. Fish. Soc.*, **98**, 762–71.

Halver, J. E., Bates, L. S. & Mertz, (1964). Protein requirements of sockeye salmon and rainbow trout (Abstr.). *Fed. Proc., Fed. Am. Soc. Exp. Biol.*, **23**, 397.

Halver, J. E., DeLong, D. C. & Mertz, E. T. (1958). Threonine and lysine requirements of chinook salmon. *Fed. Proc., Fed. Am. Soc. Exp. Biol.*, **17**, 478 (Abstr. 1873).

Halver, J. E. & Shanks, W. E. (1960). Nutrition of salmonoid fishes. VIII. Indispensable amino acids for sockeye salmon. *J. Nutr.*, **72**, 340–6.

Hamada, A. & Ida, T. (1973). Studies on specific dynamic action of fishes I. Various conditions affecting the value of measurement. *Bull. Jap. Soc. Sci. Fish.*, **39**, 1231–5.

Hamada, A., Ida, T., Tsda, T. & Kariya, T. (1975). Studies on the growth of fishes I. Maximum growth of carp. *Bull. Jap. Soc. Sci. Fish.*, **41**, 147–54.

Hamoir, G. (1955). Fish Proteins. *Adv. Protein Chem.*, **10**, 227–88.

Hanaoka, T., Furukawa, A. & Ogasawara, Y. (1948). Experiments on nutrition of fish I. Digestibility of protein in foodstuff with different nutritive values. *Bull. Jap. Soc. Sci. Fish.*, **14**, 219–22.

Harada, K. Yamada, K. (1973). Distribution of trimethylamine oxide in fishes and other aquatic animals. V. Teleosts and elasmobranchs. *J. Shimonoseki Univ. Fish.*, **22**(2), 77–94.

Harder, W. (1975). *Anatomy of fishes, part 1: Text.* Stuttgart: Schweizerbartische Verlagsbuchhan., 612 pp.

Harding, D. E., Allen, O. W. & Wilson, R. P. (1977). Sulfur amino acid requirement of channel catfish L-methionine and L-cystine. *J. Nutr.*, **107**, 2031–5.

Harper, H. A. (1971). *Review of physiological chemistry*, 13th edn. Los Altos, Calif: Lange Medical Publ., 564 pp.

Hashimoto, Y. (1953). Effect of antibiotics and vitamin B_{12} supplement on carp growth. *Bull. Jap. Soc. Sci. Fish.*, **19**, 899–904.

Hashimoto, Y., Arai, S. & Nose, T. (1970). Thiamine deficiency symptoms experimentally induced in the eel. *Bull. Jap. Soc. Sci. Fish.*, **36**, 791–7.

Haskell, D. C. (1959). Trout growth in hatcheries. *New York Fish Game J.*. **6**, 204–37.

Hastings, W. H. (1971). Study of pelleted fish foods stability in water. *Resour. Publ. US Bur. Sport Fish. Wildl.*, **102**, 75–80.

Hastings, W. H. (1973). Phase feeding for catfish. *Prog. Fish Cult.*, **35**, 195–6.

Hastings, W. & Dupree, H. K. (1969). Formula feeds for channel catfish. *Prog. Fish Cult.*, **31**, 187–96.

Hastings, W. H., Meyers, S. P. & Butler, D. P. (1971). A commercial process for water-stable fish feed. *Feedstuffs*, **43**(47), 38.

Hathaway, E. S. (1927). The relation of temperature to the quantity of food consumed by fishes. *Ecology*, **8**, 428–34.

Hayama, K. & Ikeda, S. (1972). Studies on the metabolic changes of fish caused by diet variation. I. Effect of dietary composition on the metabolic changes of carp. *Bull. Jap. Soc. Sci. Fish.*, **38**, 639–43.

Hazel, J. R. & Prosser, L. (1974). Molecular mechanisms of temperature compensation in poikilotherms. *Physiol. Rev.*, **54**, 620–77.

Heath, A. G. & Hughes, G. M. (1973). Cardiovascular and respiratory changes during heat stress in rainbow trout (*Salmo gairdneri*). *J. Exp. Biol.*, **59**, 323–38.

Heath, A. G., & Pritchard, A. W. (1965). Effects of severe hypoxia on carbohydrate energy stores and metabolism in two species of fresh-water fish. *Physiol. Zool.*, **38**, 325–34.

Henken, A. M., Kleingeld, D. W. & Tijssen, P. A. T. (1985). The effect of feeding level on apparent digestibility of dietary dry matter, crude protein and gross energy in the African catfish (*Clarias gariepinus*) (Burchell, 1822). *Aquaculture*, **51**, 1–11.

Hepher, B. (1962a). Ten years of research in fish pond fertilization in Israel I. The effect of fertilization of fish yields. *Bamidgeh*, **14**, 29–38.

Hepher, B. (1962b). Primary production in fishponds and its application to fertilization experiments. *Limnol. Oceanogr.*, **7**, 131–6.

Hepher, B. (1975). Supplementary feeding in fish culture. *Proc. Int. Cong. Nutr.*, **9** (vol. 3). 183–98.

Hepher, B. (1978). Ecological aspects of warm-water fishpond management. In S. D. Gerking (Ed.), *Ecology of freshwater fish production* pp. 447–68. Oxford: Blackwell.

Hepher, B. (1979). Supplementary diets and related problems in fish culture. In J. E. Halver & K. Tiews (Eds), *Finfish nutrition and fishfeed technology, vol. I*, pp. 343–7. Berlin: Heenemann Verlagsgesel.

Hepher, B. & Chervinski, J. (1965). Studies on carp nutrition. The influence of protein-rich diets on growth. *Bamidgeh*, **17**(2), 31–46.

Hepher, B. & Pruginin, Y. (1981). Commercial fish farming, with a special reference to fish culture in Israel. New York: Wiley Interscience, 261 pp.

Hepher, B. & Sandbank, E. (1984). The effect of phosphorus supplementation to common carp diets on fish growth. *Aquaculture*, **36**, 323–32.

Hepher, B., Chervinski, J. & Tagari, H. (1971). Studies on carp nutrition III. Experiments on the effect on fish yields of dietary protein source and concentration. *Bamidgeh*, **23**(1), 11–37.

Hepher, B., Sandbank, E. & Shelef, G. (1979). Alternative protein sources for warm-water fish diets. In J. E. Halver & K. Tiews (Eds), *Finfish nutrition and fishfeed technology, vol. I*, pp. 327–41. Berlin: Heenemann Verlagsgesel.

Hepher, B., Liao, I. C., Cheng, S. H. & Hsieh, C. S. (1983). Food utilization by tilapia – effect of diet composition feeding level and temperature on utilization efficiency for maintenance and growth. *Aquaculture*, **32**, 255–75.

Herrmann, R. B., Warren, C. E., & Doudoroff, P. (1962). Influence of oxygen concentration on the growth of juvenile coho salmon. *Trans. Am. Fish. Soc.*, **91**, 155–67.

Hettler, W. F. (1976). Influence of temperature and salinity on routine metabolic rate and growth of young Atlantic menhaden. *J. Fish. Biol.*, **8**, 55–65.

Hevesy, G., Lockner, D. & Sletten, K. (1964). Iron metabolism and erythrocyte formation in fish. *Acta Physiol. Scand.*, **60**, 256–66.

Heusner, A., Kayser, C., Marx, C., Stussi, T. & Harmelin, M. L. (1963). Relation entre le poid et la consommation d'oxygene. II. Etude intraspecifique chez le poisson. *C.R. Seances Soc. Biol. Fil.*, **157**, 654–7.

Hickling, C. F. (1966). On the feeding process in the white amur *Ctenopharyngodon idella*. *J. Zool. (Lond.)*, **148**, 404–18.

Hickling, C. F. (1970). Estuarine fish farming. *Adv. Mar. Biol.*, **8**, 119–213.

Hickman, C. P. (1959). The osmoregulatory role of the thyroid gland in the starry flounder, *Platichthys stellatus*. *Can. J. Zool.*, **37**, 997–1060.

Higashi, H., Kaneko, T., Ishii, S., Ushiyama, M. & Sugihashi, T. (1964). Effect of dietary lipids on fish reared artificially. (II) Effect of ethyl-linoleate, ethyl-linolenate and ethyl esters of polyunsaturated fatty acids on deficiency of essential fatty acids of rainbow trout. *Bitamin*, **30**(4), 271–5. Transl. Ser., Fish. Res. Bd Can., (653) 1965.

Higashi, H., Kaneko, T., Ishii, S., Ushiyama, & Sugihashi, T. (1966). Effect of ethyl linoleate, ethyl linolenate and ethyl esters of highly unsaturated fatty acids on the essential fatty acid deficiency in rainbow trout. *J. Vitaminol.* (*Kyoto*), **12**(1), 74–9.

Higgs, D. A., Markert, J. R., MacQuarrie, D. W., McBride, J. R., Dosanjh, B. S., Nichols, C. & Hoskins, G. (1979). Development of practical dry diets for coho salmon, *Oncorhynchus kisutch*, using poultry-by-product meal, feather meal, soybean meal and rapeseed meal as major protein sources. In J. E. Halver & K. Tiews (Eds), *Finfish nutrition and fishfeed technology, vol. II*, pp. 191–218. Berlin: Heenemann Verlagsgesel.

Higgs, D. A., McBride, J. R., Markert, J. R., Dosanjh, B. S., Plotnikoff, M. D. & Clarke, C. (1982). Evaluation of Tower and Candle rapeseed (Canola) meal and Bronowski rapeseed protein concentrate as protein supplements in practical dry diets for juvenile chinook salmon (*Oncorhynchus tshawytscha*). *Aquaculture*, **29**, 1–31.

Higgs, D. A., Fagerlund, U. H. M., McBride, J. R., Plotnikoff, M. D., Dosanjh, B. S., Markert, J. R. & Davidson, J. (1983). Protein quality of altex canola meal for juvenile chinook salmon (*Oncorhynchus tshawytscha*) considering dietary protein and 3, 5, 3-triiodo-1-thyronine content. *Aquaculture*, **34**, 213–38.

Higuera, M. De La, Murillo, A., Varela, G. & Zamora, S. (1977). The influence of high dietary fat levels on protein utilization by trout (*Salmo gairdneri*). *Comp. Biochem. Physiol.*, **56A**, 37–41.

Hilton, J. W. (1983). Hypervitaminosis A in rainbow trout (*Salmo gairdneri*): Toxicity signs and maximum tolerable level. *J. Nutr.*, **113**, 1737–45.

Hilton, J. W., Hodson, P. V. & Slinger, S.J. (1980). The requirement and toxicity of selenium in rainbow trout (*Salmo gairdneri*). *J. Nutr.*, **110**, 2527–35.

Hilton, J. W., Atkinson, J. L. & Slinger, S. J. (1983). Effect of increased dietary fibre on the growth of rainbow trout (*Salmo gairdneri*). *Can. J. Fish. Aquat. Sci.*, **40**, 81–5.

Hirao, S., Yamada, J. & Kikuchi, R. (1960). On improving the efficiency of feed for fish culture I. Transit and digestibility of diet in eel and rainbow trout observed by use of P^{32}. *Bull. Tokai Reg. Fish. Res. Lab.*, (7), 67–72.

Hoar, W. S. (1956). Photoperiodism and thermal resistance of goldfish. *Nature (Lond.)*, **170**,

364.
Hoar, W. S. (1975). *General and comparative physiology.* Englewood Cliffs, NJ: Prentice-Hall, 774 pp.
Hoar, W. S. & Cottle, M. K. (1952). Some effects of temperature acclimatization on the chemical constitution of goldfish tissues. *Can. J. Zool.*, **30**, 49–54.
Hochachka, P. W. (1961). Glucose and acetate metabolism in fish. *Can. J. Biochem. Physiol.*, **39**, 1937–41.
Hochachka, P. W. (1962). Thyroidal effects on pathways for carbohydrate metabolism in teleosts. *Gen. Comp. Endocrinol.*, **2**, 499–505.
Hochachka, P. W. (1967). Organization of metabolism during temperature compensation. In C.L. Prosser (Ed.), *Molecular mechanisms of temperature adaptation.* AAA Publ., 84. Washington: 177–203.
Hochachka, P. W. (1969). Intermediary metabolism in fishes. In W. S. Hoar & D. J. Randall (Eds), *Fish physiology, vol. I, Excretion, ionic regulation and metabolism*, pp. 351–389. New York: Academic Press.
Hochachka, P. W. & Hayes, F. R. (1962). The effect of temperature acclimation on pathways of glucose metabolism in the trout. *Can. J. Zool.*, **40**, 261–70.
Hochachka, P. W. & Somero, G. N. (1973). *Strategies of biochemical adaptation.* Philadelphia: Saunders Co., 358 pp.
Hofer, R. (1979a). The adaptation of digestive enzymes to temperature, season and diet in roach, *Rutilus rutilus* and rudd, *Scardinius erythrophthalmus*; proteases. *J. Fish Biol.*, **15**, 373–9.
Hofer, R. (1979b). The adaptation of digestive enzymes to temperature, season and diet in roach, *Rutilus rutilus* L. and rudd, *Scardinius erythrophthalmus* L. 1. Amylase. *J. Fish. Biol.*, **14**, 565–72.
Hofer, R. & Sturmbauer, C. (1985). Inhibition of trout and carp amylase by wheat. *Aquaculture*, **48**, 277–83.
Hogendoorn, H. (1983). Growth and production of the African catfish, *Clarias lazera* (C. & V.) III: Bioenergetic relations of body weight and feeding level. *Aquaculture*, **35**, 1–7.
Hokazono, S., Tanaka, Y., Katayama, T., Chichester, C. O. & Simpson, K. L. (1979). Intestinal transport of L-lysine in rainbow trout, *Salmo gairdneri. Bull. Jap. Soc. Sci. Fish.*, **45**, 845–8.
Holstein, B. (1975). Gastric acid secretion in a teleostean fish: A method for the continuous collection of gastric effluence from a swimming fish and its response to histamine and pentagastrin. *Acta Physiol. Scand.*, **95**, 417–23.
Holton, R. W., Blecker, H. H. & Onore, M. (1964). Effect of growth temperature on the fatty acid composition of blue-green algae. *Phytochemistry*, **3**, 595–602.
Hotta, H. Y. & Nakashima, J. (1968). Experimental study of the feeding activity of jack-mackerel *Trachurus japonicus.* (in Japanese with English summ.) *Bull. Seikai Reg. Fish. Res. Lab.*, **36**, 75–83.
Houston, A. H. & De Wilde, M. A. (1968). Thermoacclimatory variations in the haematology of the common carp, *Cyprinus carpio. J. Exp. Biol.*, **49**, 71–81.
Houston, A. H. & Rupert, R. (1976). Immediate response of the haemoglobin system of the goldfish, *Carassius auratus*, to temperature changes. *Can. J. Zool.*, **54**, 1737–41.
Hruska, V. (1961). An attempt at a direct investigation of the influence of the carp stock on the bottom fauna of two ponds. *Verh. Inter. Ver. Limnol.*, **14**, 732–6.
Hsu Y.-L. & Wu, J.-L. (1979). The relationship between feeding habits and digestive proteases of some freshwater fishes. *Bull. Inst. Zool. Acad. Sinica*, **18**(1), 45–53.
Huang, Y. (1975). Oxygen consumption of the early developmental stages of the red and common carp (*Cyprinus carpio*), the silver carp (*Hypophthalmichthys molitrix*) and the Chinese bream (*Parabramis pekinensis*), (In Chinese with English summary.) *Acta Zool. Sin.*, **21**(1), 88–96.
Huckabee, W. E. (1958). Relationship of pyruvate and lactate during anaerobic metabolism.

II. Exercise and formation of O_2 debt. *J. Clin. Invest.*, **37**, 255–63.

Hudon, B. & De La Noue, J. (1984). Influence of meal frequency on apparent nutrient digestibility in rainbow trout, *Salmo gairdneri*. *Bull. Fr. Piscic.*, **293–4**, 49–51.

Huggins, A. K., Skutsch, G. & Baldwin, E. (1969). Ornithine urea cycle enzymes in teleostean fish. *Comp. Biochem. Physiol.*, **28**, 587–602.

Hughes, G. M. & Knights, B. (1968). The effect of loading the respiratory pumps on the oxygen consumption of *Callionymus lyra*. *J. Exp. Biol.*, **49**, 603–15.

Hughes, S. G., Rumsey, G. L. & Nickum, J. G. (1981). Riboflavin requirement of fingerling rainbow trout. *Prog. Fish Cult.*, **43**, 167–72.

Huh, H. T., Calbert, H. E. & Stuiber, D. A. (1976). Effects of temperature and light on growth of yellow perch and walleye using formulated feed. *Trans. Am. Fish. Soc.*, **105**, 254–8.

Huisam, E. A. (1976). Food conversion efficiencies at maintenance and production levels for carp, *Cyprinus carpio* L., and rainbow trout, *Salmo gairdneri* Richardson. *Aquaculture*, **9**, 259–73.

Huisman, E. A., Klein-Breteler, J. G. P., Vismans, M. M. & Kanis, E. (1979). Retention of energy, protein, fat and ash in growing carp (*Cyprinus carpio* L.) under different feeding and temperature regimes, in J. E. Halver & K. Tiews (Eds), *Finfish nutrition and fish feed technology, vol. I*, pp. 175–88. Berlin: Heenemann Verlagsgesel.

Hunt, B. P. (1960). Digestion rate and food consumption of florida gar, warmouth and largemouth bass. *Trans. Am. Fish. Soc.*, **89**, 206–11.

Hunt, J. N. (1980). A possible relation between the regulation of gastric emptying and food intake. *Am. J. Physiol.*, **239**, G1–G4.

Idler, D. R. & Clemens, W. A. (1959). The energy expenditures of Fraser River sockeye salmon during the spawning migration to Chilko and Stuart Lakes. *Int. Pacif. Salmon Fish Comm. Prog. Rep*, 1–80.

Iida, A., Tayama, T., Kumai, T., Takamatsu, C. & Nishikawa, T. (1970). Studies on the nutritive value of petroleum yeast in carp and rainbow trout feeding. *Suisanzoshoku (The Aquaculture)*, **18**, 35–43.

Ikeda, S. & Sato, M. (1964). Biochemical studies on L-ascorbic acid in aquatic animals – III. Biosynthesis of L-ascorbic acid by carp. *Bull. Jap. Soc. Sci. Fish.*, **30**, 365–9.

Ikeda, Y., Ozaki, H. & Yasuda, H. (1972). Effects of potassium iodide on growth of body and scale in goldfish. *J. Tokyo Univ. Fish.*, **59**(1), 33–42.

Ikeda, Y., Ozaki, H. & Uematsu, K. (1973). Effect of enriched diet with iron in culture of yellowtail. (In Japanese with English summary.) *J. Tokyo Univ. Fish.*, **59**(2), 91–9.

Inaba, D., Ogino, C., Takamatsu, C., Sugano, S. & Hata, H. (1962). Digestibility of dietary components in fishes. I. Digestibility of dietary protein in rainbow trout. *Bull. Jap. Soc. Sci. Fish.*, **28**, 367–71.

Inaba, D., Ogino, C., Takamatsu, C., Ueda, T. & Kurokawa, K.-I. (1963). Digestibility of dietary components in fishes. II. Digestibility of dietary protein and starch in rainbow trout. *Bull. Jap. Soc. Sci. Fish.*, **29**, 242–4.

Ince, B. W. & So, S. T. C. (1984). Differential secretion of glucagon-like and somatostatin-like immunoreactivity from the perfused eel pancreas in response to D-glucose. *Gen. Comp. Endocrinol.*, **53**, 389–97.

Ince, B.W. & Thorpe, A. (1974). Effects of insulin and metabolite loading on blood metabolities in the European silver eel (*Anguilla anguilla* L.). *Gen. Comp. Endocrinol.* **23**, 460–71.

Ince, B. W. & Thorpe, A. (1976). The '*in vivo*' metabolism of ^{14}C-glucose and ^{14}C-glycine in insulin treated northern pike (*Esox lucius* L.). *Gen. Comp. Endocrinol.*, **28**, 481–6.

Ince, B. W. & Thorpe, A. (1978). Effect of insulin on plasma amino acid levels in Northern pike (*Esox lucius* L.). *J. Fish. Biol..*, **12**, 503–6.

Infante, V. O. (1974). Untersuchungen ueber stickstoffexkretion junger Karpfen (*Cyprinus carpio* L.) im Hunger und bei Futterung. *Arch. Hydrobiol.*, Suppl. **7**(2), 239–81.

Ingham, L. & Arme, C. (1977). Intestinal absorption of amino acids by rainbow trout. *J. Comp. Physiol.*, **117**, 323–34.

Inui, Y. & Yokote, M. (1974). Gluconeogenesis in the eel. *Bull. Freshwater Fish. Res. Lab., Tokyo*, **24**, 33–45.

Inui, Y. & Yokote, M. (1975a). Gluconeogenesis in the eel. II. Gluconeogenesis in the alloxanized eel. *Bull. Japan. Soc. Sci. Fish.*, **41**, 291–300.

Inui, Y. & Yokote, M. (1975b). Gluconeogenesis in the eel. III. Effects of mammalian insulin on the carbohydrate metabolism of the eel. *Bull. Jap. Soc. Sci. Fish.*, **41**, 965–72.

Inui, Y. & Yokote, M. (1975c). Gluconeogenesis in the eel. IV. Gluconeogenesis in the hydrocortisone-adminstered eel. *Bull. Jap. Soc. Sci. Fish.*, **41**, 973–81.

Inui, Y., Arai, S. & Yokote, M. (1975). Gluconeogenesis in the eel. VI. Effects of hepatectomy, alloxan and mammalian insulin on the behaviour of the plasma amino acids. *Bull. Jap. Soc. Sci. Fish.*, **41**, 1105–11.

Inui, Y., Yu, J.Y.-L. & Gorbman, A. (1978). Effect of bovine insulin on the incorporation of [^{14}C]glycine into protein and carbohydrate in liver and muscle of hagfish, *Eptatretus stouti. Gen. Comp. Endocrinol.*, **36**, 133–41.

Ionas, G. P. (1974). A calculating method for determining fat content in fish (In Russian.) *Rybn. Khoz., Moskva* (10), 50–52.

Irvine, D. G., Newman, K. & Hoar, W. S. (1957). Effects of dietary phospholipids and cholesterol on the temperature resistance of goldfish. *Can. J. Zool.*, **35**, 691–709.

Ishihara, T., Yasuda, M., Kashiwagi, S., Akiyama, M. & Yagi, M. (1974). Studies on thiaminase I in marine fish. VI. Preventive effect of thiamine on nutritional disease of yellowtail fed anchovy. *Bull. Jap. Soc. Sci. Fish.*, **40**, 775–81.

Ishiwata, N. (1968). Ecological studies on the feeding of fishes. VI. External factors affecting satiation amount. *Bull. Jap. Soc. Sci. Fish.*, **34**, 785–91.

Ishiwata, N. (1969). Ecological studies on the feeding of fishes. VII. Frequency of feeding and satiation amount. *Bull. Jap. Soc. Sci. Fish.*, **35**, 979–84.

Ishiwata, N. (1970). Food consumption. *Bull. Jap. Soc. Sci. Fish.*, **36**, 329–30.

Ishiwata, N. (1979). Ecological studies of the feeding of fishes (*Trachurus japonicus*): An estimation of daily ration in nature. (In Japanese with English summary.) *Bull. Jap. Soc. Sci. Fish.*, **45**, 561–6.

Ivlev, V. S. (1939a). The utilization of energy from oxidation of fats and carbohydrate by poikilothermic animals (In Russian.) *Bull. Mosk. Obshch. Ispy. Prir. (Biol.)*, **48**, 70–8.

Ivlev, V. S. (1939b). Energy balance in the carp. *Zool. Zhur.*, **18**, 303–18.

Ivlev, V. S. (1945). The biological productivity of waters. (In Russia.) *Usp. Sovrem. Biol.*, **19**, 98–120. Transl. Ser., Fish. Res. Bd Can. (394).

Ivlev, V. S. (1959). Estimation of evolutionary significance of energetic levels of metabolism. (In Russian.) *Zhur. Obshc. Biol.*, **20**, 94–103.

Ivlev, V. S. (1961). *Experimental ecology of the feeding of fishes*. New Haven: Yale. Univ. Press, 302 pp.

Ivleva, I. V. (1973). Quantitative correlation of temperature and respiratory rate in poikilothermic animals. *Pol. Arch. Hydrobiol.*, **20**, 283–300.

Iwata, K. (1970). Relationship between food and growth in young crucian carp, *Crassius auratus* Cuvieri, as determined by the nitrogen balance. *Jap. J. Limnol.*, **31**, 129–51.

Jackim. E. & LaRoche G. (1973). Protein synthesis in *Fundulus heteroclitus* muscle. *Comp. Biochem. Physiol.*, **44A**, 851–66.

Jackson, A. J. & Capper, B. S. (1982). Investigations into the requirements of the tilapia *Sarotherodon mossambicus* for dietary methionine, Lysine and arginine in semi-synthetic diets. *Aquaculture*, **29**, 289–97.

Jackson, A. J., Capper, B. S. & Matty, A. J. (1982). Evaluation of some plant proteins in complete diets for the tilapia *Sarotherodon mossamabicus. Aquaculture*, **27**, 97–109.

Jan, L., Hsu Y.-L. & Wu, J.-L. (1977). Nutritive values of leaf protein concentrates for grass carp *Ctenopharyngodon idellus. Bull. Inst. Zool. Acad. Sinica*, **16**(2), 91–8.

Jančařík, A. (1949). Contribution to the physiology of the digestion of the carp: the digestion of proteids. (In Czech with English summary.) *Sbor. Vys. Skoly Zemed. Brne, CSR*, **D41**, 60 pp.

Jančařík, A. (1964). Die Verdauung der Hauptnaehrstoffe beim Karpfen. *Z. Fisch.*, **12**, 602–84.

Jančařík, A. (1968). The digestion of main nutrients by carp. *Proc. FAO/EIFAC Session, 5*. F1/EIFAC 68/SC II–6., 2 pp.

Janecek, V. (1968). Effect of feeding intensity and feed composition on fat content and condition of carp. *Bull. VUR Vodnany*, **4**, 22–8.

Janeck, V. (1972). Effect of intensifying interventions on the quality of fish flesh, Part II: (a) Methods and experimental part. *Bull. VUR Vodnany*, **8**(1), 3–12.

Jankowsky, H.-D. (1964a). Die Bedeutung der Hormone fuer die Temperaturanpassung in normalen Temperaturbereich. *Helgol. Wiss. Meeresunters.*, **9**, 412–9.

Jankowsky, H.-D. (1964b). Der Einfluss des Blutes auf den Sauerstoffverbrauch des isolierten Muselgewebes von Schleien (*Tinca tinca* L.). *Zool. Anz.*, **172**, 233–9.

Jankowsky, H.-D. (1966). The effect of adaptation temperature on the metabolic level of the eel *Anguilla vulgaris* L. *Helgol. Wiss. Meeresunters.*, **13**, 402–7.

Jany, K.-D. (1976). Studies on the digestive enzymes of the stomachless bonefish *Carassius auratus gibelio* (Bloch); endopeptidases. *Comp. Biochem. Physiol.* **53B**, 31–8.

Jauncey, K. (1981). The effects of varying dietary composition on mirror carp (*Cyprinus carpio*) maintained in thermal effluents and laboratory recycling systems. In K. Tiews (Ed.), *Proc. world symposium on aquaculture in heated effluents and recirculation systems, vol. 2*, pp. 249–61. Berlin: Heenemann Verlagsgesel.

Jauncey, K. (1982). The effects of varying protein level on the growth, food conversion, protein utilization and body composition of juvenile tilapias (*Sarotherodon mossambicus*). *Aquaculture*, **27**, 43–54.

Jauncey, K. (1982b). Carp (*Cyprinus carpio* L.) nutrition – a review. In J. F. Muir & R. J. Roberts (Eds), *Recent advances in aquaculture*, pp. 215–63. London: Croom Helm.

Jayaram, M. G. & Shetty, H. P. C. (1980). Digestibility of 2 pelleted diets by *Cyprinus carpio* and *Labeo rohita*. *Mysore J. Agric. Sci.*, **14**, 578–84.

Jefferson, L. S., Exton, J. H., Butcher, R. W., Sutherland, E. W. & Park, C. R. (1968). Role of adenosine-3′,5′-monophosphate in the effects of insulin and anti-insulin serum on liver metabolism. *J. Biol. Chem.*, **243**, 1031–8.

Job, S. V. (1955). The oxygen consumption of *Salvelinus fontinalis*. *Univ. Toronto, Biol. Ser.*, **61**, Publ. Ontario Fish. Res. Lab., 73: 39 pp.

Job S. V. (1969). The respiratory metabolism of *Tilapia mossambica* (Teleostei). I. The effect of size, temperature and salinity. *Mar. Biol.*, **2**, 121–6.

Job, S. V. (1977). Laboratory studies on fish energetics and their application to aquaculture. *J. Madurai Univ.*, **6**(2), 35–42.

Jobling, M. (1981). The influence of feeding on the metabolic rates of fishes: a short review: *J. Fish Biol*, **18**, 385–400.

Jobling, M. (1983a). Growth studies with fish – overcoming the problem of size variation. *J. Fish Biol.*, **22**, 153–7.

Jobling, M. (1983b). A short review and critique of methodology used in fish growth and nutrition studies, *J. Fish Biol.*, **23**, 685–703.

Jobling, M. & Davis, P. S. (1980). Effects of feeding on the metabolic rate and specific dynamic action in plaice, *Pleuronectes platessa*. *J. Fish Biol.*, **16**, 629–38.

Johnston, I. A. (1975a). Studies on the swimming musculature of the rainbow trout. II. Muscle metabolism during severe hypoxia. *J. Fish Biol.*, **7**, 459–67.

Johnston, I. A. (1975b). Anaerobic metabolism in the carp (*Carassius carassius* L.). *Comp. Biochem. Physiol.*, **51B**, 235–41.

Johnston, I. A., Davison, W. & Goldspink, G. (1977). Energy metabolism of carp swimming muscles. *J. Comp. Physiol.*, **114B**, 203–16.

Johnston, I. A. & Goldspink, G. (1973a). Quantitative studies of muscle glycogen utilization during sustained swimming in crucian carp (*Carassius carassius* L.). *J. Exp. Biol.*, **59**, 607–15.

Johnston, I. A. & Goldspink, G. (1973b). A study of the swimming performance of the crucian carp *Carassius carassius* L. in relation to the effects of exercise and recovery on biochemical changes in the myotomal muscles and liver. *J. Fish Biol.*, **5**, 249–60.

Johnston, I. A. & Goldspink, G. (1973c). A study of glycogen and lactate in the myotomal muscles and liver of the coalfish (*Gadus virens* L.) during sustained swimming. *J. Mar. Biol. Assoc. UK*, **53**, 17–26.

Johnston, I. A. & Moon, T. W. (1980). Endurance exercise in fast and slow muscles of a teleost fish (*Pollachius virens*). *J. Comp. Physiol.*, **135**, 147–56.

Jonas, E., Ragyanszki, M., Olah, J. & Boross, L. (1983). Proteolytic digestive enzymes of carnivorous (*Silurus glanis* L.), herbivorous (*Hypophthalmichthys molitrix* Val.) and omnivorous (*Cyprinus carpio* L.) fishes. *Aquaculture*, **30**, 145–54.

Jonas, R. E. E. & Bilinski, E. (1964). Utilization of lipids by fish. III. Fatty acid oxidation by various tissues from sockeye salmon (*Oncorhynchus nerka*). *J. Fish. Res. Bd Can.*, **21**, 653–6.

Jones, D. R. (1982). Anaerobic exercise in teleost fish. *Can. J. Zool.*, **60**, 1131–4.

Jones, D. R. & Randall, D. J. (1978). The respiratory and circulatory systems during exercise. In W. S. Hoar & D. J. Randall (Eds), *Fish physiology, vol. 7*, pp. 425–501. New York: Academic Press.

Joshev, L. & Grozev, G. (1971). The effect of cobalt on the fish productivity of carp ponds. (In Bulgarian with Russian and German summaries.) *Izv. Stn. Sladkovod. Ribar. Plovdiv*, **8**, 95–101.

Joyner, C. A. (1965). Use of gelatine as food bonding agent. *Prog. Fish Cult.*, **27**, 66.

Kadmon, G., Gordin, H. & Yaron, Z. (1985). Breeding-related growth of captive *Sparus aurata* (Teleostei, Perciformes). *Aquaculture*, **46**, 299–305.

Kajak, Z. (1975). Productivity freshwater in *Polish National Committee for the International Biological Programme. Polish participation in the International Biological Programme*, pp. 182–269. Warsaw: Polish Acad. of Sci.

Kakimoto, D. & Mowlah, A. (1980). Microflora in the alimentary tract of gray mullet. VIII. Utilization of amino acids by *Vibrio* and *Enterobacter* isolates. *Mem. Fac. Fish. Kagoshima Univ.*, **29**, 349–54.

Kalač, J. (1975). Proteolytic enzymes in the pyloric caeca of mackerel (*Scomber scombrus*). *Biologia (Bratisl.)*, **30**(9), 649–58.

Kamler, E. (1972). Respiration of carp in relation to body size and temperature. *Pol. Arch. Hydrobiol.*, **19**, 325–31.

Kamler, E. (1976). Variability of respiration and body composition during early developmental stages of carp. *Pol. Arch. Hydrobiol.*, **23**, 431–85.

Kanazawa, A., Teshima, S. I., Sakamoto, M. & Awal, M. A. (1980a). Requirement of *Tilapia zillii* for essential fatty acids. *Bull. Jap. Soc. Sci. Fish.*, **46**, 1353–6.

Kanazawa, A., Teshima, S., Sakamoto, M. & Shinomiya A. (1980b). Nutritional requirement of the puffer fish: purified test diet and optiumum protein level. *Bull. Jap. Soc. Sci. Fish.*, **46**. 1357–61.

Kanazawa, A., Teshima, S. I. & Imai, K. (1980c). Biosynthesis of fatty acids in *Tilapia zillii* and the puffer fish. *Mem. Fac. Fish. Kagoshima Univ.*, **29** 313–18.

Kanungo, M. S. & Prosser, C. L. (1959a). Physiological and biochemical adaptation of goldfish to cold and warm temperatures. I. Standard and active oxygen consumption of cold and warm-acclimated goldfish at various temperatures. *J. Cell. Comp. Physiol.*, 54, 259–63.

Kanungo, M. S. & Prosser, C. L. (1959b). Physiological and biochemical adaptation of goldfish to cold and warm temperatures. II. Oxygen consumption of liver homogenate; oxygen consumption and oxidative phosphorylation of liver mitochondria. *J. Cell. Comp.*

Physiol., **54**, 265–74.

Karsinkin, G. S. (1935). Zur Physiologie der Fischernaehrung. *Verh. Int. Ver. Limnol.*, **7**(2), 398–410.

Kapoor, B. G., Smit, H. & Verighina, I. A. (1976a). The alimentary canal and digestion in teleosts. *Adv. Mar. Biol.*, **13**, 109–239.

Kapoor, B. G., Evans, H. E. & Pevzner, R. A. (1976b). The gustatory system in fish. *Adv. Mar. Biol.*, **13**, 53–108.

Kashiwada, K. & Teshima, S. (1966). Studies on the production of B vitamins by intestinal bacteria of fish. I. Nicotinic acid, pantothenic acid, and vitamin B_{12} in carp. *Bull. Jap. Soc. Sci. Fish.*, **32**, 961–6.

Katz, J. & Wood, H. G. (1960). The use of glucose-C^{14} for the evaluation of the pathways of glucose metabolism. *J. Biol. Chem.*, **235**, 2165–77.

Kausch, H. (1968). Der Einfluss der Spontanaktivitaet auf die Stoffwechselrate junger Karpfen (*Cyprinus carpio* L.) im Hunger und bei Fuetterung. *Arch. Hydrobiol., Suppl.* **33**, 263–330.

Kausch, H. (1972). Stoffwechsel und Ernaehrung der Fische. *Handb. Tierernaehr.*, **118**, 690–738. Transl. Ser., Fish. Res. Bd Can., (2489), 1973.

Kausch, H. & Ballion-Cusmano, M. F. (1976). Koerperzusammensetzung Wachstum und Nahrungsausnutzung bei Jungen Karpfen (*Cyprinus carpio* L.) unter Intensivhaltungs Bedingungen. *Arch. Hydrobiol., Suppl.* **48**, 141–8.

Kaushik, S. J. (1980). Influence of nutritional status on the daily pattern of nitrogen excretion in the carp (*Cyprinus carpio* L.) and the rainbow trout (*Salmo gairdneri* R.). *Reprod. Nutr. Develop.*, **20**, 1751–65.

Kaushik, S. J. & Luquet, P. (1980). Influence of bacterial protein incorporation and of sulfur amino acid supplementation to such diets on growth of rainbow trout, *Salmo gairdneri* Richardson. *Aquaculture*, **19**, 163–75.

Kaushik, S. J., Dabrowski, K. & Luquet, P. (1982). Pattern of nitrogen excretion and oxygen consumption during ontogenesis of common carp (*Cyprinus carpio*). *Can. J. Fish. Aquat. Sci.*, **39**, 1095–105.

Kawai, S. I. & Ikeda, S. (1971). Studies on digestive enzymes of fishes. I. Carbohydrases in digestive organs of several fishes. *Bull. Jap. Soc. Sci. Fish.*, **37**, 333–7.

Kawai, S. I. & Ikeda, S. (1972). Studies on digestive enzymes of fishes II. Effect of dietary change on the activities of digestive enzymes in carp intestine. *Bull. Jap. Soc. Sci. Fish.*, **38**, 265–70.

Keddis, M. N. (1957). On the intestinal enzymes of *Tilapia nilotica* Boul. *Proc. Egypt. Acad. Sci.*, **12**, 21–37.

Kellogg, T. F. (1975). The biliary bile acids of the channel catfish, *Ictalurus punctatus*, and the blue catfish *Ictalurus furcatus*. *Comp. Biochem. Physiol.* **50B**, 109–11.

Kellor, R. L. (1974). Defatted soy flour and grits. *J. Am. Oil Chem. Soc.*, **51**, 77A–80A.

Kelso, J. R. M. (1972). Conversion, maintenance and assimilation for walleye, *Stizostedion vitreum vitreum*, as affected by size, diet, and temperature. *J. Fish. Res. Bd. Can.*, **29**, 1181–92.

Kempinska, H. (1968). The growth of females and males of carp (*Cyprinus carpio* L.). *FAO Fish. Rep.*, **44**(4), 169–78.

Kenyon, A. J. (1967). The role of the liver in the maintenance of plasma proteins and amino acids in the eel, *Anguilla anguilla* L., with reference to amino acid deamination. *Comp. Biochem. Physiol.*, **22**, 169–75.

Kerr, S. R. (1971a). Analysis of laboratory experiments on growth efficiency of fishes. *J. Fish. Res. Bd Can.*, **28**, 801–8.

Kerr, S. R. (1971b). Prediction of fish growth efficiency in nature. *J. Fish. Res. Bd Can.*, **28**, 809–14.

Kerr, S. R. (1982). Estimating the energy budgets of actively predatory fish. *Can. J. Fish. Aquat. Sci.*, **39**, 371–9.

Ketola, H. G. (1976). Choline metabolism and nutritional requirements of lake trout (*Salvelinus namaycush*). *J. Anim. Sci.*, **43**, 474–7.

Ketola, H. G. (1979a). Influence of dietary zinc on cataracts in rainbow trout (*Salmo gairdneri*). *J. Nutr.*, **109**, 965–9.

Ketola, H. G. (1979b). Influence of dietary lysine on 'fin rot', survival, and growth of fry of rainbow trout. *Fish Health News*, **8**(2), vi.

Ketola, H. G. (1982). Amino acid nutrition of fishes: requirements and supplementation of diets. *Comp. Biochem. Physiol.*, **73B**, 17–24.

Kevern, N. R. (1966). Feeding rate of carp estimated by a radioisotopic method. *Trans. Am. Fish. Soc.*, **95**, 363–71.

Khan, R. A. & Siddiqui, A. Q. (1973). Food selection by *Labeo rohita* (Ham.) and its feeding relationship with other major carps. *Hydrobiol.*, **43**, 429–42.

Khanna, S. S. & Bhatt, S. D. (1974). Histopathology of endocrine pancreas of a fresh water fish *Clarias batrachus* (L.). VI. Effect of cobalt chloride administration. *Ver. Anat. Ges., Jena Anat. Anz. Rd.*, **136**, 157–64.

Kiermeir, A. (1940). Ueber den Blutzucker der Suesswasserfische. *Z. Vergl. Physiol.*, **27**, 460–91.

Kim, I.-B., Lee, S.-H. & Kang, S.-J. (1984). On the efficiency of soybean meal as a protein source substitute in fish feed for common carp. *Bull. Korean Fish. Soc.*, **17**(1), 55–60.

Kim. Y. K. (1974). Determination of true digestibility of dietary proteins in carp with chromic oxide containing diet. *Bull. Jap. Soc. Sci. Fish.*, **40**, 651–3.

Kimelberg, H. K. & Papahadjopoulos, D. (1972). Phospholipid requirement or ($Na^+ + K^+$) ATPase activity: Head group specificity and fatty acid fluidity. *Biochem. Biophys. Acta*, **282**, 277–92.

King, M. M., Lai, E. K. & Mckay, P. B. (1975). Singlet oxygen production associated with enzyme-catalyzed lipid peroxidation in liver microsomes. *J. Biol. Chem.* **250**, 6496.

King, R. E. (1975). The morphology of the buccal cavity of the goldfish, *Carassius auratus*. *S. Afr. J. Sci.*, **71**(6), 179–83.

Kinne, O. (1960). Growth, food intake, and food conversion in a euryplastic fish exposed to different temperatures and salinities. *Physiol. Zool.*, **33**, 288–317.

Kirchgessner, M., Kuerzinger, H. & Schwarz, F. J. (1986). Digestibility of crude nutrients in different feeds and estimation of their energy content for carp (*Cyprinus carpio* L.). *Aquaculture*, **58**, 185–94.

Kitamikado, M. & Tachino, S. (1960a). Studies on the digestive enzymes of rainbow trout. I. Carbohydrases. *Bull. Jap. Soc. Sci. Fish.*, **26**, 679–84. Transl. Ser., Fish. Res. Bd Can., (2193).

Kitamikado, M. & Tachino, S. (1960b). Studies on the digestive enzymes of rainbow trout. II. Proteases. *Bull. Jap. Soc. Sci. Fish.*, **26**, 685–90. Transl. Ser., Fish. Res. Bd Can. (2195).

Kitamikado, M. & Tachino, S. (1960c). Studies on the digestive enzymes of rainbow trout. III. Esterases. *Bull. Jap. Soc. Sci. Fish.*, **26**, 691–4. Transl. Ser., Fish. Res. Bd Can., (2194).

Kitamikado, M., Morishita, T. & Tachino, S. (1964a). Digestibility of dietary protein in rainbow trouts I. Digestibility of several dietary proteins. *Bull. Jap. Soc. Sci. Fish.*, **30**, 46–9.

Kitamikado, M., Morishita, T. & Tachino, S. (1964b). Digestibility of dietary protein in rainbow trout II. Effect of starch and oil contents in diets and size of fish. *Bull. Jap. Soc. Sci. Fish.*, **30**, 50–4.

Kitamikado, M., Takahshi, T., Noda, H., Morishita, T. & Tachino, S. (1965). Digestibility of dietary components in young yellowtail *Seriola quinqueradiata* T. and S. *Bull. Jap. Soc. Sci. Fish.*, **31**, 133–7.

Kitamura, S., Ohara, S., Suwa, T. & Nakagawa, K. (1965). Studies on vitamin requirements of rainbow trout, *Salmo gairdneri* – I. On the ascorbic acid. *Bull. Jap. Soc. Sci. Fish.*, **31**, 818–26.

Kitamura, S., Suwa, T., Ohara, S. & Nakagawa, K. (1967). Studies on vitamin requirements of rainbow trout. III. Requirement for Vit A and deficiency symptoms. *Bull. Jap. Soc. Sci. Fish.*, **33**, 1126–31.

Kitchell, J. F., & Windell, J. T. (1968). Rate of gastric digestion in pumpkinseed sunfish *Lepomis gibbosus*. *Trans. Am. Fish. Soc.*, **97**, 489–92.

Kitchell, J. F., Koonce, J. F., O'Neill, R. V., Shugart, H. H., Magnuson, J. J. & Booth, R. S. (1974). Model of fish biomass dynamics. *Trans. Am. Fish. Soc.*, **103**, 786–98.

Kitchin, S. E. & Morris, D. (1971). The effect of acclimation temperature on amino acid transport in the goldfish intestine. *Comp. Biochem. Physiol.*, **40A**, 431–43.

Kleiber, M. (1961). *The fire of life, an introduction to animal energetics*. New York: Wiley & Sons, 453 pp.

Kleiber, M. (1965). Metabolic body size. In J. H. S. Blaxter (Ed.), *Energy metabolism*, pp. 427–35. London: Academic Press.

Klein, C. & Lange, R. H. (1972). Mise en evidence par immunofluorescence de cellules secretrices de glucagon dans pancreas endocrine du poisson teleosteen *Xiphophorus helleri* H. *Histochemie*, **29**, 213–19.

Klein, R. G. & Halver, J. E. (1970). Nutrition of salmonoid fishes: Arginine and histidine requirements of chinook and coho salmon. *J. Nutr.*, **100**, 1105–9.

Klekowski, R. Z. (1970). Bioenergetic budgets and their application for estimation of production efficiency. *Pol. Arch. Hydrobiol.*, **17**, 55–80.

Klekowski, R. Z. & Duncan, A. (1975). Physiological approach to ecological energetics. In W. Godzinski, R. Z. Klekowski & A. Duncan (Eds), *Methods for ecological bioenergetics, IBP Handbook, 24*, pp. 227–57. Oxford: Blackwell Sci. Publ.

Klenk, E. & Kremer, G. (1960). Ueber die Biogenese der C_{20} und C_{22} Leberpolyenfettsaeuren bei Wirbeltieren. *Hoppe-Seyler's Z. Physiol. Chem.*, **320**, 111–25.

Kloppel, T. M. & Post, G. (1975). Histological alteration in tryptophan-deficient rainbow trout. *J. Nutr.*, **105**, 861–6.

Knapka, J. J., Barth, K. M., Brown, H. G. & Cragle, R. G. (1967). Evaluation of polyethylene chromic oxide and cerium-144 as digestibility indicators in burros. *J. Nutr.*, **92**, 79–85.

Knauthe, K. (1898). Untersuchungen ueber Verdauung und Stoffwechsel der fische. *Z. Fisch.*, **6**, 139.

Knipparth, W. G. & Mead, J. F. (1965). Influence of temperature on the fatty acid pattern of muscle and organ lipids of the rainbow trout (*Salmo gairdneri*). *Fish. Ind. Res.*, **3**, 23–7.

Knox, D., Cowey, C. B. & Adron, J. W. (1981). The effect of low dietary manganese intake on rainbow trout (*Salmo gairdneri*). *Br. J. Nutr.*, **46**, 495–501.

Kolehmainen, S. E. (1974). Daily feeding rates of bluegill (*Lepomis macrochirus*) determined by refined radioisotope method. *J. Fish. Res. Bd. Can.*, **31**, 67–74.

Kono, H. & Nose, Y. (1971). Relationship between the amount of food taken and growth in fishes. 1. Frequency of feeding for maxium daily ration. *Bull. Jap. Soc. Sci. Fish.*, **37**, 169–75.

Koops, H., Tiews, K., Beck, H. & Gropp, J. (1976). Die Verwertung von Sojaprotein durch die Regenbogenforelle (*Salmo gairdneri*). *Arch. Fischereiwiss.*, **26**, 181–91.

Korovin, V. A. (1976). Metabolic rate of the underyearling of the carp, *Cyprinus carpio*, adapted to different water temperatures. *J. Ichthyol*, **16**, 168–72.

Kosiorek, D. (1979). Changes in chemical composition and energy content of *Tubifex tubifex* (Muel). Oligochaeta, in life cycle. *Pol. Arch. Hydrobiol.*, **26**, 73–89.

Kovalskii, V. V., Krymova, R. V. & Letunova, S. L. (1965). Effect of cobalt fertilizer on cobalt migration in fish ponds. *Dokl. Akad. Sel'skokhoz. Nauk*, **10**, 24–9.

Koyama, H., Okubo, H. & Miyajima, T. (1961). Studies of fish food substituted for silkworm pupae as available foods for carp-culturing in farm ponds. II. Experiment about availability of soy-bean cake, 'Ko' meal and fish meal. *Bull. Freshwater Fish Res. Lab., Tokyo*, **11**(1), 49–55.

Krayukhin, B. V. (1962). Effect of temperature and diet on oxygen uptake rate in carp. *Akad. Nauk Ukrain. RSR*, 25–39, Referat Zhur. Biol., II 69, 1962.

Krayukhin, B. V. (1963). *Physiology of digestion in freshwater bony fish.* (In Russian.) Moscow: Akad. Nauk SSR, 139 pp.

Krebs, F. (1975). The influence of oxygen tension on the temperature adaptation in gibel carp (*Carassius auratus gibelio* Bloch). *Arch. Hydrobiol.*, **76**, 89–131.

Krebs, F. & Brandt, E. (1975). Der Einfluss der Sauerstoffspannung auf die Temperature-adaption des anaeroben Stoffwechsels im Muskelgewebe des Goldfisches (*Carassius auratus* L.). *Arch. Hydrobiol.*, **76**, 229–47.

Krishnamoorthy, R. V. & Narasimhan, T. (1972). Ascorbic acid and fat content in the red and white muscles of carp, *Catla catla*, *Comp. Biochem. Physiol.*, **43B**, 991–7.

Krishnandhi, S. & Shell, W. (1967). Utilization of soybean protein by channel catfish (*Ictalurus punctatus* Raff.). *Proc. Ann. Conf. Southeast. Assoc. Game Fish Commrs,*, **19**, 205–9.

Krogh, A. (1914). Quantitative relation between temperature and standard metabolism in animals. *Intern. Z. Phys. Chem. Biol.*, **1**. 491–508.

Krueger, F. (1963). Versuch einer mathematischen Analyse der 'Normalkurve' von Krogh. *Helgol. Wiss. Meeresunters.*, **8**, 333–56.

Krueger, F. (1966). Zur mathematischen Struktur der lebenden Substanz, dargelegt an Problem der biologischen Temperature- und Wachstums-Funktion. *Helgol. Wiss. Meeresunters.*, **14**, 302–25.

Krueger, F. (1973). Zur Mathematik des tierischen Wachstums. II. Vergleich einiger Wachstumsfunktionen. *Helgol. Wiss. Meeresunters.*, **25**, 509–50.

Krueger, H. M., Saddler, I. B., Chapman, G. A., Tinsley, I. I. & Lowry, R. R. (1968). Bioenergetics, exercise and fatty acid of fish. *Am. Zool.*, **8**, 119–29.

Kudrinskaya O. I. (1969). Metabolic rate in the larvae of pike-perch, perch, carp-bream and roach. *Hydrobiol. J.*, (Transl. Am. Fish Soc.), **5**(4), 68–72.

Kutscher, F. & Ackermann, D. (1933). The comparative biochemistry of vertebrates and invertebrates. *Ann. Rev. Biochem.*, **2**, 355–76.

Kutty, M. N. (1968). Respiratory quotients in goldfish and rainbow trout. *J. Fish. Res. Bd Can.*, **25**, 1689–728.

Kutty, M. N. (1972). Respiratory quotient and ammonia excretion in *Tilapia mossambica*. *Mar. Biol.*, **16**, 126–33.

Kutty, M. N. & Peer Mohamed, M. (1975). Metabolic adaptation of mullet *Rhinomugil corsula* (Hamilton) with special reference to energy utilization. *Aquaculture*, **5**, 253–70.

Kuzmina, V. V. (1971a). Effect of insulin on the glycemia level in freshwater bony fish. (In Russian.) *Tr. Inst. Biol. Vnutr. Vod. Akad. Nauk SSSR*, **22**, 190–7.

Kuzmina, V. V. (1971b). Effect of insulin on the glycogen level in the liver and muscles of freshwater bony fish. (In Russian.) *Tr. Inst. Biol. Vnutr. Vod. Akad. Nauk SSSR*, **22**, 197–203.

Kuzmina, V. V. (1977). Peculiarities of membrane digestion in freshwater teleosts. *J. Ichthyol.*, **17**, 99–107.

Kuzmina, V. V. & Nevalennyy, A. N. (1983). Effect of hydrogen concentration on the activity of some carbohydrases on the fish digestive tract. *J. Ichthyol.*, **23**, 114–23.

Laarman, P. W. (1969). Effects of limited food supply on growth rates of coho salmon and steelhead trout. *Trans. Am. Fish. Soc.*, **98**, 393–7.

Lakshmanan, M. A. V., Murty, D. S., Pillai, K. K. & Banerjee, S. C. (1967). On a new artificial food for carp fry. *FAO Fish. Rep.*, **44** (3), 373–87.

Lall, S. P. & Bishop F. J. (1979). Studies on the nutrient requirements of rainbow trout *Salmo gairdneri*, grown in sea water and fresh water. In T. V. R. Pillay & A. Dill (Eds), *Advances in aquaculture*, pp. 580–4. Farnham, England: Fishing News Books.

Lapkin, V. V., Poddubnyy, A. G. & Svirskiy, (1983). Thermoadaptive properties of fishes from temperate latitudes. *J. Ichthyol.*, **23**, 45–54.

Lardy, H. A. & Feldott, G. (1950). The net utilization of ammonium nitrogen by the growing rat. *J. Biol. Chem.*, **186**, 85–91.

Lassuy, D. R. (1984). Diet, intestinal morphology, and nitrogen assimilation efficiency in damselfish, *Stegastes lividus*, in Guam. *Environ. Biol. of Fishes*, **10**(3), 183–93.

Law, A. T. (1986). Digestibility of low-cost ingredients in pelleted feed by grass carp (*Ctenopharyngodon idella* C.et V.). *Aquaculture*, **51**, 97–103.

Lawrence, F. B. (1950). The digestive enzymes of the bluegill bream (*Lepomis macrochirus*). M. A. Thesis, Dept. of Zool., Auburn University, Alabama, USA.

Lebedeva, M. N., Gutveib, L. G. & Benzhitsky, A. G. (1972). The production of vitamin B-12 like substances by bacteria from alimentary canal of mediterranean fishes. *Rapp. P.-V. Réun., Comm. Int. Explor. Sci. Mer Mediterr., Monaco*, **21**(5), 245–6.

Lee, D. J. & Putnam, G. B. (1973). The response of rainbow trout to varying protein/energy ratios in a test diet. *J. Nutr.*, **103**, 916–22.

Lee, D. J., Roehm, J. N., Yu, T. C. & Sinnhuber, R. O., (1967). Effect of 3 fatty acids on the growth of rainbow trout, *Salmo gairdneri*. *J. Nutr.*, **92**, 93–8.

Lellak, J. (1957). Der Einfluss der Fresstaetikeit des Fischbestandes auf die Bodenfauna der Fischteich. *Z. Fisch., N.F.*, **6**, 621–33.

Leray, C. (1971). Experimental approaches to artificial feeding of some sea fish. In J. L. Gaudet (Ed.), *Report of the 1970 workshop on fish feed technology and nutrition, Resour. Bull., Bureau Sport Fish and Wildl., 102*, pp. 169–71.

Lesauskiene, L., Jankevicius, K. & Syvokiene, J. (1974). The role of the digestive tract microorganisms in the nutrition of pond fishes, 6. Content of free amino acids in the body of second-year fish and their synthetization by the digestive tract microorganisms. (In Russian with English summary.) *Tr. An. Lit. SSR (Biol.)*, (2), 127–135.

Lewander, K. (1976). Insulin in fish. *Zool. Revy.*, **37**(1–4), 44–6.

Lewander, K., Dave, G., Johansson, M.-L. Larsson A. & Lidman, U. (1974). Metabolic and hematological studies on the yellow and silver phases of the European eel, *Anguilla anguilla*, L. I. Carbohydrate, lipid, protein and inorganic ion metabolism. *Comp. Biochem. Physiol.*, **47B**, 571–81.

Lewis, R. W. (1962). Temperature and pressure effects on fatty acids of some marine ectotherms. *Comp. Biochem. Physiol.*, **6**, 75–89.

Liao, P. B. (1971). Water requirements of salmonids. *Prog. Fish Cult.*, **33**, 210–15.

Liebich, H. von & Mann, H. (1950). Ueber ein proteolytisches Ferment im Fischedarm und die Moeglichkeit seiner Verwendung fuer ein pharmazeutisches Praeparat. *Arch. Fischereiwiss.*, **2**, 41–9.

Lieder, U. (1964). Ueber die Moeglichkeiten der Verbesserung der biologischen Wertigkeit von Futtermittelproteinen in der Karpfernfuetterung durch Ausnutzung der Aminosaeuren – Ergaenzungswirkung. *Dtsch. Fisch. Ztg.*, **11**, 278–81.

Lieder, U. (1965a). Das Eiweiss der Nahrung der Karpfen. *Dtsch. Fisch. Ztg.*, **12**, 16–26.

Lieder, U. (1965b). Die Erscheinugen der B1 – Avitaminose bei Karpfen. *Dtsch. Fisch. Ztg.*, **12**, 264–5.

Likimani, T. A. & Wilson, R. P. (1982). Effect of diet on lipogenic enzyme activities in channel catfish hepatic and adipose tissue. *J. Nutr.*, **112**, 112–7.

Lim, C. & Lovell, R. T. (1978). Pathology of the vitamin C deficiency syndrome in channel catfish (*Ictalurus punctatus*). *J. Nutr.*, **108**, 1137–46.

Lim, C., Sukhawongs, S. & Pascual, F. P. (1979). A preliminary study on the protein requirements of *Chanos chanos* (Forskal) fry in a controlled environment. *Aquaculture*, **17**, 195–201.

Lin, H., Romsos, D. R., Tack, P. I. & Leveille, G. A. (1978). Determination of glucose utilization in coho salmon [*Oncorhynchus kisutch* (Walbaum)] with (6-^{3}H) – and (U-^{14}C)-glucose. *Comp. Biochem. Physiol.*, **59A**, 189–91.

Lin, Y., Dobbs, G. H. & Devries, A. L. (1974). Oxygen consumption and lipid content in red and white muscles of Antarctic fishes. *J. Exp. Zool.*, **189**, 379–86.

Lindsay, G. J. H. (1984a). Distribution and function of digestive tract chitinolytic enzymes in fish. *J. Fish. Biol.*, **24**, 529–36.

Lindsay, G. J. H. (1984b). Adsorption of rainbow trout (*Salmo gairdneri*) gastric lysozymes and chitinase by cellulose and chitin. *Aquaculture*, **42**, 241–6.

Lindsay, G. J. H. & Harris, J. E. (1980). Carboxymethylcellulase activity in the digestive tracts of fish. *J. Fish. Biol.*, **16**, 219–33.

Lindsay, G. J. H., Walton, M. J., Adron, J. W., Fletcher, T. C., Cho, C. Y. & Cowey, C. B. (1984). The growth of rainbow trout (*Salmo gairdneri*) given diets containing chitin and its relationship to chitinolytic enzymes and chitin digestibility. *Aquaculture*, **37**, 315–34.

Liu, D. H. W., Krueger, H. & Wang, C. (1970). Catabolic pathways for glucose in the cichlid fish *Cichlasoma bimaculatum*. *Comp. Biochem. Physiol.*, **36**, 173–81.

Love, R. M. (1970). *The chemical biology of fishes*. London: Academic Press, xv + 547 pp.

Lovell, R. T. (1972). Protein requirement of cage-cultured channel catfish. *Proc. Ann. Conf. Southeast. Assoc. Game Fish. Commrs.*, **26**, 357–60.

Lovell, R. T. (1973). Essentiality of vitamin C in feeds for intensively fed caged channel catfish. *J. Nutr.*, **103**, 134–8.

Lovell, R. T. (1975). Fish feeds and nutrition. How much protein in feeds for channel catfish? *Commer. Fish Farmer Aquacult. News*, **1**(4), 40–1.

Lovell, R. T. (1977). Feeding practices. In *Nutrition and feeding of channel catfish, Southern Cooperative Series, 218*, pp. 50–5.

Lovell, R. T. (1979). Nutritional diseases in channel catfish. In T. V. R. Pillay & A. Dill (Eds), *Advances in aquaculture*, pp. 605–9. Farnham, England: Fishing News Books.

Lovell, R. T. (1984). Use of soybean products in diets for aquaculture species. *Soybeans*, Feb. 1984, 1–6.

Lovell, R. T. & Li, Y.-P. (1978). Essentiality of vitamin D in diets of channel catfish (*Ictalurus punctatus*). *Trans. Am. Fish. Soc.*, **107**, 809–11.

Lovell, R. T., & Limsuwan, T. (1982). Intestinal synthesis and dietary nonessentiality of vitamin B12 for *Tilapia nilotica*. *Trans. Am. Fish. Soc.*, **111**, 485–90.

Lovell, R. T., Prather, E. E., Tres-Dick, J. & Chhorn, L. (1975). Effects of addition of fish meal to all-plant feeds on the dietary protein needs of channel catfish in ponds. *Proc. Ann. Conf. Southeast, Assoc. Game Fish Commrs*, **28**, 222–8.

Lovern, J. A. (1938). Fat metabolism in fishes. XIII. Factors influencing the composition of depot fat in fishes. *Biochem. J.*, **32**, 1214–24.

Lovern, J. A. (1951). The biochemistry of fish. IV. The chemistry and metabolism of fats in fish. *Biochem. Soc. Symposia*, **6**, 49–62.

Loyanich, A. A. (1974). Vitamin A dynamics in livers of artificially reared rainbow trout. (In Russian.) *Rybokhoz. Izuch. Vnutr. Vodoemov*, (12), 24–6.

Luce, W. G., Peo, E. R. & Hudman, D. B. (1967). Availability of niacin in corn and milo for swine. *J. Anim. Sci.*, **26**, 76–84.

Ludwig, B., Higgs, D. A., Fagelund, U. H. M. & McBride, J. R. (1977). A preliminary study of insulin participation in the growth regulation of coho salmon (*O. kisutch*). *Can. J. Zool.*, **55**, 1756–8.

Luehr, B. (1967). Die Futterung von Karpfen bei Intensivhaltung. Vortragsveranst. ueber neue Methoden der Fischzuechtung und Haltung in Hamburg, Feb. 1967, pp. 47–58.

Lukowsky, A., Prehn, S. & Rapoport, T. A. (1974). Biosynthesis of proinsulin in islets of Langerhans of the carp (*Cyprinus carpio*). *Biochim. Biphys. Acta*, **359**, 248–52.

Luquet, P. & Sabaut, J. J. (1973). Preliminary study on the protein requirements of the gilthead bream (*Chrysophyrs aurata*). *Stud. Rev. Gen. Fish. Counc. Medit.*, **52**, 81–9.

Lyakhnovich, V. P. (1967). Effect of yearling carp on the quantitative growth of zooplankton and benthos. *Tr. Karel. Otd. GosNIORKh.*, **5**(1), 458–65. Transl. Ser., Fish. Res. Bd Can. (2804), 1973, 12 pp.

Mabaye, A. B. E. (1971). Observation on the growth of *Tilapia mossambica* fed artificial diets. *Fish. Res. Bull. (Zambia)*, **5**, 379–96.

Machiels, M. A. M. & Henken, A. M. (1985). Growth rate, feed utilization and energy metabolism of the African catfish, *Clarias gariepinus* (*Burchell* 1822), as affected by dietary protein and energy content. *Aquaculture*. **44**, 271–84.

MacLeod, M. G. (1977). Effects of salinity on food intake, absorption and conversion in the rainbow trout, *Salmo gairdneri. Mar. Biol.*, **43**, 93–102.

Macleod, R. A., Jonas, R. E. E. & Roberts, E. (1963). Glycolytic enzymes in the tissues of a salmonoid fish (*Salmo gairdneri gairdneri*). *Can. J. Biochem. Physiol.*, **41**, 1971–81.

Majkowski, J. & Waiwood, K. G. (1981). A procedure for evaluating the food biomass consumed by a fish population. *Can. J. Fish. Aquat. Sci.*, **38**, 1199–208.

Majkowski, J. & Hearn, W. S. (1984). Comparison of three methods for estimating the food intake of a fish. *Can. J. Fish. Aquat. Sci.*, **41**, 212–5.

Matlzan, Grafin von M. (1935). Zur Ernaehrungsbiologie und Physiologie des Karpfens. *Zool. Zentr. Abt. 3*, **55**, 191–218.

Malvin, R. L., Cafruni, E. J. & Kutchai, H. (1965). Renal transport of glucose by the aglomerular fish *Lophius americanus. J. Cell. Comp. Physiol.*, **65**, 381–4.

Mann, H. (1948). Bedeutung der Zerkleinerung der Futtermittel fuer die Ausnutzung der Naehrstoffe durch Karpfen. *Allg. Fisch.-Ztg.*, **73**, 703–5.

Mann, H. (1967). The utilization of food by *Tilapia melanopleura* Dum. *FAO Fish. Rep.* **44**(3), 408–10.

Mann, H. (1968). Der Einfluss der Ernaehrung auf der Saurstoffverbrauch von Forellen. *Arch. Fischereiwiss.*, **19**, 131–3.

Mann, H. (1969). Difference in the digestibility of plant and animal protein by various kinds of fish. *FAO/EIFAC Tech. Pap.*, **9**, 194–6.

Mann, H. (1974). Eiweissbedarf und Eiweissausnutzung bei Teichfischen. *Fisch. Teichwirt*, **25**(5), 44–6.

Mann, K. H. (1965). Energy transformations by a population of fish in the river Thames. *J. Anim. Ecol.*, **34**, 253–75.

Mann, K. H. (1969). The dynamics of aquatic ecosystems. *Adv. Ecol. Res.*, **6**, 1–71.

Manoukas, A. G., Ringrose, R. C. & Teeri, A. E. (1968). The availability of of niacin in corn, soybean meal and wheat middlings for the hen. *Poultry Sci.*, **47**, 1836–42.

Marek, M. (1975). Revision of supplementary feeding tables for pondfish. *Bamidgeh*, **27**(3), 57–64.

Marek, M. & Sarig, S. (1971). Preliminary observations of superintensive fish culture in the Beith-Shean Valley in 1969–70. *Bamidgeh*, **23**(3), 93–9.

Maslennikova, N. V. (1974). The amino acid composition of some fish tissues. *J. Ichthyol.*, **14**, 1087–97.

Masoni, A. & Payan, P. (1974). Urea, insulin and para-amino-hippuric acid (PAH) excretion by the gills of the eel *Anguilla anguilla* L. *Comp. Biochem. Physiol.*, **47A**, 1241–4.

Matches, J. R., Liston, J. & Curran, D. (1974). *Clostridium perfringens* in the environment. *Appl. Microbiol.*, **28**, 655–60.

Mathavan, S., Vivekanadan, E. & Pandian, T. J. (1976). Food utilization in the fish *Tilapia mossambica* fed on plant and animal foods. *Helgol. Wiss. Meeresunters.*, **28**, 66–70.

Mathur, D. (1970). Food habits and feeding chronology of channel catfish *Ictalurus punctatus* (Refinesque) in Conowingo Reservoir. *Proc. Ann. Conf. Southeast Assoc. Game Fish Commrs*, **24**, 377–86.

Mattheis, T. (1964). Oekologie der Bakterien in Darm von Susswassernutzfischen. I. Familie Pseudomonadaceas Gattung *Pseudomonas. Z. Fisch. N.F.*, **12** 507–600.

Matthews, D. M. (1977). Protein absorption – then and now. *Gastroenterology*, **73**, 1267–79.

Matty, A. J. & Smith, P. (1978). Evaluation of a yeast, bacterium and an alga as a protein source for rainbow trout: I. Effect of protein level on growth, gross conversion efficiency and protein conversion efficiency. *Aquaculture*, **14**, 235–46.

Mazid, M. A., Tanaka, Y., Katayama, T., Simpson, K. L. & Chichester, C. O. (1978). Metabolism of amino acids in aquatic animals. III. Indispendable amino acids for *Tilapia zilli. Bull. Jap. Soc. Sci. Fish.*, **44**, 739–42.

Mazid, M. A., Tanaka, Y., Katayama, T. Rahman, M. A., Simpson, K. L. & Chichester,

C. O. (1979). Growth response of *Tilapia zilli* fingerlings fed isocaloric diets with variable protein levels. *Aquaculture*, **18**, 115–22.

McBean, R. L., Neppel, M. J. & Goldstein, L. (1966). Glutamate dehydrogenase and ammonia production in eel (*Anguilla rostrata*). *Comp. Biochem. Physiol.*, **18**, 909–20.

McCay, C. M., Tunison, A. V., Crowell, M. & Paul, H. (1936). The calcium and phosphorus content of the body of the brook trout in relation to age, growth, and food. *J. Biol. Chem.*, **114**, 259–63.

McCay, C. M., Tunison, A., Crowell, M., Tressler, D. K., MacDonald, S. P., Titcomb, J. W. & Cobb, E. W. (1931). The nutritional requirements of trout and chemical composition of the entire trout body. *Trans. Am. Fish. Soc.*, **61**, 58–79.

McCoy, K. E. M. & Wesig, P. H. (1969). Some selenium responses in the rat not related to vitamin E. *J. Nutr.*, **98**, 383–9.

McDonald, P., Edwards, R. A., & Greenhalgh, J. F. D. (1966). *Animal nutrition*. Edinburgh: Oliver & Boyd, 407 pp.

McGeachin, R. L. & Debnham, J. W. (1960). Amylase in freshwater fish. *Proc. Soc. Exp. Biol. Med.*, **103**, 814–15.

McLaren, B. A., Keller, E., O'Donnell, D. J. & Elvehjem, C. A. (1947). The nutrition of rainbow trout I. Studies on vitamin requirements. *Arch. Biochem.*, **15**, 169–78.

Mead, J. F., Kajama, M. & Reiser, R. (1960). Biogenesis of polyunsaturated acids in fish. *J. Am. Oil Chem. Soc.*, **37**, 438–40.

Medford, B. A. & Mackay W. C. (1978). Protein and lipid content of gonads, liver, and muscle of northern pike (*Esox lucius*) in relation to gonad growth. *J. Fish. Res. Bd Can.*, **35**, 213–19.

Medland, T. E. & Beamish, F. W. H. (1985). The influence of diet and fish density on apparent heat increment in rainbow trout, *Salmo gairdneri. Aquaculture*, **47**, 1–10.

Meister, R., Zwingelstein, Ç. & Jouanneteau, J. (1974). Salinité et composition en acides grasides phosphoglycerides tissulaires chez l'anguille (*Anguilla anguilla*). *Ann. Inst. Michel Pacha*, 1973(8), 58–71.

Melnick, D., Oser, B. L. & Weiss, S. (1946). Rate of enzymatic digestion of protein as a factor in nutrition. *Science*, **103**, 326–9.

Menzel, D. W. (1960). Utilization of food by a Bermuda reef fish *Epinephelus guttatus. J. Cons.*, **25**, 216–22.

Merla, G. (1966a). Untersuchungen ueber die quantitative Entwicklung der natuerlichen Nahrung des Karpfens (*Cyprinus carpio* L.) in Streck- und Abwachsteichen und ueber ihre Beziehung zum Karpfenzuwachs. *Z. Fisch., N.F.*, **14**, 161–248.

Merla, G. (1966b). Naehrtiermengen und Fischzuwachs in vier Streckteichen im Jahre 1965. *Dtsch. Fisch. Ztg.*, **13**, 51–5.

Merla, G. (1969). Availability of natural food for *Cyprinus carpio* in ponds where fish culture intensity is increasing. *FAO/EIFAC Tech. Pap.*, **9**, 131–9.

Mertz, E. T. (1968). Amino acid and protein requirements of fish. In O. W. Neuhaus & J. E. Halver (Eds). *Fish in research*, pp. 233–44. New York: Academic Press.

Mertz, E. T. (1972). The protein and amino acid needs. In J. E. Halver (Ed.), *Fish Nutrition* New York: Academic Press. pp. 105–43.

Mertz, E. T., Beeson, W. M. & Jackson, H. D. (1952). Classification of essential amino acids for the weanling pig. *Arch. Biochem. Biophys.*, **38**, 121–8.

Mertz, W. (1969). Chromium occurrence and function in biological systems. *Physiol. Rev.*, **49**, 163–239.

Mertz, W. & Schwarz, K. (1955). Impaired intravenous glucose tolerance as an early sign of dietary necrotic liver degradation. *Arch. Biochem. Biophys.*, **58**, 504–6.

Mertz, W., Toepfer, E. W., Roginski, E. E. & Polansky, M. M. (1974). Present knowledge of the role of chromium. *Fed. Proc., Fed. Am. Soc. Exp. Biol.*, **33**, 2275–9.

Meske, Ch. & Becker, K. (1981). Untersuchungen ueber den Einfluss von Futtermischungen unterschiedlichen Fett- und Proteingehaltes auf Zuwachs und Koerperzusammensetzung von Karpfen. *Informationen fuer die Fischwirtschaft*, **28**(5/6), 182–4.

Meske, Ch., Ney, K. H. & Pruss, H. D. (1977). A fish feed without fish meal, based on whey. *Fortschritte in der Tierephysiol. und Tierenaehrung*, (8). 56–70.

Meske, Ch., & Pruss, H. D. (1977). Mikroalgen als komponente von fischmehlfreiem Fischfutter. *Fortschritte in der Tierphysiol. und Tierernaehrung*, (8). 71–81.

Metailler, R., Febver, A. & Alliot, E. (1973). Preliminary notes on the essential amino-acids of the sea-bass *Dicentrarchus labrax* (Linne). *Stud. Rev., Gen. Fish. Counc. Medit.*, **52**, 91–6.

Meuwis, A. L. & Heuts, M. J. (1957). Temperature dependence of breathing rate in carp. *Biol. Bull.*, **112**, 97–107.

Micha, J.C., Dandrifosse, G. & Jeuniaux, Ch. (1973). Distribution et localisation tissulaire de la synthèse de chitinase chez les vertebres inférieurs. *Arch. Int. Physiol. Biochim.*, **81**, 439–51.

Migita, M., Hanaoka, T. & Tuzuki, K. (1937). A study of vegetable feedstuffs in fish culture. I. Nutritive value of some polysaccharides. *Imp. Fish. Exp. Sta.*, **8**, 99–177.

Miller, D. S. & Bender, A. E. (1955). The determination of the ulitization of proteins by a shortened method. *Br. J. Nutr.*, **9**, 382–8.

Millikin, M. R. (1982). Effects of dietary protein concentration on growth, feed efficiency, and body composition of age-0 stripped bass. *Trans. Am. Fish. Soc.*, **111**, 373–8.

Milstein, A., Hepher, B. & Teltsch, B. (1985a). Principal component analysis of interactions between fish species and the ecological conditions in fish ponds. I. Phytoplankton. *Aquacult. Fish. Managmt.*, **16**, 305–17.

Milstein, A., Hepher, B. & Teltsch, B. (1985b). Principal component analysis of interactions between fish species and the ecological conditions in fish ponds. II. Zooplankton. *Aquacult. Fish. Managmt.*, **16**, 319–30.

Mironova, N. V. (1976). Changes in the energy balance of *Tilapia mossambica* in relation to temperature and ration size. J. Ichthyol., **16**, 120–9.

Mishvelov, E. G. (1983). Relationship between growth rate and body water content of carp, *Cyprinus carpio* (Cyprinidae) influenced by scale cover and temperature. *J. Ichthyol.*, **23**, 144–7.

Miura, T. (1973). Computation of food consumption of a herbivorous crucian carp population in a small pond by the energy balance method. *Bull. Osaka Perf. Freshwater Fish. Exp. St.*, **1**, 109–18.

Miura, T., Suzuki, N., Nagoshi, M. & Yamamura, K. (1976). The rate of production and food consumption of the biwamusa, *Oncorhynchus rhodurus*, population in Lake Biwa. *Res. Popul. Ecol. Kyoto Univ.*, **17**, 135–54.

Moav, R. & Wohlfarth, G. W. (1968). Genetic improvement of yield in carp. *FAO Fish. Rep.*, **44**(4), 12–29.

Molnar, G., & Tolg. I. (1962a). Experiments concerning gastric digestion of pike-perch (*Lucioperca lucioperca* L.) in relation to water temperature. *Acta Biol. Acad. Sci. Hung.*, **13**(3), 231–9.

Molnar, G., & Tolg. I. (1962b). Relation between water temperature and gastric digestion in largemouth bass (*Micropterus salmoides*). *J. Fish. Res. Bd Can.*, **19**, 1005–12.

Moon., T. W. (1975). Temperature adaptation: Isozyme function and the maintenance of heterogeneity. In C. L. Market (Ed.), *Isozymes, II. Physiological function*, pp. 207–19. New York: Academic Press.

Moore, W.G. (1941). Studies on the feeding habits of fishes. *Ecology*, **22**, 91–6.

Moorthy, C. V. N., Murthy, V. K., Reddy, G. V., Haranath, V. B., Reddanna, P. & Govindappa, S. (1980). Effect of starvation and refeeding on tissue proximate analysis of hepatic and muscular tissues of *Tilapia mossambica* (Peters). *Indian J. Fish.*, **27**(1–2), 215–19.

Morgulis, S. (1918). Studies on the nutrition of fish. Experiments on brook trout. *J. Biol. Chem.*, **36**, 391–413.

Mori, T., Hashimoto, Y. & Komata, Y. (1956). B-vitamin content in the muscle of fish. *Bull. Jap. Soc. Sci. Fish.*, **21**, 1233–35.

Moriarty, D. J. W. (1973). The physiology of digestion of blue-green algae in the cichlid fish, *Tilapia nilotica*. *J. Zool.* (*Lond.*), **171**, 25–39.

Moriarty, D. J. W. (1976). Quantitative studies in bacteria and algae in the food of mullet, *Muqil cephalus C.* and in the prawn, *Metapenaeus bennette* (Racek and Dall). *J. Exp. Mar. Biol. Ecol.*, **22**, 131–43.

Moriarty, D. J. W. & Moriarty, C. M. (1973). The assimilation of carbon from phytoplankton by two herbivorous cichlid fishes: *Tilapia nilotica* and *Hoplochromis nigripinnis*. *J. Zool. (Lond.)*, **171**, 41–5.

Morishita, T., Noda, H., Kitamikado, M., Takahashi, T. & Tachino, S. (1964). The activity of the digestive enzymes in fish. *J. Fac. Fish., Univ. of Mie (Japan)*, **6**(2), 239–46.

Morita, K., Furuichi, M. & Yone, Y. (1982). Effect of carboxymethlycellulose supplemented to dextrin containing diets on growth and feed efficiency of red sea bream. *Bull. Jap. Soc. Sci. Fish.*, **48**, 1617–20.

Moss, D. D. & Scott, D. C. (1964). Respiratory metabolism of fat and lean channel catfish. *Prog. Fish Cult.*, **26**, 16–20.

Moule, M. L. & Yip, C.C. (1973). Insulin biosynthesis in the bullhead, *Ictalurus nebulosus*, and the effect of temperature. *Biochem. J.*, **134**, 753–61.

Mourik, J., Raeven, P., Steur, K. & Addink, A. (1982). Anaerobic metabolism of red skeletal muscle of goldfish. *Carassius auratus* (L.): Mitochondrial produced acetaldehyde as anaerobic electron acceptor. *FEBS Lett.*, **137**, 111–14.

Muir, B. S. (1969). Gill, dimensions as a function of fish size. *J. Fish. Res. Bd Can.*, **26**, 165–70.

Muir, B.S. & Niimi, A. J. (1972). Oxygen consumption of the euryhaline fish aholehole (*Kuhlia sandvicensis*) with reference to salinity, swimming and food consumption. *J. Fish. Res. Bd Can.*, **29**, 67–77.

Muir, B. S., Nelson, G. J. & Bridges, K. W. (1965). A method for measuring swimming speed in oxygen consumption studies on the aholehole *Kuhlia sandvicensis*. *Trans. Am. Fish. Soc.*, **94**, 378–82.

Muller-Feuga, A., Petit, J. & Sabaut, J. J. (1978). The influence of temperature and wet weight on oxygen demand of rainbow trout (*Salmo gairdneri* R.) in fresh water. *Aquaculture*, **14**, 355–64.

Mullins, L. J. (1950). Osmotic regulation in fish as studied with radio isotopes. *Acta Physiol. Scand.*, **21**, 303–14.

Murai, T. & Andrews, J. H. (1974). Interactions of dietary α-tocopherol, oxidized menhaden oil and ethoxyquin on channel catfish (*Ictalurus punctatus*). *J. Nutr.*, **104**, 1416–31.

Murai, T. & Andrews, J. W. (1975). Pantothenic acid supplementation of diets for catfish fry. *Trans. Am. Fish. Soc.*, **104**, 313–16.

Murai, T. & Andrews, J. W. (1976). Effect of frequency of feeding on growth and food conversion of channel catfish fry. *Bull. Jap. Soc. Sci. Fish.*, **42**, 159–61.

Murai, T. & Andrews, J. W. (1978a). Riboflavin requirement of channel catfish fingerlings. *J. Nutr.*, **108**, 1512–17.

Murai, T. & Andrews, J. W. (1978b). Thiamin requirement of channel catfish fingerlings. *J. Nutr.*, **108**, 176–80.

Murai, T. & Andrews, J. W. (1979). Pantothenic acid requirements of channel catfish fingerlings. *J. Nutr.*, **109**, 1140–2.

Murai, T., Akiyama, T. & Nose, T. (1981a). Use of crystalline amino-acids coated with casein in diets for carp. *Bull. Jap. Soc. Sci. Fish.*, **47**, 523–7.

Murai, T., Andrews, J. W. & Smith, R. G. (1981b). Effects of dietary copper on channel catfish. *Aquaculture*, **22**, 353–7.

Murai, T., Ogata, H. & Nose, T. (1982). Methionine coated with various materials supplemented to soybean meal diet for fingerling carp *Cyprinus carpio* and channel catfish

Ictalurus punctatus. *Bull. Jap. Soc. Sci. Fish.*, **48**, 85–8.

Murai, T., Akiyama, T. & Nose, T. (1983a). Effects of glucose chain length of various carbohydrates and frequency of feeding on their utilization by fingerling carp. *Bull. Jap. Soc. Sci. Fish.*, **49**, 1607–11.

Murai, T., Hirasawa, Y., Akiyama, T. & Nose, T. (1983b). Effects of dietary pH and electrolyte concentration on utilization of crystalline amino acids by fingerling carp. *Bull. Jap. Soc. Sci. Fish.*, **49**, 1377–80.

Murai, T., Akiyama, T. & Nose, T. (1984a). Effect of amino acid balance on efficiency in utilization of diet by fingerling carp. *Bull. Jap. Soc. Sci. Fish.*, **50**, 893–7.

Murai, T., Ogata, H., Takeuchi, T., Watanabe, T. & Nose, T. (1984b). Composition of free amino acid in excretion of carp fed amino acid diet and casein-gelatin diets. *Bull. Jap. Soc. Sci. Fish.*, **50**, 1957.

Murai, T., Ogata, H., Kosutarak, P. & Arai, S. (1986). Effects of amino acid supplementation and methanol treatment on utilization of soy flour by fingerling carp. *Aquaculture*, **56**, 197–206.

Murakami, Y. (1967). Studies on a cranial deformity in hatchery reared young carp. *Fish Pathol.*, **2**(1), 1–10.

Murat, J. & Plisetskaya, E. M. (1977). Effets du glucagon sur la glycemie, le glycogene et la glycogene-synthetase hepatique chez la carpe et la Lamprioe. *C.R. Seances Soc. Biol.*, *171*, 1302.

Murat, J. C. & Serfaty, A. (1975). Effets de l'adrenaline, du glacagon et de l'insuline sur le metabolisme glucidique de la carp. *C.R. Seances Soc. Biol.*, **169**, 228–32.

Murat, J. C., Castilla, C. & Paris, H. (1978). Inhibition of gluconeogenesis and glucagon induced hyperglycemia in carp (*Cyprinus carpio* L.). *Gen. Comp. Endocrinol.*, **34**, 243–6.

Murray, M. W., Andrews, J. W. & DeLoach, H. L. (1977). Effect of dietary lipid, dietary protein and environmental temperatures on growth, feed conversion and body composition of channel catfish. *J. Nutr.*, **107**, 272–80.

Nagai, M. & Ikeda, S. (1971a). Carbohydrate metabolism in fish – I. Effects of starvation and dietary composition on the blood glucose level and the hepatopancreatic glycogen and lipid contents in carp. *Bull. Jap. Soc. Sci. Fish.*, **37**, 404–9.

Nagai, M. & Ikeda, S. (1971b). Carbohydrate metabolism in fish – II. Effect of dietary composition on metabolism of glucose-6-^{14}C in carp. *Bull. Jap. Soc. Sci. Fish.*, **37**, 410–14.

Nagai, M. & Ikeda, S. (1972). Carbohydrate metabolism in fish – III. Effect of dietary composition on metabolism of glucose-U-^{14}C and glutamate-U-^{14}C in carp. *Bull. Jap. Soc. Sci. Fish.*, **38**, 137–43.

Nagai, M. & Ikeda, S. (1973). Carbohydrate metabolism in fish – IV. Effect of dietary composition on metabolism of acetate-U-^{14}C and L-alanine-U-^{14}C in carp, *Bull. Jap. Soc. Sci. Fish.*, **39**, 633–43.

Nagase, G. (1964). Contribution to physiology of digestion in T. *Mossambica* Peters: digestive enzymes and effects of diets on their activity. *Z. Vergl. Physiol.*, **49**, 270–84.

Nagayama, F., Oshima, H. & Umezawa, K. (1972a). Distribution of glucose-6-phosphate metabolizing enzymes in fish. *Bull. Jap. Soc. Sci. Fish.*, **38**, 589–93.

Nagayama, F., Oshima, H., Umezawa, K. & Kaiho, M. (1972b). Effect of starvation on the activities of glucose-6-phosphate metabolizing enzymes in fish. *Bull. Jap. Soc. Sci. Fish.*, **38**, 595–8.

Nagayama, F., & Saito, Y. (1968). Distribution of amylase, and glucosidase, and galactosidase in fish. *Bull. Jap. Soc. Sci. Fish*, **34**, 944–9.

Nail, M. L. (1962). The protein requirement of channel catfish *Ictalurus punctatus* (Rafinesque). *Proc. Ann. Conf. Southeast Assoc. Game Fish Commrs*, **16**, 307–16.

Nakamura, Y. (1982). Effects of dietary phosphorus and calcium contents on the absorption of phosphorus in the digestive tract of carp. *Bull. Jap. Soc. Sci. Fish.*, **48**(3), 409–13.

Nakanishi, T. & Itazawa, Y. (1974). Effect of hypoxia on the breathing rate, heart rate and rate of oxygen consumption in fishes. *Rep. Fish. Res. Lab., Kyushu Univ.*, (2), 41–52.

National Research Council (USA). (1977). *Nutrient requirements of swine*. Washington DC: National Academy of Sciences.

National Research Council (USA). (1979). *Nutrient requirements of poultry*. Washington DC: National Academy of Sciences.

National Research Council (USA). Subcommittee on coldwater fish nutrition. (1981). *Nutrient requirements of coldwater fishes*. Washington, DC: National Academy Press, Nutrient requirements of domestic animals, no. 16, 63 pp.

National Research Council (USA) (1983). *Nutrient requirements of warmwater fishes and shellfishes* (revised edition). National Academy Press, Washington; 102 pp.

Nehring, D. (1963). Verdauungsversuche an Fishchen nach der chromoxyd-indikatormethode, *Z. Fisch.*, **11**, 769–77.

Neurath, H. & Walsh, K. A. (1976). The role of proteases in biological regulation. In D. W. Ribbons & K. Brew (Eds), *Proteolysis and physiological regulation*, pp. 29–42. New York: Academic Press.

Neurath, H., Bradshaw, R. A. & Arnon, R. (1970). Homology and phylogeny of proteolytic enzmyes. In P. Desnuelle, H. Neurath & M. Ottesen (Eds), *Structure–function relationship of proteolytic enzymes*, pp. 113–37. Copenhagen: Munksgaard.

Nicolaides, N., & Woodall, A. N. (1962). Impaired pigmentation in chinook salmon fed diets deficient in essential fatty acids. *J. Nutr.*, **78**, 431–7.

Nielsen, (1973). On the density dependence of growth in soles (*Solea solea*). *Aquaculture*, **1**, 349–57.

Nightingale, J. W. (1975). Bioenergetic responses of nitrogen metabolism and respiration to various temperatures and feeding intervals in Donaldson strain rainbow trout. *Res. Fish., Coll. Fish., Univ. Wash.*, (415), 80.

Niimi, A. J. (1975). Relationship of body surface to weight in fishes. *Can. J. Zool.*, **53**, 1192–4.

Niimi, A. J. & Beamish, F. W. H. (1974). Bioenergetics and growth of largemouth bass (*Micropterus salmoides*) in relation to body weight and temperature. *Can. J. Zool.*, **52**, 447–56.

Nishihara, H. (1967). Studies on the structure of red and white fin muscles of the fish (*Carassius auratus*). *Arch. Histol. Jap.*, **28**, 425–47.

Nishizuka, Y. & Hayaishi, O. (1963). Enzymic synthesis of niacin nucleotides from 3-hydroxyanthranilic acid in mammalian liver. *J. Biol. Chem.*, **238**(1), pc483–pc485.

Noaillac-Depeyre, J. & Gas, N. (1973). Absorption of protein macromolecules by the enterocytes of the carp (*Cyprinus carpio* L.). *Z. Zellforsch.*, **146**, 525–41.

Noaillac-Depeyre, J. & Gas, N. (1974). Fat absorption by the enterocystes of the carp (*Cyprinus carpio* L.). *Cell Tissue Res.*, **155**, 353–65.

Noe, B. D. & Bauer, G. E. (1970). Evidence for glucagon biosynthesis and participation of a precursor protein in islets of the anglerfish (*Lophius americanus*). *Biol. Bull.*, **139**, 431.

Nordlie, F. (1966). Thermal acclimation and peptic digestive capacity in the black bullhead *Ictalurus melas* (Raf.). *Am. Midl. Nat.*, **75**, 416–24.

Norris, E. R. & Mathies, J. C. (1953). Preparation, properties and crystallisation of tuna pepsin. *J. Biol. Chem.*, **204**, 673–80.

Norris, J. S., Norris, D. O. & Windell, J. (1973). Effects of simulated meal size on gastric acid and pepsin secretory rate in bluegill (*Lepomis macrochirus*). *J. Fish. Res. Bd Can.*, **30**, 201–4.

Nose, T. (1960). On the digestion of food protein by goldfish (*Carassius auratus* L.) and rainbow trout (*Salmo irideus* G.). *Bull. Freshwater Fish. Res. Lab., Tokyo*, **10**, 11–22.

Nose, T. (1961). Determination of nutritive value of food protein in fish. I. On the determination of food protein utilization by carcass analysis. *Bull. Freshwater Fish. Res. Lab., Tokyo*, **11**, 29–42.

Nose, T. (1963). Determination of nutritive value of food protein in fish. 2. Effect of amino acid composition of high protein diets on growth and protein utilization of the rainbow

trout. *Bull. Freshwater Fish. Res. Lab., Tokyo*, **13**, 41–50.
Nose, T. (1964). Protein digestibility of several test diets in cray and prawn fish. *Bull. Freshwater Fish. Res. Lab., Tokyo*, **14**, 23–8.
Nose, T. (1967). Recent advances in the study of fish digestion in Japan. *FAO/EIFAC Tech. Pap.*, **3**, 83–94.
Nose, T. (1971). Determination of nutritive value of food protein in fish. III. Nutritive value of casein, whitefish meal and soybean meal in rainbow trout fingerlings. *Bull. Freshwater Fish. Res. Lab., Tokyo*, **21**, 85–98.
Nose, T. (1972). Changes in pattern of free plasma amino acid in rainbow trout after feeding. *Bull. Freshwater Fish. Res. Lab., Tokyo*, **22**, 137–44.
Nose, T. (1979). Summary report on the requirements of essential amino acids for carp. in J. E. Halver & K. Tiews (Eds), *Finfish nutrition and fishfeed technology*, pp. 145–56. Berlin: Heenemann Verlagsgesel.
Nose, T. & Arai, S. (1973). Optimum level of protein in purified diet for eel, *Anguilla japonica*. *Bull. Freshwater Fish. Res. Lab., Tokyo*, **22**, 145–55.
Nose, T., & Arai, S. (1979). Recent advances in studies on mineral nutrition of fish in Japan. In T. V. R. Pillay & A. Dill (Eds), *Advances in aquaculture*, pp. 584–90. Farnham, England: Fishing News Books.
Nose, T. & Mamiya H. (1963). Protein digestibility of flatfish meal in rainbow trout. *Bull. Freshwater Fish. Res. Lab., Tokyo*, **12**, 1–4.
Nose, T. & Toyama, K. (1966). Protein digestibility of brown fish meal in rainbow trout. *Bull. Freshwater Fish. Res. Lab., Tokyo*, **15**, 213–24.
Nose, T., Arai, S., Lee, D.-L. & Hashimoto, Y. (1974). A note on amino acids essential for growth of young carp. *Bull. Jap. Soc. Sci. Fish.*, **40**, 903–8.
Odum, W. (1970). Utilization of the direct grazing and plant detritus food chains by *Mugil cephalus*. In H. Steele (Ed.), *Marine food chains*, pp. 222–40. London: Oliver & Boyd.
Ofojekwu, P. C. & Ejike, C. (1984). Growth response and feed utilization in the tropical cichlid *Oreochromis niloticus* (Linn.) fed on cottonseed-based artificial diets. *Aquaculture*, 27–36.
Ogami, A. (1964). Digestibility of wheat containing diet for eels and its growth experiments. (In Japanese.) *Shizuoka-ken, Suisan Shikenjo Hamanako Bunjo Shiken Hokokusho*, **1**, 9–13.
Ogata, H., Arai, S. & Nose, T. (1983). Growth response of cherry salmon *Oncorhynchus masou* and a mago salmon *O. rhodurus* fry fed purified casein diets supplemented with amino acids. *Bull. Jap. Soc. Sci. Fish.*, **49**, 1381–5.
Ogino, C. (1963). Studies on the chemical composition of some natural foods of aquatic animals. *Bull. Jap. Soc. Sci. Fish.*, **29**, 459–62.
Ogino, C. (1965). B vitamin requirements of carp, *Cyprinus carpio*. I. Deficiency symptoms and requirement of vitamin B_6. *Bull. Jap. Soc. Sxi. Fish.*, **31**, 546–51.
Ogino, C. (1967). B vitamin requirements of carp. II. Requirements for riboflavin and pantothenic acid. *Bull. Jap. Soc. Sci. Fish.*, **33**, 351–4.
Ogino, C. (1980). Requirements of carp and rainbow trout for essential amino acids. *Bull. Jap. Soc. Sci. Fish.*, **46**, 171–4.
Ogino, C. & Chen, M. S. (1973a). Protein nutrition in fish. 3. Apparent and true digestibility of dietary proteins in carp. *Bull. Jap. Soc. Sci. Fish*, **39**, 649–51.
Ogino, C. & Chen, M. S. (1973b). Protein nutrition in fish. 4. Biological value of dietary proteins in carp. *Bull. Jap. Soc. Sci. Fish.*, **39**, 797–800.
Ogino, C., & Chiou, J. Y. (1976). Mineral requirements in fish. II. Magnesium requirements in fish. II. Magnesium requirement of carp. *Bull. Jap. Soc. Sci. Fish.*, **42**, 71–5.
Ogino, C., Chiou, J. Y. & Takeuchi, T. (1976). Protein nutrition in fish. VI. Effects of dietary energy sources on the utilization of proteins by rainbow trout and carp. *Bull. Jap. Soc. Sci. Fish.*, **42**, 213–18.
Ogino, C., Kakino, J. & Chen, M. S. (1973). Protein nutrition in fish. II. Determination of

metabolic fecal nitrogen and endogenous nitrogen excretions of carp. *Bull. Jap. Soc. Sci. Fish.*, **39**, 519–23.

Ogino, C. & Kamizono, M. (1975). Mineral requirements in fish. I. Effect of dietary salt mixture levels on growth, mortality, and body composition in rainbow trout and carp. *Bull. Jap. Soc. Sci, Fish.*, **41**, 429–34.

Ogino, C. & Kawasaki, H. (1980). Method for determination of nitrogen retained in the fish body by the carcass analysis. *Bull. Jap. Soc. Sci. Fish.*, **46**, 105–8.

Ogino, C., & Nanri, H. (1980). Relationship between the nutritive value of dietary proteins for rainbow trout and the essential amino acid compositions. *Bull. Jap. Soc. Sci. Fish.*, **46**, 109–12.

Ogino, C. & Saito, K. (1970). Protein nutrition in fish I. The utilization of dietary protein by young carp. *Bull. Jap. Soc. Sci. Fish.*, **36**, 250–4.

Ogino, C., Takashima, F. & Chiou, J. Y. (1978). Requirement of rainbow trout for dietary magnesium. *Bull. Jap. Soc. Sci. Fish.*, **44**, 1105–8.

Ogino, C. & Takeda, H. 1976). Mineral requirements in fish. III. Calcium and phosphorus requirements in carp. *Bull. Jap. Soc. Sci. Fish.*, **42**, 793–9.

Ogino, C., Takeuchi, L., Takeda, H. & Watanbe, T. (1979). Availability of dietary phosphorus in carp and rainbow trout. *Bull. Jap. Soc. Sci. Fish.*, **45**, 1527–32.

Ogino, C., Watanabe, T. Kakino, J., Iwanaga, N. & Mizuno, M. (1970a). B vitamin requirements of carp. III. Requirement for biotin. *Bull. Jap. Soc. Sci. Fish.*, **36**, 734–40.

Ogino, C., Uki, N., Watanabe, T., Iida, Z. & Ando, K. (1970b). B vitamin requirements of carp. IV. Requirements for choline. *Bull. Jap. Soc. Sci. Fish.*, **36**, 1140–6.

Ogino, C. & Yang, G.-Y. (1978). Requirements of rainbow trout for dietary zinc. *Bull. Jap. Soc. Sci. Fish.*, **44**, 1015–18.

Ogino, C. & Yang, G.-Y. (1979a). Requirements of rainbow trout for dietary zinc. In J. E. Halver & K. Tiews (Eds), *Finfish nutrition and fishfeed technology, vol. I*, pp. 105–11. Berlin: Heenemann Verlagsgesel.

Ogino, C. & Yang, G.-Y. (1979b). Requirement of carp for dietary zinc. *Bull. Jap. Soc. Sci. Fish.*, **45**, 967–9.

Ogino, C. & Yang, G.-Y. (1980). Requirements of carp and rainbow trout for dietary manganese and copper. *Bull. Jap. Soc. Sci. Fish.*, **46**, 455–8.

Oguri, M. (1968). Urinary constituents of snakehead fish, with special reference to urinary sugar. *Nippon Suisan Gakkaishi*, **34**(1), 6–10.

O'Hara, J. (1968). Influence of weight and temperature on the metabolic rate of sunfish. *Ecology*, **48**, 159–61.

Ojha, J. & Datta Munshi, J. S. (1975). Oxygen consumption in relation to body size and respiratory surface area of a fresh water mud-eel *Macrognathus aculeatum* (Bloch). *Indian J. Exp. Biol.*, **13**(4), 353–7.

Okoniewska, G. & Kruger, A. (1979). Passage of various food through the digestive tract of fish (*Cyprinus carpio, Ctenopharyngodon idella, Hypophthalmichthys molitrix*). *Rocz. Nauk Roln. H*, **99**(3), 179–200.

Okutani, K. & Kimata, M. (1964a). Studies on chytinolytic enzyme in digestive tracts of Japanese sea-bass *Lateolabrax japonicus*. I. *Bull. Jap. Soc. Sci. Fish.*, **30**, 262–6.

Okutani, K., & Kimata, M. (1964b). Studies on chitinolytic enzyme present in aquatic animals. 3. Distribution of chitinase in digestive organs of a few kinds of aquatic animals. *Bull. Jap. Soc. Sci. Fish.*, **30**, 574–6.

Omae, H., Suzuki, R. & Shimma, Y. (1979). Influence of single-cell protein feeds on the growth and reproductivity of carp with reference to fatty acid composition. In J. E. Halver & K. Tiews (Eds), *Finfish nutrition and fishfeed technology, vol. II*, pp. 63–73. Berlin: Heenemann Verlagsgesel.

Onishi, T., Murayama, S. & Takeuchi, M. (1973a). Sequence of digestive enzyme levels in carp after feeding. I. Amylase and protease of intestinal content, hepatopancreas and gall-bladder. *Bull. Tokai Reg. Fish. Res. Lab.*, (75), 23–32.

Onishi, T., Murayama, S. & Takeuchi, M. (1973b). Sequence of digestive level in carp after feeding. II. Protease in activated and zymogen forms of intestine, hepatopancreas, gall-bladder and spleen. *Bull. Tokai Reg. Fish. Res. Lab.*, (75), 33–8.

Onishi, T., Murayama, S. & Shibata, N. (1974). Studies on enzymes of cultivated salmonoid fishes, III. The change of the activities of several hepatic enzymes during their maturation process. *Bull. Tokai Reg. Fish. Res. Lab.*, (78) 47–57.

Ooshiro, Z. (1974). Studies on proteinase in the pyloric caeca of fishes. II. Some properties of proteinase purified from pyloric caeca of mackerel. *Bull. Jap. Soc. Sci. Fish.*, **37**, 145–8.

Oser, B. L. (1959). An integrated Essential Amino Acid index for predicting the biological value of proteins. In A. A. Albanese (Ed.), *Protein and amino acid nutrition*, pp. 281–95. New York: Academic Press.

Ottolenghi, C., Puviani, A. C., Baruffaldi, A. & Brighenti, L. (1982). 'In vivo' effects of insulin on carbohydrate metabolism of catfish (*Ictalurus melas*). *Comp. Biochem. Physiol.*, **72A**, 35–41.

Overnell, J. (1973). Digestive enzmyes of the pyloric caeca and of their associated mesentery in the cod (*Gadus morhua*). *Comp. Biochem. Physiol.*, **46B**, 519–31.

Owen, J. M., Adron, J. W., Middleton, C. & Cowey, C. B. (1975). Elongation and desaturation of dietary fatty acids in turbot *Scophthalmus maximus* L. and rainbow trout, *Salmo gairdneri* Rich. *Lipids*, **10**. 528–31.

Page, J. E., Rumsey, G. L., Riis, R. C. & Scott, M. L. (1978). Dietary sulfur requirements of fish: nutritional and pathological criteria. *Fed. Proc., Fed. Am. Soc. Exp. Biol.*, **37**, 435.

Page, J. W. & Andrews, J. W. (1973). Interaction of dietary levels of protein and energy on channel catfish. *J. Nutr.*, **103** 1139–46.

Page, J. W., Andrews, J. W., Murai, T. & Murray, M. W. (1976). Hydrogen ion concentration in the gastrointestinal tract of channel catfish. *J. Fish Biol.*, **8**, 225–8.

Palmer, D. D., Robinson, L. A. & Burrows, R. E. (1951). Feeding frequency: its role in the rearing of blue black salmon fingerlings in troughs. *Prog. Fish Cult.*, **13**, 205–12.

Palmer, T. N. & Ryman, B. E. (1972). Studies on oral glucose intolerance in fish *J. Fish Biol.*, **4**, 311–9.

Paloheimo, J. E. & Dickie, L. M. (1965). Food and growth of fishes. I. A growth curve derived from experimental data. *J. Fish. Res. Bd Can.*, **22**, 521–42.

Paloheimo, J. E. & Dickie, L. M. (1966a). Food and growth of fishes. II. Effects of food and temperature on the relation between metabolism and body weight. *J. Fish. Res. Bd Can.*, **23**, 869–908.

Paloheimo, J. E. & Dickie, L. M. (1966b). Food and growth of fishes. III. Relations among food, body size, and growth efficiency. *J. Fish. Res. Bd Can.*, **23**, 1209–48.

Pandian, T. J. (1967). Intake, digestion, absorption and conversion of food in the fishes *Megalops cyprinoides* and *Ophiocephalus striatus*. *Mar. Biol.*, **1**, 16–32.

Papaparaskeva-Papoutsoglou, E. & Alexis, M. N. (1986). Protein requirements of young grey mullet, *Mugil capito*. *Aquaculture*, **52**, 105–15.

Parker, R. R. & Larkin, P. A. (1959). A concept of growth in fishes. *J. Fish. Res. Bd Can.*, **16**, 721–45.

Patankar, V. B. (1973). Esterases in the stomach of fishes with different food habits. *Folia Histochem. Cytochem.*, **11**, 253–8.

Pauly, D. (1981). The relationships between gill surface area and growth performance in fish: a generalization of von Bertalanffy's theory of growth. *Meeresforschung – Reports on Marine Research*, **28**, 251–82.

Pavlovskii, E. N. (1958). Uptake and loss of calcium in first year carp as a function of the concentration of calcium salts in the surrounding medium. *Dokl. Akad. Nauk SSSR.* English transl: Biol. Sci. Soc., **120**(1–6), 349–52.

Peer, M. M. & Kutty, M. N. (1981). Respiratory quotient and ammonia quotient in *Tilapia mossambica* (Peters) with special reference to hypoxia and recovery. *Hydrobiologia*, **76**, 3–9.

Pegel, V. A. (1950). *Digestive physiology of fishes.* (In Russian.) Univ. Tomsk: Tomskij Gosudarstv, 199 pp.

Penczak, T., Molinski, M., Kuslo, E., Ichniowska, B. & Zalewski, M. (1976). The ecology of roach, *Rutilus rutilus* (L.) in the barbel region of the polluted Pilica River. III. Lipids, protein, total nitrogen and caloricity. *Ecol. Pol.*, **25**, 75–88.

Pequin, L. (1967). Degradation and synthesis of glutamine in carp (*Cyprinus carpio*) L. (In French.) *Arch. Sci. Physiol.*, **21**, 193–203.

Percy, W. G. & Avault, J. W. (1974). Influence of floating and sinking feeds and fingerling size on channel catfish production. *Proc. Ann. Conf. Southeast. Assoc. Game Fish Commrs.*, **27**, 500–11.

Pérès, G., Boge, G., Colin, D. & Rigal, A. (1974). Effets de la temperature sur les processes digestifs des poissons. *Rapp. P.-V. Réun., Comm. Int. Explor. Sci. Mer Mediterr., Monaco*, **22**(7), 59–60.

Pessah, E. & Powles, P. M. (1974). Effect of constant temperature on growth rates of pumpkinseed sunfish (*Lepomis gibbosus*). *J. Fish. Res. Bd Can.*, **31**, 1678–82.

Peter, R. E., Hontela, A., Cook, A. F. & Paulencu, C. R. (1978). Daily cycle in serum cortisol levels in the goldfish: effects of photoperiod, temperature, and sexual conditions. *Can. J. Zool.*, **56**, 2443–8.

Peters, D. S. & Hoss, D. E. (1974). A radio-isotopic method of measuring food evacuation time in fish. *Trans. Am. Fish. Soc.*, **103**, 626–9.

Peters, D. S., Kjelson, M. A. & Boyd, M. T. (1972). The effect of temperature on food evacuation rate in the pinfish (*Lagodon rhomboides*), spot (*Leiostomus xanthurus*) and silverside (*Menidia menidia*). *Proc. Ann. Conf. Southeast. Assoc. Game Fish Commrs.*, **26**, 637–43.

Peterson, R. H. & Anderson, J. M. (1969). Influence of temperature change on spontaneous locomotor activity and oxygen consumption of Atlantic salmon *Salmo salar*, acclimated to two temperatures. *J. Fish. Res Bd Can.*, **26**, 93–109.

Petrenko, I. N. & Karasikova, A. A. (1960). Investigations of the indicators of the amino acid complex of the Baltic herring in the compilation of short-term forecast of its catches. (In Russian.) *Tr. Vses. Nauch-issled. Inst. Morsk. Ryb. Khoz. Okeanogr.* (*VNIRO*), **42**, 189–94.

Petrusewicz, K. & MacFadyen, A. (1970). *Productivity of terrestrial animals: principles and methods, IBP Handbook 13.* Oxford: Blackwell.

Pfuderer, P., Williams, P. & Francis, A. A. (1974). Partial purification of the crowding factor from *Carassius auratus* and *Cyprinus carpio*. *J. Exp. Zool.*, **187**, 375–82.

Phillips, A. M. (1969). Nutrition, digestion, and energy utilization. In W. S. Hoar & D. J. Randall (Eds), *Fish physiology, I. Excretion, ionic regulation, and metabolism*, pp. 391–432. New York: Academic Press.

Phillips, A. M. (1972). Calorie and energy requirement. In J. E. Halver (Ed.), *Fish nutrition*, pp. 1–28. New York: Academic Press.

Phillips, A. M., Brockway, D. R., Lovelace, F. E., Podoliak, H. A. & Maxwell, J. M. (1951). The nutrition of trout. Cortland Hatchery Report 21. *Fish. Res. Bull.*, **15**, 18–22.

Phillips, A. M., Livingston, D. L. & Poston, H. A. (1966a). The effect of changes in protein quality, calorie sources, and calorie levels upon the growth and chemical composition of brook trout. *Fish. Res. Bull.*, **29**, 6–14.

Phillips, A. M., Livingston, D. L. & Poston, H. A. (1966b). Use of calorie sources by brook trout. *Prog. Fish Cult.*, **28**, 67–72.

Phillips, A. M., Lovelace, F. E., Brockway, D. R. & Balzer, G. C. (1953). The nutrition of trout. Cortland Hatchery Report 22. *Fish. Res. Bull.*, **17**, 31 pp.

Phillips, A. M., Lovelace, F. E., Podoliak, H. A., Brockway, D. R. & Balzer, G. C. (1954). The nutrition of trout. Cortland Hatchery Report 23. *Fish. Res. Bull.*, **18**, 52 pp.

Phillips, A. M., Lovelace, F. E., Podoliak, H. A., Brockway, D. R. & Balzer, G. C. (1955). The nutrition of trout. Cortland Hatchery Report 24. *Fish. Res. Bull.*, **19**, 56 pp.

Phillips, A. M., Podoliak, H. A., Brockway, D. R. & Balzer, G. C. (1956). The nutrition of trout. Cortland Hatchery Report 25. *Fish. Res. Bull.*, **20**, 61 pp.

Phillips, A. M., Podoliak, H. A., Brockway, D. R. & Vaughn, R. R. (1957). The nutrition of trout. Cortland Hatchery Report 26. *Fish. Res. Bull.*, **21**, 93 pp.

Phillips, A. M., Podoliak, H. A., Livingston, D. L., Dumas, R. F. & Hammer, G. L. (1961). Cortland Hatchery, Report 29. *Fish. Res. Bull.*, **24**, 76 pp.

Phillips, A. M., Podoliak, H. A., Livingston, D. L., Dumas, R. F. & Thoesen, R. W. (1960). The nutrition of trout. Cortland Hatchery, Report 28. *Fish. Res. Bull.*, **23**, 83 pp.

Phillips, A. M., Podoliak, H. A., Poston, H. A., Livingston, D. L., Booke, H. E. & Hammer, G. L. (1962). Cortland Hatchery, Report 30. *Fish. Res. Bull.*, **25**, 57 pp.

Phillips, A. M., Podoliak, H. A., Poston, H. A., Livingston, D. L., Booke, H. E., Pyle, E. A. & Hammer, G. L. (1963). The relationship of protein and ash in water and fat free samples of fish tissues. *Fish. Res. Bull.*, **26**, 49–51.

Phillips, A. M., Tunison, A. V. & Brockway, D. R. (1948). The utilizaton of carbohydrates by trout. *Fish. Res. Bull.*, **11**, 1–44.

Phillips, A. M., Tunison, A. V., Shaffer, H. B., White, G. K., Sullivan, M.W., Vincent, C., Brockway, D. R. & McCay, C. M. (1945). The nutrition of trout. Cortland Hatchery, Report 14. *Fish. Res. Bull.*, **8**, 31 pp.

Pieper, A. & Pfeffer, E. (1979). Carbohydrate as possible sources of dietary energy for rainbow trout (*Salmo gairdneri*, Richardson). In J. E. Halver & K. Tiews (Eds), *Finfish nutrition and fishfeed technology, vol. I*, pp. 209–19. Berlin: Heenemann Verlagsgesel.

Pierce, R. J. & Wissing, T. E. (1974). Energy cost of food utilization in the bluegill (*Lepomis macrochirus*). *Trans. Am. Fish. Soc.*, **103**, 38–45.

Pillay, T. V. R. (1953). Food, feeding habits, and alimentary tract of grey mullet. *Proc. Natl. Inst. Sci., India*, **19**, 777.

Pitcher, T. J., Magurran, A. E. & Winfield, I. J. (1982). Fish in larger shoals find food faster. *Behav. Ecol. Sociobiol.*, **10**. 149–51.

Plantikow, A. & Plantikow, H. (1985). Alanine aminopeptidase (AAP) activity in the midgut of rainbow trout (*Salmo gairdneri* R.): The influence of feed quantity and quality, temperature and osmolarity. *Aquaculture*, **48**, 261–76.

Podoliak, H. A. (1961). Relation between water temperature and metabolism of dietary phosphorus by fingerling brook trout. *Trans. Am. Fish. Soc.*, **90**, 398–403.

Polimanti, O. (1912). Untersuchungen über die Topographie der Enzyme in Magen-Darmrohr der Fische. *Biochem. Ztg.*, **38**, 113–28.

Post, G. (1975). Ration formulation for intensively reared fishes. *Proc. Int. Congr. Nutr.*, **9**(3), 199–207.

Poston, H. A. (1965). Effect of protein and calorie sources on blood serum protein of immature brown trout. *Fish. Res. Bull.*, **28**, 19–23.

Poston, H. A. & Combs, G. F. (1980). Nutritional implications of tryptophan catabolizing enzymes in several species of trout and salmon. *Proc. Soc. Exp. Biol. Med.*, **163**, 452–4.

Poston, H. A. & Dilorenzo, R. N. (1973). Tryptophan conversion to niacin in brook trout (*Salvelinus fontinalis*). *Proc. Soc. Exp. Biol. Med.*, **144**, 110–12.

Poston, H. A., Combs, G. F. & Leibovits, L. (1976). Vitamin E and selenium interrelations in the diet of Atlantic salmon (*Salmo salar*): Gross histological and biochemical deficiency signs. *J. Nutr.*, **106**, 892–904.

Poston, H. A., Riis, R. C., Rumsey, G. L. & Ketola, H. G. (1977). The effect of supplemental dietary amino acids, minerals and vitamins on salmonids fed cataractogenic diets. *Cornell Vet.*, **67**, 472–509.

Prather, E. E. & Lovell, R. T. (1973). Response of intensively fed channel catfish to diets containing various protein – energy ratios. *Proc. Ann. Conf. Southeast. Assoc. Game Fish. Commrs*, **27**, 455–9.

Precht, H. (1955). Wechselwarme Tiere und Pflanzen. In H. Precht, J. Christophersen & H. Hensel (Eds), *Temperatur und Leben*, pp. 1–77. Berlin: Springer Verlag. English translation – Temperature and life, Springer, Berlin, 1973.

Prikryl, I. & Janecek, V. (1982). The effect of food consumption on the amount of natural food in ponds stocked with carp. (In Czech with English summary.) *Bull. Vyzk. Ustav Ryb. Hydrobiol., Vodnany*, **18**(1), 20–6.

Pritchard, A. W., Florey, E. & Martin, A. W. (1958). Relationship between metabolic rate and body size in an elasmobranch (*Squalus suckleyi*) and in a teleost (*Ophiodon elongatus*). *J. Mar. Res.*, (*Sears Found.*), **17**, 403–11.

Prosser, C. L. (1955). Physiological variation in animals. *Biol. Rev..*, **30**, 229–62.

Prosser, C. L. & Brown, F. A. (1961). *Comparative animal physiology* (2nd edn). Philadelphia: Saunders, 688 pp.

Prosser, C. L., Precht, H. & Jankowsky, H.-D. (1965). Nervous control of metabolism during temperature acclimation of fish. *Naturwissenschaften*, **52**, 168–9.

Prus, T. (1970). Calorific value of animals as an element of bioenergetical investigations. *Pol. Arch. Hydrobiol.*, **17**, 183–99.

Puckett, K. J. & Dill, L. M. (1984). Cost of sustained and burst swimming of juvenile coho salmon (*Oncorhynchus kisutch*). *Can. J. Fish. Aquat. Sci.*, **41**, 1546–51.

Rackis, J. J. (1974). Biological and physiological factors in soybeans. *J. Am. Oil Chem. Soc.*, **51**, 161A-74A.

Rajbanshi, V. K. (1966). A note on the study of gustatory senses of the barbels of fresh water teleosts. *Z. Naturwiss.*, **53**(8), 208–9.

Rajbanshi, V. K. (1979). Scanning microscopical and histological studies on barbel epithelium of *Heteropneustes fossilis* (Bloch). In K. Herter (Ed.), *Zoologische Beitraege*, pp. 73–80. Berlin: Duncker and Humblot.

Ranade, S. S. & Kewalramani, H. G. (1967). Studies on the rate of food passage in the intestine of *Labeo rohita* (Ham.), *Cirrhina mrigala* (Ham.) and *Catla catla* (Ham.). *FAO Fish. Rep.*, **44**(3), 349–58.

Randall, D. J., & Shelton, G. (1963). The effects of changes in enviromental gas concentrations on the breathing and heart rate of a teleost fish (*Tinca tinca*). *Comp. Biochem. Physiol.*, **9**, 229–39.

Rantin, F. T. & Johansen, K. (1984). Responses of the teleost *Hoplias malabaricus* to hypoxia. *Environ. Biol. of Fishes*, **11**, 221–8.

Rao, G. M. M. (1968). Oxygen consumption of rainbow trout (*Salmo gairdneri*) in relation to activity and salinity. *Can. J. Zool.*, **46**, 781–6.

Rayner, M. D. & Keenan, M. J. (1967). Role of red and white muscle in the swimming of the skipjack tuna. *Nature* (*Lond.*), **214**, 392–3.

Read, C. P. (1967). Studies on membrane transport. I. A common transport system for sugars and amino acids? *Biol. Bull.*, **133**, 630–42.

Redgate, E. S. (1974). Neural control of pituitary adrenal activity in *Cyprinus carpio.*, *Genl. Comp. Endocrinol.*, **22**, 35–41.

Reece, D. L., Wesley, D. E., Jackson, G. A. J. & Dupree, H. K. (1975). A blood meal–rumen contents blend as a partial or complete substitute for fish meal in channel catfish diets. *Prog. Fish Cult.*, **37**, 15–19.

Reed, J. R. (1971). Uptake and excretion of ^{60}Co by black bullhead *Ictalurus melas* (Rafinesque). *Health Phys.*, **21**, 835–44.

Reed, J. R., Samsel, G. L., Daub, R. R. & Llewellyn, G. C. (1974). Oxidation pond algae as supplement for commercial catfish feed. *Proc. Ann. Conf. Southeast. Assoc. Game Fish Commrs.*, **27**, 465–70.

Refstie, T. & Austreng, E. (1981). Carbohydrate in rainbow trout diets. III. Growth and chemical composition of fish from different families fed four levels of carbohydrate in the diet. *Aquaculture*, **25**, 35–49.

Reichle, G. (1980). Soybean meal as a substitute for herring meal in practical diets for rainbow trout. *Prog. Fish. Cult.*, **42**, 103–6.

Reichle, G. & Wunder, W. (1974). Kann die Regenbogenforelle pflanzliches Eiweiss auswerten? *Arch. Hydrobiol.*, **74**, 468–72.

Reinitz, G. (1980). Soybean meal as a substitute for herring meal in practical diets for rainbow trout. *Prog. Fish Cult.*, **42**, 103–6.

Reinitz, G. & Hitzel, F. (1980). Formulation of practical diets for rainbow trout based on desired performance and body comparison. *Aquaculture*, **19**, 243–52.

Reinitz, G. L. & Yu, T. C. (1981). Effects of dietary lipids on growth and fatty acid composition of rainbow trout (*Salmo gairdneri*). *Aquaculture*, **22**, 359–66.

Reinitz, G. L., Orme, L. E., Lemm, C. A. & Hitzel, F. N. (1978). Influence of varying lipid concentrations with two protein concentrations in diets for rainbow trout (*Salmo gairdneri*). *Trans. Am. Fish. Soc.*, **107**, 751–4.

Richard, P. J., Berhaut, J. A. & Ceccaldi, H. J. (1982). Development of some digestive enzymes in young *Mugil capito* at various growth stages. *Biochem. Syst. Ecol.*, **10**, 185–90.

Richman, S. (1958). The transformation of energy by *Daphnia pulex*. *Ecol. Monogr.*, **28**, 273–91.

Ricker, W. E. (1979). Growth rates and models. In W. S. Hoar, D. J. Randall & J. R. Brett (Eds), *Fish physiology, vol. VIII*. pp. 677–743. New York: Academic Press.

Ringrose, R. C. (1971). Calorie-to-protein ratio for brook trout (*Salvelinus fontinalis*). *J. Fish. Res. Bd Can.*, **28**, 1113–17.

Risse, S. (1971). Enzymatic changes in the intestinal tract due to differences in the supply of amino acids in food. *Nahrung*, **15**, 295–305.

Roberts, J. L. (1964). Metabolic responses of fresh-water sunfish to seasonal photoperiods and temperatures. *Helgol. Wiss. Meeresunters.*, **9**, 459–73.

Robinson, E. H., Wilson, R. P. & Poe, W. E. (1980a). Total aromatic amino acid requirement, phenylalanine requirement and tyrosine replacement value for fingerling channel catfish. *J. Nutr.*, **110**, 1805–12.

Robinson, E. H., Poe, W. E. & Wilson, R. P. (1980b). Quantitative amino acid requirements for channel catfish. *Feedstuffs*, **52**(43), 29–36.

Robinson, E. H., Rawles, S. D., Oldenburg, P. W. & Stickney, R.R. (1984). Effects of feeding glandless or glanded cottonseed products and gossypol to *Tilapia aurea*. *Aquaculture*, **38**, 145–54.

Roginski, E. E. & Mertz, W. (1969). Effects of chromium (III) supplementation on glucose and amino acid metabolism in rats fed a low protein diet. *J. Nutr.*, **97**, 525–30.

Rose, W. C., Smith, L. C., Womack, M. & Shane, M. (1949). The utilization of the nitrogen of ammonium salts, urea, and certain other compounds in the synthesis of non-essential amino acids in vivo. *J. Biol. Chem.*, **181**, 307–16.

Rosebrough, R. W. & Steele, N. C. (1981). Effect of supplemental dietary chromium or nicotinic acid on carbohydrate metabolism during basal, starvation, and refeeding periods in poults. *Poultry Sci.*, **60**, 407–17.

Ross. B. & Jauncey, K. (1981). A radiographic estimation of the effect of temperature on gastric emptying time in *Sarotherodon niloticus* (L.) × *S. aureus* (Steindachner) hybrids. *J. Fish. Biol.*, **19**, 333–44.

Rothbard, S. (1979). Observations on the reproductive behaviour of *Tilapia zillii* and several *Sarotherodon* spp. under aquarium conditions. *Bamidgeh*, **31**(2), 35–43.

Rozin, P. & Mayer, J. (1961). Regulation of food intake in the goldfish. *Am. J. Physiol.*, **201**, 968–74.

Rumsey, G. L. & Ketola, H. G. (1975). Amino acid supplementation of casein in diets of Atlantic salmon (*Salmo salar*) fry and of soybean meal for rainbow trout (*Salmo gairdneri*) fingerlings. *J. Fish. Res. Bd Can.*, **32**, 422–6.

Rychly, J. & Marina, B. A. (1977). The ammonia excretion of trout during a 24-hour period. *Aquaculture*, **11**, 173–8.

Rychly, J. & Spannhof, L. (1979). Nitrogen balance in trout. I. Digestibility of diets containing varying levels of protein and carbohydrate. *Aquaculture*, **16**, 39–46.

Sakaguchi, H., Takeda, F. & Tange, K. (1969). Studies on vitamin requirements by yellowtail, I. Vitamin B_6 and vitamin C deficiency symptoms. *Bull. Jap. Soc. Sci. Fish.*, **35**, 1201–6.

Sakaguchi, M. & Kawai, A. (1968). Histidine metabolism in fish. III. Purification and some properties of histidine deaminase from mackerel muscle. *Bull. Jap. Soc. Sci. Fish.*, **34**, 1040–6.

Sakaguchi, M. & Kawai, A. (1970a). Change in the composition of free amino acids in carp muscle with the growth of the fish. *Mem. Res. Inst. Food Sci., Kyoto Univ.*, (31), 19–21.

Sakaguchi, M. & Kawai, A. (1970b). Histidine metabolism in fish. V. Effect of protein deficiency and fasting on the activities of histidine deaminase and urocanase in carp liver. *Bull. Jap. Soc. Sci. Fish.*, **36**, 783–7.

Sakaguchi, M. & Kawai, A. (1971). Occurrence of histidine in muscle extracts and its metabolism in fish tissues. *Kyoto Univ. Res. Inst. for Food Sci., Rep.*, **34**, 28–51.

Sakaguchi, M., Sugiyama, M., Sugiyama, T. & Kawai, A. (1970). Histidine metabolism in fish. IV. Comparative study on histidine deaminases from muscle and liver of mackerel and from bacteria. *Bull. Jap. Soc. Fish.*, **36**, 200–6.

Sakamoto, S. & Yone Y. (1973). Effect of dietary calcium/phosphorus ratio upon growth, feed efficiency and blood serum Ca and P levels in red sea bream. *Bull. Jap. Soc. Sci. Fish.*, **39**, 343–8.

Sakamoto, S. & Yone, Y. (1978a). Effect of dietary phosphorus level on chemical composition of red sea bream. *Bull. Jap. Soc. Sci. Fish.*, **44**, 227–30.

Sakamoto, S. & Yone, Y. (1978b). Iron deficiency symptoms of carp. *Bull. Jap. Soc. Sci. Fish.*, **44**, 1157–60.

Sakamoto, S. & Yone, Y. (1979). Availabilities of phosphorus compounds as dietary phosphorus sources for red sea bream. *J. Fac. Agr., Kyushu Univ.*, **23**, 177–94.

Sakata, T., Okabayashi, J. & Kakimoto, D. (1980a). Variations in the intestinal microflora of *Tilapia* reared in fresh and sea water. *Bull. Jap. Soc. Sci. Fish.*, **46**, 313–17.

Sakata, T., Sugita, H., Mitsuoka, T., Kakimoto, D. & Kadota, H. (1980b). Isolation and distribution of obligate anaerobic bacteria from the intestine of freshwater fish. *Bull. Jap. Soc. Sci. Fish.*, **46**, 1249–55.

Sakata, T., Sugita, H. & Mitsuoka, T. Kakimoto, D. & Kadota, H. (1981). Characteristics of obligate anaerobic bacteria in the intestines of freshwater fish. *Bull. Jap. Soc. Sci. Fish.*, **47**, 421–7.

Salonen, K., Sarvala, J., Hakala, I. & Viljanen, M.-L. (1976). The relation of energy and organic carbon in aquatic invertebrates. *Limnol. Oceanogr.*, **21**, 724–30.

Saltiel, A. R. & Cuatrecasas, P. (1986). Insulin stimulates the generation from hepatic plasma membranes of modulators derived from an inositol glycolipid. *Proc. Natl. Acad. Sci., USA*, **83**, 5793–7.

Saltiel, A. R., Fox, J. A., Sheline, P. & Cuatrecasas, P. (1986). Insulin-stimulated hydrolysis of a novel glycolipid generates modulators of cAMP phosphodiesterase. *Science* (*Wash.*), **233**, 967–72.

Sandbank, E. (1979). Harvesting of micro-algae from wastewater stabilization pond effluents and their utilization as a fish feed. Ph.D. Thesis, Technion – Israel Institute of Technology, Haifa, Israel. (In Hebrew with English summary.)

Sandbank, E. & Hepher, B. (1978). The utilization of microalgae as a feed for fish. *Arch. Hydrobiol.*, **11**, 108–20.

Santiago, C. B., Aldaba, M. B. & Laron, M. A. (1981). Effect of varying crude protein levels on spawning frequency and growth of *Sarotherodon niloticus* breeders. *Quarterly Res Rep. (SEAFDEC)*, **5**(4), 5–10.

Sarbahi, D. S. (1951). Studies of the digestive tracts and the digestive enzymes of the goldfish *Carassius auratus* (*Linnaeus*) and the largemouth black bass, *Micropterus salmoides* (Lacepede). *Biol. Bull.*, **100**, 244–57.

Sastry, K. V. (1972). Amino tripeptidase activity in three fishes. *Proc. Int. Acad. Sci.*, **76B**, 251–7.

Sastry, K. V. (1974). Distribution of lipase in the digestive system of two teleost fishes. *Acta Histochem.*, **48**, 320–5.

Sastry, K. V. (1975). Alkaline and acid phosphatase in the digestive system of two teleost fishes. *Anta. Anz.* **13**, 159–65.

Sastry, K. V. (1977). Peptidase activity in a few teleost fishes. *Zool. Beitr.*, **23**, 29–33.

Satia, B. P. (1974). Quantitative protein requirements of rainbow trout. *Prog. Fish Cult.*, **36**, 80–5.

Satia, B. P., Donaldson, L. R., Smith, L. S. & Nightingale, J. N. (1974). Composition of ovarian fluid and eggs of the University of Washington strain of rainbow trout (*Salmo gairdneri*). *J. Fish. Res. Bd. Can.*, **31**, 1796–9.

Sato, M., Yoshinaka, R., Yamamoto, Y. & Ikeda, S. (1978). Nonessentiality of ascorbic acid in the diet of carp. *Bull. Jap. Soc. Sci. Fish.*, **44**, 1151–6.

Satoh, S., Yamamoto, H., Takeuchi, T. & Watanabe, T. (1983a). Effects on growth and mineral composition of carp of deletion of trace elements or magnesium from fish meal diet. *Bull. Jap. Soc. Sci. Fish.*, **49**, 431–5.

Satoh, S., Yamamoto, H., Takeuchi, T. & Watanabe, T. (1983b). Effects on growth and mineral composition of rainbow trout of deletion of trace elements or magnesium from fish meal diet. *Bull. Jap. Soc. Sci. Fish.*, **49**, 425–9.

Saunders, R. L. (1963). Respiration of the Atlantic cod. *J. Fish. Res. Bd Can.*, **20**, 373–86.

Savitz, J. (1969). Effects of temperature and body weight on endogenous nitrogen excretion in the bluegill sunfish (*Lepomis macrochirus*). *J. Fish. Res. Bd Can.*, **26**, 1813–21.

Savitz, J. (1971). Effect of starvation on body protein utilization of bluegill sunfish (*Lepomis macrochirus* Rafinesque) with a calculation of caloric requirements. *Trans. Am. Fish. Soc.*, **100**, 18–21.

Schaeperclaus, W. (1961). *Lehrbuch der Teichwirtschaft.* Berlin: Paul Parey, xii + 582 pp.

Schaeperclaus, W. (1964). Ergebnisse eines Fuetterungsversuchs mit Koblatchlorid bei Karpfen und ueber die Bedeutung der Spurelemente in der Teichwirtschaft im allgemeinen. *Dtsch. Fisch. Ztg.*, **11**, 33–9.

Schaeperclaus, W. (1965). Versuche mit der Verfuetterung von Spurelementen an Karpfen. *Dtsch. Fisch. Ztg.*, **12**, 161–3.

Schaeperclaus, W. (1966a). Weitere Untersuchungen ueber Groesse und Bedeutung des Naturnahrungsanteils an der Gesamtnahrung der Karpfen bei Fuetterung mit Getreidenkoernern in Abwachsteichen. *Z. Fisch., N.F.*, **14**, 71–100.

Schaeperclaus, W. (1966b). Bedeutung des effektiven Naturzuwachses fuer die Gesamtproduktion in Karpfenteich. *Verh. Int. Ver. Limnol.*, **16**, 1415–20.

Schaeperclaus, W. (1968). Einfluss der Nahrung auf die chemische Zusammensetzung von Speisekarpfen. *Z. Fisch., N.F.*, **16**, 77–102.

Schaeperclaus, W. (1968b). Neue Methoden zur verbesserten energetischen Bewertung von Karpfenfuttermittelen. *Z. Fisch., N.F.*, **16**, 43–56.

Schalles, J. F. & Wissing, T. E. (1976). Effects of dry pellet diets on the metabolic rates of bluegill (*Lepomis macrochirus*). *J. Fish. Res. Bd Can.*, **33**, 2443–9.

Schepartz, B. (1973). *Regulation of amino acid metabolism in mammals.* Philadelphia: Saunders, 205 pp.

Scheuring, L. (1928). Beziehungen zwischen Temperatur und Verdauungs-geschwindigkeit bei Fischen. *Z. Fisch.*, **26**, 231–6.

Schindler, D. W., Clark, A. S. & Gray, J. R. (1971). Seasonal calorific values of freshwater zooplankton, as determined with Phillipson bomb calorimeter modified for small samples. *J. Fish. Res. Bd Can.*, **28**, 559–64.

Schlisio, W. & Nicolai, B. (1978). Kinetic investigations on the behaviour of free amino acids in the plasma and of two aminotransferases in the liver of rainbow trout (*Salmo gairdneri* Richardson) after feeding on a synthetic composition containing pure amino acids. *Comp. Biochem. Physiol.*, **59B**, 373–9.

Schlottke, E. (1938–9). The change in the enzyme strength in the intestine of carp during digestion. *Sitzbar. Abhandl. naturforsch. Ges. Rostock*, **7**, 27–88.

Schmeing-Engberding, F. (1953). Die Vorzugstemperaturen einiger Knochfische und ihre physiologische Bedeutung, *Z. Fisch., N.F.*, **2**, 125–55.

Schmidt-Nielsen, K. (1975). *Animal physiology, adaptation and environment*. London: Cambridge University Press, 699 pp.

Schmitz, O., Greuel, E. & Pfeffer, E. (1983). A method for determining digestibility of nutrients in eels. *Aquaculture*, **32**, 71–8.

Schmitz, O., Greuel, E. & Pfeffer, E. (1984). Digestibility of crude protein and organic matter of potential sources of dietary protein for eels (*Anguilla anguilla*, L.). *Aquaculture*, **41**, 21–30.

Schneberger, E. 1941. Fishery research in Wisconsin, *Prog. Fish Cult.*, (56). 14–17.

Schroeder, G. L. (1974). Use of fluid cowshed manure in fish ponds. *Bamidgeh*, **26**(3), 84–6.

Schroeder, G. L. (1980). The breakdown of feeding niches in fish ponds under conditions of severe competition. *Bamidgeh*, **32**(1), 20–4.

Schwarz, K. (1961). Nutritional significance of selenium (Factor 3). *Fed. Proc., Fed. Am. Soc. Exp. Biol.*, **20**, 665–73.

Schwarz, K. & Mertz, W. (1959). Chromium (III) and the glucose tolerance factor. *Arch. Biochem. Biophys.*, **85**, 292–5.

Schwarz, F. J. Zeitler, M. H. & Kirchgessner, M. (1983). Wachstum und Naehrstoffaufwand bei Karpfen (*Cyprinus carpio* L.) mit unterschiedlicher Protein- und Energieversorgung. 2. Mitteilung Gewichtsentwicklung, Futtervewertung, Protein und Energieaufwand. *Z. Tierphysiol., Tierernaehr., Futtermittelk.*, **49**(2), 88–98.

Scott, M. L., Olson, G., Krook, L. & Brown, W. R. (1967). Selenium responsive myopathies of myocardium and smooth muscle in the young poult. *J. Nutr.*, **91**, 573–83.

Seagran, H. L., Morey, D. E. & Dassow, J. A. (1954). The amino acid content of roe at different stages of maturity from five species of Pacific salmon. *J. Nutr.*, **53**, 139–49.

Sehgal, P. (1966). Food and feeding habits of *Labeo calbasu* (Hamilton). *Res. Bull. Punjab Univ.*, **17**, 251–6.

Sen, P. R. & Chatterjee, D. K. (1979). Enhancing production of Indian carp fry and fingerlings by use of growth-promoting substances. In T. V. R. Pillay & W. A. Dill (Eds), *Advances in aquaculture*, pp. 134–41. Farnham, England: Fishing News Books.

Sen, P. R., Rao, N. G. S., Ghosh, S. R. & Rout, M. (1978). Observations on the protein and carbohydrate requirements of carps. *Aquaculture*, **13**, 245–55.

Sera, H. & Ishida, I. (1972). Bacterial flora in the digestive tract of marine fish. III. Classification of isolated bacteria. *Bull. Jap. Soc. Sci. Fish.*, **38**, 853–8.

Seshadri, B. (1959). Effect of insulin injection on the amino nitrogen content of free and protein-bound amino acids in skeletal muscle in the murrel, *Ophiocephalus striatus* (Bloch). *Curr. Sci. (India)*, **28**, 121–2.

Shabalina, A. A. (1964). Effect of cobalt on the size of young carp and filamentous algae. (In Russian.) *Izv. Gos. Nauch-issled. Inst. Ozer, Rech. Ryb. Khoz. (GOSNIORKH)*, **57**, 290–4.

Shafi, M. (1980). Histophysiological studies on the peptic glands of a murrel fish *Ophiocephalus punctatus*. *Zool. Beitr.*, **26**, 203–10.

Shamberger, R. J. (1980). Trace metals in health and disease in M. A. Brewster & H. K. Natio (Eds), *Nutritional elements and clinical biochemistry*, pp. 241–75. New York: Plenum Publ. Co.

Shcherbina, M. A. (1964). Determination of the digestibility of artificial foodstuffs of pond fish with the help of inert substances. *Vopr. Ikhtiol.*, **4**, 672–8.

Shcherbina, M. A. (1975). Physiological assessment of the food value of artificial foods of fishes. *J. Ichthyol.*, **15** 303–10.

Shcherbina, M. A. & Kazlauskene, O. P. (1971). Water temperature and the digestibility of nutrient substances by carp. *Hydrobiol. Jour. (SSSR)*, **7**(3), 40–4.

Shcherbina, M. A. & Sorvachev, K. F. (1967). On the digestion of oil meals by two-year-old carp. (In Russian.) *Rybovod. Rybolov.*, **2**, 12–13.

Shcherbina, M. A., Motshulskaya, W. F. & Erman, E. Z. (1970). Study of the digestibility of feed by pond fish. I. Digestibility of peanut oil cake, pea and mixed diet by two-year-old carp. (In Russian.) *Vop. Ikhtiol.*, **10**, 876–82.

Shelbourn, J. E., Brett, J. R. & Shirahata, S. (1973). Effect of temperature and feeding regime on the specific growth rate of sockeye salmon fry (*Oncorhynchus nerka*), with a consideration of size effect. *J. Fish. Res. Bd Can.*, **30**, 1191–4.

Sherstneva, T. A. (1971). Glycemia level in carp and rainbow trout. (In Russian.) *Izv. Gos. Nauch-issled. Inst. Ozer. Rech. Ryb. Khoz.*, **75**, 136–42.

Shewan, J. M. (1961). The microbiology of sea water fish. In G. Borgstrom (Ed.), *Fish as food, vol. I*, pp. 487–560. New York: Academic Press.

Shiflett, J. M. & Haskell, B. E. (1969). Effect of vitamin B_6 deficiency on leucine transaminase activity in chick tissue. *J. Nutr.*, **98**, 420–6.

Shimeno, S., Hosokawa, H. & Takeda, M. (1979). The importance of carbohydrate in the diet of carnivorous fish. In J. E. Halver & K. Tiews (Eds), *Finfish nutrition and fishfeed technology, vol. I*, pp. 127–43. Berlin: Heenemann Verlagsgesel.

Shimeno, S. & Ikeda, S. (1967). Studies on glucose-6-phosphatase of aquatic animals. II. The enzyme activities in fish tissues. *Bull. Jap. Soc. Sci. Fish.*, **33**, 112–16.

Shimeno, S. & Takeda, M. (1972). Studies on hexose monophosphate shunt of fishes. I. Properties of hepatic glucose-6-phosphate dehydrogenase of barracuda. *Bull. Jap. Soc. Sci. Fish.*, **38**, 645–50.

Shimeno, S. & Takeda, M. (1973). Studies on hexose monophospate shunt of fishes II. Distribution of glucose-6-phosphate dehydrogenase. *Bull. Jap. Soc. Sci. Fish.*, **39**, 461–6.

Shimeno, S., Takeda, M., Takayama, S., Fukui, A., Sasaki, H. & Kajiyama, H. (1981). Adaptation of hepatopancreatic enzymes to dietary carbohydrate in carp. *Bull. Jap. Soc. Sci. Fish.*, **47**, 71–7.

Shimma, Y. & Nakada, M. (1974). Utilization of petroleum yeast for fish feed. I. Effect of supplemental Oil. *Bull. Freshwater Fish. Res. Lab., Tokyo*, **24**, 47–63.

Shoubridge, E. A. & Hochachka, P. W. (1980). Ethanol: novel end product of vertebrate anaerobic metabolism. *Science* (*Wash*), **209**, 308–9.

Shrable, J. B., Tiemeir, O. W. & Deyoe, C. W. (1969). Effects of temperature on rate of digestion by channel catfish. *Prog. Fish Cult.*, **31**, 131–8.

Siebert, G., Schmitt, A. & Bottke, I. (1965). Enzyme des Aminosaeurestoffwechsel in der Kabeljaumuskulatur. *Arch. Fischereiwiss.*, **15**, 233–44.

Siddiqui, A. Q., Siddiqui, A. H. & Ahmad, K. (1973). Free amino acid contents of the skeletal muscle of carp at juvenile and adult stages. *Comp. Biochem. Physiol.*, **44B**, 725–8.

Silk, D. B. A., Marrs, T. C., Addison, J. M., Burston, D., Clark, M. L. & Matthews, D. M. (1973). Absorption of amino acids from an amino acid mixture simulating casein and a tryptic hydrolysate of casein in man. *Clin. Sci. Mol. Med.*, **45**, 715–19.

Sin, A. W. (1973a). The dietary protein requirements for growth of young carp (*Cyprinus carpio*). *Hong Kong Fish. Bull.*, **3**, 77–81.

Sin, A. W. (1973b). The utilization of dietary protein for growth of young carp (*Cyprinus carpio*) in relation to variations in fat intake. *Hong Kong Fish. Bull.*, **3**, 83–8.

Singh, R. P., & Nose, T. (1967). Digestibility of carbohydrates in young rainbow trout. *Bull. Freshwater Fish. Res. Lab., Tokyo*, **17**, 21–5.

Sinha, G. M. (1975). A histochemical study of the mucous cells in the bucco-pharyngeal region of four Indian freshwater fishes in relation to their origin, development, occurrence and probable functions. *Acta Histochem.*, **53**, 217–23.

Sinha, G. M. & Moitra, S. K. (1975). Functional morpho-histology of the alimentary canal of an Indian freshwater major carp, *Labeo rohita* (Hamilton) during its different life-history stages. *Anat. Anz.*, **138**, 222–39.

Sitaramaiah, P. (1967). Water, nitrogen and calorific value of freshwater organisms. *J. Cons. Perm. Int. Explor. Mer*, **31**, 27–30.

Sizer, I. W. (1943). Effects of temperature on enzyme kinetics. *Adv. Enzymol.*, **3**, 35–62.

Slinger, S. J., Cho, C. Y. & Bayley, H. S. (1974). Effects of binding agents in steam-pelleted and extruded diets on fish performance and pellet durability. *Proc. Fish Feed and Nutr. Workshop (Cortland, N.Y.)*, **3**, 64–72.

Smagula, C. M. & Adelman, I. R. (1982). Day-to-day variation in food consumption by largemouth bass. *Trans. Am. Fish. Soc.*, **111**, 543–8.

Smit, H. (1967). Influence of temperature on the rate of gastric juice secretion in the brown bullhead (*Ictalurus nebulosus*). *Comp. Biochem. Physiol.*, **21**, 125–32.

Smith, B. W. & Lovell, R. T. (1971). Digestibility of nutrients in semi-purified rations by channel catfish in stainless steel troughs. *Proc. Ann. Conf. Southeast. Assoc. Game Fish Commrs*, **25**, 452–9.

Smith, B. W. & Lovell, R. T. (1973). Determination of apparent protein digestibility in feeds for channel catfish. *Trans. Am. Fish. Soc.*, **102**, 831–5.

Smith, E. L. (1951). Proteolytic enzymes. In Sumner and Myrbaeck (Eds), *The enzymes, vol. 1*, pp. 793–872. New York: Academic Press.

Smith, H. W. (1929). The excretion of ammonia and urea by the gills of fish. *J. Biol. Chem.*, **81**, 727–42.

Smith, H. W. (1936). The retention and physiological role of urea in elasmobranchii. *Biol. Rev.* **11**, 49–82.

Smith, L. S. (1980). Digestion in teleost fishes. *In Aquaculture development and coordination programme, fish feed technology* – Lectures presented at the FAO/UNDP Training Course in Fish Feed Technology, pp. 3–18. Rome: FAO.

Smith, K. L. (1973). Energy transformations by the Sargassum fish, *Histrio histrio* (L.). *J. Exp. Mar. Biol. Ecol.*, **12**, 219–27.

Smith, M. W. (1970). Selective regulation of amino acid transport by intestine of goldfish. *Comp. Biochem. Physiol.*, **35**, 387–401.

Smith, M. W. & Ellory, J. (1971). Sodium-amino acid interactions in the intestinal epithelium. *Phil. Trans. Roy. Soc. Lond. B*, **262**, 131–40.

Smith, M. W. & Kemp, P. (1971). Parallel temperature-induced changes in membrane fatty acids and in transport of amino acids by the intestine of goldfish (*Crassius auratus* L.). *Comp. Biochem. Physiol.*, **39B**, 357–65.

Smith, R. L. (1969). Intestinal amino-acid transport in the marine teleost *Haemulon plumieri*. *Comp. Biochem. Physiol.*, **30**, 1115–23.

Smith, R. R. (1971). A method for measuring digestibility and metabolizable energy of fish feeds. *Prog. Fish Cult.*, **33**, 132–4.

Smith, R. R. (1976). Metabolizable energy of feedstuffs for trout. *Feedstuffs*, **48**(23), 16–21.

Smith, R. R. (1977). Recent research involving full-fat soybean meal in salmonid diets. *Salmonid*, **1**(4), 8–18.

Smith, R. R., Rumsey, G. L. & Scott, M. L. (1978a). Net energy, maintenance requirements of salmonids as measured by direct calorimetry. Effect of body size and environmental temperature. *J. Nutr.*, **108**, 1017–24.

Smith, R. R., Rumsey, G. L. & Scott, M. L. (1978b). Heat increment associated with dietary protein, fat, carbohydrate and complete diets in salmonids: comparative energetic efficiency. *J. Nutr.*, **108**, 1025–32.

Smuths, D. B. (1935). The relation between the basal metabolism and the endogenous nitrogen metabolism, with particular reference to the estimation of the maintenance requirement of protein. *J. Nutr..*, **9**, 403–33.

Soller, M., Shchori, Y., Moav, R., Wohlfarth, G. & Lahman, M. (1965). Carp growth in brackish water. *Bamidgeh*, **17**(1), 16–23.

Solomon, D. J. & Brafield, A. E. (1972). The energetics of feeding, metabolism and growth of perch (*Perca fluviatilis* L.). *J. Anim. Ecol.*, **41**, 699–718.

Somero, G. N. (1975). The roles of isozymes in adaptation to varying temperatures. In C. L. Market (Ed.), *Isozymes II. Physiological function*, pp. 221–34. New York: Academic Press.

Somero, G. N. & Childress, J. J. (1980). A violation of the metabolism-size scaling paradigm:

activities of glycolytic enzymes in muscle increase in larger-size fish. *Physiol. Zool.*, **53**, 322–37.

Sondermann, von U., Becker, K. & Guenther, D. (1985). Untersuchungen zum oxienergetischen Aequivalent beim Spiegelkarpfen (*Cyprinus carpio* L.) im Hunger. *Z. Tierphysiol. Tierernaehr. Futtermittelk.*, **54**(4). 161–75.

Spallholz, J. E., Martin, J. L., Gerlach, M. L. & Heinzerling, R. H. (1973). Immunologic responses of mice fed diets supplemented with selenite selenium. *Proc. Soc. Exp. Biol. Med.*, **143**, 685–9.

Spannhof, L. & Kuehne, H. (1977). Untersuchungen zur Verwertung verschiedener Futtermischungen durch europaeische Aale (*Anguilla anguilla*). *Arch. Tierernaehr.*, **27**, 517–31.

Spannhof, L. & Plantikow, H. (1983). Studies on carbohydrate digestion in rainbow trout. *Aquaculture*, **30**, 95–108.

Spataru, P. & Hepher B. (1977). Common carp predating on tilapia fry in high density polyculture fishpond system. *Bamidgeh*, **29**(1), 25–8.

Spataru, P., Hepher, B. & Halevy, A. (1980). The effect of the method of supplementary feed application on the feeding habits of carp (*Cyprinus carpio* L.) with regard to natural food. *Hydrobiologia*, **72**, 171–8.

Spencer, W. P. (1939). Diurnal activity rhythms in fresh-water fishes. *Ohio J. Sci.*, **39**, 119–132.

Spinelli, J., Houle, C. R. & Wekell, J. C. (1983). The effect of phytates on the growth of rainbow trout (*Salmo gairdneri*) fed purified diets containing varying quantities of calcium and magnesium. *Aquaculture*, **30**, 71–83.

Spinelli, J., Mahnken, C. & Steinberg, M. (1979). Alternative sources of proteins for fish meal in salmonid diets. In J. E. Halver & K. Tiews (Eds), *Finfish nutrition and fishfeed technology vol. 2.*, pp. 131–42. Berlin: Heenemann Verlagsgesel.

Spoor, W. A. (1946). A quantitative study of the relationship between the activity of oxygen consumption of the goldfish, and its application to the measurement of respiratory metabolism in fishes. *Biol. Bull.*, **91**, 312–25.

Stanley, J. G. (1974). Nitrogen and phosphorus balance of grass carp, *Ctenopharyngodon idella*, fed Elodea, *Egeria densa*. *Trans. Am. Fish Soc.*, **103**, 587–92.

Staples, D. J. & Nomura, M. (1976). Influence of body size and food ration on the energy budget of rainbow trout *Salmo gairdneri* Richardson. *J. Fish Biol.*, **9**, 29–43.

Steele, N. C. & Rosebrough, R. W. (1981). Effect of trivalent chromium on hepatic lipogenesis by the turkey poult. *Poultry Sci.*, **60**, 617–22.

Steffens, W. (1964). Vergleichende anatomisch-physiologische Untersuchungen an Wild- und Teichkarpfen (*Cyprinus carpio*). *Z. Fisch., N.F..*, **12**, 725–800.

Steffens, W. & Albrecht, M.-L. (1973). Proteineinsparung durch Erhoehung des Fettanteils im Futter fuer Regenbogenforellen (*Salmo gairdneri*). *Arch. Tierernaehr.* **23**, 711–17.

Steffens, W. & Albrecht, M.-L. (1976). Untersuchungen ueber die Moeglichkeiten zur Senkung des Fishmehlanteils im Futter fuer Regenbogenforellen (*Salmo gairdneri*). *Arch. Tierernaehr.*, **26**, 285–91.

Steiner, D. F., Clark, J. L., Nolan, C., Rubenstein, A. H., Margoliash, E., Aten, B. & Over, B. (1969). Proinsulin and the biosynthesis of insulin in E. B. Astwood (Ed.), *Recent progress in hormone research, vol. 25*, Proc. of the 1968 Laurentian Hormone Conf., pp. 207–89. New York: Academic Press.

Stewart, N. E., Shumway, D. L. & Doudoroff, P. (1967). Influence of oxygen concentration on growth of juvenile largemouth bass. *J. Fish. Res. Bd Can.*, **24**, 475–94.

Stickney, R. R. & Andrews, J. W. (1971). Combined effects of dietary lipids and environmental temperature on growth, metabolism and body composition of channel catfish (*Ictalurus punctatus*). *J. Nutr.*, **101**, 1703–10.

Stickney, R. R. & Andrews, J. W. (1972). Effects of dietary lipids on growth, food conversion, lipid and fatty acid composition of channel catfish. *J. Nutr.*, **102**, 249–58.

Stickney, R. R. & Shumway, S. E. (1974). Occurrence of cellulase activity in the stomachs of fishes. *J. Fish Biol.*, **6**, 779–90.

Stirling, H. P. (1972). The proximate composition of the European bass, *Dicentrarchus labrax* (L.) from the bay of Naples. *J. Cons. Int. Explor. Mer*, **34**, 357–64.

Stirling, H. P. (1976). Effects of experimental feeding and starvation on the proximate composition of the European bass *Dicentrarchus labrax*. *Mar. Biol.*, **34**, 85–91.

Stokes, R. M. & Fromm, P. O. (1964). Glucose absorption and metabolism by the gut of rainbow trout. *Comp. Biochem. Physiol.*, **13**, 53–69.

Storebakken, T. (1985). Binders in fish feeds. I. Effect of alginate and Guam gum on growth, digestibility, feed intake and passage through the gastrointestinal tract of rainbow trout. *Aquaculture*, **47**, 11–26.

Storer, J. H. (1967). Starvation and the effects of cortisol in the goldfish (*Carassius auratus* L.). *Comp. Biochem. Physiol.*, **20**, 939–48.

Strickland, J. D. H. (1960). Measuring the production of marine phytoplankton. *Bull. Fish. Res. Bd Can.*, **122**, 1–172.

Stroganov, N. S. & Buzinova, N. S. (1969a). Enzymes of the digestive tract in grass carp I. Amylase and lipase. (In Russian.) *Vestn. Mosk. Univ., Biol. Pochvoved.*, **24**(3), 27–31.

Stroganov, N. S. & Buzinova, N. S. (1969b). Enzymatic activity of the grass carp (*Ctenopharyngodon idella*) intestinal tract. II. Proteolytic enzymes. (In Russian.) *Vestn. Mosk. Univ., Biol. Pochvoved.*, **24**(4), 3–7. Chem. Abstr., 7.

Sturmbauer, C. & Hofer, R. (1986). Compensation for amylase inhibitors in the intestine of the carp (*Cyprinus carpio*). *Aquaculture*, **52**, 31–3.

Sugita, H., Sakata, T., Ishida, Y., Deguchi, Y. & Kadota, H. (1981). Measurement of total viable counts in the gastrointestinal bacteria of freshwater fish. *Bull. Jap. Soc. Sci. Fish.*, **47**, 555.

Sugita, H., Enomoto, A. & Deguchi, Y. (1982a). Intestinal microflora in fry of *Tilapia mossambica*. *Bull. Jap. Soc. Sci. Fish.*, **48**, 875.

Sugita, H., Ishida, Y. Deguchi, Y. & Kadota, H. (1982b). Studies on the gastrointestinal bacteria of freshwater. IV. Aerobic microflora attached to wall surface in the gastrointestine of *Tilapia nilotica*. *Bull. Coll. Agr. Vet. Med. Nihon Univ.*, **39**, 212–17.

Sukhoverkhov, F. & Krymova, R. (1961). Effect of cobalt salts on the growth of carp. (In Russian.) *Rybovod. Rybolov.*, **1**, 18–19.

Sukhoverkhov, F. M., Krymova, R. V. & Farberov, V. G. (1963). The effect of cobalt upon the growth and hematological indices of carp. *Ryb. Khoz.*, **39**(8), 35–8.

Summerfelt, R. C., Mauck, P. E. & Mensinger, G. (1970). Food habits of the carp, *Cyprins carpio* L., in five Oklahoma reservoirs. *Proc. Ann. Conf. Southeast. Assoc. Game Fish Commrs.*, **24**, 352–77.

Sundnes, G. (1957). Notes on the energy metabolism of cod (*Gadus callarias* L.) and the coalfish (*Gadus virens* L.) in relation to body size. *Fiskeridirektorat skrifter, Ser. Havundersoek.*, **11**(9), 3–10.

Suppes, C., Tiemeier, O. W. & Deyoe, C. W. (1968). Seasonal variation of fat, protein and moisture in channel catfish. *Trans. Kansas Acad. Sci.*, **70**, 349–58.

Sutherland, E. W. & Cori, C.F. (1948). Influence of insulin preparation on glycogenolysis in liver slices. *J. Biol. Chem.*, **172**, 737–50.

Suyama, M. & Ogino, C. (1958). Changes in chemical composition during development of rainbow trout eggs. *Bull. Jap. Soc. Sci. Fish.*, **23**, 785–8.

Suzuki, R., Yamaguchi, M., Ito, T. & Toi, J. (1976). Difference in growth and survival in various races of common carp. *Bull. Freshwater Fish. Res. Lab., Tokyo*, **26**(2), 59–70.

Svob, T. & Kilalic, T. (1967). A contribution to comparative studies on the physiology of alimentary canal in cyprinidae. (In Serbian.) *Ribar. Jugosl.*, **22**(5), 130–2.

Swarup, C. & Goel, K. A. (1975). Histochemical study of the activity of lipase in the digestive system of some teleost fishes. *Acta Histochem.* **54**, 10–15.

Swarup, K. & Srivastav, S. P. (1982). Vitamin D_3 induced hypercalcaemia in male catfish,

Clarias batrachus. *Gen. Comp. Endocrinol.*, **46**, 271–4.

Swift, D. R. (1955). Seasonal variation in growth rate, thyroid gland activity and food reserves of brown trout (*Salmo trutta* Linn.) and its cause. *J. Exp. Biol.*, **32**, 751–64.

Swift, D. R. (1961). The annual growth-rate cycle in brown trout (*Salmo trutta* Linn.). *J. Exp. Biol.*, **38**, 595–604.

Swift, D. R. (1962). Activity cycles in the brown trout (*Salmo trutta* Lin). I. Fish feeding naturally. *Hydrobiologia*, **29**, 241–7.

Swift, D. R. (1963). Influence of oxygen concentration on growth of brown trout, *Salmo trutta* L. *Trans. Am. Fish. Soc.*, **92**, 300–1.

Swingle, H. S. (1958). Experiments on growing fingerling channel catfish to marketable size in ponds. *Proc. Southeast. Assoc. Game Fish Commrs.*, **12**, 63–72.

Swingle, H. S. (1961). Relationship of pH of pond waters to their suitability for fish culture. *Proc. Pac. Sci. Congr. (Fisheries)*, **9**, 4pp.

Syazuki, K. (1956). Studies on nutrients for carp fry. II. Digestibility of protein and percentage of ingested feed per body weight of foodstuff having various levels of protein. *J. Shimonoseki Coll. Fish.*, **6**, 109–13.

Syazuki, K., Takesue, K. & Matsui, I. (1953). Studies on the nutritive value of the blood dust of whales for breeding. I. On the breeding test for carp fingerling and the digestibility of protein of the blood dust. *J. Shimonoseki Coll. Fish.*, **3**, 193–6.

Syvokene, J. & Grigorovie, G. (1974). The role of the digestive canal microorganisms in feeding of pond fish. 7. Physiological activity of microorganisms. (In Russian with English summary.) *Tr. Akad. Nauk Lit. SSR (Biol.)*, (4), 85–92.

Tacon, A. G. J. & Beveridge, M. M. (1982). Effect of dietary trivalent chromium on rainbow trout. *Nutr. Resp. Internat.*, **25**(1), 49–56.

Tacon, A. G. J. & Rodrigues, A. M. P. (1984). Comparison of chromic oxide, crude fibre polyethylene and acid-insoluble ash as dietary markers for the estimation of apparent digestibility coefficients in rainbow trout. *Aquaculture*, **43**, 391–9.

Tacon, A. G. J., Knox, D, & Cowey, C. B. (1984). Effect of different levels of salt mixtures on growth and body composition in carp. *Bull. Jap. Soc. Sci. Fish.*, **50**, 1217–22.

Takeda, M., Shimeno, S., Hosokawa, H., Kajiyama, H. & Kaisyo, T. (1975). The effect of dietary calorie-to-protein ratio on the growth, feed conversion and body composition of young yellowtail. *Bull. Jap. Soc. Sci. Fish.*, **41**, 443–7.

Takeuchi, T., Arai, S., Watanabe, T. & Shimma, Y. (1980a). Requirement of eel *Anguilla japonica* for essential fatty acids. *Bull. Jap. Soc. Sci. Fish.*, **46**, 345–53.

Takeuchi, L., Takeuchi, T. & Ogino, C. (1980b). Riboflavin requirements in carp and rainbow trout. *Bull. Jap. Soc. Sci. Fish.*, **46**, 733–7.

Takeuchi, T., & Watanabe, T. (1977a). Requirement of carp for essential fatty acids. *Jap. Soc. Sci. Fish.*, **43**, 541–51.

Takeuchi, T., & Watanabe, T. (1977b). Dietary levels of methyl laurate and essential fatty acid requirement of rainbow trout. *Bull. Jap. Soc. Sci. Fish.*, **43**, 893–8.

Takeuchi, T. & Watanabe, T. (1978). Growth-enhancing effect of cuttlefish liver oil and short-necked clam oil on rainbow trout and their effective components. *Bull. Jap. Soc. Sci. Fish.*, **44**, 733–8.

Takeuchi, T., Watanabe, T. & Ogino, C. (1978). Optimum ratio of protein to lipid in diet of rainbow trout. *Bull. Jap. Soc. Sci. Fish.*, **44**, 683–8.

Takeuchi, T., Watanabe, T. & Ogino, C. (1979). Optimum ratio of dietary energy to protein for carp. *Bull. Jap. Soc. Sci. Fish.*, **45**, 983–7.

Takeuchi, T., Watanabe, T., Ogino, C., Saito, M., Nishimura, K. & Nose, T. (1981). A long term feeding with rainbow trout by a low protein diet with high energy value. *Bull. Jap. Soc. Sci. Fish.*, **47**, 637–43.

Talbot, C., Higgins, P. J. & A. M. Shanks, (1984). Effects of pre- and post-prandial starvation on meal size and evacuation rate of juvenile Atlantic salmon, *Salmo salar* L. *J. Fish. Biol.*, **25**, 551–60.

Talling, J. F. (1982). Utilization of solar radiation by phytoplankton. In C. Helene, M. Charlier, Th. Monteney-Garestier & B. Laustrial (Eds), *Trends in photobiology*, pp. 619–31. New York: Plenum Publ.

Tanaka, Y., Hokazono, S., Katayama, T., Simpson, K. L. & Chichester, C. O. (1977a). Metabolism of amino acids in aquatic animals I. The effect of the addition of phosphate salts, indigestible materials and algae to the diets of carp and the relationship of intestinal retention time to their growth rate. *Mem. Fac. Fish. Kagoshima Univ.*, **26**. 39–43.

Tanaka, Y., Hokazono, S., Katayama, T., Simpson, K. L. & Chichester, C. O. (1977b). Metabolism of amino acids in aquatic animals. II. The effect of an amino acid supplemented casein diet on the growth rate of carp. *Mem. Fac. Fish. Kagoshima Univ.*, **26**, 45–8.

Tandler, A. & Beamish, F. W. H. (1979). Mechanical and biochemical components of apparent specific dynamic action in largemouth bass, *Micropterus salmoides* Lacepede. *J. Fish. Biol.*, **14** 343–50.

Tandler, A. & Beamish, F. W. H. (1980). Specific dynamic action and diet in largemouth bass, *Micropterus salmoides* (Lacepede) *J. Nutr.*, **110**, 750–64.

Tandler, A. & Beamish, F. W. H. (1981). Apparent specific dynamic action (*SDA*), fish weight and level of caloric intake in largemouth bass, *Micropterus salmoides* Lacepede. *Aquaculture*, **23**, 231–42.

Tang, Y. A. (1970). Evaluation of balance between fishes and available fish foods in multispecies fish culture ponds in Taiwan. *Trans. Am. Fish. Soc.*, **99**, 708–18.

Tang, Y. A. & Hwang, T. L. (1967). Evaluation of the relative suitability of various groups of algae as food for milkfish in brackish-water ponds. *FAO Fish. Rep.*, **44**(3), 365–72.

Tashima, L. & Cahill, G. F. (1965). Fat metabolism in fish. In A. E. Renold & G. F. Cahill (Eds), *Handbook of physiology, section 5: Adipose tissue*, pp. 55–8. Washington DC: Am. Physiol. Soc.

Tashima, L. & Cahill, G. F. (1968). Effects of insulin in the toadfish, *Opsanus tau*. *Gen. Comp. Endocrinol.*, **11**, 262–71.

Tatrai, I. (1981). The nitrogen metabolism of bream, *Abramis brama* L. *Comp. Biochem. Physiol.*, **68A**, 119–21.

Teng, S.-K., Chua, T.-E. & Lim, P.-E. (1978). Preliminary observation on the dietary protein requirement of estuary grouper, *Epinephelus salmoides* Maxwell, cultured in floating net-cages. *Aquaculture*, **15**, 257–71.

Terrell, J. W. & Fox, A. C. (1974). Food habits, growth and catchability of grass carp in the absence of aquatic vegetation. *Proc. Ann. Conf. Southeast. Assoc. Game Fish Commrs.*, **28**, 251–9.

Tesch, F.W. (1977). *The eel – biology and management of anguillid eels*. London: Chapman & Hall, 434 pp.

Teshima, S., & Kashiwada, K. (1967). Studies on the production of B vitamins by intestinal bacteria of fish. III. Isolation of vitamin B_{12} synthesizing bacteria and their bacteriological properties. *Bull. Jap. Soc. Sci. Fish.*, **33**, 979–83.

Teshima, S., & Kashiwada, K. (1969). Studies on the production of B vitamins by intestinal bacteria of fish. IV. Production of nicotinic acid by intestinal bacteria of carp. *Mem. Fac. Fish. Kagoshima Univ.*, **18**, 87–91.

Teshima, S.-I., Kanazawa, A. & Sakamoto, M. (1982). Essential fatty acids of *Tilapia nilotica*. *Mem. Fac. Fish. Kagoshima Univ.*, **31**, 201–4.

Thiele, V. F. & Brin, M. (1968). Availability of vitamin B_6 vitamers fed orally to Long Evans rats as determined by tissue transaminase activity and vitamin B_6 assay. *J. Nutr.*, **94**, 237–42.

Thomas, N. W. (1970). Morphology of endocrine cells in the islet tissue of the cod *Gadus callarias*. *Acta Endocrinol.*, **63**, 679.

Thompson, J. N. & Scott, M. L. (1970). Impaired lipid and vitamin E absorption related to atrophy of the pancreas in selenium deficient chicks. *J. Nutr.*, **100**, 797–809.

Thorpe, A. (1976). Studies on the role of insulin in teleost metabolism in T. A. I. Grillo,

L. Leibson & A. Apple (Eds), *The evolution of pancreatic islets*, pp. 271–84. Oxford: Pergamon Press.

Thorpe, A. & Ince, B. W. (1974). The effects of pancreatic hormones, catecholamines and glucose loading on blood metabolites in northern pike (*Esox lucius* L.). *Gen. Comp. Endocrinol.*, **23**, 29–44.

Thorpe, A. & Ince, B. W. (1976). Plasma insulin levels in teleosts determined by a charcoal separation radioimmunoassay technique. *Gen. Comp. Endocrinol.*, **30**, 332–9.

Tiemeier, O. W. & Deyoe, C. W. (1980). Channel catfish produced in Kansas ponds for profit and pleasure. *Bull. Kansas Agric. Expt. St.*, **635**, 36 pp.

Tiemeier, O. W., Deyoe, C. W. & Weardon, S. (1965). Effects on growth of fingerling channel catfish of diets containing two energy and two protein levels. *Trans. Kansas Acad. Sci.*, **68**(4), 180–6.

Tiews, K., Koops, K., Gropp, J. & Beck, H. (1979). Compilation of fish meal free diets obtained in rainbow trout (*Salmo gairdneri*) feeding experiments at Hamburg (1970–1977/78). In J. E. Halver & K. Tiews (Eds), *Finfish nutrition and fishfeed technology, vol. II*, pp. 219–28. Berlin: Heenemann Verlagsgesel.

Toepfer, E. W., Mertz, W., Polansky, M. M., Roginski, E. E. & Wolf, W. R. (1977). Preparation of chromium-containing material of glucose tolerance activity from brewers yeast extracts and by synthesis. *J. Agric. Food Chem.*, **25**, 162–6.

Tomiyama, T. & Ohla, N. (1967). Biotin requirement of goldfish. *Bull. Jap. Soc. Sci. Fish.*, **33**, 448–52.

Tomnatik, E. N. & Batyr, A. K. (1965). The role of organic cobalt in the nutrition of fingerling carp. (In Russian.) In *Biologicheskie Resury Vodoemov Moldavii Kartya Moldovenyaske Kishiner*, (3), 54–8.

Trama, F. B. (1957). The transformation of energy by an aquatic herbivore, *Stenonema pulchellum*, Ph.D. Thesis, Univ. Michigan, Ann Arbor, Mich., USA.

Trent, L. & Hassler, W. W. (1966). Feeding behaviour of adult striped bass, *Roccus saxatilis*, in relation to stages of sexual maturity. *Chesapeake Sci.*, **7**(4), 189–92.

Trepsiene, O., Jankevicius, K. & Peciukenas, A. (1974a). Complex B vitamins in the food and some organs of naturally feeding carp, white amur and tench. I. Vitamin content in the tissues of second year fish. (In Russian with English summary.) *Tr. Akad. Nauk Lit. SSR (Biol.)*, (2), 137–43.

Trepsiene, O., Jankevicius, K. & Lubianskiene, V. (1974b). B group vitamins in the food and some organs of naturally feeding carp, white amur and tench. 2. Synthesis of vitamins by bacteria isolated from the digestive tract of 1st and 2nd year fish (In Russian with English summary.) *Tr. Akad. Nauk Lit. SSR (Biol.)* **3**(67), 129–39.

Trofimova, L. N. (1973). Dynamics of total proteolytic activity along the digestive tract in carp in relation to incubation temperature. (In Russian with English summary.) *Sb. Nauch. Tr. VNIIPRKH*, **10**, 170–81.

Tropical Fish Culture Research Institute, Malacca. (1961). Report for 1960/1.

Truscott, B. (1979). Steroid metabolism in fish. Identification of steroid moieties of hydrolyzable conjugates of cortisol in the bile of trout *Salmo gairdneri*. *Gen. Comp. Endocrinol.*, **38**, 196–206.

Trust, T. J., Bull, L. M., Currie, B. R. & Buckley, J. T. (1979). Obligate anaerobic bacteria in the gastrointestinal microflora of the grass carp (*Ctenopharyngodon idella*), goldfish (*Carassius auratus*) and rainbow trout (*Salmo gairdneri*). *J. Fish. Res. Bd. Can.*, **36**, 1174–9.

Trust, T. J. & Sparrow, R. A. H. (1974). The bacterial flora in the alimentary tract of freshwater salmonid fishes. *Can. J. Microbiol.*, **20**, 1219–28.

Tryamkina, S. P. (1973). The effect of wheat flour on the rate of digestive and metabolic processes in trout. (In Russian with English summary.) *Sb. Nauch. Tr. VNIIPRKH*, **10**, 182–91.

Tsuchiya, Y. & Kunii, K. (1960). Studies on the influence of treatments immediately after catching upon the quality of fish flesh. V. Determination of lactic acid in fish muscle. *Bull.*

Jap. Soc. Sci. Fish., **26**, 284–8.

Tsukamoto, K. (1981). Direct evidence for functional and metabolic differences between dark and ordinary muscles in free swimming yellowtail, *Seriola quinqueradiata. Bull. Jap. Soc. Sci. Fish.*, **47**, 573–5.

Tsukuda, H. (1975). Temperature dependency of the relative activities of liver lactate dehydrogenase isozymes in goldfish acclimated to different temperatures. *Comp. Biochem. Physiol.*, **52**, 343–5.

Tuba, J. & Dickie, N. (1955). The role of alkaline phosphatase in intestinal absorption. 4. The effects of various proteins on the levels of the enzyme in intestinal mucosa. *Can. J. Biochem. Physiol.*, **33**, 89–92.

Tunison, A. V., Phillips, A. M., McCay, C. M., Mitchell, C. R. & Rodgers, E. O. (1939). Carbohydrate utilization by trout. *Cortland Hatchery Report*, **8**, 9–12.

Tunison, A. V., Phillips, A. M., Brockway, D. R., Dorr, A. L., Mitchell, C. R. & McCay, C. M. (1941). The nutrition of trout. *Fish. Res. Bull.*, **1**, 7–16.

Tunison, A. V., Brockway, D. R., Maxwell, J. M., Dorr, A. L. & McCay, C. M. (1942). The nutrition of trout. *Fish. Res. Bull.*, **4**, 1–51.

Tunison, A.V., Brockway, D. R., Shaffer, H. B., Maxwell, J. M., McCay, C. M., Palm, C. E. & Webster, D. A. (1943). The nutrition of trout. *Cortland Hatchery Report 12, Fish Res. Bull.*, **5**.

Tunison, A. V., Phillips, A. M., Shaffer, H. B., Maxwell, J. M., Brockway, D. R. & McCay, C. M. (1944). The nutrition of trout, *Cortland Hatchery Report* 13., Fish. Res. Bull., **6**.

Tyler, A. V. (1970). Rates of gastric emptying in young cod. *J. Fish. Res. Bd. Can.*, **27**, 1177–89.

Uchida, N., Obata, T. & Saito, T. (1973). Occurrence of inactive precursors of proteases in chum salmon pyloric caeca. *Bull. Jap. Soc. Sci. Fish.*, **39**, 825–8.

Ufodike, E. B. C. & Matty, A. J. (1983). Growth responses and nutrient digestibility in mirror carp (*Cyprinus carpio*) fed different levels of cassava and rice. *Aquaculture*, **31**, 41–50.

Ugolev, A. M. (1960). Influence of the surface of the small intestine on enzymatic hydrolysis of starch by enzymes. *Nature (Lond.)*, **188**, 588–9.

Ugolev, A. M. (1965). Membrane (contact) digestion. *Physiol. Rev.*, **45**, 555–95.

Ulla, O. & Gjedrem, T. (1985). Number and length of pyloric caeca and their relationship to fat and protein digestibility in rainbow trout. *Aquaculture*, **47**, 105–11.

Ursin, E. (1967). A mathematical model of some aspects of fish growth, respiration and mortality. *J. Fish. Res. Bd Can.*, **24**, 2355–453.

Ushiyama, H., Fujimori, T., Shibata, T. & Yoshimura, K. (1965). Studies on carbohydrates in pyloric caeca of the salmon, *Oncorhynchus keta. Bull. Hokkaido Univ. Fish.*, **16**(3), 183–8. Transl. Ser., Fish. Res. Bd Can. (2183), 12 pp.

Utida, S. & Isono, N. (1967). Alkaline phosphatase in intestinal mucosa of the eel activity adapted to fresh-water and sea-water. *Proc. Jap. Acad.*, **43**, 782–92.

Vahl, O. & Davenport, J. (1979). Apparent specific dynamic action of food in the fish *Blennius pholis. Mar. Ecol. Prog. Ser.*, **1**, 109–13.

Van den Thillart, G. (1982). Adaptation of fish energy metabolism to hypoxia and anoxia. *Molec. Physiol.*, **2**, 49–61.

Van den Thillart, G. & Kesbeke, F. (1978). Anaerobic production of carbon dioxide and ammonia by goldfish, *Carassius auratus* (L.). *Comp. Biochem. Physiol.*, **59A**, 393–400.

Van den Thillart, G. & Verbeek, R. (1982). Substrates for anaerobic CO_2 production by the goldfish, *Carassius auratus* (L.): Decarboxylation of ^{14}C-labelled metabolites. *J. Comp. Physiol.*, **149**, 75–81.

Van den Thillart, G., Van Berge-Henegouwen, M. & Kesbeke, F. (1983). Anaerobic metabolism of goldfish, *Carassius auratus* (L.): Ethanol and CO_2 excretion rates and anoxia tolerance at 20, 10 and 5 °C. *Comp. Biochem. Physiol.*, **76A**, 295–300.

Van Dyke, J. M. & Sutton, D. L. (1977). Digestion of duckweed (*Lemna* spp.) by the grass carp (*Ctenopharyngodon idella*). *J. Fish Biol.*, **11**, 273–8.

Van Waarde, A., Van Den Thillart, G. & Kesbeke, F. (1983). Anaerobic energy metabolism of the European eel, *Anguilla anguilla* L. *J. Comp. Physiol.*, **149**, 469–75.

Vavruska, A. & Janecek, V. (1973). Dry matter, fat, nitrogenous and ash in carp meat. *Bull. Vyzk. Ustav Ryb. Hydrobiol., Vodnany*, **9**(3), 26–31.

Vellas, F. & Serfaty, A. (1974). L'ammoniaque et l'urée chez un teleostéen d'eau douce; la carpe (*Cyprinus carpio* L.). *J. Physiol. (Paris)*, **68**, 591–614.

Vens-Cappell, B. (1984). The effects of extrusion and pelleting of feed for trout on the digestibility of protein, amino-acids and energy and on feed conversion. *Aquacult. Engin.*, **3**, 71–89.

Verma, S. R., Tyagi, M. P. & Dalela, R. C. (1974). Morphological variations in the stomach of a few teleost fishes in relation to food and feeding habits. *Morphol. Jahrb.*, **120**, 367–80.

Verjbinskaya, N. A. & Savina, M. V. (1969). Regulation of carbohydrate utilization in muscles of cold-blooded animals. (In Russian.) *J. Evol. Biokh. Fiziol.*, **5**, 234–40.

Vijverberg, J. & Frank, Th.H. (1965). The chemical composition and energy content of copepodes and cladocerans in relation to size. *Freshwater Biol.*, **6**, 333–45.

Vinogradov, V. K., & Erokhina, L. V. (1961). Use of granulated food with addition of inorganic cobalt. *Tr. Vses. Nauch-issled. Inst. Prud. Ryb. Khoz.*, **10**, 24–5.

Vinogradov, V. K. & Erokhina, L. V. (1962). Use of cobalt in the breeding of carp fingerlings. (In Russian). *Trud. Vses. Nauch-issled. Inst. Prud. Ryb. Khoz.*, **11**, 18–26.

Viola, S. (1975). Experiments on nutrition of carp growing in cages. Part 2: Partial substitution of fish meal. *Bamidgeh*, **27**(2), 40–8.

Viola, S. (1981). The replacement of feedstuffs of animal origin by plant feeds in the nutrition of common carp (*Cyprinus carpio*). (In Hebrew with English summary.) Ph.D. Thesis, Technion – Israel Institute of Technology, Haifa, 122 + vi pp.

Viola, S. & Rappaport, U. (1979a). The 'extra-caloric effect' of oil in the nutrition of carp. *Bamidgeh*, **31**(3), 51–68.

Viola, S. & Rappaport, U. (1979b). Acidulated soapstocks in intensive carp diets, their effects on growth and composition. In J. E. Halver & K. Tiews (Eds), *Finfish nutrition and finfeed technology, vol. 2*, pp. 51–62. Berlin: Heenemann Verlagsgesel.

Viola, S. & Zohar, G. (1984). Nutrition studies with market size hybrids of tilapia (*Oreochromis*) in intensive culture. 3. Protein levels and sources. *Bamidgeh*, **36**(1), 3–15.

Viola, S., Rappaport, U., Arieli, Y., Amidan, G. & Mokady, S. (1981). The effects of oil-coated pellets on carp (*Cyprinus carpio*) in intensive culture. *Aquaculture* **26**, 49–65.

Viola, S., Mokady, S., Rappaport, U. & Arieli, Y. (1982). Partial and complete replacement of fishmeal by soybean meal in feeds for intensive culture of carp. *Aquaculture*, **26**, 223–36.

Viola, S., Mokady, S. & Arieli, Y. (1983). Effects of soybean processing methods on growth of carp (*Cyprinus carpio*). *Aquaculture*, **32**, 27–38.

Vonk, H. J. (1937). The specificity and collaboration of digestive enzymes in Metazoa. *Biol. Rev.*, **12**, 245–84.

Walsh, P. J., Foster, G. D. & Moon, T. W. (1983). The effects of temperature on metabolism of the American eel *Anguilla rostrata* (Le Sueur): compensation in the summer and torpor in the winter. *Physiol. Zool.*, **56**, 532–40.

Walter, E. (1934). Grundlagen der allgemeinen fischereilichen Produktions lehre einschliesslich ihrer Anwendug auf die Fuetterung. In R. Demoll & H. N. Maier (Eds), *Hanb. Binnenfisch. Mitteleurop., vol. 4(5)*, pp. 481–662. Stuttgart: Schweizerbartsche Verlag.

Walton, M. J. & Cowey, C. B. (1979). Gluconeogenesis by isolated hepatocystes from rainbow trout *Salmo gairdneri*. *Comp. Biochem. Physiol.*, **62B**, 75–9.

Walton, M. J. & Wilson, R. P. (1986). Postprandial changes in plasma and liver free amino acids of rainbow trout fed complete diets containing casein. *Aquaculture*, **51**, 105–15.

Warren, C. E. & Davis, G. E. (1967). Laboratory studies on the feeding, bioenergetics, and growth of fish. In S. D. Gerking (Ed.), *The biological basis of freshwater fish production*, pp. 175–214. Oxford: Blackwell Sci. Publ.

Watanabe, T. (1982). Lipid nutrition in fish. *Comp. Biochem. Physiol.*, **73B**, 3–15.

Watanabe, T., Ogino, C., Koshiishi, Y. & Matsunaga, T. (1974a). Requirement of rainbow trout for essential fatty acids. *Bull. Jap. Soc. Sci. Fish.*, **40**, 493–9.

Watanabe, T., Takashima, F. & Ogino, C. (1974b). Effect of dietary methyl linolenate on growth of rainbow trout. *Bull. Jap. Soc. Sci. Fish.*, **40**, 181–8.

Watanabe, T., Utsue, D., Kobayashi, I. & Ogino, C. (1975a). Effect of dietary methyl linoleate and linolenate on growth of carp. I. *Bull. Jap. Soc. Sci. Fish.*, **41**, 257–62.

Watanabe, T., Takeuchi, T. & Ogino, C. (1975b). Effect of dietary methyl linoleate and linolenate on growth of carp. II. *Bull. Jap. Soc. Sci. Fish.*, **41**, 263–9.

Watanabe, T. & Takeuchi, T. (1976). Evaluation of polloc liver oil as a supplement to diets of rainbow trout. *Bull. Jap. Soc. Sci. Fish.*, **42**, 893–906.

Watanabe, T., Takeuchi, T., Matsui, M., Ogino, C. & Kawabata, T. (1977). Effects of α-tocopherol deficiency on carp. VII. The relationship between dietary levels of linoleate and α-tocopherol requirement. *Bull. Jap. Soc. Sci. Fish.*, **43**, 935–46.

Watanabe, T., Arakawa, T., Kitajima, C., Fukusho, K. & Fujita, S. (1978a). Proximate and mineral composition of living feeds used in seed production of fish. *Bull. Jap. Soc. Sci. Fish.*, **44**, 979–84.

Watanabe, T., Arakawa, T., Kitajima, C. & Fujita, S. (1978b). Nutritional evaluation of proteins of living feeds used in seed production of fish. *Bull. Jap. Soc. Sci. Fish.*, **44**, 985–8.

Watanabe, T., Takeuchi, T. & Ogino, C. (1979). Studies on the sparing effect of lipids on dietary protein in rainbow trout (*Salmo gairdneri*). In J. E. Halver & K. Tiews (Eds), *Finfish nutrition and fishfeed technology. vol. I*, pp. 113–25. Berlin: Heenemann Verlagsgesel.

Watanabe, T., Murakami, A., Takeuchi, L., Nose, T. & Ogino, C. (1980a). Requirement of chum salmon held in freshwater for dietary phosphorus. *Bull. Jap. Soc. Sci. Fish.*, **46**, 361–7.

Watanabe, T., Takeuchi T., Murakami, A., & Ogino, C. (1980b). The availability to *Tilapia nilotica* of phosphorus in white fish meal. *Bull. Jap. Soc. Sci. Fish.*, **46**, 897–9.

Watanabe, T., Takeuchi, T. & Wada, M. (1981a). Dietary lipid levels and α-tocopherol requirement of carp. *Bull. Jap. Soc. Sci. Fish.*, **47**, 1585–90.

Watanabe, T., Takeuchi, T., Wada, A. & Uehara, R. (1981b). The relationship between dietary lipid levels and α-tocopherol requirement of rainbow trout. *Bull. Jap. Soc. Sci. Fish.*, **47**, 1463–76.

Weatherley, A. H. (1963). Notions of niche and competition among animals, with special reference to freshwater fish. *Nature (Lond.)*., **197**, 14–17.

Weatherley, A. H. & Gill, H. S. (1983). Protein, lipid, water and caloric contents of immature rainbow trout *Salmo gairdneri* Richardson growing at different rates. *J. Fish Biol.*, **23**, 653–74.

Weber, F. & Wiss, O. (1966). Interactions between vitamin E and other components of the diet. *Biblioteca Nutritio et Dieta*, **8**, 54–63.

Wendelaar Bonga, S. E., Lammers, P. I. & Van Der Meij, J. C. A. (1983). Effects of 1,25- and 24,25-dihydroxyvitamin D_3 on bone formation in the chick. *Cell Tissue Res.*, **228**, 117–26.

Wendt, C. A. G. & Saunders, R. L. (1973). Changes in carbohydrate metabolism in young Atlantic salmon in response to various forms of stress. *Spec. Publ. Ser., Int. Atl. Salmon Found.*, **4**(1), 58–82.

Werner, E. E. & Hall, D. J. (1976). Niche shifts in sunfishes, experimental evidence and significance. *Science (Wash.)*, **191**, 404–6.

West, E. S., Todd, W. R., Mason, H.S. & Van Bruggen, J. T. (1966). *Textbook of biochemistry* (4th edn). New York: Macmillan, xii + 1595 pp.

Western, J. R. H. (1971). Feeding and digestion in two cottid fishes, the freshwater *Cottus gobio* L. and the marine *Enophrys bubalis* (Euphrasen). *J. Fish Biol.*, **3**, 225–46.

Western, J. R. H. & Jennings, J. B. (1970). Histochemical demonstration of hydrochloric acid in the gastric tubules of teleost using an *in vivo* Prussian blue technique. *Comp.*

Biochem. Physiol., **35**, 879–84.

White, A., Handler, P., Smith, E. L. & Stetten, D. (1959). *Principles of biochemistry* (2nd edn). New York: McGraw-Hill Book Co., 1149 pp.

Whitmore, D. H. & Goldberg, E. (1972a). Trout intestinal alkaline phosphatases. I. Some physical–chemical characteristics. *J. Exp. Zool.*, **182**, 47–58.

Whitmore, D. H. & Goldberg, E. (1972b). Trout intestinal alkaline phosphatases II. The effect of temperature upon enzymatic activity *in vitro* and *in vivo*. *J. Exp. Zool.*, **182**, 59–68.

Wieser, W. (1973). Temperature relations of ectotherms: A speculative review. 1973. In W. Wieser (Ed.), *Effects of temperature on ectothermic organisms*, pp. 1–23. Berlin: Springer-Verlag.

Wikfors, G. H. (1986). Altering growth and gross chemical composition of two microalgal molluscan food species by varying nitrate and phosphate. *Aquaculture*, **59**, 1–14.

Wilson, R. P. (1973). Absence of ascorbic acid synthesis in channel catfish, *Ictalurus punctatus* and blue catfish. *Ictalurus furcatus*. *Comp. Biochem. Physiol.*, **46B**, 635–8.

Wilson, R. P. & Poe, W. E. (1973). Impaired collagen formation in the scorbutic channel catfish, *J. Nutr.*, **103**, 1359–64.

Wilson, R. P., Harding, D. E. & Garling, D. L. (1977). Effect of dietary pH on amino acid utilization and the lysine requirement of fingerling channel catfish. J. Nutr., **107**, 166–70.

Wilson, R. P., Allen, O. W., Robinson, E. H. & Poe, W. E. (1978). Tryptophan and threonine requirement of fingerling channel catfish. *J. Nutr.*, **108**, 1595–9.

Wilson, R. P., Poe, W. E. & Robinson, E. H. (1980). Leucine, isoleucine, valine and histidine requirements of fingerling channel catfish. *J. Nutr.*, **110**, 627–33.

Wilson, R. P., Robinson, E. H. & Poe, W. E. (1981). Apparent and true availability of amino acids in feed ingredients for channel catfish. *J. Nutr.*, **111**, 923–9.

Wilson, R. P., Galtin, D. M. & Poe, W. E. (1985). Postprandial changes in serum amino acids of channel catfish fed diets containing different levels of protein and energy. *Aquaculture*, **49**, 101–10.

Winberg, G. G. (1956). Rate of metabolism and food requirements of fishes. *Nauch. Tr. Beloruss. Gosud. Univ. imeni Lenina, Minsk*, 253 pp. Transl. Ser., Fish. Res. Bd Can., (194), 1960.

Winberg, G. G. (1961). New information on metabolic rate in fishes. *Vopr. Ikhtiol.*, **1**(1), 157–65. Transl. Ser., Fish. Res. Bd Can., (362), 1961.

Winberg, G. G. (1964). The pathways of quantitative study of food consumption and assimilation by aquatic animals. (In Russian with English summary.) *Zhur. Obshch. Biol.*, **25**, 254–66.

Winberg, G. G. (1971). *Methods for the estimation of production of aquatic animals*. London: Academic Press, 157 pp.

Winberg, G. G. & Khartova, L. E. (1953). Intensity of metabolism in young carp. *Dokl. Akad. Nauk SSSR*, **89**, 1119–22.

Windell, J. T. (1966). Rates of digestion in the bluegill sunfish. *Invest. Indiana Lakes Streams*, **7**, 185–214.

Windell, J. T. (1967). Rates of digestion in fishes. In S. D. Gerking (Ed.), *The biological basis of freshwater fish production*, pp. 151–73. Oxford: Blackwell Sci. Publ.

Windell, J. T. (1978). Estimating food consumption rates of fish populations. In T. Bagenal (Ed.), *Methods for assessment of fish production in fresh waters, IBP Handbook no. 3* (3rd edn), pp. 227–254. Oxford: Blackwell Sci. Publ.

Windell, J. T. & Norris, D. O. (1969a). Dynamics of gastric evacuation in rainbow trout, *Salmo gairdneri*. *Am. Zool.*, **9**, 584.

Windell, J. T. & Norris, D. O. (1969b). Gastric digestion and evacuation in rainbow trout. *Prog. Fish Cult.*, **31**, 20–6.

Windell, J. T., Foltz, J. W. & Sarokon, J. A. (1978). Effect of fish size, temperature and amount fed on nutrient digestibility of a pelleted diet by rainbow trout *Salmo gairdneri*. *Trans. Am. Fish. Soc.*, **107**, 613–16.

Winfree, R. A. (1979). Effects of dietary protein and energy on growth, feed conversion efficiency and body composition of *Tilapia aurea.* M.Sc. Thesis, Texas A & M Univ., Texas, USA, 47 pp. (mimeographed).

Winfree, R. A., & Stickney, R. R. (1984). Starter diets for channel catfish: effects of dietary protein on growth and carcass composition. *Prog. Fish Cult.*, **46**, 79–86.

Wissing, T. E. (1974). Energy transformation by young-of-the-year white bass *Morone chrysops* (Rafinesque) in Lake Mendota, Wisconsin. *Trans. Am. Fish. Soc.*, **103**, 32–7.

Wissing, T. E. & Hasler, A. D. (1968). Calorific values of some invertebrates in Lake Mendota, Wisconsin. *J. Fish. Res. Bd Can.*, **25**, 2515–18.

Wissing, T. E. & Hasler, A. D. (1971). Intraseasonal changes in caloric content of some freshwater invertebrates. *Ecology*, **52**, 371–73.

Wittenberger, C. (1967). On the function of the lateral red muscle of teleost fishes. *Rev. Roum. Biol., Ser. Zool.*, **12**, 139–44.

Wittenberger, C. & Diaciuc, I. V. (1965). Effort metabolism of lateral muscles in carp. *J. Fish. Res. Bd Can.*, **22**, 1397–406.

Wittenberger, C. & Giurgea, R. (1973). Transaminase activities in muscles and liver of the carp. *Rev. Roum. Biol., Ser. Zool.*, **18**, 441–4.

Wittenberger, C. & Vitca, E. (1966). Variation of the glycogen content in the lateral muscles of the carp during work performed by isolated muscles and during starvation. *Studuia Univ. Babes-Bolyai, Ser. Biologica*, (2), 117–23.

Wodtke, E. (1974). Wirkungen der Adaptationstemperatur auf den oxidativen Stoffwechsel des Aales (*Anguilla anguilla* L.). I. Leber und rote Muskulatur: Veraenderungen im Mitochondrien gehalt und in der oxidativen Kapazitaet isolierler, gekoppelter Mitochondrien. *J. Comp. Physiol.*, **91**, 309–32.

Wohlfarth, G., Lahman, M. & Moav, R. (1962). Genetic improvement of carp. IV. Leather and line carp in fishpond of Israel. Bamidgeh, **15**(1), 3–8.

Wohlfarth, G. W., Moav, R. & Hulata, G. (1983). A genotype-environment interaction for growth rate in the common carp, growing in intensively manured ponds. *Aquaculture*, **33**, 187–95.

Wolf, L. E. (1942). Fish diet disease of trout. A vitamin deficiency produced by diets containing raw fish. *N.Y. Conserv. Dept., Fish Res. Bull.*, **2**, 1–5.

Wolf, L. E. (1951). Diet experiments with trout. *Prog. Fish Cult.*, **13**, 17–20.

Wolny, P., Wojno, T. & Trzebiatowski, R. (1965). Changes in the protein, fat, water and average matter content in the body of two-year-old carp during its growth. (In Polish with English summary.) *Rocz. Nauk Roln. B*, **86**(2), 283–300.

Wood, J. D. (1958). Nitrogen excretion in some marine teleosts. *Canad. J. Biochem. Physiol.*, **36**, 1237–42.

Wood, J. D.,Duncan, D. W. & Jackson, M. (1960). Biochemical studies on sockeye salmon during spawning migration. XI. The free histidine content of the tissues. *J. Fish. Res. Bd Can.*, **17**, 347–51.

Woodall, A. N. & LaRoche, G. (1964). Nutrition of salmonoid fishes. xi. Iodide requirements for chinook salmon. *J. Nutr.*, **82**; 574–81.

Woodall, A. N., Ashley, L. M., Halver, J. E., Olcott, H. S. & van Der Veen, J. (1964). Nutrition of salmonoid fishes. xiii. The α-tocopherol requirement of chinook salmon. *J. Nutr.*, **84**, 125–35.

Worthington, R. E. & Lovell, R. T. (1973). Fatty acids of channel catfish (*Ictalurus punctatus*): variance components related to diet, replications within diets, and variability among fish. *J. Fish. Res. Bd Can.*, **30**, 1604–8.

Wu, J.-L. & Jan, L. (1977). Comparison of the nutritive value of dietary proteins in *Tilapia aurea. J. Fish. Soc. Taiwan.* **5**(2), 55–60.

Wunder, W. (1936). Physiologie der Suesswasserfische Mitteleuropas. In R. Demoll & H. N. Maier (Eds), *Handb. Binnenfisch. Mitteleurop., vol. 2B.* Stuttgart: Schweizerbartsche Verlag, 340 pp.

Wunder, W. (1937). Richtlinien fuer die Verfuetterung von Kartoffeln im Karpfenteich. *Fisch.-Ztg.*, **40**(14), 3 pp.

Wunder, W. (1955). Leistungspruefungsversuche mit einzeln markierten Karpfen verschiedener Staemme in praktischen Zuchtbetrieben. *Allg. Fisch.-Zeit.*, **80**(6/7), 8 pp.

Yamada, J., Kikuchi, R., Matsushima, M. & Ogami, H. (1962). On improving the efficiency of feed for fish culture. 2. Digestibilities of feedstuffs for rainbow trout and some trials on the improvement. *Bull. Jap. Soc. Sci. Fish.*, **28**, 905–8.

Yamada, K. & Yokote, M. (1975). Morphological analysis of mucosubstances in some epithelial tissues of the eel (*Anguilla japonica*). *Histochemistry*, **43**, 161–72.

Yamada, S., Simpson, K. L., Tanaka, Y. & Katayama, T. (1981a). Plasma amino acids changes in rainbow trout *Salmo gairdneri* forced-fed casein and a corresponding amino acid mixture. *Bull. Jap. Soc. Sci. Fish.*, **47**, 1035–40.

Yamada, S., Tanaka, Y. & Katayama, T. (1981b). Feeding experiments with carp fry fed an amino acid diet by increasing the number of feedings per day. *Bull. Jap. Soc. Sci. Fish.*, **47**, 1247.

Yamane, S. (1973a). Localization of amylase activity in digestive organs of carp determined by a substrate film method. *Bull. Jap. Soc. Sci. Fish.*, **39**, 497–504.

Yamane, S., (1973b). Localization of amylase activity in the digestive organs of the Mozambique mouth-brooder, *Tilapia mossambica*, and bluegill, *Lepomis macrochirus*, determined by starch substrate film method. *Bull. Jap. Soc. Sci. Fish.*, **39**, 595–603.

Yamashina, I. (1956). Action of enterokinase on trypsinogen. *Acta Chem. Scand.*, **10**, 739–43.

Yamauchi, T., Stegeman, J. J. & Goldberg, E. (1975). The effects of starvation and temperature acclimation on pentose phosphate pathway dehydrogenases in brook trout liver. *Arch. Biochem. Biophys.*, **167**, 13–20.

Yanase, M. (1963). Pantothenic acid and vitamin B_{12} contents assessed for the rainbow trout during maturation. *Bull. Jap. Soc. Sci. Fish.*, **29**, 1024–30.

Yarzhombek, A. A., Shcherbina, T. V., Shamakov, N. F. & Gusseynov, A. G. (1983). Specific dynamic effect of food on fish metabolism. *J. Ichthyol.*, **23**, 111–17.

Yashouv, A. (1969). The fish pond as an experimental model for study of interactions within and among fish populations. *Verh. Inter. Ver. Limnol.*, **17**, 582–93.

Yasunaga, Y. (1972). Studies on the digestive function of fishes. III. Activity of the digestive enzymes, protease and amylase of some flatfish species. *Bull. Tokai Reg. Fish. Res. Lab.*, **71**, 169–75.

Yone, Y. (1976). Nutrition studies of red sea bream. *Proc. Int. Conf. Aquacult.*, **1**, 39–64.

Yone, Y. (1979). The utilization of carbohydrate by fishes. *Proc. Japan-Soviet Joint Symp. Aquaculture*, **7** (1978), 39–48.

Yone, Y. & Fujii, M. (1975a). Studies on nutrition of red sea bream. 11. Effect of ω3 fatty acid supplements in a corn oil diet on growth rate and feed efficiency. *Bull. Jap. Soc. Sci. Fish.*, **41**, 73–7.

Yone, Y. & Fujii, M. (1975b). Studies on nutrition of red bream. 12. Effect of ω3 fatty acid supplement in corn oil diet on fatty acid composition of fish. *Bull. Jap. Soc. Sci. Fish.*, **41**, 79–86.

Yone, Y. & Toshima, N. (1979). The utilization of phosphorus in fish meal by carp and black sea bream. *Bull. Jap. Soc. Sci. Fish.*, **45**, 753–6.

Yone, Y., Sakamoto, S. & Furuichi, M. (1974). Studies on nutrition of red sea bream. 9. The basal diet for nutrition studies. *Rep. Fish. Res. Lab., Kyushu Univ.*, (2), 13–24.

Yoshida, H. & Sakurai, Y. (1984). Relationship between food consumption and growth of adult walleye pollock *Theragra chalcogramma*. *Bull. Jap. Soc. Sci. Fish.*, **50**, 763–9.

Yoshida, Y. (1970a). Studies on the efficiency of food conversion to fish body growth. I. The formulae on the efficiency of conversion. *Bull. Jap. Soc. Sci. Fish.*, **36**, 156–60.

Yoshida, Y. (1970b). Studies on the efficiency of food conversion to fish body. II. Actual application of the formulae concerning efficiency of conversion. *Bull. Jap. Soc. Sci. Fish.*,

36, 160–4. Transl. Ser., Fish. Marine Serv., Can., (3199), 1974.

Yoshinaka, R., Sato, M. & Ikeda, S. (1973). Studies on collagenase of fish. I. Existence of collagenolytic enzyme in pyloric caecae of *Seriola quinqueradiata*. *Bull. Jap. Soc. Sci. Fish.*, **39**, 275–81.

Yoshinaka, R., Sato, M. & Ikeda, S. (1981a). *In vitro*, activation of trypsinogen and chymotrypsinogen in the pancreas of catfish. *Bull. Jap. Soc. Sci. Fish.*, **47**, 1473–8.

Yoshinaka, R., Sato, M. & Ikeda, S. (1981b). Distribution of trypsin and chymotrypsin and their zymogens in digestive system of catfish. *Bull. Jap. Soc. Sci. Fish.*, **47**, 1615–18.

Yoshinaka, R., Sato, M., Morishita, J., Itoh, Y., Hujita, M. & Ikeda, S. (1984). Purification of some properties of two carboxypeptidases B from the catfish pancreas. *Bull. Jap. Soc. Sci. Fish.*, **50**, 171–22.

Yu, T. C. & Sinnhuber, R. O. (1972). Effect of dietary linolenic and docosahexaenoic acid on growth and fatty acid composition of rainbow trout (*Salmo gairdneri*). *Lipids*, **7**, 450–4.

Yu, T. C. & Sinnhuber, R. O. (1975). Effect of dietary linolenic and linoleic acids upon growth and lipid metabolism of rainbow trout (*Salmo gairdneri*). *Lipids*, **10**, 63–6.

Yu, T. C., Sinnhuber, R. O. & Putnam, G. B. (1977). Effect of dietary lipids on fatty acid composition of body lipid in rainbow trout (*Salmo gairdneri*). *Lipids*, **12**, 495–9.

Zaitschek, E. A. (1960). The respiration and the basal and active metabolism of the carp (*Cyprinus carpio* L.). (In Hebrew with English summary.) Ph.D. Thesis, Hebrew University of Jersualem, iv + 28 pp.

Zebian, M. F. & Creach, Y. (1979). Free aminated and fraction and oxidative degradation of some amino acids in carp (*Cyprinus carpio* L.): Importance of nutritional factors. In J. E. Halver & K. Tiews (Eds), *Finfish nutrition and fishfeed technology, vol. II*, pp. 531–44. Berlin: Heenemann Verlagsgesel.

Zeitler, M. H., Kirchgessner, M. & Schwarz, F. J. (1984). Effects of different protein and energy supplies on carcass composition of carp (*Cyprinus carpio* L.). *Aquaculture*, **36**, 37–48.

Zeitler, M. H., Schwarz, F.J. & Kirchgessner, M. (1983). Wachstum und Naehrstoffaufwand bei Karpfen (*Cyprinus carpio* L.) mit unterschiedlicher Protein- und Energieversorgung I. Mitteilung: Versuchsplan, Rationszusammensetzung, Naehrstoffverdaulichkeit. *Z. Tierphysiol., Tierernaehr. Futtermittelk.*, **49**, 80–7.

Zeitoun, I. H., Halver, J. E., Ullrey, D. E. & Tack, P. I. (1973). Influence of salinity on protein requirement of rainbow trout (*Salmo gairdneri*) fingerlings. *J. Fish. Res. Bd Can.*, **30**, 1867–73.

Zeitoun, I. H., Ullrey, D. E., Halver, J. E., Tack, P. I. & Magee, W. T. (1974). Influence of salinity on protein requirements of coho salmon (*Oncorhynchus kisutch*) smolts. *J. Fish. Res. Bd Can.*, **31**, 1145–8.

Zeitoun, I. H., Ullrey, D. E., Magee, W. T., Gill, J. L. & Bergen, W. G. (1976). Quantifying nutrient requirements of fish. *J. Fish. Res. Bd Can.*, **33**, 167–72.

Zintzen, H. (1972). A summary of the vitamin E/selenium problems in ruminants. *News and reviews, Hoffmann La Roch Co., Basel*, 1–18.

Zohar, G. (1986). Improved feeding chart for tilapias in high stocking rates. (In Hebrew with English summary.) *Fisheries and Fishbreeding in Israel*, **19**, 28–33.

Zorn, M. (1984). The effect of temperature on some factors of protein utilization by carp, *Cyprinus carpio*. M.Sc. Thesis, The Hebrew University, Faculty of Agriculture, Rehovot.

Zur, O. (1981). Primary production in intensive fish ponds and a complete organic carbon balance in the ponds. *Aquaculture*, **23**, 197–210.

Systematic Index

(for fish)

Subject Index

WIDENER UNIVERSITY
WOLFGRAM
LIBRARY
CHESTER, PA.